Principles of Engineering Management

Jishan He

Principles of Engineering Management

中国建筑工业出版社
CHINA ARCHITECTURE & BUILDING PRESS

Springer

Jishan He
Central South University
Changsha, Hunan, China

ISBN 978-981-99-1167-7 ISBN 978-981-99-1168-4 (eBook)
https://doi.org/10.1007/978-981-99-1168-4

Jointly published with China Architecture & Building Press
The print edition is not for sale in China (Mainland). Customers from China (Mainland) please order the print book from: China Architecture & Building Press.

B&R Book Program

This Springer imprint is published by the registered company Springer Nature Singapore Pte Ltd.
The registered company address is: 152 Beach Road, #21-01/04 Gateway East, Singapore 189721, Singapore

Contents

Chapter 1
Engineering Management Ontology

Engineering ontology is the theoretical core and fulcrum of the ternary theory of science, technology, and engineering. Engineering ontology believes that engineering has an "ontological position" and is not a derivative of science or technology. Engineering is autonomous rather than dependent or accessory. Engineering has its basis of self-existence (not a vassal), has its own rules of activities and development, and has its own goals and values [1]. In the engineering management theoretical system research, the author agrees with the ternary theory of science, technology, and engineering.

This chapter discusses science and technology as an entry point and defines engineering as a whole process activity of creating new "artificial nature" and running this "artificial nature" until its decommissioning. Scholars have different opinions on the stage division of engineering. Based on the definition of engineering, the stage division of engineering is discussed in depth. Engineering is the activity of human beings to adapt to nature, exploit nature and protect nature based on understanding and respecting nature [2]. Engineering management is the soul of engineering activities.

We have encountered such doubts more than once in engineering management research: Is engineering management a science? This chapter gives an affirmative answer to this question. This chapter takes the cognition of science, technology, and engineering as an entry point, discusses the definition of engineering management, and generalizes the definition of engineering management from general cognition to philosophical cognition. Engineering management includes engineering management science, engineering management technology, and engineering management art. Engineering management science is a knowledge system that forms an understanding of the objective laws of engineering management; Engineering management technology is a variety of methodologies applied in the engineering management process, such as coordination technology, evaluation technology, etc.; Engineering management art is the management of objects and people in the process of engineering management. The management art of managers has a very important influence on the management effect.

J. He, *Principles of Engineering Management*,
https://doi.org/10.1007/978-981-99-1168-4_1

Based on the cognition of engineering management science, technology, and art, this chapter provides a comprehensive definition of engineering from four dimensions of function, process, element, and philosophy. According to the historical facts of scientific management and project management, it is determined that they are all developed from engineering management.

Summarized and refined from the investigation and study of a large number of large-scale, super-large engineering management practices, "people-oriented, harmony between nature and human, collaborative innovation, and building harmony" are the core values of engineering management, the soul of engineering management, and one of the focus of this chapter.

1.1 Science and Technology

1.1.1 Science

1.1.1.1 The Meaning of Science

According to Shuowen Jiezi (Chinese: 説文解字), an ancient Chinese dictionary from the Han dynasty, the original meaning of the word "science (Kexue)" was "the knowledge of measurement" [3]. But from the Tang Dynasty to modern times, science was an abbreviation of the study of the Chinese Imperial Examination (科举). For example, Chen Liang, a bibliophile of the Song Dynasty, in his *Song Shu Zu Zhu Jun Zhou Gao Yao Bu Xu* (送叔祖主筠州高要簿序), the meaning of science refers to the study of imperial examinations, which is not the definition of science in the modern sense.

Chinese tradition calls all knowledge known as learning. Su Shi of the Song Dynasty said in the *Deng Zhou Xie Shang Biao*, "I am born to be dull, lacking knowledge." In ancient China, the knowledge about "natural objects" was called Wu Li (物理). Therefore, Wu Li of ancient times is equivalent to the natural science of modern times. In the Ming Dynasty, until the Sino-Japanese War of 1894–1895, the word corresponding to modern science is "Ge Zhi (格致)," the abbreviation of "Ge Wu Zhi Zhi (格物致知)," which means the knowledge gained from studying natural objects.

"Science" in English means "Knowledge" and "Learning," and in modern times, it gradually refers to the learning of nature. "Science" was introduced to Japan during the Meiji era. It was translated as "科学Kexue" in Chinese in the Western Zhou Dynasty. Kang Youwei listed *the Introduction of Science* and *The Principle of Science* in his published book *Ri Ben Shu Mu Zhi*, who is considered the first Chinese who introduces the word "Kexue" with modern meaning. After the Revolution of 1911, the use of the word "Kexue" became more widespread and finally replaced "Ge Zhi."

In the modern era, "science" is a frequent term, and people often blurt it out. However, it is not easy to answer when asked what science is. When talking about science, people often consider that science is sublime, absolutely correct, sacred, and even mysterious. It seems that science already has a straightforward definition. Many scholars have made efforts to do it, but they are not very successful. The book's authors do not intend to discuss the definition of science too much. However, in engineering and engineering management research, the concept of science needs to be involved.

From many pieces of literature, the author agrees with the explanation of "science" in *Modern Chinese Dictionaries* 《现代汉语词典》: Science is a disciplinary knowledge system that studies the objective laws of nature, society, and thinking [4].

Although this definition is somewhat sketchy, it is simple and straightforward, and it points out the essential characteristics of science in general: the disciplinary knowledge system about objective laws. As for the requirements of this "system" to be considered science, it is somewhat complicated and is not the goal of this book. Although there are no strict requirements, it is not difficult to judge which knowledge systems belong to science under the premise of conventions. Some common sense "systems," theology, or fortune-telling, should not be recognized as a science.

Science in *Encyclopedia Britannica* means: Science involves all kinds of intellectual activities in the material world and its various phenomena and requires unbiased observations and systematic experiments. In general, science involves the pursuit of knowledge, including the pursuit of various universal truths or various basic laws [5]. The above explanation mainly shows that science is an intellectual activity, and the aim of this activity is to pursue the truth. While the *Modern Chinese Dictionary* explains what science is.

The Scientific Theory in *Encyclopedia Britannica* is described as follows: "Systematic conceptual structure of a broad field conceived by human imagination. It includes a system of empirical laws on the inherent regularity of objects and events; These objects and events can be either observed or assumed; The structures proposed by these laws are designed to explain these things in a scientific and rational way. In order to explain what is being experienced, scientists use (1) careful observation or experiment, (2) reporting the various regularities they found, and (3) a systematic description of the theory. If these rules' statements are precise, they can be used as a law of experience to express the relationship of the observed things or features. So, when the law of experience satisfies the curiosity of scientists by revealing a kind of rationality in things, scientists propose a systematic outline or scientific theory to provide an acceptable explanation for why these laws are obtained" [6].

This indicates that scientific theory is an interpretation or assumption of things, and it is a human understanding of objective things. This is consistent with the statement that "Science is a disciplinary knowledge system that reflects the objective laws of nature, society, and thinking."

The author believes that the definition of science in the *Modern Chinese Dictionary* can be further summarized. From a philosophical point of view, science can be expressed as Science is a disciplinary knowledge system about the objective laws of the world. The "world" here includes the objective and subjective worlds, including nature, society, thinking, etc.

1.1.1.2 Natural Science and Social Science

Based on the previous discussion, the following extensions can be made:

1. **The disciplinary knowledge system of the objective law of nature is the natural science**

The task of natural science is to discover the nature and laws of natural phenomena. Its disciplines include physics, chemistry, biology, astronomy, and earth science. Mathematics is their working language.

Western natural philosophy refers to the philosophical thoughts formed by human beings thinking about the natural world they face. It includes the relationship between nature and human beings, the relationship between natural and artificial nature, the most basic laws of nature, etc. The general philosophy of nature includes natural science. In ancient times, the distinction between philosophy and science was not noticeable. Not only did the natural sciences contain natural philosophy, but the two usually did not distinguish. For example, Newton's most famous book is called *The Mathematical Principle of Natural Philosophy.*

2. **The disciplinary knowledge system of the objective law of society is the social science**

Social science is the science about the nature of social things and their laws. Natural science is usually objective, and social science has different standpoints. When holding a correct viewpoint, the nature and laws of social things reflected in social science achievements are also objective; but on the contrary, it is uncertain. The generalized "social science" includes the humanities, a general designation for social science and humanities.

1.1.1.3 Scientific Theory is Constantly Developing and Perfecting

Science is a knowledge system about the objective laws of things, and it is human cognition of the world. The objective law is objective and unique, but human cognition is influenced by the time and space environment. The understanding of the objective law needs to be gradually deepened. Therefore, a scientific theory is in a situation of inexhaustible development and change. In addition, in the process of development, some misunderstandings may occur for various reasons. The well-known Dalton atomic theory and Einstein et al.'s EPR paradox about quantum mechanics are vivid examples of the development of science.

Dalton is a chemist. He analyzed the facts of chemical experiments. On the basis of inheriting the ancient Greek naive atomism and Newtonian particles, in 1803, he first explicitly stated that chemical elements are composed of atoms [7]. At the same time, he believes that the quality of atoms is one of the basic characteristics of elements, and the atoms of the same element are not only of the same nature but of the same quality; the atoms of different elements have different quality; when atoms of different elements react chemically, each other's atoms are combined in a simple integer ratio. Dalton also proposed a relative comparison method to obtain the atomic weight of various elements and published the first atomic scale. A large number of experimental results confirm his above theory. However, while proposing the concept of atoms, he believes that atoms are the smallest particles that cannot be subdivided.

Dalton's atomic theory is not a hypothesis, not a philosophical theory. It is based on 150 years of chemical experiments and the epoch-making scientific theory of the substance microstructure. Although it contains errors that the atom is impossible to be divided, it cannot be denied that Dalton's atomism is a great scientific theory. It is the most advanced understanding of substance in that era.

Einstein has always questioned the completeness of quantum mechanics theory. In the 47th issue of the American *Physical Review* in 1935, Einstein, Podolsky, and Rosen published a paper entitled *Can Quantum–Mechanical Description of Physical Reality Be Considered Complete?* Einstein acknowledged the glorious achievements of quantum mechanics, but he denied the complementary principles by Bohr et al. EPR is the initials of the last names of the three physicists Einstein, Podolsky, and Rosen. This paper is called the EPR Paradox, and it is a paradox they put forward to demonstrate the incompleteness of quantum mechanics. In the same year, the 48th issue of the American *Physical Review* published another paper with the same title: *Can Quantum–Mechanical Description of Physical Reality be Considered Complete?* by Bohr. Bohr defends the completeness of quantum mechanics theory and believes that the microscopic system and the measuring instrument constitute a whole. In this way, EPR correlation can be reasonably explained in the range of quantum mechanics. The debate has continued, and since then, quantum mechanics has basically evolved along the route of Boer et al. and has achieved remarkable achievements. But we can't simply think that Einstein et al. are losers and Bohr wins. The EPR paradox proposed by Einstein et al. promotes deeper research and stimulates the new theory of quantum mechanics, even the formation and development of a new school of thought.

The story of Einstein's public apology at the age of 70 is particularly touching. Einstein adheres to the concept of the static universe model. However, the research results of American scholar Friedman et al. have obtained a dynamic solution, thinking that the universe is in a state of uniform expansion or contraction, not static. Einstein still insists on his static view of the universe and constantly criticizes and ridicules the stupidity of Friedman et al. Later, some scholars came up with dynamic solutions, especially the observation by American astronomer Hubble, who discovered that planets are moving away from the earth. This important discovery also supports the dynamic universe model. Einstein realized that he was wrong, but he did not publicly acknowledge it. On his 70th birthday, he wrote to the newspaper that he had done many wrong things in science, hurt many people, and solemnly apologized to Friedman et al.

In addition, Einstein spent ten years in his later years seeking a unified field theory that could include universal gravitation and electromagnetic phenomena, but it was not successful. He published a new unified field theory paper in 1950, but physicists consider that it is impossible. It also shows how hard the process of scientific exploration is.

These examples vividly and fully demonstrate that scientific theory is constantly evolving.

1.1.1.4 Summary

Through the above definitions and examples related to science, the summary is as follows:

(1) The law of movement of things is objective and unique.
(2) Science is a knowledge system about the objective law of things and is the human cognition of the world. However, human understanding is affected by the time and space environment, so the scientific theory is also constantly developing and changing. The core of scientific activities is to "discover" objective laws. The role of scientists in various scientific activities is to pursue the truth and discover the world that people have not yet cognized. People can only constantly discover and approach the truth but not exhaust the truth. This kind of activity has no end, and it is an eternal process. The enduring debate between Einstein and Bohr itself is an outstanding and typical example of the development of science.
(3) Science (scientific theory) does not mean correct or entirely correct. It is not strange because science is a knowledge system about the objective law of the world and people's cognition has a limitation; it's possible to have some incorrect parts. Due to the limited knowledge of human beings and the limitations of the environment, the world is often recognized from a certain angle or a certain aspect in constructing the knowledge system. Therefore, a scientific theory is usually incomplete.

Engineering management research is also like this. The author tries to summarize and sublimate the existing cognition of engineering management itself and build a theoretical engineering management system. However, this book is only a phased cognition about engineering management science at a certain depth and breadth.

1.1.2 Technology

1.1.2.1 The Meaning of Technology

Technology is a specialized area of human activity. From the early days of mankind, technology, the universe, nature, and the social environment have always been the four major environmental factors of human life and have largely changed social features over thousands of years [8].

The word technology is a combination of Greek Techne and Logos, which means the discussion of plastic arts and applied technology. When it first appeared in the UK in the seventeenth century, it only referred to various application techniques. In the nineteenth century, it was defined as the general designation of practical art. And at that time, "practical art," "applied science," and "engineering" usually referred to the technology defined currently. By the twentieth century, the meaning of technology has gradually expanded, involving tools, machines, and their methods of use and processes. After World War II, technology was defined as "a means or activity by which humans change or control the objective environment." Technical science includes traditional engineering disciplines, agricultural disciplines, and modern disciplines such as space, computing, and automation. Humans have produced technology in the process of manufacturing tools, and the most remarkable feature of modern technology is its combination with science [9]. Technology is essential "all the means and methods humans use to expand their muscles, feelings, and wisdom when interacting with nature, also playing an important role in creating cultural values" [10].

The demand for human activities drives technology to innovate. Technological innovation definitely promotes scientific research; scientific research conversely affects technological innovation. The two complement each other. In ancient times, the upper-class mastered science, and technology was mainly mastered by craftsmen and workers. It seems that science is nobler than technology. With the development of the economy and social exchanges, the role of technology in social production and life has become more prominent, and it has its due status and position in modern society.

However, the impression that technology is lower than science still exists. It is often considered that technology is created and developed due to the creation and development of science. Technology is the application of science. It is even considered that technology is an appendage of science. It is not. When there was no corresponding scientific theory in ancient times, people invented various techniques from the practice of production and life, not because of the application of scientific theory. Archimedes has a famous saying: "Give me a lever long enough and a fulcrum on which to place it, and I shall move the world!" Now it is widely accepted that he demonstrated the principle of leverage. However, far before he was born, the Chinese and the Egyptians knew how to use leverage to move heavy objects without the lever theory. In fact, Archimedes was inspired by the ancient Egyptians to use leverage to develop the principle of leverage. Therefore, technology is not simply a product of science. Their development has its own path. Modern technology and science are very closely integrated. On the one hand, people often study the corresponding scientific principles when they are engaged in technological innovation; on the other hand, all of the contemporary significant scientific discoveries require technical support. For example, without the corresponding microscopic techniques and techniques for observing long distances, there is no way to proceed with the scientific discovery of the microscopic world or the macroscopic universe. Therefore, science, technology, and engineering are ternary.

1.1.2.2 Four Characteristics of Technological Innovation

Technological innovation activities are characterized by diversity, selectivity, and phase, that is, four characteristics.

Innovation is the core of technological activities. Innovation includes invention and innovation. The invention refers to the research of a technology that has not been done before. Innovation is based on the main body of the original technology, improving several technical aspects. Both inventions and innovations are carried out by the demand of life and production activities and therefore have obvious utilitarian purposes. There are also a few technical inventions and innovations that are due to personal interests, hobbies, or curiosities and have no specific utilitarian purpose. There are many ways and means of technological innovation. Therefore, a particular problem can be solved by many technologies, that is, the diversity of technologies. In real life and production activities, there is often only one needed technology to solve a specific problem, and it must be selected from a variety of technologies. This kind of selection is based on economics, convenience, and social acceptability, resulting in selectivity. Any type of technology must constantly improve with the development of the times. There is no end to technological innovation activities; that is, no best, only better. This is the phased character of technical activity.

Simple mineral water bottles can be used as an example to demonstrate the characteristics of innovation, diversity, selectivity, and phase of the technology. Besides widely used plastic, glasses, metal, and ceramic can also be used, that is diversity. Plastic bottles are widely used nowadays because of their light weight, low price, convenient transportation, and convenient use. But plastic bottle also has drawbacks. For example, once the mineral water is stored for too long or the environmental temperature is too high, a considerable amount of plasticizer may enter the mineral water, adversely affecting our health. Therefore, the use of plastic bottles is a compromise at the current stage. With the continuous development of technology, it is highly probable to invent a new mineral water container that is light, inexpensive, durable, hygienic, and healthy.

1.2 Engineering

1.2.1 The Definition of Engineering

Looking at engineering from different perspectives will have different understandings. Engineering in *Encyclopedia Britannica* means: "A specialized technique that is applying scientific knowledge to enable natural resources to provide the best services to humans." The designer of engineering is called an engineer. A scientist's duty is how to recognize, but the duty of an engineer is how to realize [11]. Some literature also considers engineering as the application of science.

In fact, engineering is prior to science from the perspective of the history of human development. Before they had corresponding scientific knowledge, humans could build houses to shelter from the wind and rain and construct roads and bridges to facilitate travel. Although engineering can apply the necessary scientific principle, it cannot be simply defined as the application of science. Actually, the primary character of engineering is integrating and applicating a variety of technologies. The former president of the National Academy of Engineering, Bill Wulf, once said: "Science is the knowledge about 'What is it,' engineering is the knowledge about 'What to do'" [12].

From the perspective of engineering science, this book defines engineering as follows:

Engineering is the whole process of human beings for survival and development, to achieve specific purposes, to effectively use resources, to systematically integrate and innovate technologies, to create new "artificial nature," to run this "artificial nature" until the "artificial natural" decommissioning. In general, engineering has technical integration and industry relevance. In addition, creating new "artificial nature" and changing "natural objects" traits are complementary.

Engineering is a practical activity for the benefit of mankind, a knowledge system involving scientific technology; Engineer is a specific occupation of creating a new world; Engineering culture is the oldest but most energetic "third culture" that is different from science culture and human culture, a culture of creation and work that is closely related to human material life.

The definition of science in Sect. 1.1.1 can be extended to engineering science:

The disciplinary knowledge system about the objective law of engineering is engineering science

The Chinese Academy of Engineering, the United Nations Educational, Scientific and Cultural Organization (UNESCO), and the International Academy of Engineering and Technology Sciences (CAETS) held two international conferences on engineering science and technology in Shanghai and Beijing in 2000 and 2014, respectively. This shows that engineering science is widely recognized internationally.

As an independent discipline, engineering includes engineering science, engineering technology, and engineering management.

In order to better understand the concepts of engineering, the following demonstration is required.

1.2.1.1 About Artificial Nature

Engineering is a kind of activity; an activity carried out by humans to achieve a specific goal. In the construction phase of engineering, the most important feature of engineering activity is to create new "artificial nature," that is, to create objects or to change the traits of "natural objects" to achieve specific goals through activities such as creating or changing traits of objects.

"Natural nature" refers to the part of nature that has not yet been influenced by human practice. "Artificial Nature" refers to the objects and environment created by human beings in the process of utilizing nature in order to maintain their own survival and development requirements.

"Artificial Nature" can be divided into two categories: (1) artificial natural objects, the artificial existences that are made by people using natural objects, such as houses, roads, bridge tunnels, and other examples such as various metal and non-metal materials, various finished products, including items for the basic necessities of life, and various kinds of machines and tools, etc. (2) artificial natural world, that is, artificial ecosystem, such as large area of artificially planted forests, pastures, urban ecosystems, etc. There is no strict boundary between artificial natural objects and the artificial natural world. Especially the scale of modern engineering is getting bigger and bigger, and the original artificial natural objects will also play the role of the artificial natural world. For example, the hydraulic engineering built in the Three Gorges is an artificial natural object, but it has a significant influence on the environment in the Three Gorges Dam area and the reservoir area, and it plays the role of the artificial natural world.

Artificial nature is produced from natural nature and exists in natural nature. It is restricted by the laws of natural nature evolution and accepts the test of natural laws of nature.

During the process of research, different names have been used for the products of engineering, such as "artificial existence," "existence," "artifacts," "artificial nature," etc. These terms can express the meaning of engineering products. The "artificial nature" is used uniformly in this book because "artificial nature" is mainly used to express this meaning in natural philosophy. Natural philosophy is the predecessor of modern natural science. Natural philosophy mainly considers the philosophical problems of nature that human beings face, including the relationship between nature and human beings, the relationship between artificial nature and natural nature, the most fundamental laws of nature, etc.

1.2.1.2 The Life Cycle of Engineering

All Engineering can be divided into three periods: the construction period, the operation period, and decommissioning period (Fig. 1.1).

Although engineering can be divided into the above three periods in general, engineering varies significantly due to the variety of engineering types. For example, the Three Gorges Project's construction phase mainly includes the construction of dams, power generation systems, and shipping systems. When these were completed, the construction phase of the Three Gorges Project was completed. The next phase, the operation phase, is crucial because the goal of construction is operation. The quality of operation is the evaluation of each work in the construction phase. The main tasks of the Three Gorges Project are flood control, power generation, and shipping, so in the operation phase, it is necessary to ensure the smooth process of the three main tasks. Of course, there are maintenance, repair, or technical transformation works

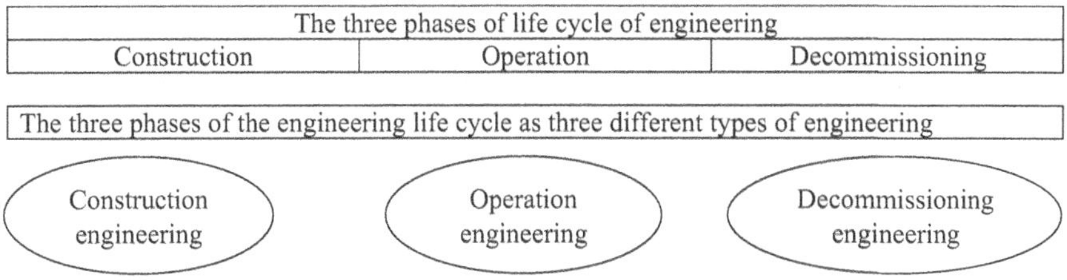

Fig. 1.1 The life cycle of engineering

during the operation phase to protect and enhance its functions. Things always have a lifetime. After a few years, perhaps a hundred years later, the Three Gorges Project will inevitably enter the final phase, that is, the decommissioning phase. Although it will be a long time before the Three Gorges Project decommission, it will happen. As the engineering product ages, degenerates, and can no longer continue to perform its original function, it must face its decommissioning phase.

Some scholars advocate that the engineering phase is mainly identified as the construction phase. That is because, after the completion of the construction phase, it is the operation phase. At this phase, the staff, the content of the work, and the management methods are very different, and the decommissioning phase is even more different. Therefore, some scholars believe that construction, operation, and decommissioning phases should be a kind of engineering. This view also has its reasons. For example, building an aircraft carrier is a large project. The aircraft carrier is delivered to use after it is built. It becomes the airport and combat system at sea. The operation phase and construction phase of the aircraft carrier is totally different. When the aircraft carrier must be retired, it can drive back to a certain place for decommissioning. Decommissioning engineering can also be carried out in different ways, either by disassembling them and recovering valuable substances; that is to say, after the decommissioning phase, the aircraft carrier disappears. It can also be parked in a certain place and reused as a sightseeing spot. It has no military combat mission, but it plays its value and role as a tourist attraction for civilian use. Another example is mining engineering. After the ore is exhausted, the reclamation process is a kind of decommissioning engineering. However, many mines are located in scenic spots, and people have transformed them into tourist attractions. This is also a way of decommissioning. Another example is China's two bombs and one satellite project. The two bombs include the atomic bomb and the hydrogen bomb. Making atomic bombs and hydrogen bombs is huge engineering. After being manufactured, the operation is mainly reflected in the aspects of storage, maintenance, and detonation. Once detonated, it is not like other engineering bodies that must be dismantled or destroyed because the detonation process is destroyed. Like the manned spaceflight project, from the construction of the manned spacecraft-spacecraft to its successful launch, this is its construction period. After launching, the operation from the air to the recovery is the operation period. The spacecraft, after recycling, can be used as a research sample for further improvement or for the public to visit; this is its decommissioning period. Another example is the artificial rain project, which emits the corresponding ice material so that the water in the air condenses into the rain. The condensed rainfall is a very short operation phase.

Since the founding of New China, especially in the 30 years since the reform and opening up, China has completed a large number of engineering projects. Society and builders are mainly focused on how to complete these projects. The theory and practices of engineering management are also concentrated primarily on the construction phase of engineering. The projects finished construction are all in operation, so there are so many practices in terms of operation. Some buildings have reached the end of their service time, or there is no way to repair them, or repair is not worthwhile, then they will be destroyed by directional blasting. It is the final decommissioning phase of the project, but the number of decommissioning and destruction is not much. In fact, many large-scale engineering projects in China have not been built for a long time and are currently in their youth of operation phase. There is still less experience in how to decommission after many years in the future. Especially for artificial nature like nuclear power plants, its decommissioning is very complicated. At present, China has little experience.

Considering the whole life cycle of engineering (construction, operation, and decommissioning) as one project, or dividing construction, operation, and decommissioning into three projects, both of these two ways are available, as long as the concept is clear. Some literature advocates treating engineering from the whole life cycle perspective but only defines engineering from the construction perspective, which requires coordination.

The definition of engineering in this book is a whole process activity that is to create new "artificial nature," operate this "artificial nature," until decommissioning this "artificial nature," which is based on the perspective of the entire life cycle. Currently, engineering projects in China are mainly in the construction stage and operation stages. Some building demolitions are decommissioning projects which are generally done manually or by bulldozers, forklifts, etc. For larger buildings, directional blasting is often used for demolition. Due to the author's information, the book basically does not list the cases for decommissioning period, which needs to be supplemented when reprinting.

1.2.1.3 Life Cycle Assessment

Life cycle assessment (LCA) is a systematic method for quantitative evaluation of a product or a service system according to the input and output of substances and energy in its whole life cycle and environmental impact. It has been widely used.

The cycle of LCA means cycle and repeats, which is true for many products, but the engineering project is generally not repeated. Therefore, in the book, when applying the whole life cycle theory to engineering, it is not necessary to use the "whole life cycle" but the "whole life period", which means the whole life of the construction, operation, and decommissioning of the project. That is the whole life assessment.

1.2.1.4 Create Artificial Nature and Change the Characteristics of Natural Objects

Creating artificial nature and changing the characteristics of natural objects, the two often complement and accompany each other.

The Three Gorges Project and Qinghai-Tibet Railway Project are great works of artificial nature created by mankind, while the artificial rainfall project is mainly to change the characteristics of objects. However, the Three Gorges Project can also change the characteristics of objects, such as cement becoming concrete. The artificial rainfall project also creates artificial nature, that is, the new artificial nature—rain. Metallurgical engineering created new artificial nature–metals, and changed the properties of ores as natural objects, making them metal and slag.

1.2.1.5 Technology Integration and Industry Relevance

Technology integration refers to the integration of engineering performance in terms of related or series of technologies to form a specific form of technology integration. Engineering is not a simple addition of various technologies but an orderly integration of various technologies for specific goals based on specific laws or rules.

Industrial relevance is derived from the inseparability between engineering, products, as well as enterprises, and the connotation of engineering is often associated with specific products, specific enterprises, or specific industries. Engineering activities and industrial activities often have inseparable internal relations. It also shows that all professional fields related to industrial activities can become a particular engineering field.

1.2.1.6 Engineering Activities and Engineering Products

Engineering plays a very active role in human society's development. The term engineering is widely used in various language activities, and it is often borrowed or extended in literary arts and media language. For example, the appearance and application of "five-one projects," "ideological construction projects," etc. However, according to the above-mentioned definition of engineering, these concepts in daily languages are not the engineering of engineering science or engineering management science.

The construction process and its products are often referred to as engineering in daily languages. For example, someone stood on the Three Gorges Dam and admired the grand project. The project they referred to here is the magnificent Three Gorges Dam. Engineering refers to a specific process, not the product of a particular project or its consequences. In other words, the outcomes of the engineering, such as the Three Gorges Project, are different from the project itself (the process of building the Three Gorges Project).

1.2.2 Classification of Engineering

1.2.2.1 The Basic Classification of Engineering

Projects such as the construction of buildings and water conservancy occurred early, were wide-ranging, and were easy to be perceived by people. Most people think of civil engineering construction when they mention engineering and even only regard civil engineering as engineering. However, the concept of modern engineering is broad and includes various engineering, including civil engineering. The scope of engineering is extensive, so it is not easy to classify it. To facilitate discussion, engineering in this book is temporarily classified as follows:

Construction and Infrastructure Engineering: Urban Construction Engineering, Transportation Engineering, Energy Engineering, Communication Engineering, Medicine and Public Health Engineering, Public Safety Engineering;
Process and manufacturing engineering: manufacturing engineering, chemical engineering, metallurgical engineering;
Exploration and mining engineering: oil exploration and mining engineering, natural gas exploration and mining engineering, metal ore exploration and mining engineering, coal exploration and mining engineering, and other non-metallic exploration and mining engineering;
Planting, biological and environmental engineering: agricultural engineering, forestry engineering, environmental engineering;
Defense engineering: aviation engineering, aerospace engineering, ship engineering, island reef engineering, weapon engineering, electronic countermeasure engineering, etc.

This classification is rough. For example, transportation engineering also includes road engineering, bridge engineering, tunnel engineering, dam engineering, port engineering, etc.

1.2.2.2 The Examples for Engineering Generality

The generality of engineering has existed since ancient times. The Chinese Couplet Society has created a couplet for the Chinese Academy of Engineering, which has made a good refinement of China's ancient and modern engineering achievements. The content is as follows:

Back to the history of five thousand years, four inventions, *Nine Chapters of Arithmetic* vibrate the first voice, look up *Tian Gong Kai Wu*, *Sheng Nong Ben Cao*, arch bridge, canal, the Dujiangyan, the Great Wall. Splendid! Shuojin Zhengu Shu Jiazhen, Zhan Jinglun Pin Tufeng.

Along 80,000 miles of a spring breeze, one satellite and two bombs are developed by self-reliance, see the dam, the Qinghai-Tibet Railway, the West Gas, the South

Water, manned space flight, and Lunar exploration, the hybrid rice. Brilliant! Qiushi Chuangxin Tian Guoyu, Xing Keji Jing Tenglong.

The comments on this couplet:

The right scroll counts ancient Chinese engineering and technology achievements from the history of five thousand years. "Four inventions" refer to papermaking, printing, the compass, and gunpowder, which are the significant contribution of the Chinese people to world civilization. The original author of the *Nine Chapters of Arithmetic* has been challenging to verify, and it is said that Li Shou created it during the Yellow Emperor period. Liu Wei of Liang Dynasty in the Southern Dynasties said in *Wen Xin Diao Long · Shi Ji*: The ancient study of "*Nine Chapters of Arithmetic*" has accumulated algorithms; it is called techniques. "*Tian Gong Kai Wu" and* "*Sheng Nong Ben Cao"* are the ancient engineering masterpieces of the Ming Dynasty Song Yingxing *Tian Gong Kai Wu*, the Ming Dynasty Li Shizhen *Compendium of Materia Medica,* respectively. The "arch bridge," "canal," "Dujiangyan," and "Great Wall" are great ancient works such as the Zhaozhou Bridge, the Grand Canal, the Dujiangyan Irrigation System, and the Great Wall. The "first voice" is a prestige that makes people shocked. "Jinglun" refers to the ambition and ability to govern the country. *The Book of Rites· the Doctrine of the Mean* said: Only the world is sincere; we can manage the world's great fundamentals, establish the fundamental laws of the World, and master the world's knowledge. "Tufeng" was explained in *Miscellany of the Western Capital* Volume II: "Yang Xiong writes *Tai Xuan Jing* dreaming of phoenix." Later, "Tufeng" means the beauty of the text or character.

The left scroll, "along 80,000 miles of a spring breeze," means the spring of science and technology, which shows the great achievements of modern engineering technology in China. "One satellite" and "two bombs" refer to satellites, nuclear bombs (including atomic bombs and hydrogen bombs), and missiles. The "dam" means the Three Gorges Project. The "Qinghai-Tibet Railway" means the smooth operation of the Qinghai-Tibet Railway. The "West Gas" and the "South Water" refer to the West–East Gas Pipeline Project and the South-to-North Water Transfer Project. "Manned space flight and Lunar exploration" means manned space project and lunar exploration project. The "hybrid rice" refers to the hybrid rice project. "Self-reliance" means self-reliance, relying on your own strength to get things done. "Tenglong" means riding the dragon. Xiang Liu in Han Dynasty said in *Shuoyuan · Shuo Cong:* "Riding Dragon on the cloud and raises."

The last sentences of the right and left scrolls, "Zhan Jinglun Pin Tufeng" and "Xing Keji Jing Tenglong" fully express the pride and sense of responsibility of the Chinese people and engineers and scientific workers, embody the concept of scientific development, and have the sense of the times that science and technology is the primarily productive force.

Engineering in contemporary China is very extensive. The Chinese Engineering Science and Technology Achievements Exhibition was held on the 20th anniversary of the founding of the Chinese Academy of Engineering. The main topic of the exhibition echoes the contents of the couplets mentioned above, as listed below:

1. **Two bombs and one satellite project**

The "two bombs and one satellite" refers to the atomic, hydrogen, and artificial satellite. On October 16, 1964, China's first atomic bomb exploded successfully. On April 24, 1970, China's first artificial satellite was successfully launched.

At that time, the country's economy and technology base was weak, and working conditions were difficult; the country became self-reliant and broke through cutting-edge technology in a short period with less investment. The "Two Bombs and One Satellite" project reflects the development level of China's economy, science, and technology, societal and military capabilities at that time. It is an important symbol of the country's comprehensive national strength and a remarkable achievement in national defense engineering.

2. **Manned space flight and lunar exploration project**

The Long March II F launch vehicle (CZ-2F) is the main application of the Chinese manned spaceflight project. The rocket consists of four liquid boosters, a core first-level rocket, a secondary core rocket, a fairing, and an escape tower. It was the rocket with the highest mass and the most extended length of all the launch vehicles in China, 62 m high, and weighed 464 tons.

China's manned spaceflight project has been implemented since the 1990s and has established a three-step strategy. In June 2013, the completion of the rendezvous and docking mission between the Shenzhou-10 and the Tiangong-1 marked the completion of the first phase of the second step of the Chinese manned spaceflight project.

The "three-step" strategy of China's lunar exploration is also called the "Chang E Plan," including "around," "fall," and "back" three steps, all of which are unmanned detection. The significance of this is that after completing these three steps, China has the ability to research and implement the strategy of a manned landing on the moon.

3. **Yangtze Three Gorges Project**

The Three Gorges Dam is located in Sandouping Town, Yichang City, Hubei Province. As the regional crust rises, the Yangtze River flows strongly and cuts down to form the Three Gorges of the Yangtze River, with rapid currents and abundant water resources. The idea of building a dam in the Three Gorges of the Yangtze River can be traced back to the Sun Yat-sen period. After repeated consideration by Chairman Mao Zedong and other national leaders, in March 1992, the engineering proposal was submitted to the Fifth Session of the Seventh National People's Congress for consideration and obtained approval. The Three Gorges Hydropower Station was officially started in 1994 and began to store water on the afternoon of June 1, 2003. It was completed in 2009 which is currently the largest water conservancy project in the world, with three major benefits: flood control, power generation, and shipping.

4. **High-speed railway project**

The railway is the main artery of the Chinese national economy and has made important contributions to economic construction. In recent years, China's high-speed railway has developed rapidly. By the end of 2013, China had built 39 high-speed railways, including the Beijing-Tianjin inter-city railway and Beijing-Guangzhou high-speed railway, Beijing-Shanghai high-speed railway, etc., and completed a mileage of 110,000 km. At present, a high-speed railway with a top speed of 350 km/h has been designed, which runs at a speed of 300 km/h. The high-speed railway with a top speed of 250 km/h has been designed to operate at a speed of 200 km/h.

5. **Qinghai-Tibet railway project**

The Qinghai-Tibet Railway is known as the "Tianlu," a landmark project for implementing the strategy of developing the western region. It is one of the four major projects of China in the new century. It starts east from Xining City in Qinghai and south to Lhasa, Tibet, with a total length of 1956 km. Most areas are cold and oxygen-poor, the environment is harsh, and the project faces many technical difficulties in the world, such as permafrost, ecological fragility, and bad weather. The construction of the Qinghai-Tibet Railway has taken comprehensive measures, such as the bridge and the ventilation pipeline foundation, to solve the problem of frozen soil. During the project, the ecological balance of the area was ensured by the approach of restoring 1 m^2 of vegetation per damaged 1 m^2 of vegetation. 33 Tibetan antelope passages were also established for the passage of Tibetan antelope and yellow sheep.

6. **Bridge projects**

China's bridges have a long history, dating back to the Yin and Shang Dynasties. The famous bridges include Zhaozhou Bridge, Lugou Bridge, etc. In 1957, the first Yangtze River Bridge, Wuhan Yangtze River Bridge was built. In the past thirty years, more than 70 bridges have been built on the Yangtze River. Today, China's bridge construction ranks first in the world. Existing bridge types include the arch bridge (Chongqing Chaotianmen Yangtze River Bridge), Liangqiao (Shibanpo Yangtze River Bridge), cable-stayed bridge (Sutong Bridge), suspension bridge (Hong Kong Tsing Ma Bridge), cross-sea bridge (Hangzhou Bay Bridge), etc.

7. **High-performance computer engineering**

In 1946, the world's first digital electronic computer was born. Since then, human life has undergone tremendous changes. In 1983, China's first supercomputer, "Yinhe" was successfully developed, and its speed can reach more than 100 million operations per second. At present, domestic high-performance computers represented by "Shenwei," "Dawn" and "Tianhe" have created "China Speed" in terms of computing speed and product performance, which has dramatically enhanced national competitiveness. In 2010, China's first petascale computer, "Tianhe No. 1," became the world's fastest supercomputer at a sustained rate of 2566 trillion per second. In 2013, the Tianhe No. 2 supercomputer system was successfully developed. It ranked first in the world in the 2013 international supercomputer top 500 list.

8. **Liaoning ship engineering**

The Liaoning ship was the first aircraft carrier of the Chinese People's Liberation Army Navy that could carry fixed-wing aircraft. It is 309.3 m long, draught 10.5 m, and can carry more than 50 aircraft, of which 26 are J-15. In November 2013, the Liaoning ship sailed from Qingdao to the South China Sea for a comprehensive marine exercise, marking the beginning of the capability of the maritime formation battle group.

9. **Early warning aircraft engineering**

The early warning aircraft is known as the aircraft carrier's clairvoyant. It mainly comprises aircraft carrier shipborne early warning aircraft and reconnaissance satellites. It is an information-based special military aircraft that integrates electronics, information, aviation, and other fields. China's existing self-developed early warning aircraft models include Air Police 2000 and Air Police 200.

10. **Jiaolong manned submersible project**

The Jiaolong Manned Submersible is independently developed by China and is mainly used for seabed resource development and seabed-specific operations which have four characters: first, the dive depth reaches 7000 m; the second is the hovering positioning capability, which provides a reliable guarantee for the high-precision operation of the submersible; the third is the advanced underwater acoustic communication and seabed micro-topography detection capability; the fourth, equipped with a variety of high-performance work tools. The successful development of the "Jiaolong" manned submersible has enhanced China's international influence in the field of deep-sea technology and enhanced the confidence of Chinese deep-sea researchers.

11. **Oil and gas exploration and development engineering**

After more than 60 years of hard work, China's oil and gas exploration methods and technologies, refining, and major petrochemical products have been fully developed. The crude oil processing technology has been fully grasped and reached an advanced level in the world.

12. **Deep-sea drilling platform engineering**

"Offshore Oil 981" is China's first self-designed and constructed sixth-generation deep-water semi-submersible drilling platform. The maximum operating depth is 3000 m, and the maximum drilling depth is 10,000 m, indicating that China has established independent research and development capabilities in marine engineering equipment and has strong international competitiveness.

13. **Ultra-high voltage (UHV) technology**

UHV transmission technology refers to AC transmission engineering and related technologies with voltage levels of 1000 kV and above, featuring long-distance, large capacity, low loss, and economic performance. China's energy and power load distribution is uneven, the western region is rich in energy, and the eastern economy is economically well developed. More than two-thirds of the country's electricity load is concentrated in the east of the Beijing-Guangzhou Railway. Existing UHV transmission technologies cannot meet the needs of future power growth. Therefore, it is necessary to accelerate grid development and technological innovation.

14. **Hybrid rice engineering**

Food is essential for people. The staple food of more than half of the world's population is rice-based. According to experts' estimates, in order to meet the needs, global rice production should increase by 60% by 2030 compared to 1995. China is a populous country and a grain power. Internationally, hybrid rice is called "Oriental Magic Rice" and has greatly contributed to China's food security and world grain production.

Since the 1960s, scientists and technicians led by Academician Yuan Longping have developed three-line method hybrid rice, two-line method hybrid rice, super hybrid rice, and other varieties through years of breeding. In 2000, the super hybrid rice reached a yield of 10.5 tons per hectare; in 2012, the yield of the super hybrid rice reached 13.5 tons per hectare.

15. **Medical engineering**

Since the founding of New China, the health level of urban and rural residents has been greatly improved. In the early days of the founding of New China, China's average life expectancy was 35 years. In 2010, China's average life expectancy was 74.83 years. At the same time, China's maternal and infant mortality rates have fallen sharply. With the continuous development of science and technology, the medical level is constantly innovating. The national physical quality is also enhanced continuously. The nation's overall health is greatly improved, and the life happiness index is increasing day by day.

Since the founding of the People's Republic of China, especially after the reform and opening up, in this ancient land of China, the most significant engineering projects in the history of mankind have been carried out, which has brought significant changes to the entire country. The above 15 major projects are only typical representatives, but they have not yet included large-scale municipal construction projects in all levels of towns and villages in the country.

1.2.3 Engineering Technology

Engineering technology is "the means and methods for human beings to create, control and apply artificial natural systems by relying on natural laws and the material, energy, and information of nature in order to meet the needs of society." [11] Engineering technology has both natural and social attributes.

Construction engineering technology is the oldest engineering technology. In the past, people often regarded engineering technology as only construction engineering technology. A broad concept of engineering technology has been formed with the different purposes of human beings using nature and the different methods used.

Generalized engineering technology includes many fields, such as hydraulic engineering technology, power engineering technology, metallurgical engineering technology, material engineering technology, mining engineering technology, etc. In recent decades, with the comprehensive development of science, technology, and engineering, the concepts, means, and methods of engineering technology have penetrated into all aspects of human activities, resulting in information engineering technology, genetic engineering technology, system engineering technology, satellite engineering technology and so on. Military engineering technology is the integration of engineering technology in different industries. Engineering technology has broken through the scope of industrial production technology and is closely related to people's production and life.

To further illustrate the generality of engineering technology, here are two examples of authoritative activities. In the twenty-first century, the US National Academy of Engineering, the American Federation of Engineers' Associations, and the National Engineers Weekly, plus 27 institutes/associations, selected the most remarkable engineering techniques of the twentieth century. Twenty engineering technologies were selected: electrification, automotive, aircraft, water and distribution systems, electronics, radio and television, agricultural mechanization, computers, telephones, air conditioning and refrigeration, highways, spacecraft, the Internet, imaging technology, appliances, medical technology, petroleum and petrochemical technologies, laser and fiber optic technology, nuclear technology, and high-performance materials [12] The Chinese Academy of Engineering also organized a group of academicians, and together with 24 associations such as the China Aerospace Society, 8 industry associations, 14 national ministries and commissions, and large enterprises and institutions, selected China's major engineering and technological achievements in the twentieth century. The selected 25 engineering technologies are: two bombs and one satellite, Chinese character information processing and printing revolution, oil, crop yield increasing technology, infectious disease prevention, electrification, river governance and development, railway, ship, steel, family planning, telecommunications engineering, geology exploration and resource extraction, livestock and poultry aquaculture technology, broadcasting and television, computer, highway, major mechanical technical equipment, aviation engineering, inorganic chemicals, surgical treatment, development and application of rare metals and advanced materials, urbanization, light industry and textiles, and coal mining project [13].

The innovation, diversity, selectivity, and phase of technology are discussed from the perspective of innovation in Sect. 1.1.2.2. Engineering also has these four characteristics. Engineering technology is the means and method adopted by human beings for the purpose of survival and development, and it must be feasible from the application point of view. Feasibility requires engineering technology to be practical, economical, mature, and integrated.

1.2.3.1 Feasibility

Engineering technology is to complete the engineering service. In the process of achieving the goal, it will be bound by the engineering project itself and objective conditions. The engineering itself mainly has project establishment, scale, and progress. Objective conditions include natural and social conditions, such as the environment, financial capabilities, materials, equipment, etc. As far as engineering technology is concerned, in the engineering demonstration and design stage, it is necessary to consider the needs and possibilities of national economic and social development and form various options. After analyzing and evaluating each option, selecting technologies that meet both the requirements and the corresponding constraints is considered feasible.

The feasibility of engineering technology does not depend entirely on the advancement or complexity of the technology itself but also on the time of the project, geography, society, economy, environment, etc. Therefore, for a certain technology, feasibility is not a fixed concept. The same technology is not feasible in one project and may be feasible in another project; it may be feasible in the same engineering phase and may not be feasible in another phase. Many factors determine the technical feasibility of engineering, mainly including practicality, economy, maturity, and integration.

1.2.3.2 Practicality

Any technology that uses its natural rule to achieve its use in a certain form of matter has its own characteristics and application conditions. Moreover, similar technologies each have their advantages and disadvantages and must be traded off in engineering activities. Based on its specific use, it's necessary to minimize the shortcomings' impact and maximize its advantages. The technology that deviates from engineering use, no matter how advanced it is, is meaningless. Therefore, engineering technology must be practical; otherwise, it will have no vitality.

1.2.3.3 Economics

Engineering is governed by natural laws and governed by social laws, especially economic laws. To apply a certain technology to engineering, it must first be feasible from a technical point of view. This is the most basic, but not the only one. Engineering technology must promote economic and social development as an important task and achieve the unity of technological advancement and economic efficiency. This requires that good economic benefits be realized so that the materialized form of engineering technology is artificial nature benefitted in terms of social, economic aspects. Otherwise, it will be uncompetitive and challenging to sustain or develop rapidly. The technological advancement and poor economy of the Iridium is the best example.

The original design spreads 77 satellites evenly on seven orbits. The communication system formed is similar to the 77 electrons distribution of the element Iridium atom, so it is called the "Iridium" communication system. It can make communications anywhere on the planet unimpeded, thus starting a new era of personal satellite communications. Iridium communication is technologically advanced, but its price is ten times higher than regular satellite communication. Due to various economic reasons, the Iridium project has quickly bankrupted with more than $4 billion in debt.

1.2.3.4 Maturity

There is often a misconception that engineering technology should adopt the most advanced technology, but this is not the case. In order to ensure security and reliable continuous operation after the completion of the project, engineering technology must first be reliable and safe, that is, mature. Or, to be safe and reliable, adopt advanced technology, not necessarily the most advanced technology. Whether or not to use the most advanced technology, and prove its safety and reliability is important. Its economic factors must also be considered.

1.2.3.5 Integration

Engineering technology is often the integration of multiple existing technologies. For example, civil engineering projects integrate concrete technology, wood technology, metal materials technology, plumbing technology, electrical technology, lighting technology, air conditioning technology, information technology, etc. With the development and progress of modern engineering technology, integration is becoming more and more apparent. Integration is the integration of multiple technologies in an orderly fashion rather than a simple overlay.

1.2.4 Engineering Education

At the end of this section, it is necessary to introduce STEM briefly. STEM is a general term for science, technology, engineering, and mathematics. "STEM education" refers to education in science, technology, engineering, and mathematics. On January 31, 2006, US President Bush announced the "American Competitiveness Initiative (ACI)" in his State of the Union Address, which proposes that cultivating talents with STEM literacy is one of the educational goals in the era of the knowledge economy. It is called the key to global competitiveness. Since then, STEM has become a popular term, and the US government has continuously increased investment in STEM education and strengthened the cultivation of students' science and technology literacy.

Engineering practice is the key to the development of engineering and technical talents. The development experience in the world shows that it is impossible to produce engineering science and technology talents without the work positions provided by engineering practice. This is also the fundamental rule for the growth of engineering science and technology talents. For example, the atomic bomb project in 1941 in the United States cost $2 billion, and 100,000 scientific and technical personnel participated in the project. The development of the atomic bomb was successful and dropped on Hiroshima and Nagasaki, which gave the Allied troops a decisive victory. In the 1960s, the US government invested $25 billion in Apollo's moon landing project, which lasted 11 years, more than 20,000 companies nationwide, more than 120 universities and research institutions, and hundreds of thousands of people participated in this project. In 1969, Apollo successfully achieved the feat of human landing on the moon for the first time, resulting in a leading position in the US aerospace industry and substantial economic benefits. The Information Highway Program in 1993 made the information industry one of the four pillar industries in the United States today. In the 1980s, the US auto industry was strongly impacted by Japan. At that time, the government launched the Next Generation Automotive Cooperation Program, which successively proposed two slogans: "Developing technology for the benefit of the United States" and "Technology is the engine of the economy." In September 1993, there were eight government departments, three major automobile companies (GM, Ford, Chrysler), and 453 units in 38 states (colleges, national laboratories, technology suppliers, etc.) jointly participating in the "governmental, production, learning, research" joint project. There are also nuclear development programs, TMD programs, etc., which have trained and created many engineering and scientific talents.

However, due to the economic slowdown and the difference in income and working conditions, top high school graduates in the US choose to major in economics, law, medicine, and other related disciplines instead of STEM. The percentage of the number of doctoral students in science and engineering in the United States fell from 50% in the 1970s to 15% in 2010.

At the beginning of the founding of New China, in addition to the "two bombs and one satellite" project, China has also built 12,000 tons of free-forging hydraulic presses, nine major pieces of equipment, annual output of 1.5 million tons of steel complex sets of equipment in Panzhihua, and the full set of production equipment by the second automobile manufacturing plant. After the reform and opening up in the late 1970s, China took the development of major equipment as a national goal. It took about ten years to manufacture key equipment of 300 and 600 MW thermal power units, slab casting machines, and heavy-duty trains for coal transportation units. The technology has achieved made-in-China, breakthroughs in coal magnetic pulse vibration unloading technology and shallow draft coal transportation ship. China has developed many products and core technologies with independent intellectual property rights in the hybrid rice project, the Three Gorges Project, the Qinghai-Tibet Railway, the military industry, etc. A large number of engineering and technical talents have been trained.

China's unique conditions are different from many other countries in the world. For example, good student sources, large scales of student populations, and a broad job market. More than 50% of bachelor's degrees in China are awarded in science, technology, engineering, and mathematics, compared to only 17% in the US. It can be expected that in the next 15–20 years, a large number of engineering constructions in China will cultivate a large number of world-class engineering and scientific talents who will become a beautiful landscape in the world. By then, many large-scale projects worldwide will be presided over by Chinese engineers and engineering teams. In fact, Chinese companies have won bids for construction in many large-scale international projects such as hydropower stations, high-speed railways, highways, and large buildings. Not only that, but the author also believes that by the 100th anniversary of the founding of New China, China's overall engineering technology level will rank among the best in the world.

1.3 Engineering Management

1.3.1 The Meaning of Engineering Management

Engineering management refers to the management of engineering activities.

From the perspective of ontology, "engineering management" establishes the "main body" position of engineering. In engineering management, management is attached to engineering. Naturally, no engineering, and there is no engineering management. But this does not mean that management is not important. Modern engineering, especially large-scale engineering, cannot carry out scientific engineering activities without scientific management.

First of all, to understand the relationship between engineering and management, we must take engineering as the main body. Secondly, management knowledge can not be copied; it is based on the actual activities of engineering. Thirdly, the management theory and method are not simply copied and applied but combined with the reality of the engineering, based on digestion and absorption, try to develop and upgrade, adapt it to the needs of the engineering, and integrate it with the engineering to form a whole. That is a new form of existence—engineering management.

The definition of science in Sect. 1.1.1 can also be extended as follows:

The knowledge system about the objective law of engineering management is engineering management science.

From this point of view, engineering management science is an independent science. Moreover, because engineering and engineering management involve engineering, society, nature, and many other aspects, it is also a comprehensive science. From the discussion below, we will see "Engineering Management" as an independent discipline containing science, technology, and art.

The International Conference on Engineering Science and Technology 2014, jointly organized by the Chinese Academy of Engineering, UNESCO, and the International Council of Engineering and Technology Sciences (CAETS), was held in Beijing on June 3, 2014. There are nine parallel sessions in this conference, and the session on Engineering Philosophy and Engineering Management is one of them. It indicates that the international engineering community recognizes Engineering Science and Engineering Management Science [14].

It is not an easy task to establish a complete definition of engineering management. In 2006, Jishan He et al. cited the ideas of Henri Fayol (1844–1925) in the study of "engineering management connotation and definition," considered the importance of decision-making in engineering management, and defined engineering management as "the decision, planning, organization, command, coordination, and control of the engineering" [15]. Similarly, the American Engineering Management Institute defined engineering management in 2012 as "an art and science that plans, organizes, distributes, directs, and controls activities which have technical elements." And they believe that: engineering management is a bridge between engineering and management disciplines [16]. Such definitions are all discussed from the functions of engineering management and belong to the one-dimensional definition.

In order to understand engineering management more comprehensively, this book constitutes a four-dimensional definition of engineering management from functions, processes, elements, and philosophy [17].

(1) From the perspective of engineering management functions, engineering management refers to the decision-making, planning, organization, command, coordination, and control of the engineering;
(2) From the perspective of the engineering process, engineering management refers to the management of the whole process of the engineering, that is, the pre-project assessment, decision-making, design, medium-term implementation, operation after completion, until the management of decommissioning;
(3) From the perspective of engineering management elements, engineering management is the integrated management of organizing, quality, cost, construction period, occupational health and safety, environmental protection, resources, contracts, risks, technology, information, culture, etc. in engineering activities;
(4) From a philosophical perspective, engineering management is about the status and role of people in engineering activities, the relationship between people, people and engineering, engineering and society, engineering and nature, and the interaction of science, technology, and art.

In this definition, engineering management integrates science, technology, and art.

1.3.1.1 Engineering Management Science

In management study, many of the original scattered contents have been refined into a scientific knowledge system. For example, Taylor's scientific management has developed into industrial engineering. Another example is that human-oriented management has evolved into behavioral science. For another example, applying the quantitative study to management has evolved into operations research, and so on. It can be seen that the content of management itself can be developed into a scientific knowledge system, which is the scientific nature of management [18]. From the perspective of engineering management, the understanding of objective laws such as engineering decision-making, engineering implementation, engineering operation, and engineering decommissioning belongs to the category of engineering management science. In addition, the basic connotation of engineering management science can be understood at three levels as follows: The first level, when the engineering management activities make full use of quantitative analysis such as mathematical methods, construct many engineering management activities into a mathematical model, look for the variables in the function, try to control this variable to obtain the mathematical solution, and then apply it to engineering management practice, it becomes engineering management science, which is the narrowest understanding of engineering management science. The second level, based on the discussion of the nature, functions, means, processes, and methods of general management, engineering management science integrates engineering practice, plus mathematics, statistical methods, and modern scientific methods such as the use of computers and networks to seek systematic understanding. The third level, engineering management science, has evolved into a comprehensive discipline; it focuses on the purpose and characteristics of management, applying knowledge from philosophy, sociology, history, law, political science, mathematics, economics, and other subject knowledge, including natural sciences, social sciences, and thinking sciences. Now, with the increasing scale of engineering as well as the gradual increase of the level and difficulty, engineering management science has gradually risen to the level of a comprehensive discipline.

1.3.1.2 Engineering Management Technology

There are many engineering management technologies, such as engineering target control technology, engineering information technology, engineering financial management technology, engineering risk management technology, engineering assessment technology, etc., all of which belong to engineering management technology.

1.3.1.3 Engineering Management Art

Although engineering management is scientific, engineering management in behavior aspects is also artistic. Engineering management art can have a significant impact on the role of engineering management science and technology. The word art here is not the same as the daily understanding of art, such as stage art, painting, or calligraphy. The art of engineering management refers to the morality, charm, style, ways, methods, etc., of being an engineering manager. In his book *The Art of Leadership and the Wisdom of Management*, Carnegie summarizes leadership art into five major areas: reconciliation skills, communication skills, persuasion skills, work skills, and eloquence skills [19].

Some scholars believe that "science can't reflect all the engineering management behaviors. The engineering management content or behavior that science can't reflect is called the artistic nature of management. The artistry of management can have multiple performances, but engineering management needs to rely on people's intuition, experience, and insight. In many cases, intuition, experience, and insight are difficult to express in words, so a system of knowledge cannot be formed. But intuition, experience, and insight are very flexible and highly creative, which is the subtlety of the artistry of engineering management. Second, some problems in the engineering management process are difficult to quantify accurately and objectively, unable to use mathematics to formulate, and cannot follow fixed procedures. For example, human latent energy and behavior, accidental changes in people and things, random feedback between various factors in the management system, various accidents, etc., all of which depend on the artistic nature of management. Third, some management problems can be quantified, modeled, and procedurally processed. However, due to the limitation of people's cognitive ability and cognitive tools, the quantified patterns of things and procedures determined and reflected by people also have limitations. This limitation can only be compensated by the artistic nature of management" [18]. In this discussion, the author sums up the three aspects of engineering management that are currently difficult to form a systematic understanding of and believes that it can only be expressed in terms of artistry.

The above facts exist, but the conclusions need to be discussed. As mentioned in Sect. 1.1.1.1, science is a disciplinary knowledge system that reflects the objective laws of nature, society, and thinking [4]. First of all, "human intuition, experience, and insight" belong to the category of thinking. Therefore, the disciplinary knowledge system that reflects the objective laws of "human intuition, experience, and insight" belongs to thinking science. However, currently, the research is not enough in this area. Secondly, for "some problems in the engineering management process are difficult to accurately and objectively quantify," it is necessary to clarify a question: Is it called science that must have a model, calculation, and accurate quantification? Actually, this is a misunderstanding. Any model simplifies objective things, an approximate expression, a research tool, not science itself. It cannot be said that the expression without the "mathematical model" is not science. Even for physics and chemistry with a large number of formulas, there are many theorems and axioms that have no formulas but are expressed in narratives. The book, *the Art of Leadership and the Wisdom of Management* by Carnegie, is a scientific exploration of management art. The author believes that management art is mainly in the category of thinking, and its objective knowledge system belongs to management art science. Art itself contains techniques such as painting, performance, eloquence, etc. "Science, technology" in the expression "engineering management science, technology, and art" is relatively narrow and does not include "art science and technology."

Although engineering management has various techniques and corresponding regulations, the importance of management art is beyond doubt in practice. Because engineering management manages not only things, but a more critical object is people with ideas. Ideas can change with time and space. A rule or procedure will go smoothly if people can agree and accept it happily. Sometimes when the project encounters unprecedented difficulties and problems, a moving talk by the manager may arouse the enthusiasm of the masses. They may work together to solve the problem.

1.3.1.4 Basic Characteristics of Engineering Management

Unlike general management, engineering management is the management of various engineering projects with technical integration and industry-related characteristics. In general, engineering management has the basic characteristics of systems, comprehensiveness, and complexity [15].

(1) Engineering management is systematic management. In theory, the systematic expression of engineering management is an orderly integration of various technologies to achieve specific goals, that is, the organic integration of various engineering components and the coordination of various engineering subsystems to achieve the overall goal of the engineering. In the practice of modern engineering management, the application of system theory and system thinking is indispensable, which is the essence of engineering management thinking.
(2) Engineering management is a comprehensive management. Because engineering requires the organic integration of technology, engineering is often associated with specific artificial nature, specific products, and specific groups. Therefore, any form of engineering management must be comprehensive, considering different technical coordination and industrial characteristics. In addition, the comprehensiveness of engineering management is also reflected in the effectiveness of the various resource utilization required for the realization of the engineering goals and the coordination between the engineering management entity and the engineering management environment.
(3) Engineering management is complexity management. In general, engineering consists of multiple parts and multiple organizational participation; therefore, engineering management is highly complex and requires multidisciplinary knowledge to solve problems. Since engineering itself has many unknown factors, and each aspect often has uncertainty, it is necessary to organically organize people with different experiences and from different organizations in a specific organization to achieve expected aims under various constraints. Therefore, the complexity of engineering management is much higher than general production management.

When understanding the status and role of engineering management, it reminds us to quote Jiang Zemin's speech at the 2000 Shanghai International Engineering Technology Conference, that is, "Engineering science and technology have always played an engine role in promoting the progress of human civilization" [20]. Since engineering management plays a leading role in engineering activities, we can say that engineering management is the energy and power to promote the engine of human civilization.

1.3.2 Scientific Management and Project Management All Originated from Engineering Management

Due to the large scale and deep impression of civil engineering projects, people only regard the management of civil engineering as engineering management for a long time. For this reason, many cases of engineering management in the past have not been included in the scope of engineering management. After examining the history of scientific management and project management, it can be clearly seen that their origins are closely related to engineering management.

1.3.2.1 Scientific Management Stems from Engineering Management

Frederick Winslow Taylor (1856–1915), the famous American management scientist is known as "the father of scientific management," entered Philadelphia's Leadville Steel Company at 22. At the age of 28, he was the chief engineer of the steel company. In 1898, he joined the Bethlehem Steel Company for management research. He has been engaged in technical and management work in steel companies for a long time, and his working experiences are the practical basis for his research on scientific management. His main achievement in his life is how industrial engineering management can improve production efficiency. The famous tests of "moving pig iron" and "iron shovel" were carried out by Taylor during his employment at Bethlehem Steel Company in 1898. In 1911, based on the practice of metallurgical engineering management in his early years, he published *The Principles of Scientific Management*, which laid the foundation for scientific management [21].

In addition, the scientific management concept created by Taylor does not only refer to part-time learning and reasonable factory organization. His main idea was that engineers should strive for professional autonomy within the enterprise, to a certain degree, enjoying superior status that the enterprise does not affect. He tried to re-establish the traditional autonomy of engineer entrepreneurs and the image of independent engineers in the nineteenth century under the economic environment characterized by large enterprises in the twentieth century [22].

As one of the main representatives of classical management theory, Henri Fayol was committed to mining coal mines when he was young. At the age of 30, he became a person in charge of a group of mines. At the age of 40, he became the CEO of the coal and iron joint ventures. And then, he worked for more than 30 years until he retired at the age of 77. Fayol is the founder of the management process theory, and his main achievement is also based on engineering management. Fayol believes that the functions of management are planning, organizing, commanding, coordinating, and controlling. His main work is *General and Industrial Management* [21].

Peter F. Drucker (1909–2005), known as the father of modern management, was hired by General Motors, the world's largest company, in 1942. At the site of such a giant manufacturing engineering, he studied the internal management structure. In 1946, based on his thoughts, he wrote a famous book, *Concept of the Corporation*. The book makes a groundbreaking discussion on engineering organization, "telling how people with different skills and knowledge can work together in a large organization." Drucker first proposed the concept of "organization" and laid the foundation for the organizational theory.

Taylor's management practice is metallurgical engineering, Fayol's management practice is mining engineering, and Drucker's management practice is manufacturing engineering. Metallurgical engineering, mining engineering, and manufacturing engineering are all engineering disciplines. It can be said that they are the pioneers of engineering management.

1.3.2.2 Project Management Stems from Engineering Management

Project management is a widely used and successful management method. Its early breakthrough and famous ground-breaking cases belong to engineering management. Project management originated from engineering management and applied to engineering management in its development process.

The Louisville chemical plant in the US must be continuously produced day and night. In order to carry out a comprehensive overhaul, it is necessary to arrange a certain period to stop production every year. Since the time management technique "Critical Path Method (CPM)" was invented in 1957, the maintenance time was reduced from 125 to 78 h, saving 38%. Now, the "Critical Path Method" is still used in project management.

The US Navy began developing Polaris missiles in 1958, a substantial military engineering project. Based on the "Critical Path Method" technique, they used the "three-valued weighting" method to plan and design, formed Program/Project Evaluation and Review Technique (PERT), modelized the relationship between the project tasks and shortened two years of the time of completion. It took only four years to complete the project scheduled in 6 years, saving more than 33% of the time.

Apollo moon landing project, with 20,000 companies involved, 400,000 people, 7 million parts, and a cost of $30 billion, was a famous mega project in the 1960s. It used Network Planning Technology to ensure that it could be successfully completed.

By the 1970s, the scale of construction was even larger, the project type and external environment became more and more complex, and the application of project management gradually expanded from military engineering and aerospace engineering to construction engineering, water conservancy engineering, electric power engineering, petrochemical engineering, and other fields. As a successful management method, project management is no longer focused only on engineering management. It has become a critical integrated management method for many government departments and large enterprises in the US and Europe. Even the World Bank manages each loan as a project.

1.4 The Core Values of Engineering Management

China has achieved remarkable achievements with a large number of major engineering projects represented by the two bombs and one satellite project, the manned spaceflight project, the Three Gorges project, and the Qinghai-Tibet Railway project. The management practice of this engineering is itself a very rich encyclopedia of Chinese engineering management. Through systematic investigation and research, the author extracts **the core values of engineering management**.

People-oriented, harmony between nature and human, collaborative innovation, building harmony [17].

"People-oriented, harmony between nature and human" is traditional Chinese philosophy's earliest and longest-standing philosophical proposition. It has a very rich connotation and is considered to be the main point of Chinese traditional culture. People-oriented, harmony between nature and humans, as the main point of traditional culture in China, will inevitably have a subtle influence on all Chinese and will inevitably be integrated into the concept of engineering management.

1.4.1 The Origin of "People-Oriented."

"Oriented" has two meanings: one is the source, and the other is fundamentality. The "oriented" in "people-oriented" is fundamental, the root and foundation of things, and the most important part.

Among people, gods, and things, ancient China has long recognized that people are the most important. *Shangshu·Taishishang* describes that King Wu of Zhou said in the oath of the meeting of the feudal lords in Mengjin, Henan, that heaven and earth are the parents of all things, and people are the spirit of all things.

The "people-oriented" formulation was first seen in *Guanzi*. In the article of Bayan of Guanzi, it said that "To be an overlord, it must be people-oriented; Only by straightening out people can the country be consolidated, otherwise the country will be in danger." The book *Guanzi* is the collection of thoughts and speeches of Guan Zhong (about 723 BC–645 BC), who was the prime minister of Qi State. It was collected and edited by his students and successors [23]. "*Bayan*" is the suggestion that Guan Zhong made to Duke Huan of Qi outlining how to be an overlord. Guan Zhong's people-oriented; it is based on the people.

People-oriented thinking of Confucianism is consistent. The tenth chapter of the *Analects of Confucius* said the horse shed caught fire. Confucius retreated and asked, "Is anyone hurt?" without asking about the loss of the horse. This text is a very prominent expression of people-oriented thinking.

"Introduction to Chinese Culture" takes "people-oriented" as the main item of the four major points of Chinese traditional culture, and the other three are "the harmony between nature and human, vigorous and promising, and respecting harmony and balance" [24].

In ancient China, there was no difference between "people" and "citizens," and the basis of people is also the basis of the citizen. But as a philosophical value, there is a difference between the basis of people and citizens. People are relative to things and God. People-oriented is about the relationship between people and things, between people and God. Compared with things and God, people are fundamental, and people are more important. A citizen is relative to the king, relative to the government, or relative to the official. In Jin Xin Zhang Ju Xia, Mencius said the citizens are important, the country is second, and the king is minor. This is an important idea put forward by Mencius, which means that the citizens are the first, the country is second, and the king is at the end. Because with the citizen, it is possible to establish a country; it is necessary to have a "king" with the citizen. This is a classic expression of the Confucian political philosophy of the citizen-centered.

The Third Plenary Session of the 16th Communist Party of China (CPC) Central Committee clearly stated that "the people-oriented principle should be adhered to, and a comprehensive, coordinated and sustainable development concept should be established to promote the comprehensive development of the economy, society and people." The 18th National Congress of the CPC put forward "promoting people's all-round development" and "adhering to people-oriented," and clearly put people-oriented as the highest value orientation of development. Adhering to the people-centered principle is consistent with the fundamental purpose of serving the people wholeheartedly of CPC and the demands of representing the fundamental interests of the overwhelming majority of the Chinese people. It is to respect people, understand people, care for people, that is, to constantly meet the overall needs of people and promote people's comprehensive development as the fundamental starting point for development.

The Western humanistic management philosophy is based on humanistic philosophy. "Humanism" stems from humanistic thinking. The main object of the study of scholars in ancient Greece was the natural world, which focused on the common origin of all things in the world and placed human problems in second place. In the fifth century BC, Protagoras proposed that "human beings are the scale of all things," which opened the prelude to people's own exploration of human beings. Since then, through the rational and irrational confrontation between modern humanism and contemporary humanism, humanism's development has peaked. Although humanism was born earlier, due to the tension between labor and management, coupled with absenteeism and disciplinary relaxation, employees are regarded as the production factors of the organization. The organization aims to obtain benefits, and the rigid and strict system is used to improve efficiency and form a kind of management style of ignoring people. In the 1970s, employees were regarded as independent bodies alongside the organization, striving to achieve the goals in which individual employee goals are aligned with the overall goals of the organization, respecting their own values and development goals, and creating a broad space for development for individuals, at the same time achieving organizational prosperity. "People-oriented" management means taking people as the starting point and center in the management process, respecting personality, establishing the status of employees, caring for employee development, focusing on employee training, highlighting flexible management, motivating people's initiative, enthusiasm and creativity, and expanding management activities to achieve win–win development of people and enterprises.

1.4.2 "People-Oriented" in Engineering Management

The value orientation of people-oriented in engineering management emphasizes that people are the ultimate goal of engineering activities. The ultimate goal of engineering construction is to affirm the main position and role of people in engineering activities and respect people, rely on people, and protect people in engineering activities. The proposition that people are the fundamental purpose of engineering activities answers why engineering activities are carried out, and the engineering activities are "for whom"? And the claiming that people are the fundamental driving force of engineering activities answers the question of how to carry out engineering activities, and engineering activities "depending on whom"? "For whom" and "depends on whom" are inseparable. Everything is for people, everything depends on people, and the unity of the two constitutes the complete content of people-oriented engineering activities.

The purpose of engineering construction is for people; that is, people-oriented is to make the achievements of engineering construction benefit people and promote the overall development of people. Putting people-oriented in engineering management is relative to project-oriented. It is a negation and transcendence of the development mode of "material-oriented" that only pursues the enterprise or local interests and only focuses on engineering and ignores people. In July 2013, the project to build "Sky City"—the world's tallest building in Changsha, was signed in Wangcheng District, Changsha. This news has caused heated discussions around the world. At that time, it announced an investment of 9 billion yuan. In 7 months, it will build a 208-story, 838-m-high "sky city" in Changsha using "building blocks." Its height will exceed that of Burj Khalifa in Dubai. Sky City can provide residents with all living services such as residence, education, work, medical care, shopping, entertainment, etc. Unless you are on a business trip or traveling, you can live in the building without going out [25]. In fact, Changsha still has a lot of room for construction, and it is still far from being needed to build such high-rise buildings to alleviate the crowding. Moreover, once the high-rise building is completed, water supply, fire-fighting, rescue facilities, etc., will bring many new burdens to society.

What needs to be emphasized is that people-oriented is a way of thinking which requires people to develop human perspectives and provide humanized services when analyzing, considering, and solving all engineering problems—confirming people's subject status and goal status in the practice of engineering. Reflecting the value of respecting people, liberating people, relying on people, for people, and shaping people should be available in engineering practice [26].

Trying to build the Changsha air city is just an extreme case. Moreover, this project has not been fully implemented due to the intervention of the administration. A large number of engineering projects in China reflect people-oriented thinking. These kinds of cases are too numerous to enumerate. Here we only mentioned examples such as the Yangtze River Three Gorges Hydropower-complex Project, the railway speed-up project, the high-speed railway, and the Shendong coal mine project.

The Yangtze River Three Gorges Hydropower-complex Project is called for short as the Three Gorges Project. The Three Gorges include Qutang Gorge, Wu Gorge, and Xiling Gorge. The poet Li Bai once wrote a very famous poems: "Leaving at dawn the White Emperor City in the cloud; I've sailed a thousand miles through Gorges in a day. Apes cry load on riverbanks, my skiff has left ten thousand mountains far away", which describes the rush water flowing of the Three Gorges. The Three Gorges Hydropower Station dam is 185 m high, with a water storage height of 175 m and a reservoir length of more than 600 km. The construction of the Three Gorges Project has three major goals: flood control, power generation, and shipping. Its primary aim is flood control. The flood control capacity reserved for reservoir operation is 22.15 billion cubic meters. Reservoir flood control can reduce flood peak flow to 27,000–33,000 cubic meters per second. It is the world's largest water conservancy project, effectively protecting people's lives and property. In addition to flood control, the dam is equipped with 32 hydropower units with a capacity of 700,000 kW. It is the world's largest hydropower station with the largest installed capacity, and its power generation efficiency is enormous. The annual carbon dioxide generated by coal combustion is reduced by more than 85 million tons. These fully reflect the goal of building the Three Gorges Project is people-oriented [27].

At 0:00 on April 1, 1997, China Railway's first large-scale speed-adjusting and mapping project was fully implemented, which opened the prelude to speeding up the railway. On the world's busiest railway line, China had implemented large-scale speed-up under the conditions of passenger-cargo collinear operation and mixed-class trains. Moreover, China pays more attention to improving the average travel speed rather than simply pursuing the improvement of the running speed of a specific section. This requires optimizing transportation organization, compressing operation time, and selecting the best train arrival and departure times to maximize the time-saving needs of passengers [28]. For this speed-adjusting and mapping project, the train's maximum running speed reached 140 km. The national railway passenger train travel speed increased from 48.1 km per hour in 1993 to 54.9 km per hour; the express trains and the one day's (start at dawn and arrive at dusk) trains were opened for the first time, which greatly facilitated the travel of people.

The high-speed railway station is no longer just the station for railway passengers but an integrated transportation hub that connects the high-speed rail with the city's transportation system, such as buses, subways, taxis, long-distance buses, and even aviation, realizing a "zero transfer." These are examples of people-oriented engineering.

Confucius believes that the essence of management is that "cultivate oneself to appease others," people are the center, and making people satisfied is the ultimate goal. Human-centered management is different from the management mode of using people as tools and means but based on profoundly understanding the role of people in social and economic activities, highlighting the position of people in management, and realizing people-centered management. Shendong Coal Group follows this management philosophy, clarifies the main position of employees in management, invigorates the spirit of employees, standardizes the behavior of employees, establishes a scientific management system centered on employees, fully motivates the enthusiasm of employees, encourages the company to be full of vitality and hope, vigorously promotes the successful completion of the company's various reforms, and realizes the rapid, sustained and healthy development [29]. Moreover, many underground positions in Shendong coal mines, hazardous jobs, are fully mechanized and automated and are visualized and remotely operated on the ground, achieving "unattended." This fully reflects the people-oriented concept.

The purpose of engineering construction is people-oriented. People-oriented in engineering management reflects the status and role of people in engineering activities. Many concepts related to engineering management, such as engineers, engineering ethics, engineering organization, engineering safety, engineering innovation, and engineering culture, all run through the people-oriented red line.

1.4.2.1 Engineer—The Soul of Engineering Activities

Engineers are considered to be a career that has lasted for 6000 years, and the group of engineers has made outstanding contributions to the long history of technological advancement in human civilization. The concepts "Engineer" and "Engine" appeared in the Middle Ages in Britain. The *Oxford English Dictionary* recorded the word "Engineer" to refer to military engineers for the first time in 1300. In addition, the term is often used to refer to inventors, designers, cartographers, and authors [30]. The oldest civilization in the world is China and the countries of the ancient Near East. It has created ancient civilizations, built large cities and classic buildings, and has reasonable operational procedures, economic management under the bureaucracy, and the use of characters. These technical experts who solve practical engineering problems are recorded in the ancient Eastern literature by specific occupational titles, such as architects, river supervisors, etc., or by a generalized concept of "wise men." It can be seen that technicians are also at the intersection of theory and practice in ancient Eastern civilization. Early "engineers" worked in areas such as construction, mining, infrastructure, measurement, military, shipbuilding, transportation, and water conservancy. In these areas, specific tasks such as design, production, planning, management, and development have formed different professional groups, stipulating different job responsibilities, resulting in various occupational titles, which approximate with contemporary "engineers" and "technicians" [31].

In today's society, engineers are the soul of engineering activities, important creators of new productive forces, and active pioneers of emerging industries. The thinking and inspiration of engineering activities are mostly from engineers. Engineers are also the main carrier of engineering ethics and are responsible for engineering activities affecting society, the economy, and the environment. Engineers should have good professional ethics, innovative ideas and firmly master modern design, construction technology, and engineering management technology. The primary obligation of the engineer is to be loyal to the client or employer and the public's health, well-being, and safety. Engineering activities must rely on engineers and workers. They are required to realize that the prerequisite for enjoying the results of the engineering project is first to create the results and be aware of contributing to the engineering project based on their own capabilities. At the 2000 International Engineering Science and Technology Conference, Jiang Zemin pointed out that "all achievements China has made in modernization construction are inseparable from the tremendous support of engineering technology". The Chinese government and people highly value the contributions made by Chinese engineers. They are proud of many internationally renowned engineers represented by Zhan Tianyou, Mao Yisheng, Li Siguang, Qian Sanqiang, and Qian Xuesen" [19]. Since the reform and opening up, China has emerged many outstanding engineers represented by Lu Youjia and Sun Yongfu in many major engineering projects represented by the Three Gorges Project, Qinghai-Tibet Railway, Aerospace, Daqing Oilfield, and municipal projects throughout the country. These engineers have made outstanding contributions to China's modernization development.

1.4.2.2 Engineering Ethics—The Constitution of Engineering Activities

Although the definition of engineering ethics is not completely consistent, there are at least two aspects from the perspective of engineering ethics. On the one hand, engineering activities are a kind of social practice activities, and engineering ethics refers to the study of the moral values involved in engineering practice; on the other hand, as a sacred profession, engineers themselves should have their own unique professional ethics. Whether as practical or professional, engineering ethics has its normative and descriptive dimensions. The priority of engineering ethics is "engineering benefits mankind" [32], which is the same meaning as the concept of "people-oriented." Engineering ethics emphasizes the loyalty, honesty, responsibility, and team spirit of engineers. The broad responsibility should include the responsibility of engineering to society and nature, namely environmental protection and green engineering. Generalized engineering also includes bioengineering, information engineering, aerospace engineering, etc. In these engineering projects, engineering ethics should be extended to issues such as cloning human moral principles, network ethics, and space ethics. The core of the above engineering ethics is to embody the concept of being people-oriented and the harmony between nature and humans.

1.4.2.3 Engineering Organization—The Main Theme of People-Oriented Engineering Activities

The engineering organization is to organize the engineering personnel into an organic whole in an optimal way under the premise of respecting people to facilitate orderly engineering. The engineering organization exerts the overall effectiveness of the organization and exerts the individual's subjective dynamic performance. There is no doubt that engineering organizations are the central theme of engineering activities, and the core of engineering organizations is people-oriented. The goal of an engineering organization is "for people." Without the concept and thought of "for people," it is difficult to formulate optimized goals. Of course, this is limited to serving customers and users and considering the interests of engineering workers, providing a standard operating environment, doing a good job in safety management, and making reasonable arrangements for labor insurance and medical insurance. Engineering organizations also need to "manage people," using various management functions, including planning, organizing, commanding, motivating, etc., to manage engineering operations personnel. Engineering organizations also need to "rely on people," that is, those who are excellent in organizational skills, engineering practice, professional theoretical knowledge, and professional ethics. And, in engineering activities, in the relationship between people and people, we must emphasize justice, that is, respect the ability and contribution of elite groups and the basic needs, legitimate rights, and independent personality of vulnerable groups. It is necessary to provide a good interpersonal environment for all engineering people and create conditions for continuous harmony and win–win between people involved in engineering.

Qinshan Phase II Nuclear Power Plant has established and improved its organizational structure, management procedures, and guarantee system, achieved full coverage in the fields of nuclear safety supervision, quality assurance, environmental protection, radiation protection, industrial safety, occupational health, fire protection, and emergency management, and strictly implemented supervision and management before, during and after the event. The Qinshan Phase II Nuclear Power Plant organization implements the second-level management, company-level, and department-level management. The management functions on engineering design, procurement, construction, and functions about planning and contract control are all handled by separate management offices. These management offices are managed by a number of deputy general managers and deputy chief engineers or directly by the general manager. At the same time, Qinshan Phase II also set up a quality assurance department and a finance department independent of the project management and control department, directly under the management of the general manager [33]. The entire engineering organization is in order. This kind of engineering organization mode of Qinshan Phase II nuclear power plant is conducive to mastering the initiative of design autonomy and localization of equipment. It is flexible for purchasing equipment, conducive to obtaining technology transfer, facilitating interface coordination and unified command, and reducing the total cost of the engineering.

1.4.2.4 Engineering Safety is the Basic Ethics of Engineering Activities, and It is also an Essential Embodiment of People-Oriented Engineering Activities

Ensuring engineering safety is the primary bottom line of people-oriented engineering management. Engineering safety includes the safety of people and things, and the most important is the safety of people. From the procedures to the various facilities, it is necessary to ensure that the engineering personnel is not harmed; and the accidents that damage the engineering project (such as fires, collapses, etc.) cannot occur. Engineering safety should be positive to ensure that the engineering personnel is not physically harmed and to ensure their mental health. It is necessary to take care of engineering personnel in all aspects, actively adopt health care measures, and carry out various recreational activities to ensure their physical and mental health. In terms of engineering safety, safety education and warning content should be full of humanity, actively create a civilized construction, occupational safety, and healthy environment, and establish a clear and visual image. The content and scope of safety management for engineering are not limited to engineering construction areas, and the safety management of non-engineering areas needs to be emphasized and strengthened. The construction of the Qinghai-Tibet Railway fully considers the safety and physical health of employees. Since the construction of the Qinghai-Tibet Railway needs to face the problems of low temperature, lack of oxygen, dryness, wind, radiation, plague, etc., the following measures have been taken to ensure the health and safety of employees: establish a health protection system, build a high-altitude disease prevention and treatment system, jointly prevent the occurrence of plague epidemics, strengthen the management of high-pressure, explosive and flammable equipment and goods, and comprehensively manage traffic accidents, etc. [34]. For the problem of lack of oxygen, 17 oxygen-making stations and 25 hyperbaric oxygen chambers had been established along the Qinghai-Tibet Railway. The average mandatory oxygen intake per employee per day for 40,000 employees was not less than two hours. At the world's highest altitude, Fenghuoshan tunnel with an altitude of 4905 m, the China Railway 20th Bureau had developed oxygen-making equipment of 24 cubic meters per hour. The tunnel was filled with oxygen, and the increase in oxygen content of the air in the tunnel was equivalent to a reduction of 1200 m in altitude. During the five-year construction period of the Qinghai-Tibet Railway, tent hospitals could be seen everywhere on the construction site, and a total of 530,000 patients were treated. Among them, 470 cases of high-altitude cerebral edema and 931 cases of high-altitude pulmonary edema were effectively treated, and no deaths from high-altitude sickness occurred.

For the prevention of plague, they implemented a policy of "three noes" and "three reports," that is, not hunting for epidemic animals, not eating epidemic animals, not carrying the animal and product out of the epidemic areas; reporting the dead rat otter, reporting the suspected plague patient, and reporting unidentified patients with high fever and acute death. At the same time, they also established a plague isolation ward, established a monitoring and medical system, and trained professional rat protection personnel. During the construction of the Qinghai-Tibet Railway, no plague epidemic occurred. These measures fully guaranteed the safety and health of people in the construction of the Qinghai-Tibet Railway.

The fundamental mission of the Shenzhou spacecraft is to transport the astronauts safely into space and ensure they return to the ground safely. Guaranteeing the safety of astronauts is the first mission of the Shenzhou spacecraft. Therefore, the design of the Shenzhou spacecraft must implement the "people-oriented" concept. This "people-oriented" design concept reflects the difficulties and levels of the Shenzhou spacecraft project. The main challenges: (1) Shenzhou spacecraft is the first manned spacecraft in China, which factors must be considered in relation to people? (2) What are the physical and psychological characteristics of astronauts in a weightless environment? How is their ability to activity? (3) Reliability and security are not the same concepts, and sometimes they are contradictory. "Reliability is not necessarily safe, and safety must be reliable." Therefore, security analysis must be performed simultaneously for reliability design, and sometimes trade-off analysis must be performed to reduce risk. The theme of implementing the whole process of the development and testing of the Shenzhou spacecraft is carrying out the "people-oriented" design concept and ensuring the safety of astronauts. Through systematic analysis and testing, the spacecraft system had carried out a large number of research projects related to people as (1) astronaut medical topics; (2) ergonomics design and evaluation topics; (3) landing impact test verification topics; (4) astronaut manual control system design and verification topics; (5) independent emergency return design and verification topics, etc. According to this design and development of the Shenzhou-5 spacecraft, China's first astronaut Yang Liwei was successfully sent to space and returned safely [35].

1.4.2.5 Engineering Culture—Humanization of Engineering Activities

Engineering culture refers to the humanization of engineering activities, which is the embodiment of people-oriented in engineering. Each engineering project has its own specific environmental conditions and historical traditions, resulting in unique philosophical beliefs, ideologies, values, and behaviors. Therefore, each engineering project has its own unique engineering culture. Engineering culture is a kind of branch culture formed in engineering management practice under a specific cultural background. It is an applied culture closely integrated with engineering management practice. Culture is the foundation, and engineering is the platform. The role of culture is enormous; it can penetrate all aspects of engineering management. For example, advocating an innovation culture can make engineering innovation activities more in-depth; advocating a safety culture can make the space and time of engineering activities more consciously strengthen security measures, etc. Therefore, the construction of engineering culture in engineering management is an important guarantee for condensing engineering teams, improving the engineering management level, and promoting the successful completion of engineering. Strengthening the construction of engineering culture and giving play to the role of cultural orientation will help reduce accidents and improve engineers' enthusiasm and quality of work, thereby enhancing the overall benefits of the project. It is generally believed that engineering culture is the behavioral orientation formed by the main body of engineering to achieve the engineering project's goals. If this behavioral orientation deviates from the inherent laws of engineering activities, it may form a cultural atmosphere that is not conducive to the smooth development of engineering.

Engineering culture has another connotation: the culture and art embodied in the engineering achievements itself. Just because engineering works for people, the engineering results should reflect the pleasant feelings of people enjoying this achievement. For example, the Great Hall of the People and the History Museum in Beijing reflect the solemn, plain, and dignified Chinese tradition. In front of them, they can naturally inspire people's national pride and patriotic enthusiasm; any mosque must reflect Islamic culture and be loved by people of all nationalities.

China's manned spaceflight project has always adhered to the spirit of "being brave in innovation, unity and cooperation, scientific truth-seeking, people-oriented, and patriotic dedication." Aerospace engineering is one of the most innovative areas. For China's manned spaceflight project, innovation is both a driving force and a fundamental way out. The aerospace industry has the characteristics of high investment, high risk, intensive technology, and complex system. Scientific truth-seeking is the eternal theme of the aerospace industry and a guarantee for the successful completion of the aerospace mission. In June 2013, Shenzhou-10 was launched in Jiuquan, and it was docked with the target aircraft Tiangong-1. The whole process flew in orbit for 15 days, and the combination with Tiangong-1 flew in space for 12 days. The successful launch of Shenzhou-10 marks the application stage of China's manned spaceflight project, and manned spaceflight has reached a new level. The astronaut system is the first system in the eight systems of China's manned spaceflight project. In order to make astronauts fly safer, ride more comfortably, and operate more conveniently, the humanized design concept runs through the entire process of design, development, and production of spacecraft, rockets, and other aerospace products, which shows that manned spaceflight is people-oriented. From the moment of its inception, China's aerospace engineering bears the mission of the country and the dignity of the nation. It embodies the ardent expectations of hundreds of millions of people. The spirit of aerospace engineering vividly interprets the national spirit with patriotism as the core and the spirit of the times with reform and innovation as the core. It is the concrete embodiment of the socialist core value system in the aerospace field [36]. This culture's guidance ensures that Chinese aerospace engineers continue to make progress and innovate.

1.4.3 The Origin of "the Harmony Between Nature and Human"

"The harmony between nature and human" or "the harmony between heaven and human" is one of the fundamental concepts of Chinese classical philosophy and one of the most significant differences between Chinese and Western philosophy. The relationship between heaven (nature) and humans is an important topic of traditional Chinese philosophy. In the long Chinese intellectual history, different schools and eras, even the same scholars at different periods, have different connotations about heaven (nature) and humans. Although it is impossible to equate "heaven" and "human" with nature and humanity, in general, it contains the theory of the relationship between humans and nature.

In ancient Chinese, the "heaven" in the proposition of "the combination of heaven and man" refers to the natural world, sometimes called "heaven," sometimes refers to "heaven and earth" when it is called "heaven" alone. "Human" is a moral subject and responsible subject and bears the responsibility and mission for all things in nature. The moral subject is a harmonious coexistence relationship. "Harmony" is a positive "combination." What is a "combination"? How to combine? The first is assisting the transforming and nourishing powers of heaven and earth. If you can do this, you may, with heaven and earth, form a trinity. Second, people should take the initiative in front of nature. "Harmony" must also bless humans. Therefore, the "harmony" of "harmony between nature and human" is not human-centered, and human uses not everything in nature. It is also not to embed humans in nature without moving but is the dialectical unity of humans and nature in the process of mutual interaction and dedication. Humans and nature are the two focal points of the ellipse, which together form nature [37].

The mainstream Confucian theory of heaven and humanity is "the unity of nature and human," but the ideological connotations of the various representatives are different. *Liji·Zhongyong* said: sincerity is the way of heaven, to think how to be sincere is the way of man. It means that people must carry forward the moral of "honesty" before being consistent with heaven. Mencius said in *Jinxin*: he who has exhausted all his mental constitution knows his nature. Knowing his heart, he knows heaven. Dong Zhongshu was a famous Confucian in the Han Dynasty. He proposed in the *Chunqiufanlu·shenchaminghao* that "the correlation between heaven and human becomes one." Dong Zhongshu also put forward the "heaven-human induction" theory. Dong Zhongshu's "Heaven" and "Human" are generally the relationship between humans and nature. However, according to his "heaven-human induction" theory, he thinks that "Heaven" is a god with a natural appearance, willpower, and intent. If a person does something bad, heaven will drop disaster and punish this. Although he proposed that "the separation of heaven and human" was to persuade the Emperor Wu of the Han dynasty, he wanted him to follow the mandate of heaven, comply with the popular wishes of the people, and implement benevolent administration. Otherwise, he will be punished by heaven. The intention is good, but it is mixed with superstition.

As the philosophical concept of "the harmony between nature and humans" was first formally proposed by the Northern Song Dynasty Confucianism Zhang Zai in the seventeenth chapter of Zhengmeng. After criticizing the imaginary thoughts of Buddhism, he showed that Confucianism is "a Confucian scholar is sincere because of his understanding, and he achieves understanding because of his sincerity. That is why heaven and human are united as one". The above contents are all described from the moral and ideological level, and the concept of human is regarded as the attribute of "heaven." Therefore, "heaven" here refers to the concept of heaven and does not contain the meaning of the relationship between nature and humans.

Xun Kuang (known as Xuncius) proposes "the separation between heaven and human" in his book *Tianlun*. According to Li's Genpan's research [38], Xuncius's "heaven" includes both the "stars and the sun, the sun and the moon, the four seasons, the yin and yang, the wind and rain" and other natural materials and phenomena, as well as the organs and emotions of life (human natural feelings, human facial features), the relatively "human" refers to the activities of the human subject. Therefore, this "relationship between heaven and human" theory of Xuncius is roughly equivalent to the relationship between nature and humans. Xuncius believes that "heaven" and "human" each have their laws of operation. The difference is that "heaven" has no willpower and goal, while "human" has will and intent. And "heaven" can not consciously give "human" blessings and disasters, "human" also can not affect the operation of "heaven." Although "human" can't affect the operation of "heaven," it can be used to "adapt the law of heaven and make use of it," "All things live in harmony and grow with nourishments." It means that it is not because of "the separation between heaven and human" that humans and nature are in a state of opposition but that humans, heaven, and the earth are in a harmonious whole. Although a human cannot change the operation of heaven, they can master and use it for the benefit of humanity based on the law of process. The author believes that although Xuncius says "the separation between heaven and human," its essence is the harmony of nature and humans.

Taoism also advocates the harmony of nature and humans. Laozi said, "therefore the Tao is great; Heaven is great; Earth is great, and the(sage) king is also great. In the universe, there are four that are great, and the(sage) king is one of them. Man takes his law from the Earth; the Earth takes its law from Heaven; Heaven takes it from the Tao. The law of the Tao is being what it is" (*Tao-Te Ching* Chap. 25). He juxtaposes heaven, earth, and humans with the Tao to show the consistency and connectedness between humans and nature. Zhuangzi's *Uniformity Theory* further puts forward "everything as the same," "nature and mankind coexist and unite into the whole." Laozi said, "thus the sage helps the natural development of all things and does not dare to act" (Tao-Te Ching Chap. 64), which means that humans must obey the laws of nature and cannot do whatever they want. In terms of farming, Zhuangzi said in *Zeyang*: "in plowing my corn fields, I left the clods unbroken, and my compensation was in the rough unsatisfactory crops; and in weeding, I destroyed and tore up (many good plants), and my compensation was in the scantiness of my harvests. In subsequent years I changed my methods, plowing deeply and carefully covering up the seed, and my harvests were rich and abundant so that all the year

I had more than I could eat". This means that if it is roughly cultivated, there is no harvest. Once finely cultivated, it will be bumper harvested. It's explained that if humans are kind to nature, nature will return benefits to humans; otherwise, humans will be punished.

Mr. Zhang Dainian pointed out that "a distinctive theory of the relationship between man and nature in Chinese philosophy is the theory of the harmony between human and nature" [39]. Mr. Ji Xianlin's explanation of "the harmony between nature and humans" is: heaven is nature; man is human; unity is mutual understanding and friendship" [40]. He also believes that the harmony between humans and nature is the most significant contribution of Chinese culture to humankind. The harmony of humans and nature is the unity of humans and nature, peaceful coexistence, adapting to nature, not conquering and being conquered.

The famous British scholar Joseph Needham also had a very wonderful discussion about this: "Chinese thinkers don't believe that there is a God who manages the universe, but rather think about it from the 'heaven' of non-human aspect. Non-human actually means 'heaven' or many 'heaven,' but here is best translated into 'the order of the universe'. Similarly, the Tao is a 'natural order'. Therefore, in China's ancient worldview, humans are not seen as the master of the universe prepared by the creator for enjoyment. From the early days, there has been a concept of a natural layer in which people are seen as the highest form of life but never given any privilege to do whatever they want. The universe is not designed to meet the needs of humans. The role of the human in the universe is "to help the transformation and parenting process of heaven and earth." This is why people often say that humans form a trinity with heaven and earth. People should explore the way of nature or compete with nature but be consistent with it when it meets its inevitable basic laws. It's like having three levels of their own organization, such as the famous narrative 'at the right time; in the right place; with the right people. Therefore, the keyword is always 'harmony.' Ancient Chinese seek order and harmony in the entire natural world and regard this as the ideal of all human relations." [41]. The world of human life is composed of three parts: nature, people, and society. The new concept of people-oriented development is fundamentally seeking the harmonious development of the relationship between humans and nature, between humans and society, and between people. The 18th National Congress of the Communist Party of China pointed out that building a socialist harmonious society, socialist ecological civilization, promoting the all-round development of people, and proposing the goal of "promoting a harmonious world" [42].

1.4.4 The Harmony of Nature and Humans in Engineering Management

Nature and humans represent two aspects of the contradiction between all things. "The harmony of nature and humans in engineering management" refers to the harmonious unity of engineering and people, engineering and society, and engineering and nature. It does not only refer to the engineering environment; others, such as engineering decision-making, engineering economics, engineering quality, and engineering art, are all on the premise of being people-oriented through the line of harmony between humans and nature.

1.4.4.1 The Engineering Environment Includes the Social and Natural Environment of Engineering, Which is a Comprehensive Reflection of the Integration of Nature and Humans in Engineering Activities

The core of the engineering environment is the harmony between engineering and society and the harmony between engineering and nature. Therefore, it is necessary to implement a sustainable development strategy. While emphasizing the development theme and encouraging economic growth, we should recognize that sustainable development should be based on protecting nature and society and coordinate with resource sustainability and ecological carrying capacity. When creating and pursuing development and consumption, we must satisfy contemporary people's needs without harming future generations. Solving the problems of engineering development and environmental protection and mutual coordination with society is the key to achieving the harmony of nature and humans in engineering activities,

The Qinghai-Tibet Plateau is a place where natural ecology is very fragile. The Chinese Party Central Committee and the State Council have clearly stated that "the construction of the Qinghai-Tibet Railway should cherish the plateau ecological environment." Protecting the ecological environment is China's basic national policy. Throughout the entire process of construction of the Qinghai-Tibet Railway project, environmental protection laws and regulations have been implemented, such as *the Environmental Protection Law*, *the Soil and Water Conservation Law*, and *the Wildlife Protection Law*. An environmental protection policy of "prevention first, protection priority, the equal importance of development and protection" was established. And ecological facilities and the main engineering project keep the "three simultaneous" environmental protection work principles of simultaneous design, simultaneous construction, and simultaneous production. The overall goal of environmental protection is proposed to ensure that the water quality of the river source is not polluted, the migration of wild animals is not affected, the permafrost environment, vegetation and wetland environment, and the landscape on both sides of the railway are not damaged and strive to build an eco-friendly railway with plateau characteristics. The Qinghai-Tibet Railway Construction Department signed the first environmental responsibility document in the history of China's railway construction with the governments of Qinghai Province and the Tibet Autonomous Region, which established legal awareness and strengthened legal supervision. In order to restore the vegetation on the railway land, the researchers actively researched vegetation restoration and reconstruction in the plateau permafrost region. Using advanced technology, the survival rate of plants is over 70%, which is more than double the natural survival rate. The Qinghai-Tibet Railway built the wildlife channel for the first time in China's railway construction history. The Qinghai-Tibet Railway General Command monitoring shows that the Tibetan antelope has adapted to the artificially created migratory environment, and many Tibetan antelopes migrate freely through the wildlife channel [43]. The Qinghai-Tibet Railway effectively protects the ecological environment during construction and deals with the relationship between engineering and nature.

When the harmonious development of the engineering environment reaches a certain height, it will produce engineering art. Engineering art stems from the fact that engineering is the essence of serving people. Since engineering has to enter people's vision, it is necessary to have an artistic embodiment. Engineering art can also be regarded as part of the engineering environment. The highest representation of engineering art is the harmonious unity of engineering and society, engineering, and nature.

1.4.4.2 Engineering Decision-Making—The Comprehensive Embodiment of People-Oriented and the Harmony Between Nature and Human

Engineering decision-making refers to the decision-makers of the engineering establishing an overall plan for the proposed project and analyzing, comparing, and judging through different project construction plans. The choice of implementation is mainline throughout the planning stage. Engineering decision-making, especially major engineering decisions concerning the national economy and the people's livelihood, generally depends on politicians but must be based on science while ensuring democracy. Major engineering decision-making is a complicated process. Therefore, politicians need to weigh the pros and cons, draw on the advantages and avoid disadvantages, respond to all possible challenges and problems, and resolutely abandon the thoughts of "biting off more than one can chew" and being "eager for quick success and instant benefit." It is necessary to research the necessity and feasibility of the engineering project, fully embodying the people-oriented and the harmony between nature and humans of the project during the process. Under this premise, to make scientific decision-making, establish an argumentation mechanism that accommodates multiple opinions, prevent the decision-making from leaving the rational track through democratic means, and realize that listening to both sides and will be enlightened.

Some people think that China's most enormous waste comes from mistakes in strategic decision-making. According to the statistics of the World Bank, during the period from the "Seventh Five-Year Plan" to the "Ninth Five-Year Plan" period, China's investment decision-making error rate is about 30%, capital waste and economic losses are between 400 and 500 billion Yuan. It is critical to highlight scientific and correct engineering decisions.

During China's major engineering decision-making, the mistake of the Sanmenxia project decision-making has brought us many inspirations. The Sanmenxia project started in 1957 and was the first high dam reservoir built in New China. It shows that the Chinese people have tamed the rivers and achieved some practical effects in power generation but ignored its negative impact on the ecological environment and made some mistakes in engineering design. The political significance was excessively pursued in the engineering decision-making stage of the Sanmenxia project, so the project design neglected the objective facts and violated the objective laws. This project was one of the key projects aided by the former Soviet Union. It relied too much on foreign technical experts in decision-making and had insufficient consideration of the actual situation of sediment and ecological environment in the Yellow River Basin. So that the Sanmenxia Dam was renovated twice, adjusted the operation mode three times, and caused the Weihe River flood in 2003 [44]. The mistakes in engineering decision-making caused economic losses and brought serious negative impacts on society and the lives of local residents.

The Three Gorges Project has experienced more than 70 years of extraordinary history, from its initial conception, investigation, planning, and demonstration to its official start-up. In 1924, Mr. Sun Yat-sen first proposed the idea of building the Three Gorges Dam. Since the 1950s, the central leaders have visited the Three Gorges several times and organized expert groups to conduct repeated investigations. On April 3, 1992, the Fifth Session of the Seventh National People's Congress passed *the Resolution of the Three Gorges Project* due to 1767 votes in favor, 171 votes against, 664 abstentions, 25 people did not vote, and nearly one-thirds objected or abstained. The construction began in 1994, and water storage began in 2003. It was completed in 2009. The correct decision of the Three Gorges Project brings three major benefits, namely flood control, power generation, and shipping. Flood control is considered to be the core benefit of the Three Gorges Project. The normal water level of the Three Gorges Reservoir is 175 m, and the effective flood control capacity is 2.215×10^{10} m^3. It provides effective protection for flood control in Jingjiang and has a huge flood control effect on the middle and lower reaches of the Yangtze River. Hydropower development in the Three Gorges is an important milestone for China's sustainable development, especially clean energy development. The Three Gorges Hydropower Station's installed capacity is 1.82×107 kW, and the annual average power generation is 8.47×10^{10} kWh, which will generate huge power benefits. The Three Gorges Project is located at the junction of the upper reaches and the middle reaches of the Yangtze River, which is a unique geographical position. It can channel the river's upper reaches from Sandouping to Chongqing, and it can also increase the dry season flow of the middle reaches of the Yangtze River after the Gezhouba Water Control Project.

1.4.4.3 Engineering Economy is the Bridge and Link Between Engineering and Society

The implementation of any engineering project must be economically acceptable to society. The engineering economy is to obtain the greatest engineering benefits from limited resources. It is the comparison of the effects, costs, and losses in social practice with technology. It is a comparative analysis of certain useful results and the resource costs and loss, that is, the evaluation of economic effects. The dialectical relationship between engineering technology and the economy is the unity of opposites. The economy is the goal of engineering and technological progress. Engineering technology is the means to achieve economic goals and the driving force for economic development. But at the same time, engineering technology and the economy still have mutual constraints and contradictory aspects. The use of engineering technology increases the interest in engineering, and advanced technology is not necessarily economically reasonable. Technology that is not economical is not applicable. It is necessary to study which technologies and how are economically applicable. Therefore, engineering economic analysis focuses on engineering foresight, which is a systematic evaluation of engineering economic activities. Of course, meeting comparable conditions is the premise of comparing technical solutions. It is necessary to meet the comparability requirement of relevant use-value, input-related costs, time, evaluation parameters, and other factors. The judgment of the dialectical unity of engineering technology and economy requires us to deal with the relationship between technology, economy, environment and society, apply the knowledge of relevant disciplines to solve the economic problems encountered in technical practice, analyze using a large volume of data, especially the estimation and judgment beforehand, and pay attention to the balance of the system.

The South-to-North Water Diversion Project is a major strategic project to alleviate the severe shortage of water resources in northern China. China has the climatic characteristics of floods in the south and drought in the north. The South-to-North Water Diversion Project can significantly alleviate the severe shortage of water resources in northern China through the rational allocation of water resources across the basin and promote the coordinated development of the economy, resources, environment, society, and population in the north and the south. The project meets the water needs of cities and industries and is generally more economical and easier to operate than redistributing irrigation water. It can save costs and create a more excellent value. The South-to-North Water Diversion Project can improve the natural environment, especially the water resources conditions in the northern region, enhance the carrying capacity of water resources, improve the efficiency of resource allocation, and promote the strategic adjustment of the economic structure. It is of strategic importance to expand domestic demand, maintain rapid economic growth across the country, and achieve nationwide structural upgrading and sustainable development of the economic and social environment. The project promotes potential productivity by improving water resource conditions and resulting in real economic growth. Through the establishment of a new operation mechanism of the South-to-North Water Diversion Project, the water-receiving areas will be strengthened to increase water conservation and pollution control, and the ecological environment of the Huang-Huai-Hai area will be gradually improved. The project can make the northern part of China a water-saving and anti-pollution society with rational allocation of water resources, guaranteed water supply, and a good water environment [45]. It can effectively solve the water quality problems caused by natural causes in some areas in the north, such as high fluoride water, brackish water, and other natural water problems containing harmful substances harmful to the human body, and improve the quality of local rural drinking water. It is conducive to alleviating the constraints of water shortage on urbanization development in the northern region, promoting local urbanization, improving the urban ecological environment and natural landscape, meeting the requirements of people's living standards, and improving the quality of the surrounding ecological environment.

1.4.4.4 Engineering Quality—The Pass Accepted by Society And Is also the Basic Requirement and Moral Bottom Line of Engineering Activities

The general view of engineering quality is to assess the final quality of artificial nature and include the overall benefits of organization, economy, safety, society, environment, etc., in the whole process of engineering activities. The two concepts of quality and harmony are material and include spiritual and cultural aspects; Not only personal, product, but also social. Quality reflects the level of product development, demonstrates the level of government management, and reflects the degree of industrialization, modernization, and social harmony of a country. Engineering quality condenses the comprehensive quality of the engineering enterprise, including the realm of the engineering-related leaders, the quality of the team, the level of technology, the strengths and weaknesses of the corporate culture, and the level of government supervision construction behavior. The quality of engineering is related to all aspects of society, all levels, and various stages, reaching into every family and social group, directly affecting all aspects of people's lives. Engineering quality is inseparable from the harmony of society. It directly affects a series of problems, such as economic development benefits, public safety, social operation efficiency, people's life happiness index, and trust in the government. The supervision of engineering quality is the unshirkable duty and obligation of the government. It is both responsible to the people and responsible to the public social safety. Therefore, as an engineering quality supervision worker, efforts should be made to promote the engineering quality management laws and regulations system and make them more sound, the engineering quality management system and mechanism more perfect and coordinated, the engineering quality responsibility further strengthened and implemented, the quality of the construction engineering guaranteed throughout its life, and the overall quality level and the satisfaction of the people significantly improved.

Wenchuan earthquake, direct economic losses amounted to 845.1 billion yuan. Among them, the loss of schools, hospitals, and other non-residential houses accounted for 20.4%. The loss of housing for civil houses and urban residents accounted for 27.4%, accounting for nearly half of all losses [46]. The important reason for the huge number of casualties in the Wenchuan earthquake was the collapse of many houses, which fully reflected the unreasonable layout and form of the building and the low construction quality. Therefore, after the earthquake, the voice for investigation and accountability for the quality of buildings in the disaster area was very strong. For houses were designed according to seismic codes and had good construction quality, most of them had been cracked without a collapse in the high-intensity areas, and the damage was mostly light in low-intensity areas. However, a large number of collapsed houses in the Wenchuan earthquake reflected the quality of the engineering construction, such as scamp work and stint material, not building according to seismic requirements, and unreasonable operation, which made the damage of earthquake worse.

1.4.5 Collaborative Innovation

Engineering innovation is the best embodiment of giving full play to people's subjective initiative in engineering activities. Engineering innovation is also the main battlefield for innovation activities and building a national innovation system. The process of engineering innovation is the process of constantly breaking through barriers and avoiding traps. The driving force behind engineering innovation comes from being people-oriented and harmonious between nature and humans. People-oriented requirements are more humane in order to serve humans better. The harmony between nature and humans requires good, fast, and economic engineering. These requirements have prompted engineering to continue to innovate. The combination of technological innovation and management innovation, the success of binary innovation, determines the success or failure of engineering innovation. Management innovation refers to the process by which an organization forms creative ideas and translates them into useful products, services, or methods of operation. Technological innovation and management innovation are the unified boosters of engineering progress. Technology is productivity, and management is production relations. The two complement each other and dialectically exist in the development of engineering. Therefore, it is necessary to simultaneously carry out technological innovation and management innovation and combine them under the premise of being people-oriented. Modern engineering, especially large-scale engineering, involves many units and complicated engineering procedures. Some major innovation activities involve many aspects, including technological innovation and management innovation and innovative research units and implementation units for innovation. Therefore, collaborative innovation is the inevitable course of engineering innovation. There is a comprehensive and in-depth discussion in Sect. 7.1.2 of Chap. 7.

Construction of the Qinghai-Tibet Railway began in the 1950s. The former Ministry of Railways was in a leading position in the technological innovation process of the Qinghai-Tibet Railway project. They fully recognized the "three major problems" in constructing the Qinghai-Tibet Railway: permafrost, plateau lack of oxygen, and ecological fragility, and overcame them one by one through technological innovation. In the innovative practice of Qinghai-Tibet Railway project management, the railway project first implemented the legal person responsibility system and proposed five primary control goals for the particular requirements of the project: namely project quality, environmental protection, health and safety, construction period, and investment scale, and realized them through the decomposition of responsibility, the implementation of the target, the participation of personnel, and the step-by-step control of the process. The engineering quality management with a responsibility system as the core implemented the full-time supervision system of environmental protection management and highlighted the critical control period and investment management. Based on the overall technological innovation plan, the pre-station project of the Qinghai-Tibet Railway was divided into 33 sections, and the post-station project was divided into 17 sections, and then public bidding was carried out. The former Ministry of Railways arranged 88 major projects and more than 120 scientific research sub-projects, conducted targeted scientific research experiments in conjunction with the actual situation, and carried out more than 80 scientific research projects on frozen soil issues [34]. The engineering innovation of the Qinghai-Tibet Railway indicated that scientific research played a leading role, ensured the smooth completion of the project, produced a large number of independent scientific and technological achievements, and created more than 100 world firsts.

1.4.6 Building Harmony

Building harmony is the ultimate goal of an engineering project. Whether in the construction, operational, or decommissioning of engineering activities, it is necessary to build harmony. As far as the construction phase is concerned, the construction of any engineering project will break the original balance. The larger the scale of the engineering project, the greater its impact on the environment and society, and the greater the impact on the balance between the original environment and society. A new balance will be constructed after the construction of the engineering is completed. If this new balance is better than the original balance, harmony is built. Otherwise, although the creation process is completed, the project is not well completed from a systems point of view. The Three Gorges Project not only builds a shipping and power hub, but more importantly, it plays a role in flood control and carries out ecological management and restoration. This makes the Three Gorges Project harmoniously exist with nature and society, that is, constructing harmony. Building harmony is the ultimate and the highest goal of engineering activities.

The meaning of building harmony is diverse, and engineering must be in harmony with society and nature. Harmony with society includes economic, political, and cultural aspects. Any engineering project, it must be accepted by society in terms of economics. All sectors of society must agree that the funds for investing in this project are appropriate. However, for various reasons, some projects do not achieve harmony. The Boston Central Tunnel Project is a well-known case. Engineering construction is vital to solving traffic problems such as congestion in the central area of Boston. However, the budget for 1983 was $2.3 billion. By 2005, the actual cost reached $14.7 billion, causing massive controversy in parliament and society. The Scottish Parliament Building is a project known for extending the construction period and increasing investment. In 1997, the budget was 40 million pounds. In 2004, it increased to 430 million pounds. It was completed in 2007 and finally cost

414.4 million pounds, ten times more than the budget. In terms of cultural harmony, Zhejiang Jinhua Architectural Art Park is a negative example. The original intention of building Jinhua Architectural Art Park is to be a good place for people to relax, entertain and appreciate the art of architecture. However, due to site selection, especially because the park's connotation is not culturally recognized by the public, few people come to the park, even on holidays. The park started with a high profile, and it became a ruin after a few years. The Qinghai-Tibet Railway noticed the impact on the environment from the beginning and paid attention to environmental protection and restoration in design. Its 33 billion investment had 1.2 billion for effective restoration and management of the environment, making it a successful example of harmony with nature.

References

1. Yin, R., & Li, B. (2013). Understanding of engineering ontology. *Studies in Dialectics of Nature, 29*, 43–48.
2. Pang, X., & Li, G. (2016). Xi Jinping emphasized during the inspection in Qinghai: Respecting nature and adapting to nature to protect nature and resolutely building a national ecological security barrier. *People's Daily*, 2016-08-25 (1). http://paper.people.com.cn/rmrb/html/2016-08/25/nbs.D110000renmrb_01.htm
3. Xu, S. (2014). *The origin of Chinese characters*. Line Pack Book Publishing House.
4. Dictionary Editing, Institute of Linguistics, Chinese Academy of Social Sciences. (1978). Modern Chinese dictionary. The Commercial Press.
5. British Encyclopedia Editorial Department. (1999). Encyclopedia Britannica International Chinese edition. Encyclopedia Publishing House, 15, 157.
6. British Encyclopedia Editorial Department. (1999). Encyclopedia Britannica International Chinese edition. Encyclopedia Publishing House, 15, 158.
7. Zhu, G. (1999). *Concise world science and technology celebrity encyclopedia* (p. 170). China Science and Technology Press.
8. British Encyclopedia Editorial Department. (1987). *Encyclopedia Britannica* (p. 233). Danqing Book Co., Ltd.
9. British Encyclopedia Editorial Department. (1999). Encyclopedia Britannica International Chinese edition. Encyclopedia Publishing House, 15, 485.
10. Laszlo, E. (2001). *The multicultural planet-Report of the UNESCO international panel of experts* (p. 216). Social Science Literature Publishing.
11. British Encyclopedia Editorial Department. (1987). Encyclopedia Britannica. Danqing Book Co., Ltd., 5, 413.
12. Armstrong, N. (2002). Engineer century. National academy of engineering. The greatest engineering achievements of the 20th century. Jinan University Press, 9.
13. Chang, P. (2002). *Major engineering achievements in China in the 20th century*. Jinan University Press.
14. He, J. (2016). *Engineering philosophy and engineering management* (p. 3). Higher Education Press.
15. He, J., Chen, X., & Hong, K. (2006). Engineering management. *Engineering Science, 7*(10), 5–10.
16. Shah, H. (Ed.). (2012). *A guide to the engineering management body of knowledge*. American Society for Engineering Management, 3.
17. He, J. (2013). On the core of engineering management theory. *Engineering Sciences, 14*(11), 4–11.

18. Deng, R. (1995). On the science and art of management. *Operation and Management*, 11–15
19. Carnegie, D. (2014). *Carnegie leadership art and management skills*. Tianjin Academy of Social Sciences Press.
20. Jiang, Z. (2000). Jiang Zemin's speech at the international engineering technology conference. http://www.gmw.cn/0lgmrb/2000-10/12/GB/10%5E18571%5E0%5EGMA1-106.htm
21. He, J., & Chen, X. (2003). *Management science: Historical evolution, current situation and development trend*. Hunan People's Publishing House.
22. Kaiser, W. (2008). *History of engineers—A career that lasts for six thousand years* (p. 173). Higher Education Press.
23. Dai, W. (1936). *Universal library: Guanzi* (p. 8). The Commercial Press.
24. Zhang, D. (1994). *Introduction to Chinese culture*. Beijing Normal University Press.
25. Liu, T. (2014). Yuanda Group's world's tallest building became a big puddle suspected to be a scam. http://finance.china.com.cn/industry/estate/20140710/2530108.shtml
26. Gong, J., & Dong, B. (2015). The relationship between the people-centeredness of the scientific concept of development and the humanism of the West. *Youth Years, 21*, 210.
27. Lu, Y., & Shang, C. (2016). Ecological environment impacts and countermeasures of the Three Gorges Project in the Yangtze river. In J. He (Ed.), *Engineering philosophy and engineering management* (pp. 9–21). Higher Education Press.
28. Fu, Z. (2004). *Exploration and practice of China railway reform and development* (p. 575). China Railway Publishing House.
29. Wang, A. (2006). *The transformation of traditional industry-thinking about the rapid development of Shendong* (pp. 69–70). China Science and Technology Press.
30. Kaiser, W. (2008). *History of engineers—A career that lasts for six thousand years* (p. 8). Higher Education Press.
31. Kaiser, W. (2008). *History of engineers—A career that lasts for six thousand years* (p. 129). Higher Education Press.
32. Xiao, P. (2009). *Introduction to engineering ethics*. Peking University Press.
33. Guang, G., & Yang, L. (2012). *Qinshan nuclear power phase ii expansion project management practice (comprehensive)*. Atomic Energy Press.
34. Wei, Y., Yang, H., & Liu, J. (2003). Preliminary study on the particularity and safety guarantee system of Qinghai-Tibet railway transport organization. *Chinese Journal of Safety Science, 3*, 22–26.
35. Yuan, J. (2006). Shenzhou spacecraft system engineering management. *Mechanical Industry Press, 1*, 396.
36. Ma, X. (2012). Practice the aerospace spirit and continue to create miracles. *Chinese Spaceflight, 2*, 7–8.
37. Qiao, Q. (2015). Inherit traditional cultural wisdom and promote the construction of ecological civilization. *Guangming Daily*, 2015-08-01. http://epaper.gmw.cn/gmrb/html/2015-01/01/nbs.D110000gmrb_01.htm
38. Li, G. (1999). The theory of harmony between humans and nature and the theory of three talents—Why should we discuss the relationship between heaven and human in the history of the Chinese economy? In *Proceedings of the symposium on the relationship between heaven and man in Chinese economy history*. Institute of Economics, Chinese Academy of Social Sciences, 12, 01.
39. Zhang, D. (1982). *The outline of Chinese philosophy* (p. 181). China Social Science Press.
40. Ji, X. (1993). A new interpretation of the harmony between humans and nature. *Traditional Culture and Modernization, 1*, 14.
41. Needham, J. (1986). *History and human values: A Chinese perspective for world science and technology*. Liaoning Science and Technology Press.
42. Hu, J. (2012). Unswervingly marching along the road of socialism with Chinese characteristics and striving for the comprehensive construction of a well-off society: A report at the 18th National Congress of the Communist Party of China. National People's Congress.
43. Sun, Y. (2004). *Railway construction management theory* (p. 143). China Railway Publishing House.

44. Lu, G., Fu, C., & Ji, X. (2010). Institutions of major national engineering decision-making mechanisms. *Scientific and Technological Progress and Countermeasures, 6*, 81–85.
45. Ma, J., & Zheng, C. (2010). Evaluation of water resources and social economic harmony in the water-receiving area of the east-central part of the south-to-north water transfer project. *China's Population: Resources and the Environment, 11*, 36–41.
46. Ding, L., & Chen, X. (2009). Viewing China's construction quality risk sharing mechanism from Wenchuan earthquake disaster. *Scientific and Technological Progress and Countermeasures, 21*, 113–117.

Chapter 2
Engineering Management Epistemology

As an important part of dialectical materialism, epistemology is a scientific understanding theory about the source of human cognition, cognitive ability, cognitive form, cognitive process, and understanding of truth. First of all, it is knowable. It considers that the objective material world is knowable. People can recognize the phenomena of the material world and understand its essence through phenomena. Second, its basic premise is reflection theory. The material world exists independently without relying on the subjective will of human beings. The human consciousness is the function of the human brain and the reflection of the material world. It must adhere to the materialist understanding route from object to feeling and think and draw a demarcation line with the idealism route from thinking and feeling to object. Third, it is a theory of practice. The dialectics of cognition is manifested in the relationship between cognition and practice. Cognition comes from practice and then turns to guide practice and serve for practice. In the process of understanding, people's understanding of the world is not completed once but a process of repeated and infinite deepening.

According to the basic theory of dialectical materialism epistemology, engineering management epistemology can be understood as the reflection of engineering management subject on process management objects. It contains a concrete and abstract logical evolution process. That is to say; scientific engineering management generally needs to go through three stages: "sensual concrete, abstract analysis, and rational concrete." The "sensual concrete" stage is the cognitive recognition of engineering management, showing the concern for various figurative things in engineering construction. The "abstract analysis" stage shows the analysis and synthesis of engineering management, which is the beginning of metaphysical thinking in the specific perceptual behavior of engineering management. It is an intermediary from external behavior to the intrinsic nature. After the "abstract analysis," it will inevitably rise to a higher "rational concrete" stage. At this time, people have completed the combination of representation and internal relations for engineering management. They have risen to understand and grasp the holistic and procedural essential aspects, realizing the transition of engineering management practice.

J. He, *Principles of Engineering Management*,
https://doi.org/10.1007/978-981-99-1168-4_2

This chapter takes the historical evolution of engineering management as the longitude line and the development of engineering management as the latitude line, trying to construct the engineering management epistemology network, outlining the development of engineering management through the analysis of ancient engineering management, modern engineering management, and contemporary engineering management theory and practice. On this basis, through the multi-dimensional perspective of engineering management theory and practice changes, this chapter excavates the objective laws and essential criteria of engineering management. It reveals the essence of engineering management understanding. Traditional agricultural social engineering management and industrial social engineering management at that time were constrained by productivity and technological levels. In general, there are inherent boundaries. But the distinctive features of modern engineering are based on high and new technology, with innovation as the driving force, integrating various resources, emerging technologies, and ideas, and developing in a technology-intensive and knowledge-intensive manner. The history of the theory and practice of engineering management in China presents a noticeable spiral developmental dialectical developmental trait.

2.1 Historical Evolution of China's Engineering Management Theory and Practice

The engineering world is a humanized world. The engineering management theory is based on human beings as the main body of management, with the construction, operation, and decommissioning of artificial nature as the management object, with the planning, organization, and control of each stage as the management carrier at improving its effectiveness and efficiency. The starting and the foothold point of engineering management theory point to engineering management practices. Engineering management practice is based on existing objective things, applying engineering management theory to engineering planning, design, investment, construction, operation, and decommissioning, promoting the theory from the potential productivity to the actual productivity to achieve the goal during the process of the subject acting on the object.

2.1.1 Cyclic Evolution of Engineering Management Theory and Practice

The modern engineering management process follows the dialectical path from practice to recognition, re-practice, and re-recognition. The theory and practice of engineering management continue to realize the interaction, develop synergistically, and advance cyclically in the two-way interactive reconstruction and construction so that the two can reach a new higher level of unification under the new historical conditions and environment. It is also the true meaning of the materialist dialectics process theory. Therefore, it is necessary to carry out abstract analysis on specific engineering management experience and then dialectically integrate it, raise it to the height of philosophical thinking, and extract some universality and regularity [1].

First, engineering management practice is the realistic basis of engineering management theory. Engineering management theory is developed and summarized from practical engineering management practices. It can be said that there is no engineering management theory without engineering management practice. Primitive humans used wood as a nest, excavated soil as a cave, cut wood as a stick, and polish stones as a tool. All of these are necessary survival engineering practices. During the practice of survival, engineering activities, production activities, and continuation of life need to be "primitively integrated." The activity itself has some "priori spontaneous" factors, in which management factors are available, but obviously, there is no clear theoretical form. With the accumulation of human productivity and the improvement of living skills, the content of life has become wider, the standard of living has also improved, and the elements of the spirit have been incorporated at the same time as material enjoyment. A slightly larger ancient project has begun to appear. These projects have a certain complexity compared with the original survival engineering activities. Its normal operation calls for the birth of engineering management theory. From the perspective of management history, the emergence of management as a social activity is also a realistic need after the emergence of the factory. When the factory faces various internal and external relationships that need to be coordinated, various problems require specialized personnel to solve, forcing the early factory owners to start consciously aware of the importance and necessity of management. Many scholars have begun to research and develop management as an independent knowledge area and even an independent knowledge system. In this regard, Mr. Taylor, the founder of scientific management, put forward the idea that "management practice precedes theory." He attaches great importance to scientific investigations, research, and experiments, and he strongly hopes to improve and reform things according to objective facts. The work quota theory, the standardization principle, and the theory of differential piece rate system advocated by him are all products of management practice. He once said that all the people involved in scientific management that I know are prepared to abandon any method and theory at any time and support other better methods and theories that can be found. Therefore, it is the enlightenment of history and the pursuit of every theoretical worker not to satisfy with the status quo of the theory, not satisfied with the theoretical achievements that have been achieved, continuously combine practice, constantly correct the original theory in practice, and create a new theory that fits. Accordingly, as the primary task of engineering management theory, based on the complete, comprehensive, and systematic analysis and investigation of existing engineering management practices, the basic concepts, and basic principles should be abstracted according to engineering practice, find out the intrinsic links between them, and then ultimately generate engineering management theory with logical integrity.

Second, the engineering management theory created in the engineering management practice has a methodological significance once it is formed. The practical factors in the creation process internally determine its guiding significance for future engineering management practice. From the point of view of dialectical materialism epistemology, the practice needs to be guided by correct understanding. Practice without theory guidance is blind practice. Proper theoretical guidance will make the practice go smoothly and achieve the expected effect. When the wrong theory guides practice, it will have a negative and even destructive effect on practice, making the practice fail. Theory, as the "opposite" of practice, is not only the "conceptuality" of theory and the "materiality" of practice, but also is the "ideality" of theory and the "reality" of practice. Human is a reality, but people are always dissatisfied with the reality of their existence and always demand to turn reality into an ideal reality. The theory transcends practice with its ideal world picture and idealistic purpose requirements and promotes the self-transcendence of practice. The reason why the theory can "rebut" the practice and promote the self-transcendence of practice is that the theory itself has three characteristics: first, the theory has "upward compatibility," that is, the theory is the accumulation and crystallization of the history of human cognition, so it can reflect on the practical activities of reality by "establishing theoretical thinking based on the history and achievements." Second, the theory has the "inclusiveness of the times," that is, the theory is "the era in thought," Thus, it can critically reflect on practical activities and normatively correct practical activities with the grasp of the universality, essence, and regularity of the times. Third, the theory has the "system of concept," that is, the theory is the logical system of concepts, so it can comprehensively observe the practical activities in the mutual definition and mutual understanding of concepts and guide the practice activities to achieve self-transcendence [2]. Any human practice, including today's engineering practice, is different from instinctive animal activity in that humans plan future behaviors and possible outcomes in a theoretical model before implementing activities. The complexity and high-tech nature of modern engineering practice activities determine that it has more selectivity and higher risk. If it is slightly negligent, it will bring irreparable and devastating damage. These determine the orientation, prediction, and promotion of scientific engineering management theory and become more and more important in future engineering management practices. Today, we look back at some of the engineering projects of humans in the twentieth century and still have a lot of emotions. Due to the lack of the multi-faceted value elements of engineering construction in the theoretical guidance of the early theory of decision-making, some projects failed to achieve the expected comprehensive utilization benefits but brought a series of ecological and social problems. It can be seen that the correctness of the theoretical guidance before the implementation of project construction, whether it is perfect or not, has "a priori" determined its future destiny [3].

Third, engineering management theory and engineering management practice are spirally developed and cyclically advanced in continuous testing and summarization. Various conditions restrict human beings' understanding of the nature and laws of things, and it is impossible to achieve them one after another. To achieve a relatively correct understanding, it is necessary to go through many repetitions of practice, understanding, re-practice, and re-recognition. This is mainly because: first, in the process of engineering construction, the understanding of the object being transformed by the project manager as the main body is limited by the development process of the object itself and the degree of its representation. Any object that the project faces as a system has many aspects and multiple layers of characteristics. The interaction between many aspects and layers makes things constantly change and develop. It is always in a dynamic process. This also causes the exposure of the nature and laws of things to manifest itself as a process. People also need to spend a certain amount of time mastering the essence and laws. When the phenomenon of the object of engineering construction fails to be fully presented to people, people's understanding is easily obscured by the illusion of the object, which may result in an incorrect or one-sided understanding of the essential characteristics of the object. Second, during the process of engineering construction, as the subject of the project, the manager's understanding of the object is also limited by historical conditions and technological level. The engineering construction of any era is based on the specific level of productivity and the level of social consciousness. People's ability to understand objects must be restricted by the conditions of the times. It is difficult to grasp the essence through phenomena and form a more complete and comprehensive understanding of the object. At the same time, people's understanding and transformation of objects will be influenced by factors such as politics and culture, which will make the process of understanding lose its objectivity. Third, during the process of engineering construction, the state of knowledge of the object is also limited by the project manager's own factors. These restrictions include the scope of human practice, knowledge level, cognitive ability, practical ability, position, viewpoint, method, and physiological quality. People can only constantly break these restrictions in the advancement of engineering management practice, and their understanding can be continuously surpassed. Then engineering management theory can continue to develop and advance.

Based on the above discussion, we can see that the relationship between engineering management theory and engineering management practice is not a simple closed relationship of "practice—understanding—practice." It is the relationship of an infinite loop of "practice—understanding—re-practice—re-understanding." And this infinity is not a simple circular loop but a spiraling ascending process. At the same time, during this ascending cycle, people construct, operate, and decommission artificial nature in engineering practice activities with subjective intentions and plans, and undoubtedly have integrated the value factors of the subject into the understanding. That is to say, the engineering practice often shows the intention and the plan first, indicating that the project has been branded with the value of the subject before the specific implementation, which has already reflected the unity of truth and value. The fundamental meaning of this unity is, in the final analysis, the realization of the subjective and objective unity that people are eagerly awaiting during the process of the rise of the theory and practice of engineering management.

The process of human civilization and social development is accompanied by a large number of engineering management practices. The Dujiangyan, after two thousand years of wind and frost, still benefits people; the Great Wall that stretches for thousands of miles; these great projects show the greatness of the nation and the Chinese civilization. A large number of major modern projects represented by the Qinghai-Tibet Railway are a model for people to follow the rules of engineering management consciously. The Qinghai-Tibet Railway is the highest-altitude road in the world. Under extremely difficult and harsh conditions, the Qinghai-Tibet Railway builders have overcome the three major world problems of "permafrost, cold and oxygen deficiency, and ecological fragility." They fully achieved the project management goal of "people-oriented, the harmony of nature and human," and built a human—engineering—environment coexistence harmonious picture, in which the Tibetan antelopes run leisurely on the side of the Qinghai-Tibet Railway, and the Lhasa Railway Station and the Potala Palace complement each other.

China's engineering management has undergone a long history of development. During the process of gradual accumulation, refinement, and development, from experience to science, from tradition to modernity [4–6]. The historical development of engineering management in China can be divided into ancient, modern, and contemporary engineering management.

2.1.2 *Ancient Engineering Management in China*

2.1.2.1 Ancient Engineering Management Thoughts and Systems in China

1. **Simple ancient engineering management thought**

The activities of initial human use and creation of things indicate the origin of engineering activities [7, 8]. The characteristic of engineering in ancient China is distinct. The momentum and art were inclusive, and the connection between engineering and nature echoed, condensing the intelligence of the ancient people and depositing a large number of rich and valuable thoughts and experiences, which led to the simple ancient Engineering management thought [9, 10].

The engineering design in ancient times mainly relied on people's subjective feelings and cognition and took people as the center to present the vital idea of "people-oriented" [9, 11]. For example, the dragon and phoenix carvings used in the palace buildings and the decoration of "luck, prosperity, longevity, and happiness" indicate the meaning of "good luck." These come from both real life and emotions, adding to the pursuit of ideals and beauty.

In ancient times, engineering and nature have been skillfully combined. For example, the Great Wall construction experience of "building a fortress due to the steep terrain", the classical garden construction method of "holding the thing to express will, using the view to develop feelings," and the feng shui of the palace buildings of "harmony with heaven, earth, and feng shui," all of which emphasize the adherence to the laws of nature and the laws, all having a common idea of "harmony between nature and human."

The comprehensive systems idea embodied in ancient engineering seems to be commendable even today. Although the idea that "the whole is greater than the sum of the parts" was recognized in the fourth century BC, the overall system thinking is also reflected in the ancient engineering construction in China [12]. For example, the Dujiangyan Water Conservancy Project consists of three major projects: the Minjiang Fish Mouth Water Diversion Project, the Flying Sand Weir spilling and desilting Project, and the Bottle-Neck Channel Water Diversion Project, which constitutes the system's water conservancy and irrigation system. In addition, "Ding Wei made a palace, achieving three things in one stroke" is a model of the application of systemic thinking in ancient Chinese engineering. In April of the Northern Song Dynasty Dazhong Xiangfu, the second year (AD 1009), the construction task of Yuqing Zhaoying Palace was undertaken by Ding Wei, and its scale was enormous, including 2610 buildings. At the same time, in addition to the funding problem, there are three difficulties: first, the construction of the palace requires soil, but because the capital city has less space, it takes a lot of labor to go to the countryside to borrow soil; second, because of lacking building materials, and it also takes work to transport from the Bianhe River and terminal handling; finally, the garbage of bricks and other wastes generated by the construction needs to be cleared from the capital city. Therefore, Ding Wei developed a careful construction plan: first, digging many deep trenches around the construction site to outside the city. The excavated soil can be used for the new soil required for construction, thus solving the soil problem. Second, leading the water of Bianhe to these deep trenches, so that wooden rafts or ships could transport building materials such as wood and stone to solve the problem of material transportation. Third, after the materials are transported, drain the water in these deep trenches, and bury the waste left in the site into the trenches to fill the deep trenches into flat land. In this way, the original construction time of 15 years was completed in only seven years. It saved time and cost and made the construction of the project orderly, and didn't cause excessive interference with the traffic and life in the capital city.

Ancient engineering pays attention to overall optimization and also includes a simple engineering management thought. For example, in the Song Dynasty, the hydraulic worker Gao Chao proposed a new "three-section pressure broom method" different from the traditional "complete broom tamping method" in the Yellow River dike project (broom is the material used to protect the wall of the dyke when floods), bundled with branches, stalks, stones, etc. "Helongmen" is a key step in the blockage of the breach. Due to the deep-water flow and the violent water, it failed to use the 60-step broom for closure. Therefore, combined with water potential, materials, technology, and other factors, it was proposed to shorten the 60-step broom to 3 sections, 20 steps in each section, connect each other with ropes, and press down the brooms by section, i.e., using the first section to weak the water potential. After being fixed, the second section was pressed down, and then the third section was pressed down. Finally, this kind of labor-saving optimization method successfully blocked the breach.

2. **Ancient construction organization management**

In ancient times, due to the low level of productivity and the small scale of engineering construction, the construction and management methods were relatively simple, and the builders built all. The project builder was responsible for the funds, materials and drawings, architectural design, and application. Work and management were concentrated on the builders themselves, and hired craftsmen could complete the project. Government projects (such as palace-style buildings) are large in scale and complex in structure, and construction costs involve treasury expenses. It is necessary to set up supporting management measures and systems [13]. For example, the official engineering system established in ancient China was a civil construction and operation services system for the government, the royal court, and the government. It was the product of the centralized management of ancient China, mainly adopting (quasi) militarized management means to ensure the completion of large-scale projects based on expected quality in the short term. The state organized and implemented the system and was designated and supervised by the court or local government. The person in charge of the project construction was usually the official or military leader. The temporary management organization was set up, which would be revoked after the completion of the project. For example, the Dujiangyan project was built by Taishou Li Bing; general Meng Tian and Meng Yi built the Great Wall of the Qin Dynasty.

During the slave society, the official of engineering was the bureaucrat who managed the craftsman. For example, the highest official of engineering in the Zhou Dynasty is "Sikong." The feudal society adjusted and perfected the official engineering system and set it as a part of the official office of the project construction agency. Its functions include engineering design, recruitment, and collection, standard-setting, production, and management. For example, in the Qin Dynasty, there was a "Jiang Zuo Shao Fu" responsible for constructing civil engineering. The Han Dynasty had "Jiang Zuo Da Jiang," which was responsible for the urban planning and design of the early capital city. After the Sui Dynasty, there were "Jiang Zuo Jian" and "Gong Bu," which the supervisors were called "Jiang Zuo Jian Cheng" and "Gong Bu Shang Shu." The Qing Dynasty retained the "Gong Bu" and set up a "Nei Wu Fu" to be responsible for the construction and operation of temples and governmental offices. Through the system of collective powers of the official engineering departments, many large-scale projects were successfully built and operated by the court or the military. This model lasted for a long time and reflected the characteristics of China's traditional culture, political system, and economic system.

3. **Implementation of ancient construction process control**

In terms of standard preparation, ancient engineering construction has formulated corresponding norms, standards, and procedures. Examples include the construction system *Kao Gong Ji* in the Spring and Autumn Period and the Warring States Period in China, the standardized collection of *Zhu Cheng Fa Shi* and later *Mu Jing* and *Ying Zao Fa Shi* in the Northern Song Dynasty Yuanfeng Period, *He Fang Tong Yi* in Yuan Dynasty, and *Gong Bu Gong Cheng Zuo Fa Ze Li* in the Qing Dynasty. *Ying Zao Fa*

Shi advocates that "division is dedicated to the profession" and thus proposes more than ten professional divisions to complete different professional-specific projects [14]. *Gong Bu Gong Cheng Zuo Fa Ze Li* unifies the size and quantity of some official building components.

In terms of resource allocation, the construction of large-scale ancient projects requires a lot of workforces and material resources. Therefore, it is necessary to make overall arrangements to ensure the realization of the project. For example, *The Art of War* says, "before the war, the operational plan had to be carefully discussed, and the favorable conditions and adverse conditions had to be fully estimated, so they can win." Although this applies to military warfare, it also needs to be strategized, carefully analyzed, and carefully planned in constructing large-scale projects [10]. During the Great Wall construction process, various methods were used for material transportation, and materials were transmitted in the form of teams in difficult walking places. In the winter, the pavement was made icy by pouring water to facilitate the pushing and pulling of stone. In the deep valley, the material was pulled by the "flying basket," all of which saved human resources and time [6].

Quality and safety control also have ancient characteristics. For example, in *The Annals of Lv Buwei,* Lv Buwei is very concerned about product quality in the Warring States Period. He proposes that the name of the craftsman should be engraved on the product to facilitate accountability. In the process of building the Great Wall, the responsibility system management method was also adopted. The stone inscription was used to record the construction location of the project. The names of supervisors, stonemasons, and carpenters were also engraved to facilitate accountability. At the same time, a quality assessment system was set up. The bow and arrow were used to shoot the wall at a certain distance. If the arrow did not penetrate the wall, it indicated that the wall was qualified in terms of quality; otherwise, it would be reworked. During the construction of up to 360-foot wooden towers in the Northern Song Dynasty, each layer was constructed with curtains to prevent the construction objects from damaging passers-by. This type of safety management is still in use today.

In terms of cost control, due to the complex structure of large-scale projects in ancient times, the resources were expensive, and the technology was not advanced enough, so more attention was paid to the consumption of materials and the calculation of costs. For example, the *Ying Zao Fa Shi* of the Song Dynasty counts the quantity of materials used in various types of work to form a quota basis to facilitate the subsequent use of materials [15]. In the *Gong Bu Gong Cheng Zuo Fa Ze Li,* promulgated in the Qing Dynasty, many calculation methods for materials were recorded, and detailed specifications for materials were compiled, called *Ying Zao Suan Li* [11].

China's ancient engineering construction has accumulated thousands of years of engineering wisdom. Engineering construction contains the idea of people-oriented harmony of nature and humans, and has accumulated a lot of spontaneous and simple engineering experience. These ideas, experiences, and practices have profoundly affected the construction and development of follow-up projects and significantly promoted the civilization process of human society.

2.1.2.2 Typical Practice Cases of Ancient Engineering Management in China

1. **Ling Qu**

Ling Qu is located in Xing'an county, 60 km northeast of Guilin. It is a complete ancient water conservancy project in the world. It is as famous as Sichuan Dujiangyan and Shanxi Zhengguoqu. They are called "the three major water conservancy projects of the Qin Dynasty." Mr. Guo Moruo called: "Echoes to the Great Wall in the north, the wonders of the world." The total length of the canal is 37 km. It was built in the thirty-third year of Emperor Qin Shihuang (214 BC) and consisted of Huazui, Daxiaotianping, Nanqu, Beiqu Xieshui Tianping, and Doumen. The design of the canal is scientific and exquisitely constructed. The Huazui separates the Xiangjiang River from 30 and 70% two-part. Thirty part of the water flows southward into the Lijiang River. Seventy part of the water flows northward into the Xiangjiang River, connecting the Yangtze River and the Pearl River.

After Qin Shihuang unified the six northern countries, he launched a large-scale military conquest in Zhejiang, Fujian, Guangdong, and Baiyue located in Guangxi, in 221 BC. His army won on the battlefield but struggled for three years in the Guangdong-Guangxi region. He did not make any achievements because the terrain of Guangxi was not able to supply transportation supplies. So, improving and ensuring transportation supply had become the key to the success of the war. Emperor Qin Shihuang strategized and ordered Shi Lu to cut through mountains to build channels. Through accurate calculation, Shi Lu finally dug the Ling Qu in Xing'an, miraculously connecting the Yangtze River and the Pearl River water system. The reinforcements and supplies were continuously transported to the front line. It promoted the development of the war and finally officially conquered the vast areas of Lingnan, letting these areas enter the territory of the Qin dynasty, which played an essential role in the unification of China by Qin Shihuang.

2. **The Great Wall**

The Great Wall is one of the wonders of world engineering architecture. The Great Wall has been regarded as a symbol of ancient Chinese civilization and is famous globally.

The Great Wall has a history of more than 2000 years, and its construction began in the Warring States Period. At that time, the State Qin, State Zhao, State Wei, State Qi, State Yan, State Chu, and other principalities built the Great Wall to defend against the invasion of the northern nomadic people. After the State Qin merged with the six states, in order to prevent the raids of the Xiongnu in the north, in 213 BC, a huge project to build the Great Wall was launched, and the original Great Walls of State Qin, State Yan, State Zhao, and State Wei were connected and expanded. A total of 300,000 migrant workers were employed in the entire project, which took more than ten years to complete. It was built west from Gansu Linzhao (now Minxian county), along the Yellow River to the Inner Mongolia Linhe, north to Yinshan, south to Yanmenguan in Shanxi, and east to Liaodong, with a total length of more than 3000 km.

In addition to rebuilding the Great Wall of the Qin Dynasty, the Han Dynasty built the Great Wall of Shuofang in the south of Hetao, Inner Mongolia, and the Great Wall of the western part of Liangzhou. The Great Wall of the western part of Liangzhou is from the north at Juyanhai, Inner Mongolia (now in the territory of Ejinaqi), along the Eji River, through Gansu Jinta, west to Anxi, Dunhuang, Yumenguan into Xinjiang. The entire Great Wall is "one beacon fire in every five miles, one pier in every ten miles, one fort in every thirty miles, and one city in every one hundred miles," constituting a strict defense system. After the Han Dynasty, the Northern Wei Dynasty, the Northern Qi Dynasty, the Sui Dynasty, the Jin Dynasty, and other dynasties constructed the Great Wall. In the Ming Dynasty, a comprehensive restoration was carried out. The Great Wall built in the Ming Dynasty starts from Jiayuguan in the west and reaches the Yalu River in the east, about 6700 km. The entire re-repair process lasted for more than 100 years, showing the vastness and arduousness of the project. The Great Wall east of Shanxi was made of rammed earth inside, and the outside was made of masonry. The Great Wall west of Shanxi was built with rammed earth. There are many passes buildings on the entire Great Wall, which are built on steep terrain. The famous ones include Jiayuguan, Juyongguan, and Shanhaiguan. Jiayuguan is a complete place in the existing Great Wall passes. It was built in the 5th year of Ming Hongwu (AD 1372). It is magnificent, well-organized, and rigorous in structure. It has the title of "the world's most majestic pass." According to legend, during the construction of this pass, the design and construction skills were superb, but the calculation of the materials was also very accurate. After the completion of the pass, there was only one brick left. This brick was placed on a small building by the descendants as a memorial. The Great Wall that people have seen so far is mainly rebuilt in the Ming Dynasty. The former Great Wall has only a few relics left.

The Great Wall stretches over 10,000 miles across the steep terrain of mountains, rapids, and valleys. The arduousness of the project is unimaginable. It shows the majestic spirit and ingenuity of the Chinese people. It also reflects the superb level of measurement, planning and design, building technology, and engineering management of ancient China.

3. **Dujiangyan**

Dujiangyan is located near Chengdu. The Dujiangyan Water Conservancy Project is located in the west of Dujiangyan City, Sichuan Province. It is the world's oldest and longest-lasting water conservancy project characterized by dam-free water diversion.

The project comprises the Fish Mouth Water-Dividing Embankment, the Flying Sand Weir spilling and desilting Project, and the Bottle-Neck Channel Water Diversion Project. It scientifically solves the problems of automatic diversion of river water, automatic sand discharge, control of influent flow, etc., eliminates flooding, and makes the western Sichuan Plain the "land of abundance," where people determine water and drought. The Minjiang River is a large tributary of the upper reaches of the Yangtze River, originating from the high mountain areas of northern Sichuan

Province. Before the completion of Dujiangyan, whenever the spring and summer flash floods broke out, the water rushed down. The narrow river channel often caused floods, and when the flood receded, it left thousands of miles of sand and stone. The Yulei Mountain on the east bank of the Minjiang River hindered the eastward flow of the river, causing drought in the east and flooding in the west.

In the fifty-first year of Qin Zhaoxuan (256 BC), Shujun Taishou of Qin Li Bing and his son learned the experience of water control of the predecessors, led the local people, and presided over the construction of the famous Dujiangyan Water Conservancy Project. The overall plan of Dujiangyan is to divide the water flow of the Minjiang River into two, one of which is introduced into the Chengdu Plain so that it not only diverts floods and reduces disasters but also diverts the water to the fields and changes it into benefits. The main project includes the Fish Mouth Water-Dividing Embankment, the Flying Sand Weir spilling and desilting Project, and the Bottle-Neck Channel Water Diversion Project.

First of all, Li Bing and his son invited many farmers who have experience in water control to conduct on-the-spot investigations on terrain and water conditions and determined to cut through Yulei Mountain to divert water. Since no gunpowder was invented at the time, Li Bing burned the stone with fire, causing the rock to burst. Finally, a mountain pass with a width of 20 m, a height of 40 m, and a length of 80 m were drilled in Yulei Mountain. Because the shape resembles a bottle mouth, it is named "Bottle-Neck Channel," and the stone pile separated from the Yulei Mountain is called "Lidui."

After completing the Bottle-Neck Channel project, although it played a role in diversion and irrigation, due to the high terrain in the east, the river is difficult to flow into the Bottle-Neck Channel. Li Bing and his son led the public to build a water diversion in the upper reaches of the Minjiang River and the center of the river, which are not far from Yulei Mountain. They used large bamboo cages filled with pebbles to form a narrow island shaped like a fish mouth in the center of the river. The fish mouth divides the raging Minjiang River into the outer and inner rivers, the outer river discharges floods, and the inner river flows into the Chengdu Plain through the Bottle-Neck Channel.

To further play the role of flood diversion and disaster mitigation, a 200 m long spillway was built into the outer river between the fish mouth water diversion and the Lidui to ensure that the inner river has no disaster. In front of the spillway, there is a bend, and the river water forms a circulation. When the river exceeds the dome, the mud that is entrained in flood will flow to the outer river so that it will not foul the inner river and the Bottle-Neck Channel, so it is named "Feishayan."

To observe and control the amount of water in the inner river, Li Bing carved three stone portraits. He put them in the water to determine the water level by "the water level will not flood the feet during the drought, the water level will not flood the shoulders during the flood." He also carved stone horses and put them in the center of the river as the standard for cleaning the river beach at the minimum amount of water per year.

The three major parts of Dujiangyan scientifically solved the problem of automatic diversion of river water, automatic sand discharge, control of influent flow, etc., eliminating flooding. Since then, the Chengdu Plain has become a land of abundance and prosperity, where people determine water and drought. The project is still working today, and the current irrigated area is over 10 million mu (per mu = 0.0667 ha).

4. **Beijing-Hangzhou Grand Canal**

The world-famous Grand Canal is the earliest and longest artificial river in the world. The Grand Canal starts from Beijing in the north and reaches Hangzhou in the south. It flows through Beijing, Hebei, Tianjin, Shandong, Jiangsu, and Zhejiang provinces. It communicates with the five major river systems of Haihe, Yellow River, Huaihe, Yangtze River, and Qiantang River. The total length is 1794 km. In the history of the Chinese nation, the canal has made tremendous contributions to the development of north–south traffic and communication and the economy and culture.

The Beijing-Hangzhou Grand Canal was built in 486 BC and was opened to traffic in 1293. It lasted for 1779 years. In the long years, it mainly experienced three major constructions.

The first time was in the late Spring and Autumn periods of the fifth century BC. At that time, Fuchai, the king of the Wu State, who ruled the downstream area of the Yangtze River, attacked the Qi State to the north in order to compete for the dominance of the Central and organized migrant workers to excavate the canal from Yangzhou to the northeast and through the Sheyang Lake to Huai'an into the Huaihe River (the current Li canal). Because of passing through Hancheng city, it was named "Hangou." It is 170 km long and brings the Yangtze River water to the Huaihe River, making it the earliest section of the Grand Canal.

The second time was in the early seventh century when the Sui Dynasty unified the country, and the capital was determined as Luoyang. In order to control the vast areas in the south of the Yangtze River and transport the rich materials of the Yangtze River Delta region to Luoyang, in 603 AD, Emperor Yang of Sui ordered to excavate Yongji Canal from Luoyang via Linqing, Shandong to Hebei Yuanjun (now southwest of Beijing) for about 1000 km. He also ordered to dig Tongluo Canal from Luoyang to Jiangsu Qingjiang (Huaiyin) about 1000 km in 605 AD. In 610 AD., the "Jiangnan Canal" with a length of about 400 km from Zhenjiang to Zhejiang Hangzhou (the foreign trade port at the time) was excavated. At the same time, the Hangou was renovated. In this way, a river with a total length of more than 1700 km between Luoyang and Hangzhou can be available for shipping.

In the late thirteenth century, the third time was after the Yuan Dynasty set Beijing as the capital. In order to connect the north and the south, no longer detour to Luoyang, the Yuan Dynasty spent ten years and successively excavated the "Luozhou River" and "Huitong River" to connect the natural rivers and lakes between Tianjin and Jiangsu Qingjiang. The south of the Qingjiang River is connected to the Hangou and Jiangnan Canals to Hangzhou. The original canal was abandoned between Beijing and Tianjin, and the new Tonghui River was excavated. In this way, the new Grand Canal is more than 900 km shorter than the Grand Canal detouring Luoyang.

2.1.3 Modern Engineering Management in China

Modern times mainly refer to the stage after the Opium War to the founding of New China. After the Opium War, China's traditional building construction and organization management methods underwent earth-shaking changes [10]. In the late Qing Dynasty, the Ministry of Industry was abolished, and the system of official engineering offices disappeared along with the feudal system. At this time, Western engineering methods gradually infiltrated China, setting off a revolution in engineering management in the modern era.

2.1.3.1 Thoughts and Systems of Modern Engineering Management in China

1. **The thoughts of modern engineering management in China**

From 1840 to 1949, China was in a period of drastic changes. The development process in all aspects of engineering management at this stage also showed the nature and characteristics of the drastic changes. The close-door policy of the Qing Dynasty blocked the introduction of Western engineering management concepts and practical management models. Until the middle of the nineteenth century, except for a few western-style construction projects such as Beijing Xiyang Building in the Old Summer Palace, Guangzhou "Thirteen Factories," and churches in individual places, during this period, all aspects of Chinese engineering management basically followed the ancient Chinese engineering management. After the Opium War, especially after the Revolution of 1911, along with Western learning, Western engineering management thoughts and practice began to enter China. They merged with Chinese culture, which profoundly affected China. These changes were prominently manifested in modern Chinese construction project management systems and practices.

Modern Chinese architecture has two systems: old and new. The old architecture system is a continuation of the original traditional architecture system. It follows the old functional layout, technical system, and style. However, due to the influence of the new architecture system, there are also some partial changes. The new architecture system includes new buildings introduced from the West and developed by China itself, with modern new functions, new technologies, and new styles. Even those architectures introduced from the West integrated different degrees of Chinese characteristics. In terms of quantity, the old architecture system still occupies an advantage. The old architecture system still dominates the vast rural areas, towns, small and medium-sized cities, and even the old areas of big cities. A large number of residential houses and other folk buildings maintained the traditional characteristics of local conditions and materials, although modern materials, structures, and decorations had been applied partially. From the perspective of the development trend of architecture, the mainstream of modern architecture in China was the new architecture system. It is particularly worth mentioning that Mr. Liang Sicheng, a master of architecture from the history of modern architecture, was one of the pioneers in studying Chinese architectural history. From 1931 to 1945, he and his colleagues in China investigated more than 2000 ancient buildings and cultural relics in 15 provinces and accumulated a large amount of information. According to this information and materials, the first Chinese-authored *Chinese Architecture History* was completed in 1943. This is the first time that the characteristics of ancient Chinese architecture and its development process were systematically discussed.

2. **Institutional setup and regulation development of modern engineering management in China**

After the Opium War, modern Western capitalism's engineering construction and production characteristics invaded China's trade port cities, most typical in Shanghai. It had a far-reaching influence on the engineering construction management mode and the corresponding regulations in Shanghai. It was different from the engineering construction organization structure of old official offices, artisans, and civil service in ancient times. Western capitalist characteristics have taken shape in modern engineering construction and organization management in China.

In 1854, Britain, American, and French consulates elected an administrative committee in the concession in Shanghai and subsequently changed to the municipal council, which was the Bureau of Industry. Its functions were mainly the preparation and revision of rules and regulations, drawing review, permit approval and issuance, etc. At the same time, the Bureau of Industry had a work department to handle the construction and management of municipal projects in the concession. The municipal engineering construction management system born from the concession gradually affected the country from the 1860s. Beijing, Tianjin, and other cities followed the example and set up municipal departments, such as the work bureau, to issue and implement corresponding rules and regulations, accelerate the construction of urban infrastructure projects, and promote the modernization process in China. For example, the Qing government promulgated the *Township Local Autonomy Charter* in 1909. The distinction between "city" and "township" for the first time made clear; it was the beginning of China's urban construction system. The municipal administration was gradually implemented and established during the period of the Republic of China [16]. In 1921, the *Municipal Self-Government System* and the *Municipal Self-Government Enforcement Rules* were promulgated. At the same time, influenced by the modern Western "autonomous city" concept, the administration and legislation were separated, and a municipal organization composed of the autonomous association, the municipal administration, and the councils were developed, particularly typical in the construction of the ancient city of Beijing. 1933–1935 was the municipal planning period of Beijing. During the period, 22 municipal engineering construction regulations were promulgated and implemented, which included six related to real estate, seven about road traffic, four related to construction projects (including project contracting, bidding, material procurement, etc.), and five related to environmental sanitation [17]. It can be seen that its urban construction and management and associated affairs were based on specific laws and regulations.

After a long period of exploration and practice, the National Government promulgated the *Architectural Law* in 1938, which is the first national construction management regulation with modern significance in the history of China. Subsequently, the construction industry management rules, such as *the Architectural Management Rules* and *Management Industry Rules*, and related technical specifications, namely *the Technical Rules for Construction*, were compiled. In addition, the government construction management organization was unified and systematized, including the central government construction and planning agency, the provincial construction department, and the municipal work bureau.

3. **Implementation of modern engineering management in China**

Ancient artisans were both designers and builders. In the early days of the emergence of architects in the West, they also presided over the architectural design and participated in engineering construction. After the Opium War in 1840, modern Western architecture influenced traditional Chinese construction methods. Also, it introduced the identity of Western architects into China, which broke the system of traditional Chinese artisans responsible for design and building, separating engineering design and construction. This gradually formed a workshop-style management mode. The construction plant (i.e., engineering contracting enterprise) was developed and participated in the bidding competition in the modern construction market. At the same time, there were strict legal procedures for the establishment of the construction plant. First, the Bureau of Industry reviews the qualification and then registers at the business administration department. In addition, the construction plant can be divided into four grades, A, B, C, and D, according to capital, representative education, business scope, and contracted project scale. For example, Yang Ruitai Construction Factory, founded by the plasterer Yang Sisheng in 1880, is the first engineering contractor enterprise established by the Chinese in Shanghai. In 1893, Yang Ruitai Construction Factory won the bid and built the Jianghaiguan phase II Building, the largest Western-style building at that time. With the success of China's construction plant model, it was promoted and developed accordingly, and many construction plants, such as Zhang Yutai and Yu Hongji, gradually emerged.

The introduction and influence of modern Western construction methods also involve project bidding mode. In 1864, the construction of the French consulate first time applied bidding. China did not adapt at first, and no construction plant participated in the bidding. Until the Jianghaiguan phase II project bidding in 1891, the Yang Ruitai Construction Plant participated in the bidding, and then the bidding was gradually accepted by the Chinese plants. For example, the Bank of Dehua in 1903, Ailiyuan in 1904, the German General Assembly and the Huizhong Hotel in 1906, and the Tianxiang Yanghang Building in 1916 were all built by the local construction plants that won the bid.

Western modern building technology, engineering structures, building materials, etc., were also introduced in China, such as installing elevators in Shanghai Huizhong Hotel in 1906. This kind of equipment was first used in the United States in 1887. In 1908, the Shanghai Delvfeng Company adopted the steel-reinforced concrete frame structure technology, which was first applied to the Montmartre Church in Paris in

1894. The steel structure was first applied to the Shanghai Electric Company in 1882. The cement was first applied to the Shanghai Waterworks in 1883. The reinforced concrete was first used in the British Shanghai General Assembly Building in 1903. The air-conditioning equipment was first used in HSBC in 1923. This reduced the gap between China's construction industry and Western developed countries, learned and absorbed Western modern construction technology, and promoted the development of China's construction industry development.

2.1.3.2 Typical Case of Modern Engineering Management in China

The period of the Republic of China was a period in which China's engineering construction and management moved from tradition to modernization. New concepts and systems were implemented in many engineering construction and management processes [18], involving railway engineering, bridge engineering, and water conservancy engineering.

1. **Railway Engineering**

The Beijing-Zhangjiakou Railway (Beijing to Zhangjiakou), built by Zhan Tianyou in 1909, has an important position in modern Chinese engineering construction history.

The Beijing-Zhangjiao Railway is about 200 km long. It was built in September 1905 and completed in 1909. It is the first railway in China that was built and operated by China itself without foreign funds and related personnel. The Beijing-Zhangjiao Railway connects Beijing Fengtai and Zhangjiakou in Hebei Province, passing through Juyongguan, Badaling, Shacheng, and Xuanhua, and there are so many steep mountains. The terrain is complex, and the bridges and tunnels are numerous. Therefore, the project is arduous. The planned construction time of the Beijing-Zhangjiao Railway was six years. Under the joint efforts and hard work of Zhan Tianyou and the construction workers, it was finally completed two years ahead of schedule. At the same time, it saved 356,774 taels compared with the budgeted work. Not only is the total construction cost far less than the price offered by foreign contractors, but the quality of the project construction is excellent.

Many technical problems were solved in the implementation of the Beijing-Zhangjiakou Railway. For example, both the north and the south simultaneously dig into the middle point of the tunnel. But because the tunnel is too long, it was considered to cutting two vertical wells in the middle and then cutting in the opposite direction, thereby forming six working faces to work together. The project utilized the natural topography of the Donggou of Qinglong Bridge adopted a zigzag line, and combined the slope of 33.33 ‰ to avoid excessive tunnel excavation. The project used the Mallet compound locomotive (0-6-6-0), which is lighter (weighing only 136 tons), flexible, and can pass a small curve radius. The project added an automatic hook invented by American Jenny to each car to combine it into a solid whole to ensure safety when climbing. A large number of concrete arch bridges were used in the project to collect materials on-site and save labor costs. In addition, Zhan Tianyou advocates engineering standardization. The Beijing-Zhangjiao Railway Engineering

Standard Map is the first set of railway engineering standard maps in China, including 49 standards for bridges, culverts, tracks, lines, caves, garages, etc., which guarantees the quality of the project and provides a reference to other construction of the railway.

2. **Bridge Engineering**

Taking the Qiantang River Bridge as an example, the Qiantang River Bridge is the first modern railway and highway bridge designed and presided over by the famous bridge expert Mao Yisheng. The bridge is a two-story steel truss bridge with a total length of 1453 m and a width of 9.1 m. It is a milestone in the history of Chinese railway bridges.

The Qiantang River Bridge began preparations in 1933 and officially started in April 1935. At the beginning of the design, to facilitate transportation needs, take into account military defenses, maximizing capital conservation, the design scheme was optimized considering the technical feasibility and economic advantages. At the same time, it attracted professional construction teams through open tendering and investment promotion and solved mechanical equipment problems. In addition, in order to complete the construction of the project with high quality and speed, the operation idea was that "go up and down, and consider both water and land," i.e., started at the same time and moved parallel. For projects related to the top and bottom, did the bottom part first and then the top part. Not only must the project on the water be checked, but also the underwater foundation project should be checked one by one. During the implementation of the Qiantang River Bridge project, the foundation and the pier were started simultaneously. The pier and the steel beam were started at the same time. The submerged construction can be observed on the spot using a caisson to ensure the construction quality.

The most serious difficulty in the construction process is the foundation project. The most severe problem is quicksand. In order to overcome the challenges of the foundation project, solve the key issues such as "how to pile, how to build bridge piers, how to erect steel beams," etc., they use the water of Qiantang River to overcome the quicksand, adjust to local conditions, and finally successfully overcome many difficulties via a series of technological innovations such as shooting water method for piling, caisson method for bridge pier, floating transportation method for steel beam, and nesting crate method for repairing bridge piers.

3. **Water conservancy engineering**

Traditional water conservancy projects in China were generally organized and managed by state and county officials or local people, lacking professional water management institutions. In contrast, the construction of water conservancy projects during the period of the Republic of China showed many different contents in terms of systems, concepts, and methods. There is an example of the construction of water conservancy projects in Jiangxi Province. In 1928, the Jiangxi Provincial Government established the Provincial Water Conservancy Bureau, set up the chief engineer position, survey office, engineering office, and design survey team, and set up special agencies or personnel to manage some important projects. After completing the irrigation project of the Wan'an Canal, the Provincial Water Conservancy Bureau established the Wan'an Canal Management Office and founded township water conservancy associations in more than ten counties and cities, such as Guixi and Nankang. With the introduction of Western science and technology and the return of personnel studied in Western countries, several water conservancy experts, such as Li Yizhi and Zheng Zhaojing, started the era of water conservancy engineering technicians presiding over carrying out water conservancy construction. The chief engineer hired water conservancy experts, and hydrological, meteorological, hydraulic, and other stations in various regions of the province were set up to measure hydrology, meteorology, hydropower, and general navigation channels in various regions of the province. These stations measure rainfall, rainy days, temperature, etc., in various regions and draw charts according to statistics applied in the construction of many rural reservoirs and hydropower projects. Since then, the Provincial Water Resources Bureau has formulated a detailed water conservancy construction plan and has drafted the province's water conservancy six-year plan. For example, building the Tangba Reservoir in the hilly areas, dredging the rivers in the upper reaches of the Ganjiang River, and building large reservoirs in various counties in phases.

The influx of modern Western architectural thoughts has broken the traditional Chinese way of building construction. Drawing on and learning from the experience of engineering construction in the West promoted the process of modernization engineering in China, which was a significant transition period for modern engineering management.

2.1.4 Modern Engineering Management Theory and Practice in China

From 1949 to the present, China's modern engineering management can be divided into two periods: planned economy and reform and opening up, with the Third Plenary Session of the Eleventh Central Committee of China in 1978. Due to political and economic background changes, the engineering management systems of these two periods showed different characteristics.

2.1.4.1 Engineering Management During the Planned Economy Period

1. **Engineering management theory in the planned economy period**

The outstanding achievements of the engineering management theory in the planned economy era are the "system theory" and "double methods" that were nurtured and grown in various construction practices during this period.

(1) **Systems theory**

Bertalanffy first proposed the concept of general systems theory in a lecture at a philosophy graduate class at the University of Chicago. After summarizing and updating, he published *General System Theory* in 1945. One of the sources of its systems thought is engineering technology. It mentions that "the development of technology makes people no longer follow a single machine but thinks according to the 'system,' and the 'systems approach' becomes a necessity." This discussion of the origin of engineering technology promotes the germination and formation of systems thinking by Chinese scientist Qian Xuesen. One of the sources of Qian Xuesen's system thinking is engineering technology, while another source is technical science. The books on general systems theory, cybernetics, and information theory were first published in the 1940s. At that time, Qian Xuesen, who was trying to explore the forefront of technical science, was most concerned with cybernetics. Since a critical technology for the development of rocket missiles is the automatic control of the engine and the projectile, the significance of focusing on cybernetics is obvious. Influenced by the mathematician Wiener's "Control Theory" in 1948, coupled with research and practice in aerospace science and technology, Qian Xuesen first created engineering control theory in 1954 using control system theory and methods and applied it to the area of engineering technology [19], especially in the successful application of major national defense projects in China. In the 1960s, China's "two bombs and one satellite" project, the first atomic bomb explosion, the hydrogen bomb air-explosion test, and the satellite launch were all guided by cybernetics.

Then, based on the comprehensive study of Marxist philosophy and Mao Zedong's philosophical works, scientists represented by Qian Xuesen implemented research on systems science based on practice. For example, *Organization Management Technology—System Engineering,* published in 1978, is an important milestone in China's systems science field [20]. *The Systematic Thought and Systematic Engineering,* published in 1980, embodies its systems science and philosophical thinking. Then turned to the exploration and creation of systems sciences. In 1986, tried to integrate classical systems sciences such as dissipative structure theory and synergy. In 1989, *A New Field of Science: Open Complex Giant System and Its Methodology* was published in the journal of *Nature.* It defines the concept of open complex giant systems and points out the integrated integration methodology used to deal with such systems. The thought of systems theory has important methodological significance for China's engineering management practice at that time and even later. It is also a model for the concrete application of the basic principles of materialist dialectics in the real world.

(2) **Double methods**

"Double methods" refers to the "optimization method" that improves the production process and quality and the "overall planning method" that deals with production organization and management issues.

In 1958, China's outstanding mathematician and educator, Hua Luogeng, led a large number of mathematicians to go out of school and go to industrial and agricultural production enterprises to seek practical applications of linear programming. However, because linear programming and other complicated methods require complicated calculations, it is difficult to carry out large-scale promotion, and the "linear programming" movement gradually cools down. In the early 1960s, based on the introduction, absorption, and improvement of CPM (Critical Path Method) and PERT (Program Evaluation and Review Technique), and based on Chairman Mao Zedong's guiding ideology of "overall planning" and "singling out the principal contradictions," Professor Hua Luogeng proposed "overall planning method." At the same time, the optimization method proposed by the international community has also received attention. Professor Hua Luogeng put forward the "optimization method."

Subsequently, the promotion pilot work was carried out. For example, in February 1965, the pilot project was carried out at the Beijing 774 factory. Although it did not achieve the expected results, it gained valuable experience. That is, "the overall planning method must be applied to one-time projects"; in June of the same year, Professor Hua Luogeng published the *Overall Planning Method* in the *People's Daily* and started overall planning method classes in Beijing and Nanjing. In March 1970, Premier Zhou Enlai pointed out that he would support Professor Hua Luogeng to experiment with his overall planning method. In April of the same year, Professor Hua Luogeng gave a presentation to the State Council and ministry leaders to introduce the overall planning method and optimization method, which caused significant repercussions. Subsequently, he was invited to the Shanghai Refinery to promote the "double method" and pilot work. The use of the overall planning method reduced the duration of the phenol refining and expansion project from 30 to 5 days. The use of the optimization method reduces the solidification temperature of the military 605 pour point depressant from − 38 to − 41 °C. And the pilot promotion of the electronics industry and the chemical industry has also achieved gratifying results. Therefore, since 1972, Professor Hua Luogeng has established the "promotion of optimization method and overall planning method team." This team went to 28 provinces, municipalities, and autonomous regions to carry out large-scale promotion work of "double methods" and has led the masses to participate in scientific experiment activities and achieved remarkable results.

After the founding of New China, China began to implement the planned economic system, shouldering the burden of restoring and developing the national economy and experienced the following periods: the recovery period of the national economy (1949–1952), the first five-year plan (1953–1957), Great Leap Forward, the period of national economic adjustment (1958–1965), and the period of the Cultural Revolution (1966–1976). During these periods, many basic construction projects were implemented in full swing using "double methods." That is, using concentrated financial resources to do important things to obtain a series of significant achievements. This promoted the development of China's engineering management practice and laid a solid foundation for the development of engineering management theory in China.

2. **Organizational model of engineering construction in the planned economy period**

The construction organization model mainly includes the construction enterprise self-operated mode, infrastructure management department, and engineering command headquarters during the planned economy period.

(1) **Construction enterprise self-operated mode**

At the beginning of the founding of New China, the design and construction strengths were weak and scattered. Many projects were arranged for the design and construct personnel by the construction enterprises, and then they purchased the related materials and equipment to start the project construction. The self-operated model of the construction enterprise came into being. The construction enterprise of this mode has two functions, construction and production, and these two functions are closely integrated. Although the construction and production efficiency cannot be accurately estimated and planned, it can reduce the conflicts between construction and production, flexibly mobilize existing resources, carry out production while construction, improve construction speed and efficiency, and thus improve investment efficiency.

(2) **Infrastructure management department mode**

At the beginning of the founding of New China, the infrastructure management department mode was also fostered. That is to say, setting up the infrastructure management department for the substantive work of engineering construction under the administrative units that carry out the daily administrative work. Some organizations with more projects also adopted the infrastructure management department mode. Although this mode may be due to the unpaid nature of government investment, these administrative departments responsible for project management may try their best to obtain investment, increase budgets, and maximize obtaining construction funds [21]. But the mode is fast, flexible, and targeted and can quickly meet the needs of the administrative departments [22].

(3) **Engineering command headquarters mode**

The engineering command headquarters mode appeared after 1958. Usually, the government convened construction developers, design entities, construction contractors, etc., to set up temporary headquarters or preparatory offices to separate construction and production functions. The temporary headquarters presided over the design, procurement, and construction management during the project's construction until the completion of the project. The headquarters was formed with the project's construction; it was dissolved with the completion of the project [23]. This led to the defect that is difficult to accumulate in management experience. But this mode can use powerful administrative means to reconcile the relationship between the parties and timely mobilize the design, construction, materials, and equipment resources required for the project's construction, which plays a vital role in developing the capital construction work.

3. **Typical engineering construction during the planned economy period**

Starting from the "First Five-Year Plan," China carried out large-scale economic construction intending to increase industrial growth, catch up with the UK and the US, prioritize the development of heavy industry, and establish the foundation of national industrialization. Outstanding achievements have been made in infrastructure construction, such as the successful production of the first batch of domestically produced vehicles—FAW Jiefang, the opening of the Kangzang and Qinghai-Tibet highways, and the completion of the Wuhan Yangtze River Bridge.

(1) **156 engineering projects**

On February 14, 1950, China and the former Soviet Union signed the 30-year *Sino-Soviet Treaty of Friendship, Alliance and Mutual Assistance* and *the Agreement on the Loan of the Soviet Union to the People's Republic of China.* At the same time, an agreement was signed to assist the construction and transformation of 50 large enterprises by the former Soviet Union, which was later changed to 47, including 10 coal mines, 11 power stations, 3 steel companies, 3 non-metal enterprises, and 5 chemical companies, 7 mechanical companies, 7 defense companies, and 1 paper company [23]. In May 1953, representatives of the two governments agreed to 91 new aid projects, of which defense industry and related projects accounted for a certain proportion. In August 1954, the former Soviet government expressed its willingness to provide equipment and other assistance to 15 Chinese defense enterprises and arranged defense content among 14 aided industrial enterprises. On March 28, 1955, China and the former Soviet Union signed an agreement on the aid of the construction of industrial projects, which belonged to the fields of national defense, shipbuilding, raw materials, and manufacturing.

The 156 engineering projects were mainly distributed in Harbin, Qiqihar, Jilin, Changchun, Shenyang, Fushun, Baotou, Xi'an, Luoyang, Taiyuan, Lanzhou, Chengdu, Wuhan, Zhuzhou, and other cities, mainly concentrated in the Northeast, changing the past 70% of the industry enterprises focused on the coastal layout. Among them, 106 were civil industrial enterprises, 50 were located in the northeast, 32 were located in the eastern area, 44 were defense enterprises, and 35 were located in the central and western regions, 21 of which were arranged in the provinces of Sichuan and Shaanxi. This is to use resources nearby; that is, metallurgical chemical companies are arranged in mineral resources and energy-rich areas, machinery industry is placed in the vicinity of raw material production areas. It is conducive to changing the face of economically backward areas, while at the same time, the new enterprises are placed in the rear for military needs.

Few of these 156 projects were put into operation during the "First Five-Year Plan" period, and most of them were completed and put into operation during the "Second Five-Year Plan" period. From 1951 to 1956, 26 enterprises were completed and put into operation, 31 were partially completed, and 17 separate workshops and factories were completed and put into operation [24]. Twenty-seven of the 64 projects aided by Eastern Europe countries were totally or partially put into production [20], involving aircraft, automobiles, new machine tools, power generation equipment, metallurgical equipment, and other manufacturing enterprises, as well as smelting enterprises such as alloy steel and non-ferrous metals. The 156 projects significantly improved China's industrial manufacturing capacity and made outstanding contributions to the new industrial production capacity and the completion of the First Five-Year Plan.

(2) **Missile, nuclear bomb, and satellite development project**

During the development process of the Two Bombs and One Satellite Project, the pre-research was adopted for scientific decision-making, and the concept of concurrent engineering was infiltrated. It used the management mode of two command lines to organize, manage, coordinate, and cooperate with all related departments and administrative departments, centralizing commanding and operating plan review technology for scheduling, resource allocation, and cost optimization [25, 26].

(a) **Scientific decision**

In the 1950s and 1960s, scientists and leaders in charge of developing the Two Bombs and One Satellite Project in China noticed the importance of pre-research and put it into practice. Since December 1960, China has established a neutron physics group and begun to develop hydrogen bomb research. More than 40 people had participated in pre-research and had made preliminary explorations on the principle of hydrogen bombs, proposing possible technical approaches and models. The theoretical pre-research on hydrogen bombs won the time for China's explosion of hydrogen bombs within two years and eight months after the atomic bombing. During the 20 years from 1956 to 1980, China's missiles leaped from short-range to intercontinental, which was inseparable from theoretical pre-research.

In 1955 and 1956, China made the decision to develop "two bombs." In 1958, China allocated 200 million yuan for the development of satellites. After China entered a three-year economic hardship period because the importance of the "two bombs" was greater than that of satellites, the central government decided to adjust the space technology research mission in 1959. First, developing sounding rockets, building space environment simulation laboratories, and developing ground tracking measurement equipment. This guaranteed the demand for the development of "two bombs" and achieved gradual progress in satellite research. In the early 1960s, a series of achievements were made in individual research and experimental equipment for space science and technology, which laid a good foundation for the comprehensive development of artificial satellites in 1965.

(b) **Organization management**

The development of "two bombs and one satellite" in China embodied the concept of "parallelism," which can be described as the Chinese parallel engineering method. The use of two "three combinations" combines scientific research units, production units, and users, combined with scientific and technological personnel, workers, and leading cadres. One of the ways to realize the three combinations of scientific research, production, and the user was the combination of factories and institutes. The nuclear weapons research institute was a typical example of the combination of factories and institutes, which involves all aspects of raw material production, theoretical design, atomic bomb manufacturing, and nuclear weapons testing. In 1960, the Nuclear Weapons Research Institute established scientific research and experimental buildings, processing workshops, and other related facilities. It established the theory department, the experimental department, the design department, and the production department (including 13 research laboratories). In October of the same year, it was changed to six research laboratories and one processing workshop. In 1962, it was reorganized into four ministry-level organizations.

The two administrative and technical command line management systems formed within the missile and satellite development system played an important role. In 1962, the medium-range missile failed due to overall design mistakes, so the chief designer system was established. In the same year, *the Provisional Regulations of the Fifth Institute of the Ministry of National Defense* were formulated to define two-line command management clearly. Each model was equipped with a designer system that included the chief designer responsible for the model, the lead designer responsible for the subsystem, and the lead designer responsible for the individual product. With designers at all levels as the core, supplemented by technical leaders at all levels, they constituted the technical command system for model development, which was responsible for the design, technical decision-making, and coordination in development. At the same time, the administrative leadership, the planning management department, and the command and scheduling department at all levels constituted the administrative command system, which was responsible for the formation of the team, organization and cooperation, planning and scheduling, post-guarantee, and ideological and political work. This two-line management model was also effective in intercontinental rockets, communications engineering, and submerged rockets of the late 1970s and early 1980s.

(c) **Command coordination**

In 1962, Chinese scientist Qian Xuesen proposed a trial program evaluation and review technique (PERT) in the planning and technical management department. In 1963, the Ministry of Defense Fifth Institute tried the PERT management method in the ground computer manufacturing process of the remote rocket guidance system and found that the power supply was a short-term product in the design and manufacturing process. It also played an important role in developing China's intercontinental rockets, communications engineering, and submarine rockets. In November 1962, China's atomic bomb development entered a crucial stage. The central government set up a special committee of fifteen people to focus on command and coordination within a wide range of cutting-edge tasks. After the successful development of the atomic bomb, it transferred to focus on missiles, satellites, and nuclear submarines in 1965. From 1962 to 1974, the Central Committee held more than 40 meetings, organized hundreds of ministries, commissions, bureau-level units, 26 provinces, municipalities, autonomous regions, and thousands of factories, universities, institutes, and military units to carry out the division of labor and cooperation, commanded all involving organizations to contribute to the development of "two bombs and one satellite."

(3) **Bridge engineering**

Both the Wuhan Yangtze River Bridge and the Nanjing Yangtze River Bridge are the railway and highway bridges on the Yangtze River in China. They were all born in the challenging environment of construction funds and technical difficulties.

(a) **Wuhan Yangtze River Bridge**

Wuhan Yangtze River Bridge is the first Yangtze River bridge in New China and also the first railway and highway bridge built on the Yangtze River. It was built between 1955 and 1957. The main bridge is a double-story trussed steel girder bridge. The upper level is a highway bridge, of which the roadway is 18 m wide, and the sidewalk is 2.25 m wide on each side. The lower level is a double-track railway bridge. During this period, due to the concentrated investment of construction funds, national finance was strained. How to find a balance between beautiful design and economy and practicality was a difficult issue for the construction of the Wuhan Yangtze River Bridge. Therefore, when designing the Wuhan Yangtze River Bridge, the various design schemes were compared, and the double-eave roof of the reinforced concrete structure was selected as the spired bridgehead.

During the construction period, due to the wide surface of the Yangtze River and deep-water flow, the water depth during the rising season was more than 40 m, the high and low water level difference was up to 19 m, and the high-water level period lasted for 8 months every year. The sediment cover of the river bottom was different in-depth, reaching a depth of more than 30 m. The rock formation below the cover layer had a complex geological structure, which brought great difficulties to underwater foundation construction. Through continuous experimentation and exploration, they boldly abandoned the deep-water foundation pressure caisson method. They innovated to propose a deep-water pier-building "pipe string drilling method" to expand the diameter of the pier foundation pile. This method opened up the basic structural model for the large-span beam piers that resist large horizontal forces in the deep-water rapids of the big river near the harbor.

In addition to the longitudinal transport corridors for trains, cars, and sidewalks, there are elevators and stairs on both sides of the bridge for pedestrians. The upper layer of the approach bridge is in the form of a continuous arch. The bridgeheads on both sides were arranged with Snake Mountain and Guishan Park to carry out appropriate greening. Using the Snake Mountain terrain at the bridgehead on Wuchang shore to set a spacious platform on the railway surface for visitors to rest and enjoy the views of the river. In addition, the Wuhan Yangtze River Bridge is also closely connected with the surrounding business, driving the surrounding area into a dense area of people's activities, harmoniously coexisting with the surrounding ecology and living environment.

(b) **Nanjing Yangtze River Bridge**

Nanjing Yangtze River Bridge is China's first self-designed, self-constructed railway and highway double-decker dual-purpose bridge constructed with domestic materials [27]. The railway bridge has a total length of 6772 m, a width of 14 m, and a double lane. The total length of the highway bridge is 4588 m, the width is 19.5 m, and the length of the main bridge is 1577 m with four lanes. There are 262 holes in the bridge, including 10 holes in the main bridge and 252 in the approach bridge. It started construction in January 1960 with a total investment of 287 million yuan. It was completed and opened to traffic in December 1968. In October 1958, the Ministry of Railways, Jiangsu Province, and Nanjing Municipality jointly formed the "Nanjing Yangtze River Bridge Construction Committee." In November 1959, the "Nanjing Yangtze River Bridge Engineering Headquarters" was established. In January 1960, the bridge was officially constructed. The steel used was the Angang 16 manganese steel successfully made by China own. The seamless railway was first used, and the high-strength bolts were used to replace rivets. The bridge's roadway was first used with fly ash ceramsite lightweight concrete, reducing the steel used for the steel beam.

Due to the bridge site's complex natural and geographical conditions, the river surface is wide, 400 m wider than the Yangtze River Bridge site, and the deepest water depth is 60 m. The geology is complex; the fracture zone of the rock formation and the strength difference are very different. Therefore, ensuring that the nine bridge piers of 70–80 m high were stable as rocks was a technical issue. The bridge engineering bureau integrated the strengths of various basic structures and through practical experience, creatively adopted four methods: heavy concrete caisson, steel sheet pile cofferdam string, steel caisson plus pipe string, and floating reinforced concrete caisson, so that the piers were firmly fixed on the rock formation at the bottom of the river.

Nanjing Yangtze River Bridge is also unique in architectural art, with rich characteristics of the times. The most eye-catching is the "Red Flag" style adopted by the North and South Bridgeheads. The junction between the main bridge and the approach bridge was a duplex fort, which consisted of a big fort and a small one. The big fort consists of two towers and a hall. Each tower is about 70 m high and has ten floors. The ground floor is connected to the hall. In addition, elevators and stairs lead to the railway, highway bridge, and overlooking platform. Reliefs such as "Long live the united people in the world" are engraved around the bridgehead.

(4) **Railway Engineering**

At the beginning of the founding of New China, China had fewer railways with low standards and uneven distribution. The central government put forward the goal of "dividing the gap and quickly building the road network" and implementing the "first build and then complete" construction concept. The Tianlan Railway (Tianshui to Lanzhou) has a total length of 354.3 km and started construction in May 1946. However, by 1949, only some earthwork and tunnel projects were completed. The Northwest Railway Main Line Engineering Bureau was established in 1950 to be responsible for the construction of the Tianlan Railway. Following the construction concept of "first build and then complete," the construction of the route and the bridge were considered, and the engineering train was first opened to solve the problem of material transportation difficulties, reduce construction costs, and speed up the construction progress.

This construction concept has played a positive role in promoting the first five-year plan of China's national economic development. During the five years from 1953 to 1957, China concentrated on building railways, saved initial investment as much as possible, vigorously promoted the construction of remote areas, industrial and mining enterprises, and forest railways, carried out technical transformation and reconstruction of major trunk lines, accelerated the supporting structure of railway projects, and focused on meeting the urgent needs of national economic development for railway transportation. Therefore, the railways in China had a robust, comprehensive transportation capacity.

In 1958, during the "Great Leap Forward" period, the upsurge of "the whole people building the railway" occurred. It emphasized that the facilities and equipment directly related to railway transportation should be solid and reliable, while other facilities could be simple to save money, reflecting the construction concept of "consolidation of essentials and simplification of trifles." This kind of concept emphasized the construction of "more and fast" and divided the systematic and complete project. It was challenging to achieve the purpose of comprehensively improving transportation capacity.

In 1964, in the southwestern third-line construction, it was proposed that "the railways of Chengkun, Sichuan-Guizhou, and Guikun should be built quickly." Because the southwestern three-line projects are mountainous railways, the geology and topography along the line are complex, and there are many river tunnels along the river. There are many bridges across deep valleys and caves. Under the guidance of the concept of "consolidation of essentials and simplification of trifles," the investment was mainly used for the main project. Due to ignoring the ancillary works, the roadbed subgrade waterproof and drainage and slope protection works of the mountain railways were inherently insufficient, leaving hidden dangers for the railway operation, resulting in the frequent occurrence of disasters such as landslides and mudslides, and even shutdowns. In fact, the investment spent on the remediation of disasters may be more than the investment saved during construction. Although the practice of "the whole people building the railway" was vigorous, some "reckless" works were not ideal. Railway engineering is complex and systematic, requiring scientific management concepts and systematic arrangement. It requires carefully carrying out survey and design, research and demonstration, rationally arranging investment and construction, and scientifically managing to achieve the expected goals and results.

At this stage, mainly influenced by China's economic development, the concept of using centralized financial resources to do important things had achieved a series of progress and achievements. The measures and practices adopted following the background of the times had promoted the development of engineering management practices; the rise of relevant theories and methods at home and abroad had also laid a good foundation for the development of engineering management in China.

2.1.4.2 Engineering Management During the Reform and Opening-Up Period

After the reform and opening up (from 1978 to the present), with the advancement of the socialist market economic system and rapid economic growth, and scientific and technological progress, the reform of the construction management system and the implementation of project management concepts and methods had been implemented. These promote the improvement of China's engineering management theory and practice [8] and make the engineering management concept tend to be scientific, the method tends to be specialized, and the value tends to be diversified.

1. **Engineering Management Theory and Discipline during the Reform and Opening-up Period**

During the period of reform and opening-up, China's engineering management theories and disciplines have become increasingly mature and perfect. With the development of the social economy and the advancement of science and technology, engineering management has become increasingly important, engineering practice has become increasingly rich, theoretical research has become increasingly deep, and the discipline status of engineering management has been continuously recognized. The emphasis on engineering management in developed countries are reflected in many aspects, such as the establishment of academic institutions and professional associations. Some national engineering academies have established engineering management-related departments. For example, in the Department of Electronic Engineering, the Department of Economics, and the Department of Education and Research Policy at the National Academy of Engineering in the US, there is a sub-area of management in these departments, and the academicians are not only from academic organizations but also from the business community. Academicians of the National Academy of Engineering in the US are involved in business and academia. The French Academy of Science and Technology and the Royal Academy of Engineering in the UK are mainly composed of engineers. At the same time, the American Society of Civil Engineers (ASCE) was founded in 1852, and the journals published under its name cover the field of engineering management, such as the *Journal of Construction Engineering and Management* and *Journal of Management in Engineering;* American Society for Engineering Management (ASEM) was founded in 1979, which is mainly responsible for the preparation and revision of the "Engineering Management Knowledge System." The publications include *the Guide to the Engineering Management Body of Knowledge* and *The Engineering Management Handbook.*

China's development of engineering management disciplines has changed from simple construction and civil engineering management to generalized engineering management. In the early 1980s, Chinese universities set up engineering management majors one after another. Subsequently, the Ministry of Education made many adjustments to *The Undergraduate Majors Catalogue of Universities* and added the majors of "Building Management Engineering" and "Basic Construction Management Engineering" in the second revision in 1989. The third revision was carried out in 1993, adding the majors of "Management Engineering" and "Real Estate Management." In 1998, the former management engineering (construction management engineering direction), real estate management, foreign construction engineering management, and international engineering management were merged into engineering management, and at the same time, established them under management engineering and the first-level engineering subjects [28, 29]. The adjustment of major disciplines is not limited to the Ministry of Education and higher education institutions. In 2000, the Chinese Academy of Engineering established the Department of Engineering Management, which means that the domestic engineering community and academia are concerned and identify the status of the engineering management discipline. From 2007 to 2016, the Chinese Academy of Engineering successfully held the China Engineering Management Forum ten times. The theme of the forum covers harmonious innovation, scientific development, western development, the rise of the central region, the One Belt and One Road, infrastructure construction, etc., which not only impels the in-depth discussion and exchange of engineering management in China but also promotes the innovation and progress of theoretical education. In addition, in 2008, the Department of Management Science in the National Natural Science Foundation of China set up the "G0118 Engineering Management" code in the "Eleventh Five-Year" project funding, encouraging and advocating scholars to conduct in-depth research around China's engineering construction. In 2010, the Master of Engineering Management (MEM) degree was approved and set up, reflecting the urgent need for engineering management talents in China's contemporary engineering construction and, at the same time, bringing China's engineering management talent training to a new level.

2. **Construction management system during the period of reform and opening-up**

After the reform and opening-up, China's engineering construction management system has undergone a significant transformation. A series of construction management systems, such as a bidding system, contract management system, construction project supervision system, and project legal person responsibility system, have been implemented.

(1) **Bidding system**

In 1979, China first tried the bidding system in the construction and installation market in some areas. In June 1983, the Ministry of Urban and Rural Construction and Environmental Protection issued *the Trial Procedures of Bidding for Construction and Installation Projects* and implemented the bidding contract system in engineering construction. In 1984, the State Council's *Interim Provisions on Several Issues Concerning the Reform of the Construction Industry and Capital Construction Management System* pointed out that "the engineering project bidding system should be carried out vigorously, and we must reform the old method of simply using administrative means to distribute construction tasks and implement the bidding system." In the same year, the *Report on the Work of the Government* of the Second Meeting of the Sixth National People's Congress clearly stated that the bidding contract system should be implemented in engineering construction. In 1985, the State Planning Commission and the Ministry of Urban and Rural Construction and Environmental Protection jointly promulgated *the Interim Provisions on Tendering for Construction Projects Bidding*. In 1992, the Ministry of Construction promulgated *the Measures for the Administration of Engineering Construction* and *Construction Bidding* and *the Detailed Procedures for Notarization*. In September 1999, *the Bidding Law of the People's Republic of China* was officially promulgated and implemented on January 1, 2001.

(2) **Contract management system**

After the reform and opening up, the Chinese construction market faced fierce competition. It was required to improve the engineering management level; engineering contract management gradually attracted attention. The introduction and application of FIDIC contract terms in the Lubuge Water Conservancy Project in 1982 and the Jingjintang Expressway project in 1984 reflected the development of China's engineering construction contract management. In March 1999, the Second Session of the Ninth National People's Congress passed the promulgation of *the Contract Law of the People's Republic of China*. In December of the same year, the Ministry of Construction and the State Administration for Industry and Commerce issued *the Model Text of Contracts for Construction Engineering*, which is the contractual text for the various public buildings, private houses, industrial plants, transportation facilities, and lines, pipelines construction, and equipment installation. With the rapid economic growth and the expansion of engineering scope, the contract management system has been promoted and applied in engineering practice.

(3) **Construction project supervision system**

China began the construction project supervision system in 1988. At the end of 1996, the Ministry of Construction issued *the Construction Project Supervision Regulations*, marking the beginning of China's construction project supervision system based on national conditions. *The Building Law of the People's Republic of China* promulgated in 1997 clearly stipulates that "the State implements the supervision system for construction projects." This makes the construction project supervision system enter the stage of comprehensive nationwide implementation, and the legal status of the supervision system is clearly defined by the law. At the end of 2000, the Ministry of Construction issued the *Code of Construction Project Management* to standardize the project construction supervision system gradually.

(4) **Project legal person responsibility system**

In November 1992, the State Planning Commission issued *the Interim Provisions on Implementing the Owner's Responsibility System for Construction Projects*. It is required from 1992 onwards, the project owner responsibility system will be implemented for new construction projects and basic construction projects for a state-owned unit that carries out preliminary work. In 1996, the State Planning Commission formulated and promulgated *the Interim Regulations on the Implementation of the Legal Person Responsibility System for Construction Projects* and required the state-owned units that operate large- and medium-sized capital construction projects to establish project legal persons and implement the project legal person responsibility system. It means that the original project owner responsibility system is changed to the project legal person responsibility system, which stipulates that the project legal person is responsible for the project planning, fund raising, construction and implementation, production and operation, repayment of debts, and asset value maintaining and appreciation, a whole process responsibility.

3. **Engineering Management Mode in the Period of Reform and Opening-up**
(1) **Project financing model**

After the reform and opening up, with the rapid development of the social economy, project investment is getting larger and larger. The single fund-raising method and financing channel cannot meet the needs of project construction. It is necessary to comprehensively adopt various financing methods to raise the necessary construction funds from different channels. The company financing model is the most common financing mode. During the project's development, for a long time, the project sponsors adopted the corporate financing model. The most prominent feature of this model is that it is simple and clear, and easy to operate. The company established by the project sponsor is the legal project person, directly financing from the financial institution and repaying according to the agreement.

In the 1980s, China began to pilot and adopt the BOT model (Build-Operate-Transfer). In the 1990s, the Chinese government began to attach importance to the BOT model and selected more power plants and waterworks projects as pilot projects for the BOT model to promote further. In January 1995, the Ministry of Foreign Trade and Economic Cooperation issued *the Notice on the Issue of Attracting Foreign Investment through BOT*. In August of the same year, the State Planning Commission, the Ministry of Electric Power, and the Ministry of Communications jointly promulgated *the Notice on Issues of the Trial Approval Management of the Foreign Investment Concession Projects* to ensure that the pilot work was carried out in an orderly manner. Due to the different types of infrastructure, investment and financing returns, and project property rights, there are many variations, including the BOOT (Build-Own-Operate-Transfer) model that emphasizes project ownership, the BT (Build-Transfer) model that doesn't include the operational phase, the BLT (Build-Lease-Transfer) model that is similar to financial leasing, and so on.

In the early twenty-first century, a large number of PPP (Public-Private-Partnerships) projects began to appear in China. That is, governments, for-profit companies, and non-profit companies work together for a project to share responsibility and financing risks. For example, in August 2002, Shenzhen broke the government monopoly of municipal public facilities and transferred some shares of state-owned enterprises such as Energy Group, Gas Group, and Public Transportation Group by bidding. In September of the same year, the Chengdu Municipal Public Utilities Bureau auctioned the franchise rights of six bus lines and transferred domestic bus lines for a fee. In 2003, 17 private enterprises in Zhejiang Province set up five investment companies to undertake the Hangzhou Bay Bridge and used private capital to build major infrastructure projects. The provincial government promulgated *the*

Opinions on Further Expanding Private Capital and clearly pointed out that coastal railways, thermal power plants, and other projects would continue to attract private capital. In 2008, the construction and operation mode of the Beijing Olympic venue—National Stadium, also adopted the PPP franchise model. In addition, the PPP model was also adopted in the Beijing Metro Line 4 in 2009 and the Shenzhen Metro Line 4 in 2011. Many other optimized and innovative project financing models are adopted, such as PFI mode (Private-Finance-Initiative), ABS mode (Asset-Backed-Security), and so on.

(2) **Organizational management model**

In 1996, with reference to the experience of state-owned enterprise reform and the established principles and models of corporate legal persons, the State Planning Commission issued *the Interim Provisions on the Implementation of Corporate Responsibility System for Construction Projects,* which made China's organizational management model move from the traditional model to the project legal person responsibility system mode. The engineering command headquarters mode with planned economic characteristics cannot gradually adapt to market economy laws. The combination with the project legal person responsibility system not only retains the advantages of the original decision-making and command but also establishes the responsibility and relationship of the project to ensure the investment return and expected efficiency. At the same time, with the further development of China's social economy and technology, the scale of engineering construction is getting bigger and bigger, and the technology is more and more complicated. Implementing project management by project legal person mode cannot meet the demand for engineering management specialization. Combined with China's national conditions, the organization and management model of the project legal person plus the project management consulting company emerged. That is, the project legal person acts as the owner, only makes recommendations for the project, and makes major decision-making. The project management consulting company provides the owner with comprehensive and full-process services. The common PM mode (Project Management), CM mode (Configuration Management), etc., are all organizational management modes of project legal person plus project management consulting company.

The agent construction system is a model for the problems of multi-integration of "construction, supervision, and practical" in the Chinese government investment project management. It means that by setting up a professional construction agency, the agency is responsible for the project (or providing consulting services), working in the early and implementation phases of the relevant construction project. As early as 1993, Xiamen City began to entrust the municipal financial investment and financing of social construction projects to some powerful professional companies, which replaced the owners to implement construction management, which can be described as the prototype of China's "agent construction system." In 1998, Chongqing, Qingdao, and other cities began to implement the pilot work. In 2004, China officially issued *the Decision of the State Council on Investment System Reform*, proposing "accelerating the implementation of the agent construction system for non-operating government investment projects, that is, through the method of bidding, selecting a specialized project management unit responsible for construction and implementation, which strictly controls project investment, quality, and construction period, and hands over to the user after completion acceptance." As of December 2005, about 47 provinces, autonomous regions, municipalities directly under the Central Government, cities under separate state planning, and cities at the sub-province level have successively tried to implement the agent construction system. On this basis, Shanghai, Shenzhen, and Beijing all have created a mode of agent construction system that conforms to the characteristics of their cities.

(3) **Engineering project contracting mode**

The traditional contracting mode DBB (Design-Bid-Build) is the most common engineering management mode at present. The World Bank, the Asian Development Bank loan project and the engineering project using the International Federation of Consulting Engineers' civil engineering contract conditions adopt this model. The owner and the contractor sign this mode. The owner contacts the contractor through the representative responsible for project management or authorizes the supervision engineer to manage.

In 2003, the Ministry of Construction issued *the Guiding Opinions on Cultivating Development Project General Contracting and Engineering Project Enterprise Management*, pointing out that DB (Design-Build) mode is a form of general contracting of engineering. The general contracting enterprise shall undertake the engineering design and construction following the contract and be fully responsible for the quality, safety, construction period, and cost of the contracted project. Actually, the DB model was adopted in the Shuangchongqiao project of Liuzhou City as early as 2000, but it failed due to laws and regulations, personnel quality, and other reasons.

EPC (Engineering Procurement Construction) mode, also known as turnkey mode, is another form of engineering general contracting. During the period from 1984 to 2000, the Ministry of Construction, the State Planning Commission, the Ministry of Finance, and other relevant departments of the State Council successively issued *the Notice on the Relevant Issues Concerning the Pilot Project of the Design Unit for the General Contracting of the Project, the Regulations on the Management of the General Contracting Qualifications of the Design Units, the Opinions on Vigorously Developing Foreign Contracted Projects*, and other documents. Many design units have transformed from design-based contractors to engineering companies with full functions such as design, procurement, construction management, etc.

In addition, *the Guiding Opinions on Cultivating Development Project General Contracting and Engineering Project Management Enterprises* indicates that the specific project management methods include PM mode (Project Management) and PMC mode (Project Management Contractor). In 2003, China Huanqiu Engineering Company and Vietnam Chemical Corporation signed a contract for the construction and management contracting project (PMC) of the Haiphong Diamine Phosphate Project. The project plans to invest about US$180 million, which is the largest investment project in Vietnam to date, and also is the first PMC contract signed by the Chinese company overseas.

4. **Major project construction during the reform period**

Since the twenty-first century, the field of engineering management has been expanding. The scope of application has gradually extended from the fields of architecture and national defense to aviation, machinery, petrochemicals, information engineering, etc. A large number of valuable, successful experiences and practices have been produced in engineering practice in various industries, such as the High-speed railway, Shenzhou spacecraft project, Three Gorges Project, West–East Gas Transmission, etc.

(1) **Manned spaceflight engineering**

In 1985, the Commission of Science, Technology and Industry for National Defense and the Ministry of Space proposed to the Central Committee that manned space flight should be the next step in China's space development. In 1986, manned space flight was included in the 863 Program. In September 1992, the central government approved *the Asking for Instructions for the Development of China's Manned Spacecraft Project*, which clarified the development guidelines, strategies, mission objectives, and overall concept of the manned spaceflight project. It proposed the first step of the manned spacecraft, including four major tasks, seven major systems, and proposals for funding, schedule, organization, and management [30].

China's manned spaceflight project is implemented under the direct leadership of the Central Committee. It is composed of the General Armament Department, the National Defense Science and Technology Commission, the Chinese Academy of Sciences, and the China Aerospace Science and Technology Corporation, which is an inter-departmental, cross-industry, highly centralized, and unified organizational management system. At the professional level, the project consists of seven major systems, including general engineering and astronauts, spacecraft applications, manned spacecraft, launch vehicles, launch sites, measurement and control communications, landing sites, and their respective sub-systems. At the management level, according to the nature of the mission, there are two management modes: the flight mission period and the no-flight mission period. In this way, the two command lines of the chief commander and the chief designer are vertically connected from top to bottom, the manned space engineering offices at all levels are managed horizontally, and the post and responsibility are fixed at all levels, weaving into a matrix organization system and network.

The core of the manned space engineering planning system is comprehensive planning, supporting management, interface coordination, node control, and bottleneck breakthrough. The planning system is particularly important in the implementation of special management. The manned spaceflight project is based on the key engineering and short-term projects, adopting a reasonable parallel and cross-arrangement method, formulating the planning process and node plan according to the technical process, forming a network flow chart, then decomposing them in layers, and implementing them to each system until stand-alone device.

The complexity of the manned spaceflight project, the long development cycle, and the uncertainty in the development process has made the project's overall design and each system valued. The project is based on the development goals of the central decision-making and adopts a three-cabin spaceship program conducive to the full completion of essential tasks. It fully draws on the mature technology of the aerospace model and reflects the characteristics of China. In addition, according to the development requirement, the overall project and the overall design department of each system set the technical requirements and basic plans for each development stage, clarify the technical process, and formulate the signs for completing the tasks so that the entire project has clear requirements before the start of each development stage. The process has a technical procedure that can be followed. After the completion of the development, the completion mark is used as the inspection evaluation standard.

The biggest difference in manned spaceflight engineering is in manned. It requires the safety of astronauts to be placed first, and improving the safety and reliability of engineering is the core of engineering quality. Therefore, the system development, the quality of the whole machine development, the quality of the collaborative supporting products, and the quality of the engineering hardware and software products are paid attention to. On the one hand, focusing on the "head" (leader and management authority), and on the other hand, focusing on the "source" (components, raw materials, design, and technology) to implement quality control points to each system, each unit, and each job position. At the same time, the main body of the quality consciousness of "manned" and "people-oriented" was established, and the quality control system such as *Astronaut Safety Work Guide* and *First Manned Space Flight Release Guideline* were formulated, adhering to the quality issue one-vote veto system, progress obeying quality, and focusing the quality throughout the entire process and all involving personnel comprehensively.

(2) **Energy engineering**

(a) **Petrochemical engineering**

Sinopec is China's largest petrochemical-integrated energy and chemical company. It combines project management theory with China's petrochemical engineering construction practice. It explores and creates a construction management model suitable for China's major petrochemical engineering project management, namely IPMT + EPC + engineering supervision management. IPMT (Integrated Project Management Team) is the project integration management group, and EPC (Engineering Procurement Construction) is design, procurement, construction project general contracting.

This management mode was first adopted in the Shanghai Secco 900,000 tons/year ethylene project jointly invested by Sinopec and BP. The Chinese and foreign parties each hold 50% of the shares. The "PMC + EPC" management mode was adopted in the initial stage of construction, and the contradiction was prominent during the implementation process. For example, in terms of contract management, the entire project contract's segmentation caused contractors and patentees to ask for an exorbitant price, resulting in a long-lasting business negotiation. In terms of organization and management, all parties act in their way. It is challenging to coordinate the whole process, etc. In response to these contradictions and problems,

Sinopec repeatedly communicated with the senior management of BP and finally adopted the management mode of "IMPT + EPC + engineering project supervision." The core of this management model lies in the three-tier organizational structure: the first layer is the decision-making layer, composed of members of the IPMT director group, who authorize decision-making on the major and key issues of the construction projects. The second layer is the management layer, composed of members of the IPMT project management department. It mainly undertakes coordination with EPC units and supervisory units and implements HSE, quality, schedule, cost, and effective contract execution. The third level is the executive level, which consists of pre-project consultants, EPC contractors, and supervises contractors to perform specific project management and construction tasks [31]. The matrix structure of IPMT is mainly based on vertical project management, and horizontal management is mainly based on professional management, which is primarily composed of professionals engaged in engineering management in the petrochemical construction industry. This integrated management model strengthens the engineering construction concept of "focusing on technological transformation, eliminating 'bottleneck' constraints," and promoting technological innovation. Compared with the Nanjing Yangba 600,000 tons/year ethylene project adopting the "PMC + EPC" management mode, the SECCO project adopting the new construction management mode saved investment, saved IPMT and PMC management costs and shortened the construction period. Subsequently, this model was popular and applied in major petrochemical projects such as Hainan Refining Project, Qingdao 10 million tons/year refining project, Tianjin 1 million tons/year ethylene project, Zhenhai 1 million tons/year ethylene project.

(b) **Oil and gas field engineering**

The Sulige gas field is the largest natural gas field discovered in China. It is located in Erdos, Inner Mongolia, from the north to the Aobaojiahan, south to Dingbian, east to Taolimiao, and west to Etuokeqianqi. The exploration area is 4×10^4 km^2, and the natural gas resources are nearly 4×10^{12} m^3 [32]. The Sulige gas field adopts a cooperative development mode, forming a "5 + 1" cooperative development model involving Changqing Petroleum Exploration Bureau, Liaohe Petroleum Exploration Bureau, Sichuan Petroleum Administration Bureau, Dagang Oilfield Group Corporation, North China Petroleum Administration Corporation, and Changqing Oilfield Company. During the development and construction of the Sulige gas field, valuable engineering construction management experience was gradually summarized.

Firstly, a standardized design was carried out. According to the functions and processes of the ground well station, a set of general, standardized, relatively stable guiding and operational documents suitable for ground construction were designed to achieve uniform process flow, unified layout, unified module division, unified installation size, unified model specifications, and unified supporting standards. The core was the generalization of process flow, standardization of well station plane, finalization of process equipment, modularization of installation pre-configuration, unified construction standards, and digitalization of production management. Secondly, modular construction was carried out. According to the division of each process link of the oil and gas station, different single-unit equipment and different-scale processing modules were shaped and designed. Based on the module design, the prefabrication was carried out according to the single module. Finally, digital management was an information management method that integrated information technology and automatic control technology and adopted an "integrated work platform." Its core was an "intelligent, digital, modular" development and production management system. That was, "automatic data entry, automatic generation of programs, automatic abnormal alarm, automatic operation control, single well electronic patrol, and automatic data sharing," which was mainly composed of the data transmission system, remote switch well system, gas well production dynamic prediction system, and production management system. The entire process of gas field development, construction, and production management have realized integrated management based on information technology.

(c) **Coal mining project**

In coal mining area, we take Shendong Coal Mine as an example. Shendong Coalfield is located at the intersection of the northern part of Yulin Prefecture in Shaanxi Province and the southern part of Erdos in Inner Mongolia. It is the largest coalfield in China and has a large coal-bearing area and abundant reserves. The coalfield has good storage conditions and is suitable for mechanized mining. It has good coal and high-quality thermal coal with low sulfur, low ash, medium, and high calorific value, which is perfect coal for metallurgical injection, gasification, and liquefaction [33]. In the early 1990s, in order to solve the situation that coal was in short supply, China decided to build Shenhua Project. The main goal was to build Shendong Coalfield into a Twenty-first-century coal base in China.

The Shendong mining area adopted a rapid well construction strategy to simplify the development mode, shorten the construction period, and build one to put into production, achieving low input, high output, and rolling development. The project researched and adopted the best production process and technical equipment according to the resource situation. It optimized the fee quota, reduced investment costs, and minimized non-production personnel and facilities. By implementing the rapid well construction strategy, the construction period of mines and coal washing plants was surprisingly shortened. The process of coal mining was mainly characterized by large-scale production, production technology and equipment modernization, team specialization, and management Informatization. It was reflected in the scientific development concept as the guide, the concept of innovation as the soul, and the management innovation and technological innovation as the core, to realize the harmonious development of economy, ecology, and society.

(3) **Water conservancy projects**

The Three Gorges Project is a large-scale multi-functional comprehensive project. In the fifth session of the Seventh National People's Congress in 1992, the project proposal was passed, and the research demonstration and decision-making procedures for nearly half a century were completed and transferred to the implementation stage [34]. The project was officially started on December 14, 1994. The dam, power station, and navigation buildings were gradually built in three phases.

The construction period lasted for 17 years, and the dynamic investment was 180 billion yuan. The investment was large, the cycle was long, and the scale was grand. All of which were unprecedented. At the beginning of the project construction, the construction management system adopting the project owner responsibility system and the national macro-control system was determined. The state adopted the role of macro-control and supervision. The project was implemented through the contracting bidding system centered on the legal person responsibility system, the engineering supervision system, and the contract management system.

The State Council established the Three Gorges Project Construction Committee (hereinafter referred to as the "Committee"), with the Premier of the State Council as the director of the committee, the relevant ministries and commissions of the State Council, and relevant provincial and municipal leaders as members became the highest decision making body in the construction of the Three Gorges Project. The organization is responsible for the decision-making of major issues in the construction of the Three Gorges Project, the coordination of all parties concerned, the allocation and supervision of macro resources, and the combination of market operation and government macro-control. The construction of the Three Gorges Project has caused all or part of 13 cities and counties to be flooded, with more than 1 million dynamic immigrants. In order to complete the immigration work of the Three Gorges Project, the Committee has set up an immigration bureau to formulate immigration policies and coordinate resettlement work. At the same time, in order to ensure the smooth implementation of the Three Gorges Project and prevent all violations, the Committee established the Supervision Bureau and the China Yangtze Three Gorges Project Development Corporation (hereinafter referred to as "China Three Gorges Corporation").

As the legal person of the Three Gorges Project, China Three Gorges Corporation fully undertakes the planning, financing, construction, operation, loan repayment, asset preservation, and value-added work of the Three Gorges Project. According to the characteristics of the Three Gorges Project, China Three Gorges Corporation has established a three-level, two-in-one construction management organization system, namely, decision-making level, management level, and executive level, focusing on engineering construction management and the combination of modern enterprise system centered on management [35]. The decision-making level is composed of the leaders of the China Three Gorges Corporation. The management level is an integrated engineering management department involving planning, finance, construction management, technology management, materials, and equipment, etc. The executive level mainly refers to the on-site representative of the legal project person.

Since the start of the Three Gorges Project, it has successfully overcome various technical problems such as slope stability for 175 m vertical high slope excavation, high-strength concrete pouring, interception, and deep-water cofferdam construction. These engineering and technical issues are inseparable from scientific decision-making. Whether in the design or construction process, the Three Gorges Project reflects collective wisdom and innovative ideas. For example, a transportation plan combining a high-level fully enclosed mountain quasi-level 1 road of nearly 30 km and the Yangtze River water transport solution solved the problem of external traffic.

The project adopted the artificial sand scheme and abandoned the limestone sand making plan 30 km away from the dam site, used the Xia'anxi granite material yard 10 km away from the dam site with the mica content lower than the national standard, thus obtaining relatively high-quality sand material, which guaranteed the quality of dam concrete. The project chose a continuous conveying and pouring scheme that had not been adopted on a large scale at home and abroad. Through the belt conveyor, elevated jack-up belt trestle bridge, and tower belt conveyor, concrete produced from the concrete factory is directly transported to the construction site for pouring. In 2000, the scheme successfully created a world record of 5.428 million cubic meters of concrete per year, thus ensuring the quality and progress of dam concrete pouring.

The Three Gorges Project involves multiple disciplines and requires the cooperation of various disciplines. The introduction and adoption of information technology are essential. In combination with the actual situation of the Three Gorges Project's construction, the project's informatization was realized through the overall planning and step-by-step implementation mode. Through the development and construction of three stages information management system, including the Three Gorges Project Management System (TGPMS), the Construction Applied Power Generation Control and Management System (EPMS), and the group level informatization, an all-round standardized and efficient information management system was formed, realized standardized management across departments, regions and all directions, and played an important role in the progress, quality, safety and investment control of the Three Gorges Project.

(4) **Railway Engineering**

(a) **High-speed railway**

The high-speed railway is a new thing that emerged with the needs of social development in the middle of the twentieth century and is an important achievement of scientific and technological progress. China's high-speed railway started late. It has experienced more than ten years of repeated discussion on whether China needs to develop a high-speed railway, whether it can build a high-speed railway, and how to build a high-speed railway. In 1993, the State Science and Technology Commission, the State Planning Commission, the Economic and Trade Commission, the Commission for Economic Reform, and the Ministry of Railways organized 47 units and more than 120 experts to participate in the "Preliminary Study on Major Technical and Economic Issues of the Beijing-Shanghai High-speed Railway," and further deepened their understanding. In June 1994, the Central Finance and Economics Leading Group held a meeting. The Ministry of Railways reported the *Proposal on the Pre-feasibility Study of the Beijing-Shanghai High-speed Railway*, which was approved by the central government leaders. At the same time, the Ministry of Railways organized scientific research, design, universities, and other units to carry out special research on high-speed railway basic theory, key technologies, construction and operation management modes, set up a high-speed railway overall technology research group at the Railway Science Research Institute, and arranged the third and the forth railway survey and design institutes to conduct research on the Beijing-Shanghai railway line direction and key engineering projects, which provided a scientific basis for the design and construction specifications of high-speed railways in China, and also prepared for the import of key technologies.

From 1997 to 2007, the Ministry of Railways successfully organized six significant speedups for busy main railway lines, increasing the maximum operating speed of existing lines from 120 km/h to 140–160 km/h, and some segments of main railway lines were increased to 200–250 km/h in 2007. The train speed of Qinshen Railway was over 160 km/h during the trial operation at the end of 2002, which explored the design and construction of the high-speed railway. The pre-constructed Shanhaiguan-Suizhong comprehensive test segment was 66.8 km long, providing a practical site for the high-speed train test. On August 1, 2008, the Beijing-Tianjin inter-city railway was opened and operated. The maximum speed of the train reached 350 km/h, marking the start of the era of the high-speed railway in China.

Based on its own national conditions, China's high-speed railways have learned from foreign experiences and based on its own independent innovations to solve the critical technologies of high-speed railways and established a technological innovation system based on railway authorities as to the leadership, enterprises as the mainstay, market-oriented, and the interaction of production, education, and research. China has independently researched a series of high-speed rail technologies, such as box-type simplified supported beams, subgrade settlement control, ballastless track, one-time laying seamless railway technology, and train operation control technology. On April 1, 2004, the State Council listened to the report of the Ministry of Railways on improving the equipment level of railway rolling stock and put forward the basic principles of "import advanced technology, joint design and production, and building Chinese brands." It clearly supports the main factories, determining the project operation mode of introducing a small number of foreign components, partial parts domestic assembly, and large-scale domestic manufacturing. This quickly mastered the key technologies of foreign transfer and absorbed and re-innovated on this basis. The "Harmony" CRH (China Railway High-speed) is a model for China's railway importing, digestion, absorption, and innovation.

A high-speed railway is a complex giant system [36]. China's high-speed railway system includes subsystems such as public works, traction power supply, communication signals, EMUs, and transportation management. The subsystems are related and interact with each other. It is necessary to study not only the expertise of each subsystem but also the interface ties between different subsystems or between the system and the external environment, the main parameters and functional coordination of different subsystem devices, namely system interface technology management, such as civil engineering and communication signals, traction power supply, EMU interface relationship, EMU and traction power supply, communication signals, train control interface, etc. After summarizing and refining, the Ministry of Railways promulgated a series of high-speed railway design and construction specifications and professional technical standards, forming a complete set of technical documents of 200–250 km/h and 300–350 km/h, and established a relatively complete high-speed railway technical system in China. These are the results of system integration and innovation, with distinctive characteristics of China's high-speed railway.

(b) **Plateau railway**

The Golmud-Lhasa section of the Qinghai-Tibet Railway (hereinafter referred to as the "Gera Section") is the world's highest-altitude and longest-permanent permafrost railway. The total length of the line is 1142 km (including 1110 km of the new mainline and 32 km of reconstruction line). There are 34 stations in total [37]. Construction started on June 29, 2001. In August 2005, the entire line foundation, tunnel, bridge and culvert, and other offline projects were basically completed. On July 1, 2006, the trial operation was opened to traffic. The process from the construction of the Qinghai-Tibet Railway to the opening of the whole line solved the three global engineering problems of "permafrost, ecological fragility, high and cold and oxygen deficiency" it also accumulated valuable experience in building railways in plateau permafrost regions.

The Qinghai-Tibet Railway is a typical public welfare construction project, and the state arranges all investments (75% of which is the state monetary fund and 25% is the railway construction fund). The Qinghai-Tibet Railway Company, which the State Council approved as a legal project person, is different from the traditional "construction headquarters" and different from limited liability companies or joint-stock limited companies. This is a useful exploration of institutional innovation and management innovation. The Qinghai-Tibet Railway Company is directly responsible for construction management, which manages construction and operation. This eliminates the disadvantages of separation of construction and management so that construction and operation are closely linked. It is also managed and supervised by the government department to control investment in engineering projects effectively. At the same time, the Qinghai-Tibet Railway Construction Headquarters was established in Golmud. The Party Working Committee was set up to absorb the leaders of the main construction enterprises as members of the Party Working Committee. The construction units without administrative affiliation were unified and led by the Party Working Committee, forming a management structure mode that not only manages construction but also manages the construction team and not only manages engineering projects but also manages ideological politics.

During the construction of the Qinghai-Tibet Railway, the construction policy of "striving for dedication, relying on science and technology, protecting health, protecting the environment, and striving for the best" reflects the new concept of the railway construction in the new era, namely "People-oriented, serving the transportation, consolidation of essentials and simplification of trifles, systems optimization, focusing on development." It not only reflects the spirit of hard work and tireless work but also reflects the advanced power of science and technology. It not only ensures the construction task to complete on time and with high quality but also reflects humanistic care to ensure the health and safety of the participants. It promotes all the projects and achieves the unification of economic, social, and environmental benefits. Only by relying on hard work and dedication as the basis, relying on scientific and technological progress as the driving force, preserving the health of personnel as the premise, and protecting the ecological environment as the responsibility can the project finally achieve the goal of striving for the best.

The construction of the Qinghai-Tibet Railway is an ambitious systematic project. Based on the comprehensive analysis and repeated demonstration of the objective conditions and influencing factors of the construction of the Qinghai-Tibet Railway, a set of target systems for near-term and far-term convergence and overall coordination has been formulated. The overall goal and each sub-goal target are decomposed and implemented. Focusing on the objectives of health safety, environmental protection, quality control, construction period control, investment control, etc., a goal management system of "top-down expansion, bottom-up guarantee, full participation, all-round implementation, and full process control" has been established. The research and development of the Qinghai-Tibet Railway Construction Project Management Information System have been carried out to ensure the smooth flow of information and resource sharing so that the work of the Qinghai-Tibet Railway is in a standardized, orderly and controllable manner. It also actively explored the quality-environment-occupational health and safety integrated management system, integrating the elements of quality, environment, occupational health, and safety and incorporating the requirements of integrated management into the engineering contract.

From the simple engineering management thoughts that sprouted in ancient times, to the preliminary exploration of the modern Western engineering construction methods, to the learning absorption and exploration practice of engineering management in the planned economy period, and finally to the innovation of contemporary engineering management concepts and models after the reform and opening up, the above time trajectory outlines the development of engineering management. As the saying goes, "review the past to understand the present, reality does not exist without history" clarifying the historical evolution of engineering management is to follow the basic law of dialectical materialism epistemology, summarizing the understanding process of engineering management from the historical development of engineering management practice. It provides not only rich material for theoretical research but also offers valuable experience for engineering management practice.

2.2 Engineering Management Theory System and Its Development

The history of engineering management outlines the practice track of engineering management development, revealing that engineering management is the product of social development and actual needs. During the process of engineering management practice, through continuous summarization and refinement, people systematically develop a theory on engineering management experience and further guide practice, gradually forming the theory and method of engineering management in the cognitive cycle of practice-theory–practice again. A large amount of literature shows that people's understanding of engineering management is gradually deepening, and rich research results and theoretical progress have been made in the aspects of engineering management connotation and scope, engineering management ideas and concepts, engineering management methods and means, and engineering management theory system framework. And this will continue to develop with society and science and technology development.

2.2.1 The Connotation and Characteristics of Engineering Management

Engineering management involves a wide range of engineering fields. With the continuous development of various advanced technologies, the content of engineering management is constantly updated and expanded. Therefore, the understanding and thinking of the connotation and scope of engineering management also show a state of "a hundred schools of thought contend, a hundred flowers bloom."

2.2.1.1 The Connotation of Engineering Management

From the perspective of historical evolution, the object of engineering management has experienced a transformation from simple civil engineering to engineering in a broader sense. Engineering management has the characteristics of a multidisciplinary intersection and is closely related to the social, economic, and natural environment. It is in dynamic development, and its connotation is constantly enriched and improved. Correctly defining and understanding "engineering management" is beneficial to promote the healthy development of China's engineering management theory and practice.

Relevant research institutions explain engineering management from different perspectives. For example, the American Society for Engineering Management (ASEM) defines engineering management from a scientific perspective. It considers engineering management "an activity of science and art with technical components for planning, organization, resource allocation, command, and control." The Institute of Electrical and Electronics Engineers (IEEE) defines engineering management as "a discipline on developing and implementing strategic and tactical decisions about various technologies and their interrelationships." Many scholars [38–40] highlight the role and status of engineers in engineering management. Rothman [41] proposes that environmental management, human resource management, etc. are complementary to engineering management content. Ferris and Cook [42] consider engineering management a multidisciplinary collaborative optimization problem, including project teams, team management, scheduling, and finance. It is necessary to consider these factors to establish a unified framework platform. As the understanding of engineering management continues to deepen, understanding its characteristics is becoming clearer, and there is an interactive relationship between engineering management and technological, economic, political, and environments. Closely related to the environment can achieve sustainable and effective engineering management [43, 44].

Domestic scholars also explored the concept and connotation of engineering management from different perspectives. Zhang [45] discusses the importance of engineering management from the connotation of engineering and the progress of engineering, pointing out that correct decision-making is the decisive factor for the success of engineering construction. He et al. [46] understand engineering management broadly and believe that its fields include construction project implementation management, new product development, and production management, and technology innovation management. He et al. [47] summarize the characteristics of engineering management disciplines and construct a framework of engineering management theory.

Since the concept of "engineering" is closely related to technology and industry, "engineering management" is related to technology and industry. Engineering management is different from general form management. It is the management of a specific form of technology integration by engineering managers in typical industrial environments. It is oriented to particular objects, particular forms of decision-making, planning, organization, command, coordination, and control work [46].

2.2.1.2 The Essential Characteristics of Engineering Management

The essential characteristics of engineering management are the integration and unification of valued, materialized, and intelligent features [48].

(1) Materialized characteristics. Engineering management has a clear value orientation to create a more valuable world through creative activities. For example, building railways and water conservancy facilities aim to achieve that people can live and work in peace and contentment. Engineering management activities should conform to objective laws and be carried out in a collaborative environment of nature and society. For example, the Qinghai-Tibet Railway transformed the original intention of harmonious coexistence between man and nature into environmental protection measures during construction and operation. It embodied the concepts of "people-oriented" and "harmony between nature and humans" in terms of goal realization and law compliance. Under the guidance of such a concept, the construction workforce from all aspects has achieved the goal of building a first-class railway on the plateau. Under the premise of observing objective laws, engineering management uses different technical methods to achieve the goal, reflecting the essential characteristics of engineering management: objectives and laws are unified, but technology and methods are different.

(2) Intelligent characteristics. Information is ubiquitous, but only through the processing of the subject to form the information accepted by the subject is it worthwhile. A large volume of information will be generated in the engineering management activities. The intelligent processing of all kinds of information is the prerequisite for the management activities of the main body of engineering management activities. Under the guidance of the systems thinking mode,

the main body of the engineering management activities organizes the scattered resource elements to achieve the engineering objectives. For example, the construction of the Qinghai-Tibet Railway faces three major problems: frozen soil, anoxic and ecological fragility. In order to solve the three major problems, the Qinghai-Tibet Railway project comprehensively integrates various resource elements, including science, technology, and society.

Technology is an important factor in whether an engineering project can be established. For engineering management, the rational use of technology to achieve goal optimization is an important issue. Through technology integration, existing mature technologies can be rationally deployed to achieve the ultimate goal.

(3) Value characteristics. Dujiangyan Project is a typical ancient engineering model following the harmony of humans, engineering, and nature. The design and construction process fully conform to and utilize nature, ensuring the sustainability of this project. As the world's first railway with the highest altitude, the Qinghai-Tibet Railway fully protects the ecological environment and rare species along the line, making the scenery along the line and the Qinghai-Tibet Railway enhance each other's beauty.

The project's ultimate goal is for people, and people with subjective initiative are the fundamental driving force of engineering activities. People form a close relationship with the environment through engineering creation activities, and the human-engineering-environment relationship is highlighted in all stages of engineering activities. Whether the human-engineering-environment relationship can be properly handled directly affects the realization of the final value of the project. To understand the engineering relationship from this perspective, the essence is to coordinate and balance the human-engineering-nature relationship in engineering creation activities and enhance the engineering value.

In summary, domestic and foreign scholars' understanding of the connotation and characteristics of engineering management is mainly from the concept and connotation of engineering and the concept, connotation, and characteristics of engineering management. Clarifying the various concepts of engineering and engineering management, as well as the relationship between these concepts, is conducive to revealing the essential characteristics of engineering management, thus promoting the in-depth study of the engineering management theory system.

2.2.2 *Project Management Thoughts and Ideas*

The engineering management practice has a long history in China. The typical early projects have accumulated rich experience and understanding of engineering management practice and reflected the management thought at that time but did not form systematic engineering management thought. In the 1980s, Qian Xuesen proposed the idea of system engineering and integrated integration. Scholars began to think about engineering management with reference to the ideas and concepts of systems thinking, philosophical thoughts, ethical concepts, and sustainable concepts, successfully applied to engineering management practices, and gradually formed the ideas and concepts of engineering management such as systems thinking, engineering philosophy, engineering ethics, and engineering sustainability.

2.2.2.1 Systems Thinking

Qian Xuesen's *On System Engineering,* published in 1983, proposed system ideas and system analytic methods. Based on aerospace engineering research, proposed the concept of an open complex giant system and its methodology—comprehensive integration method. It provides a basis for engineering management theory and points out new development ideas for studying large-scale complex projects. People began to use systems thinking to solve engineering management problems. Scholars believe that the use of systems thinking is the foundation of engineering management. Engineering management is a complex ecosystem related to society, the economy, and nature. It has system-level complexity [49–51]. On this basis, some scholars explore integrated management methodology with Chinese scenarios for large-scale complex projects based on the management thinking of complex systems [52]. From the aspects of integrity, openness, dynamics, hierarchy, and adaptability, they analyze the complexity of large-scale engineering projects and propose effective large-system theory [53].

2.2.2.2 Engineering Philosophy and Engineering Management Philosophy

With the continuous development of engineering practice and the increasing influence of engineering, foreign philosophers began to make an epistemological analysis of the engineering practice process. More and more engineers also reflected on their work results from the philosophical level. Many research findings on engineering philosophy have emerged from the perspectives of ontology, epistemology, methodology, and value theory [54–56]. Since then, domestic scholars and experts have begun to study and analyze various philosophical issues in the engineering process. Li [57] puts forward the "three-dimensional theory" on science, technology, and engineering and considers "three-dimensional theory" as the foundation and theoretical premise of engineering philosophy research. He points out that engineering philosophy should be a discipline parallel with the philosophy of science and the philosophy of technology. Xu [58] believes that there must be profound and complicated philosophical problems in major engineering problems, engineering needs philosophical support, and engineers need to have philosophical thinking. Wang et al. [59] believe that a new era of new engineering concepts should be established, including engineering concepts with sustainable development connotation and sustainable development benefits, engineering dialectical views, engineering systems concepts, engineering values, and engineering ecological views. Du [60] proposes that the research on engineering philosophy should involve the whole process of engineering practice activities, including engineering research, demonstration, and engineering decision-making. Fu [61] believes that engineering philosophy should focus on engineering and environment, engineering and human, engineering and culture, engineering progress and quality, and cost. Based on the perspective of engineering philosophy, Lu [62] proposes examining and evaluating contemporary engineering development in China with a dynamic and harmonious engineering ecological concept. Yin et al. [8] emphasize in the *Engineering Philosophy* (Second Edition) the need to conduct in-depth research on engineering thinking and knowledge and believe that engineering thinking penetrates and runs through all aspects and processes of engineering activities. The basic task and basic content of engineering thinking are to raise and solve engineering problems. Engineering thinking is closely related to engineering knowledge and has value content and will factors. With the deepening of engineering philosophy research, relevant scholars carry out a philosophical reflection on engineering management and, on this basis, guide engineering practice and development. He and Wang [63] discriminate the essential characteristics of engineering from the aspects of materiality, variability, and time and space of engineering. On this basis, they discuss the philosophical connotation of engineering management regarding the concepts, values, methodology, organization, and innovation, compare and analyze the categories and concerns of engineering philosophy and management philosophy, point out the necessity of upgrading engineering management philosophy, and discuss the six themes and three major changes of engineering management in the perspective of philosophy.

2.2.2.3 Engineering Harmonious Management

Xi et al. [64] proposed the theory of harmonious management and applied it to the development and management of water conservancy and hydropower projects. Wu and Cheng [65] raise harmonious management to the methodological level, believe that it should run through the full understanding of engineering management, and propose the harmonious management mode of engineering projects. Based on the theory of harmonious management, Li [66] identifies and analyzes the HSE harmony theme of engineering projects and constructs the HSE management system of engineering projects from three aspects: values, culture, and incentives. He [67] believes that the core of improving engineering management effectiveness lies in the interactive development of harmony and innovation based on improving engineering harmony and innovation. Zhang et al. [68] use the correlation function in extenics to establish a matter-element model for the harmonious management of construction projects to conduct a harmonious analysis of project incompatibility problems. Yang and Hou [69] discuss the harmonious management of government investment projects from the aspects of organizational structure design, workflow management, harmonious system construction, and public participation mechanism construction. Qiu et al. [70] obtain the equilibrium optimization point of the three construction projects by constructing the utility function model of progress, cost, and quality, which is the harmonious theme of the engineering construction project; and constructing a harmonious control model among progress, cost, and quality under the guidance of harmonious management theory.

2.2.2.4 Sustainable Engineering

The concept of engineering sustainability is proposed under the condition of an "unbalanced supply and demand of urban infrastructure" [71]. Since then, the issue of sustainable development of projects has gradually received attention. Chen [72] proposes the concept and indicator framework of sustainable development of construction projects and indicates the relationship between engineering project management and sustainable development theory and practice is an important concept and cognitive conversion. Liu and Li [73] propose the sustainability evaluation of major engineering projects from necessity evaluation, economics, technology, science and technology, and environmental impact assessment. Zheng [74] establishes a qualitative indicator system for the sustainability of construction projects and introduces the ecological footprint method and energy analysis method to calculate environmental compatibility and economic rationality. Feng et al. [75] propose that the scientific development concept should guide water resources development. Engineering and design management should be carried out in the water resources engineering project in line with the concept of harmony between humans and nature. Based on the scope of ecological economics and the systematic analysis of the connotation of sustainable construction of construction projects, Zhang et al. [76] establish a sustainable construction system for construction projects from the three dimensions of engineering major, engineering life cycle, and engineering ecology, and conduct an empirical analysis of the engineering ecological dimension using the ecological efficiency measurement tool of ecological economics.

With the further cross-infiltration of engineering management and related disciplines, people continue to introduce new management ideas and concepts in engineering management practice, making engineering management thinking move from experience to science and gradually forming engineering management thoughts with engineering philosophy, engineering management philosophy, engineering ethics, engineering harmonious management, and engineering sustainability as the main contents.

2.2.3 Engineering Management Methods and Means

Domestic and foreign scholars creatively use the general theory and methods of management and economics in engineering management practice for specific engineering or technologies, systematically summarize new ideas and methods in engineering decision-making, planning, organization, coordination, and control, and provide further guidance and apply them in new engineering management practices.

2.2.3.1 Engineering Decision

In the narrow sense, the "decision" in engineering management refers to the risk investment decision of the engineering project. In a broad sense, it includes investment decisions, planning decisions, design decisions, implementation decisions, management decisions, etc. The decision process includes finding the problem, determining the target, evaluation criteria, program development, program selection, and program implementation. Decision-making in modern science can effectively evaluate and judge engineering technology, the economy, and the environment through various technologies such as mathematical modeling and simulation. The following are some representative views put forward by some scholars from different perspectives.

Denton et al. [77] analyze market risk issues in the energy industry, using real options models and stochastic optimization techniques to manage market risks. Chinowsky [78] proposes a five-stage value engineering model, using construction simulation technology and cycle operation network to reduce subjectivity further, and using value engineering to save cost and time better. Liu [79] applies the decision theory and method to the engineering goal decision-making practice and discusses the fuzzy risk decision-making of water resources engineering investment and multi-objective group decision-making of water resources planning management. Guo et al. [80] propose a new mechanism for integrating data mining technology and engineering decision support technology because of the inherent shortcomings of traditional engineering decision support systems (EDSS) in terms of knowledge acquisition and scope application. Lei et al. [81] point out that large-scale engineering decision-making involves multiple decision-making bodies and multi-level decision-making objectives and propose an interactive multi-attribute group decision-making method for large-scale engineering projects based on the principle of relative entropy. Ren et al. [82] believe that the giant project is an effective large system and study the basic principles, decision-making methods, and evaluation of giant project decision-making. At present, from the perspective of the research scope, scholars' knowledge of engineering decision-making methods and means mainly focuses on investment risk analysis in construction plans, expected risk analysis in engineering subcontracting collaboration, etc. From the perspective of research methods, investment decision analysis is greatly affected by uncertain factors; experts focus on mathematical modeling by introducing methods such as real options, simulation techniques, multi-objective optimization, and data mining related to statistics and operations research disciplines. Other scholars have studied the properties of engineering decision management and the principles of decision-making and made relevant suggestions.

2.2.3.2 Engineering Plan

The "plan" in engineering management has two meanings: one refers to the engineering project's long-term planning, including engineering site design, project design, and product design; another refers to the short-term plan during the implementation of the project. Nie [83] proposes an improvement plan for construction enterprise engineering project planning from planning system, coordination, and control process, planning management informatization, and computer aids. AbouRizk et al. [84] summarizes the latest research progress in the application simulation theory in the civil engineering design process, analyzes the key factors in the simulation of building engineering management, and proposes that the long-term simulation technology of building design can realize the visual integrated management function in project planning and control. Based on the complex product manufacturing engineering work and breakdown structure theory, Liang [85] designs the conflict coordination mechanism for planning and managing multi-project resources. Based on multi-project management theory, Wang [86] researches the planning management methods of planning management, planning implementation management, and planning assessment management in the owner's project management.

The engineering plan has two levels of meaning: long-term planning of the project and short-term planning during the implementation of the project. At present, scholars have researched two levels of engineering plans. The influencing factors in engineering design, the application of simulation technology in engineering planning and design, and the multi-project planning method have received extensive attention.

2.2.3.3 Engineering Organization

"Organization" in engineering management refers that in the process of engineering implementation, the combination and classification of various necessary business activities to realize the engineering management objectives and plans, clarifying the functions and responsibilities of management personnel, stipulating the coordination relationship between the upper and lower sides, and at the same time, carrying out the process of adjusting the structure constantly.

The project involves many stakeholders, including the government, owners, contractors, designers, supervisors, materials and equipment suppliers, financial institutions, scientific research institutions, users, and local residents. How to optimize the allocation of resources of various organizations and achieve management goals efficiently is the primary problem to be solved by engineering management organization innovation. Traditional organizational management theory tends to think that organization is a closed system, focusing on how to ensure high productivity internally, but lacks consideration of the relationship between the organization and the external environment. This is not conducive to engineering decision-making and implementers to accurately perceive the external environment characteristics to let the organization respond timely to changes in the external environment [87]. The engineering management organization model should gradually evolve the way of dealing with stakeholders from notification to providing information, consulting, participation, and establishing partnerships to promote resources and information sharing and achieve mutual benefits. Many scholars have done a lot of research on the behavior of engineering organization groups from the perspective of building the main body alliance [88], partner selection [89], benefits and risk allocation mechanism [90], which involve game theory, evaluation methods, etc. [91, 92] Chinowsky et al. [43] analyze the importance of mutual trust and good communication between different departments in the process of engineering management, and study the social network model in building management. The results show that the model can strengthen the organization and collaboration between various project elements to create an efficient project team.

At present, the key research directions for engineering organization collaboration mainly include: from the perspective of industrial chain integration and cooperation, studying the means and ways of coordination and integration between organizations and the impact of engineering on society, economy, and environment [93, 94]; from the perspective of informatization, through the development and application of information management systems to promote information sharing among participants of various projects and to improve the efficiency of collaborative work of all parties [94]; from the perspective of learning and innovation, strengthening the knowledge sharing among project parties and the integration of internal and external knowledge, forming a platform for technology management and innovation, and promoting independent research and development technology to solve crucial and challenging problems in engineering practice [95].

It can be seen from the above literature that organizational efficiency is the core of engineering organization research. With the understanding of engineering organizations, the focus of engineering organization management extends from the internal resource allocation of the organization to the association between the organizational system and the external environment, and the organizational model shifts from traditional notification to organizational group behavior and organizational synergy. However, the existing researches are too much from the perspective of the direct stakeholders of the project, and there is a lack of a more profound understanding of the interaction mechanism between individual behavior and group behavior and group psychology and group behavior.

2.2.3.4 Engineering Coordination

The "coordination" of engineering management includes the effective communication and coordination between different links or stages of the project, between departments or agencies, and the optimization and configuration of engineering resources. Cong [96] studies the classification, content, and method of construction project organizational coordination. Wang et al. [2–97] study the coordination mechanism of the schedule and propose the procedures and techniques for the coordinated management of the project management during the planning phase. Guo et al. [98] research project coordination management from stakeholders' perspectives, coordination of management information systems, and coordination of management culture.

Resource optimization configuration plays a very important role in engineering coordination management. Zhong et al. [99] use a discrete event simulation system combined with three commonly used heuristic algorithms to simulate and optimize a large hydropower station's underground cavern group construction. The results show that the genetic algorithm can better solve the resource balance problem in the complex engineering construction process. Behzadan et al. [100] use GPS and wireless positioning technology to provide corresponding on-site information according to the role of the field personnel, and their location at the construction site and use augmented reality technology to display this information in real-time in a visual way.

The above literature found that scholars have studied the connotation, content, influencing factors, and methods of engineering coordination management. Resource optimization configuration is one of the critical contents of coordinated management research. Many scholars have researched the realization conditions of multi-project resource dynamic optimization configuration, the algorithm of resource optimization configuration, simulation technology, and the application of information technology in resource optimization configuration.

2.2.3.5 Engineering Control

"Control" in engineering management refers to the detection, measurement, and evaluation of various indicators and elements, such as project quality, safety, schedule, investment, risk, etc., during the process of engineering project implementation to ensure the realization of the project's expected goals, and propose corrective measures in a targeted manner.

In engineering quality control, Feng and Yang [101] establish a gray swing model of the quality control system of the water conservancy and hydropower project, analyze the stability of the operation mechanism of the quality control system, and predict the future state of the quality control system. Wang et al. [102], through the analysis of the quality management practices of the Three Gorges Project, point out that the Three Gorges Project has established a relatively complete quality assurance system. Yuan and Yan [103] propose applying the multivariate graph method to analyze the causes of quality variation on construction sites and carry out quality control. Sun [37] researches on the engineering management practice of Qinghai-Tibet Railway and points out that strict engineering quality management organization, perfect quality management rules and regulations, and a project quality management system with a responsibility system as the core are the key to the success of Qinghai-Tibet Railway quality management. Zhong and Ding [104] apply the concept of knowledge management to construction quality control and develop a construction quality control and accident handling system (CQIS) that integrates building quality control knowledge management and decision support functions. Wang and Xie [105] conduct a questionnaire survey to analyze China's significant project quality management status quo and the main factors affecting quality management. They believe that the engineering quality responsibility system plays a good role, but there are still some shortcomings.

In terms of safety management, Fang et al. [106, 107] use the combination of qualitative and quantitative methods to conduct in-depth research on the safety assessment of the work environment at the construction site, the relationship between safety input indicators and performance, the causes of construction casualties, the impact of industry safety climate on safety behavior, the construction safety risk probability assessment, and the cognitive causes of unsafe behavior of workers. Zhou et al. [108] analyzed the security control black box, information island, and knowledge loss and built a security-integrated control model for knowledge applications. Aiming at the problem of emergency response evaluation of urban public places, Chen et al. [109] propose a three-dimensional emergency capability model based on the emergency risk, emergency capability time, and emergency capability space and put forward an analytics method for the indicator system based on the emergency phase specified in the *Emergency Response Law*. Some scholars have used fuzzy comprehensive judgment, dynamic evaluation, and grey theory to analyze and study the construction site's safety management, evaluation, and prediction.

In terms of schedule control, Wang [110] proposes preparing the overall schedule, the organization plan of tracking management, and the successful application of the PSWS solution on the network platform in practice. Regarding the lack of linkages between the actual progress management and expense management in project management, Hui et al. [111] use the earned value management method and apply the SIMULINK toolbox in MATLAB software to carry out dynamic simulation for joint control of schedule and cost of the engineering project, and put forward the idea of joint control of schedule and cost. Hu et al. [112] introduce the parameters of Baosteel's progress control—the comprehensive progress rate. They establish a gray GM (1, 1) model for progress prediction and carry out the model generalization test based on the actual total progress rate. Based on field research and coding analysis methods, Wu et al. [113] apply the fuzzy comprehensive evaluation method to carry out project progress evaluation. In view of the difficulties in the construction of the Shanghai World Expo project, Zhu et al. [114] analyze the key issues affecting the realization of the progress target through a series of management measures, including the dynamic preparation and adjustment of the schedule, the development of a comprehensive progress assessment method, the dynamic adjustment of the organizational structure, and the active coordination of the progress work to ensure the realization of the progress objectives.

In terms of investment control, Feng et al. [115] introduce the main structure and mathematical model of the civil engineering cost estimation system based on case-based reasoning (CBR) using the theory of fuzzy mathematics. Hu et al. [116] consider that the effect of project investment control is directly related to people. Combined with the three assumptions about human behavior in the new institutional economics, they carry out theoretical analysis regarding the role of the system in the investment control of the Three Gorges Project. Lu et al. [117] construct a BP neural network model to predict project cost and optimize the list pricing model to complete project cost control.

In risk management, Liu and Han [118] combine the risk characteristics of large-scale engineering construction projects, establish a fuzzy influence diagram method based on fuzzy sets and influence diagrams, and conduct applied research for project risk analysis. From the perspective of neoclassical economics, based on the divergence of interests, He and Fu [119] explain the economics of the owners and contractors in the decision-making of engineering quality risk management and propose measures to prevent engineering quality risks from the perspective of economics. Tang et al. [120] created a project risk management model based on the partnership model. Based on the case analysis of underground engineering accidents in China, Wang and Zhang [121] proposes to use information technology and displacement distribution principle to strengthen construction safety control. Wang et al. [122] analyze the characteristics of simultaneous construction and dynamic risk management of rail transit stations, form the dynamic risk management idea of station foundation pit construction during the same period, and establish a dynamic risk management model and implementation process. Zhao and Yin [123] construct a hierarchical structure model of engineering project risk-sharing factors through the ISM method and further reveal the mechanism of action between risk-sharing factors.

In engineering practice, it is necessary to realize the goal of engineering control through informatization and standardized management means. Yao [124] proposes that civil engineering informatization should be implemented from the aspects of design, construction, and control. Based on IFC and 4D, Zhang et al. [125] built a 4D project management system to manage schedules and resources dynamically. Zhu [126] believes that the development of information technology has changed the engineering management organization and thus impacts engineering management ideas and methods. Starting from the theory of the complete life cycle, Liu and Sun [127] summarizes the connotation of engineering management informatization into four aspects: operation management, partner collaboration, public service, and integration innovation, and discuss the differences and connections between indication system of engineering management informatization and enterprise management informatization. Wang et al. [128] discuss the standardization management techniques of significant projects from the aspects of programmatic, contractualization, formatting, and informatization. Wang and Whu [129] fully apply the optimization of information technology such as BIM and BLM to conduct full lifecycle management of project management.

In quality control, scholars conduct quality control research from the aspects of quality control methods, reliability of quality control systems, quality management system construction, and policies and regulations affecting quality management. In safety management, the research priorities are safety assessment, safety risk assessment, safety behavior, safety integrated control, and emergency management. In terms of scheduling, dynamic tracking management of schedules, joint control of schedules and costs, schedule control parameters and forecasting, and factors affecting schedules have received wide attention. Investment control research focuses on the improvement of estimation methods and the prediction of the project cost. The influencing factors of engineering investment are also concerned by scholars. In risk management, the main aspects of scholars' research include the causes of risk, the dynamic management model of risk, risk-sharing, and the application of information technology. In addition, engineering management informatization has also received extensive attention. Scholars mainly research informatization objectives, connotations, application practices, and the impact of informatization on engineering management.

2.2.4 Engineering Management Theory System Framework

With the accumulation of experience in engineering construction in China, the rapid development of various related disciplines and the active exploration of engineering management in the practical and theoretical fields have formed many management ideas, concepts, methods, and means suitable for China's national conditions. However, a complete theory has not yet been formed, which has influenced the development of engineering management disciplines and the role of practical guidance. Therefore, scholars began to summarize and enhance engineering management experience, sort out the scattered application theory, and carry out research on the framework of the engineering management theory system suitable for China's national conditions.

2.2.4.1 Understanding of the Theoretical Framework of Engineering Management

On the basis of defining the concept of engineering and engineering management and analyzing the characteristics of engineering management disciplines and the theoretical origins of the discipline, He et al. [47] construct the framework of the engineering management theory system under the guidance of engineering philosophy. Liu [130] breaks through the framework of engineering investment, construction general contracting, and construction management theory in foreign engineering management theory and proposes the engineering management theory system with the engineering life cycle as the mainline, including engineering planning management theory, engineering design management theory, engineering investment management theory, engineering construction management theory, and engineering use management theory. Wang et al. [131]] analyze the evolution of engineering thinking under the direction of value thinking in engineering practice, propose the dynamic model of modern engineering management theory under engineering thinking mode, and construct theoretical model of modern engineering management based on engineering investment management theory, engineering transaction management theory, and engineering project management theory. Then, based on the analysis of the engineering management knowledge development dimension, engineering management knowledge innovation model, and structural model, they propose the engineering management knowledge system framework under engineering thinking mode, including basic knowledge system framework and application knowledge system framework. Wang and Wang [132] analyze the logical starting point of the construction of the engineering management theory system, expound the practical starting point theory and environmental starting point theory of engineering management theory system, construct the theoretical system framework of "ontology + extension" structure, and explain the connotation and relationship of the various parts. He [133] constructs a theoretical system of engineering management from a theoretical module, activity module, and application module.

2.2.4.2 Conceptual Model of Engineering Management Theory System

The engineering management theory system is the systemization of engineering management theory. It is necessary to clarify the theoretical content of engineering management covered, reveal the inherent logical structure and hierarchical relationship between various constituent elements, and clarify the relationship and context of engineering management theory. The conceptual model of the engineering management theory system is shown in Fig. 2.1.

The conceptual model of the engineering management theory system is rooted in China's unique national conditions and cultural backgrounds; as a large tree, its soil is China's rich engineering management practice and related policies and systems. As an interdisciplinary subject, the relationship between engineering management discipline and other related disciplines is complex, forming the basic theory of engineering management in the intersection and blending, forming the foundation of the big tree. The subject theory is the backbone and supports the theoretical system of engineering management. The application theory of work management as its branches, the branches extended, represent specific application techniques and methods [134].

The basic engineering management theory is a theoretical principle with stability, fundamentality and universality, which plays a fundamental role in engineering, management, economics, engineering philosophy, sociology, art, law, etc.

The main body of engineering management mainly includes engineering ontology, epistemology, methodology, decision-making, organization, value, innovation, environment, humanities, and ethics. From the perspective of ontology, "engineering management" is to establish the "main body" position of the project. Engineering

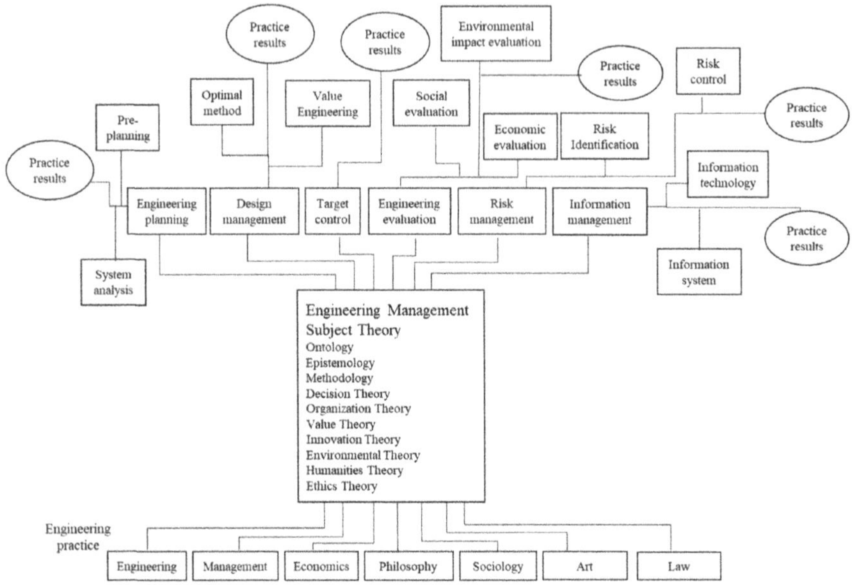

Fig. 2.1 Conceptual model of engineering management theory system

management depends on engineering and, at the same time, occupies an important position in engineering activities. Engineering epistemology and methodology represent the understanding of engineering management knowledge systems and the organizing of engineering management methods. Engineering decision-making mainly refers to the methods and ideas of decision-making in engineering management activities. Engineering organization mainly refers to the organization management methods, principles, and laws in engineering management activities. The engineering value relates primarily to the economic problems, quality, and safety issues involved in realizing and improving value in engineering management activities. Engineering innovation mainly refers to binary innovation, namely management innovation and technological innovation. The engineering environment mainly refers to the interaction analysis between engineering management activities and the social and natural environment. Engineering humanities mainly refers to engineering ethics, engineering culture and engineering art, and the study of the status and roles of engineers, ethical issues in engineering, function and modeling of engineering culture, and artistic expression of engineering.

Engineering management application technology can be divided into engineering planning technology, engineering design management technology, target control technology, engineering evaluation technology, information management technology, etc. For example, target control technology includes cost control, schedule control, quality control, etc. Engineering evaluation includes economic evaluation, social evaluation, environmental impact evaluation, etc. Information management technology involves information technology and information systems.

In recent years, scholars have paid more attention to the construction of engineering management theory systems. The construction of an engineering management theory system helps clarify the relationship between various engineering management concepts, reveal its essential characteristics and principles, produce the key propositions and axioms necessary for the development of engineering management theory and practice, and promote the enrichment and perfection of engineering management theory.

References

1. Xu, K. (2008). Development of engineering philosophy to implement scientific development. *Journal of Beijing Normal University (Social Science Edition), 1*, 90–92.
2. Sun, Z. (2010). The dialectical relationship between theory and practice. *Xinhua Digest, 4*, 41–43.
3. Lu, Y. (2005). Understanding hydropower engineering from the perspective of philosophy. *China Three Gorges, 2*, 4–8.
4. He, J., Wang, M., & Wang, Q. (2013). *Current status and development of China's engineering management*. Higher Education Press.
5. Wang, Y. (2013). Introduction to engineering management. Xi'an Jiaotong University Press.
6. Ren, H., & Chen, Y. (2007). *Introduction to engineering management*. China Construction Industry Press.
7. Harms, A. A., Baetz, B. W., & Volti, R. (2004). Engineering in time: The systematics of engineering history and its contemporary context. Imperial College Press.
8. Yin, R., Wang, Y., Li, B., et al. (2013). *Engineering philosophy* (2nd ed.). Higher Education Press.
9. He, J. (2013). The characteristics and management thoughts of ancient Chinese engineering architecture. *Engineering Science, 10*, 4–9.
10. Cheng, H. (2007). *Introduction to engineering management*. China Construction Industry Press.
11. Zhu, H. (2013). Analyzing the far-reaching influence of ancient Chinese management thoughts. *Manager, 07*, 104–105.
12. Zheng, J., & Wang, M. (2014). The historical evolution of China's engineering management. *Science and Technology Management Research, 23*, 245–250.
13. Fu, R. (1990). The official system and engineering supervisor of ancient China. *Building Economy, 11*, 30–32.
14. He, C. (2011). Project management thought in ancient construction engineering. *Construction Enterprise Management, 12*, 100–102.
15. Wang, M., Yao, J., & Lu, W. (2012). Investigation on the measurement and valuation of ancient Chinese city project. *Journal of China Science and Technology History, 02*, 204–221.
16. Sun, X. (2009). Beijing city management system and municipal construction during the Republic of China. *Journal of Beijing Institute of Civil Engineering and Architecture, 3*, 51–54.
17. Beijing Archives Editorial Board. (2002). *Beijing archives historical materials*. Xinhua Press.
18. Liao, Y., & Jin, S. (2008). An attempt to modernize water conservancy in the Republic of China—Take Jiangxi province in the period of 1932–1942 as an example. *Theory Monthly, 05*, 45–46.
19. Sheng, Z., Zhang, J., Du, J., et al. (2009). *Social science computing experiment theory and application*. Shanghai Sanlian Bookstore.
20. Miao, D. (2010). System scientist Qian Xuesen. *Journal of Liaodong University (Social Science Edition), 03*, 16–22.

21. Lu, R., Wang, M., Luo, J., et al. (2009). *Government investment project management model and general contract management practice*. China Construction Industry Press.
22. Zhang, W., & Zhu, H. (2005). Overview of government project management mode and reform direction. *Construction Economy, 07*, 19–23.
23. Zhang, W., & Zhu, H. (2007). Institutional changes in the management of government investment projects in China. *Journal of Civil Engineering, 05*, 79–84.
24. Li, D. (2002). *Beijing and Moscow: From the alliance to confrontation*. Guangxi Normal University Press.
25. Liu, Y. (2002). Successful experience at the management level of the "two bombs and one satellite" project. *Aerospace Industry Management, 2*(7), 23–26.
26. Liu, Y. (2002). Successful experience at the management level of the "two bombs and one satellite" project. *Aerospace Industry Management, 1*(6), 13–16.
27. Zhang, G. (2009). Can't erase the memory--The documentary of Nanjing Yangtze River bridge construction. *New Long March (Party Edition), 05*, 62–64.
28. Wang, Y., & Wang, N. (2006). Current status and development of engineering management discipline in China. *China Engineering Science, 03*, 11–17.
29. Ren, H., Zhu, X., & Gu, X. (2001). Development prospect of engineering management major. *Higher Architecture Education, 02*, 32–34.
30. Hu, S., & Zhang, Q. (2004). China manned space engineering—A model of successful practice in system engineering. *China Aerospace, 10*, 4–7.
31. Wang, J. (2007). Innovation of Sinopec petrochemical major project management model. *Sinopec, 07*, 45–49.
32. Ran, X. (2008). Engineering management innovation in the development and construction of Changqing oil and gas field. *China Engineering Science., 12*, 58–62.
33. Ye, Q. (2003). Shenhua engineering management model. *Journal of China Coal Society, 02*, 113–116.
34. Lu, Y. (2002). The practice of the construction project management of the Three Gorges Project. *China Three Gorges, 09*, 3–4.
35. He, G., & Chen, W. (2001). Summary of construction management experience of the Three Gorges Project in the past eight years. *China Three Gorges, 08*, 1–6.
36. Sun, Y. (2009). The road to success of China's high-speed railway. *Journal of the China Railway Society, 06*, 135–139.
37. Sun, Y. (2005). Innovation and practice of Qinghai-Tibet railway construction management. *Management World, 3*, 1–6.
38. Thamhain, H. J. (1992). Engineering management: Managing effectively in technology-based organizations. Wiley-Interscience.
39. Lock, D. (1993). *Handbook of engineering management*. Butterworth-Heinemann.
40. Lannes, W. J. (2001). What is engineering management? *IEEE Transactions on Engineering Management, 48*(1), 107–115.
41. Rothman, J. (1999). Successful engineering management: seven lessons learned. In *Proceedings of the technical program. ELECTRO 99 (Cat. No. 99CH36350)* (pp. 277–281). IEEE.
42. Ferris, T. L., & Cook, S. C. (2005, September). Away from a single theory of engineering management. In *Proceedings 2005 IEEE international engineering management conference, 2005* (Vol. 1, pp. 271–275). IEEE.
43. Chinowsky, P. S., Diekmann, J., & O'Brien, J. (2009). Project organizations as social networks. *Journal of Construction Engineering and Management, 136*(4), 452–458.
44. Xu, J., & Li, Z. (2012). A review on ecological engineering based engineering management. *Omega, 40*(3), 368–378.
45. Zhang, S. (2002). The scope of project management and the importance of project management. *Journal of Wuhan University of Technology (Information and Management Engineering Edition), 24*(3), 7–10.
46. He, J., Chen, X., & Hong, K. (2005). On engineering management. *China Engineering Science, 7*(10), 5–10.

47. He, J., Wang, M., & Wang, Q. (2009). Analysis of engineering management theory and system construction. *Science and Technology Progress and Policy, 26*(21), 1–4.
48. Zhu, W., Wang, M., & Zheng, J. (2015). Philosophical thinking of engineering management. *Theory Monthly, 8*, 47–51.
49. Knoepfel, H. (1983). Systematic project management. *International Journal of Project Management, 1*(4), 234–241.
50. Saynisch, M. (1983). Project management system for a large international project. *International Journal of Project Management, 1*(2), 115–121.
51. Payne, J. H. (1995). Management of multiple simultaneous projects: A state-of-the-art review. *International Journal of Project Management, 13*(3), 163–168.
52. Sheng, Z., You, Q., & Li, Q. (2008). Methodology and method of large-scale complex engineering management: Integrated management-taking Sutong bridge as an example. *Science and Technology Progress and Countermeasures, 25*(10), 193–197.
53. Ren, H., Zeng, D., Zhang, W., et al. (2012). Effective large-scale system of giant projects. *China Engineering Science, 12*, 75–80.
54. Mitcham, C. (1991). Book review-philosophy of technology.
55. Mitcham, C. (2000). Metaphysics, epistemology, and technology.
56. Bucciarelli, L. L. (1994). *Designing engineers*. MIT Press.
57. Li, B. (2002). Strive to open up the field of engineering philosophy. *Research on Dialectics of Nature, 18*(7), 36–39.
58. Xu, K. (2004). Establishing a new concept of engineering and promoting the new development of productivity. *Engineering Research—Interdisciplinary Engineering, 1*, 4–8.
59. Wang, Y., & Wang, H. (2005). Contemporary engineering concept and building a harmonious society. *Engineering Research—Interdisciplinary Vision, 02*, 26–28.
60. Du, X. (2005). Some thoughts on engineering philosophy. *China Engineering Science, 7*(2), 24–25.
61. Fu, Z. (2007). Philosophical thinking on railway speed-up project in China. *Journal of Zhejiang University (Humanities and Social Sciences), 37*(3), 5–14.
62. Lu, Y., & Zhang, Z. (2009). Construction of the ecology of engineering with dynamic harmony and the original ecology concept criticism. *Engineering Research-Interdisciplinary Perspectives on Engineering, 1*(4), 346–353.
63. He, J., & Wang, M. (2008). Philosophical thinking on engineering and engineering management. *China Engineering Science, 10*(3), 9–12.
64. Xi, Y., & Shang, Y. (2001). Harmonious management thought and contemporary harmonious management theory. *Journal of Xi'an Jiaotong University (Social Science Edition), 21*(3), 23–26.
65. Wu, W., Hou, Y., & Cheng, H. (2007). Engineering project management from the perspective of harmonious management theory. *Civil Engineering and Environmental Engineering, 29*(4), 129–132.
66. Li, J. (2007). Research on HSE management system design and performance based on harmonious management theory. Tianjin University.
67. He, J., Wang, J., & Yu, Z. (2008). Research the interaction relationship between engineering harmony and engineering innovation. *Engineering Science, 10*(12), 4–9.
68. Zhang, W., & Yang, B. (2001). Research on harmonious management of construction engineering projects. In *International conference on engineering and business management* (pp. 1186–1188).
69. Yang, W., & Hou, X. (2011). Harmonious management of government investment projects from the perspective of partnering model. *Economist, 1*, 29–30.
70. Qiu, X., Zhang, L., & Zong, H. (2012). Harmony theme optimization control of engineering construction projects based on harmonious management theory. *Project Management Technology, 5*, 90–93.
71. Carter, R. C., Tyrrel, S. F., & Howsam, P. (1999). The impact and sustainability of community water supply and sanitation programmes in developing countries. *Water and Environment Journal, 13*(4), 292–296.

72. Chen, G., & Cheng, H. (2004). Research on the whole life target system of construction projects. *Journal of Civil Engineering, 37*(10), 87–91.
73. Liu, Y., & Li, X. (2004). Project sustainability research. *China Engineering Consulting, 1*, 13–14.
74. Zheng, X. (2005). Research on sustainability and evaluation of construction projects. Chongqing University.
75. Feng, J., & Shao, X. (2009). The connotation and level of harmony in water resources engineering. *Journal of Water Resources and Water Engineering, 9*(5), 19–23.
76. Zhang, J., & Wang, L. (2012). Research on sustainable construction system of building engineering from an ecological economics perspective. *Construction Economy, 10*, 5–10.
77. Denton, M., Palmer, A., Masiello, R., & Skantze, P. (2003). Managing market risk in energy. *IEEE Transactions on Power Systems, 18*(2), 494–502.
78. Chinowsky, P. S. (2009). Construction planning, programming and control. *Construction Management & Economics, 27*(12), 1269–1270.
79. Liu, N. (2006). Research on engineering target decision-making. China Water Resources and Hydropower Press.
80. Guo, J., Zhang, J., Hu, Z., et al. (2006). Intelligent engineering decision platform based on data mining. *Journal of Harbin Institute of Technology, 38*(9), 1518–1522.
81. Lei, L., Zhou, J., & Li, M. (2011). Research on interactive multi-attribute group decision-making method for large-scale engineering projects based on relative entropy principle. *China Soft Science, 2*, 166–175.
82. Ren, H., Ma, X., & Jin, H. (2012). Research on the evaluation of giant project input based on decision-making "core three principles." *Science & Technology Progress and Policy, 29*(18), 81–85.
83. Nie, Z. (2005). Improvement of project management of construction enterprises. Zhengzhou University.
84. AbouRizk, S., Mohamed, Y., Taghaddos, H., Saba, F., & Hague, S. (2010, December). Developing complex distributed simulations for industrial plant construction using high-level architecture. In: *Proceedings of the winter simulation conference* (pp. 3177–3188). Winter Simulation Conference.
85. Liang, B. (2012). Research on multi-project planning and scheduling method for complex products based on critical chain. Jiangsu University of Science and Technology.
86. Wang, J. (2012). Research on the development planning of commercial real estate projects. *Anhui Architecture, 19*(2), 213–213.
87. Nadler, D., Tushman, M., Tushman, M. L., & Nadler, M. B. (1997). Competing by design: The power of organizational architecture. Oxford University Press.
88. Sheng, F., Qi, A., & Wang, J. (2008). Research on stakeholder game and partnership relationship management model of government investment projects. *Productivity Research, 7*, 86–88.
89. Liu, S., & Dong, Y. (2007). Virtual enterprise design partner selection model and real application. *Industrial Engineering, 10*(5), 78–84.
90. Hao, H., & Zheng, P. (2005). The risk distribution mechanism of supply chain partners based on Shapley value. *Journal of Harbin Institute of Technology (Social Science Edition), 7*(5), 71–75.
91. Sun, D., & Ye, F. (2001). Research on negotiation model of dynamic alliance benefit distribution. *Research Management, 22*(2), 91–95.
92. Wan, J., Li, M., & Kou, J. (2002). Study the influence of distribution mechanism in supply chain on Bullwhip effect. *Journal of Systems Engineering, 17*(4), 340–348.
93. Eriksson, P. E., & Pesämaa, O. (2013). Buyer-supplier integration in project-based industries. *Journal of Business & Industrial Marketing, 28*(1), 29–40.
94. Eriksson, P. E., & Westerberg, M. (2011). Effects of cooperative procurement procedures on construction project performance: A conceptual framework. *International Journal of Project Management, 29*(2), 197–208.

95. Bossink, B. A. (2004). Managing drivers of innovation in construction networks. *Journal of Construction Engineering and Management, 130*(3), 337–345.
96. Cong, P. (2003). *Construction project management specification* Series Lecture Chapter 13 Construction project organization and coordination. *Construction Technology, 4*, 58–59.
97. Wang, Y., Li, Y., & Cheng, H. (2007). Mechanism and design of coordination management of engineering projects in the planning stage. *Construction Technology, 36*(7), 90–92.
98. Guo, F., Xu, H., Qu, G., et al. (2011). Innovation and practice research on coordination management of construction engineering supply chain partners—Take Shenzhen metro line 5 project as an example. *Science and Technology Progress and Countermeasures, 28*(13), 26–31.
99. Zhong, D., Yang, X., & Li, J. (2005). Research on simulation optimization of TBM construction in a long tunnel. *Water Resources and Hydropower Technology, 36*(4), 85–87.
100. Behzadan, A. H., Iqbal, A., & Kamat, V. R. (2011, December). A collaborative augmented reality-based modeling environment for construction engineering and management education. In *Proceedings of the 2011 winter simulation conference (WSC)* (pp. 3568–3576). IEEE.
101. Feng, J., Yang, J., et al. (1999). Gray swing model of quality control system for water conservancy and hydropower projects. *Journal of Hohai University (Natural Sciences), 27*(5), 57–62.
102. Wang, J. (2001). Three Gorges project and several rock mechanics problems. *Journal of Rock Mechanics and Engineering, 20*(5), 597–602.
103. Yuan, J., & Yan, G. (2003). Quality control method of highway subgrade engineering based on multiple variation graph method. *Highway, 7*, 108–112.
104. Zhong, B., Ding, L., & Zhou, W. (2005). Research on knowledge flow analysis and knowledge integration of large-scale construction engineering. *Science & Technology Progress and Countermeasures, 22*(10), 120–122.
105. Xie, H., & Wang, M. (2010). Analysis of the status quo and problems of the work in the decision-making stage of major engineering projects in China. *China Engineering Science, 12*(1), 18–23.
106. Zhou, Q., & Fang, D. (2009). An empirical study on the influence mechanism of construction industry safety atmosphere on safety behavior. *Journal of Civil Engineering, 11*, 129–132.
107. Qiang, M., Fang, D., Xiao, H., et al. (2004). Safety investment and performance research of construction projects. *Journal of Civil Engineering, 37*(11), 101–107.
108. Zhou, C., Ding, L., Zhou, Y., et al. (2010). Research on safety integrated control process of metro engineering construction for knowledge reuse. *Construction Technology, 39*(9), 17–20.
109. Chen, T., Yuan, H., & Su, G. (2012). Three-dimensional analysis framework of emergency response capability in urban public places. *China Emergency Management, 2*, 28–31.
110. Wang, G. (2003). Compilation and tracking management of the overall progress outline of large-scale construction projects. *Journal of Tongji University, 31*(11), 1359–1363.
111. Hui, Y., Li, H., & Ning, F. (2007). Simulation research on project progress—Cost joint control system. *Journal of Xi'an University of Science and Technology, 27*(3), 498–501.
112. Hu, C., Liang, S., Wang, J., et al. (2007). GM (1,1) schedule control model based on comprehensive progress rate and its application. *Journal of Xi'an University of Architecture & Technology (Natural Science Edition), 39*(4), 468–473.
113. Wu, Q., Gao, S., & Zhou, W. (2011). A comprehensive evaluation of schedule control of construction projects. *Journal of Xi'an University of Science and Technology, 31*(4), 412–419.
114. Zhu, Y., Liu, G., et al. (2010). Research progress control and coordination of Shanghai world expo project construction. *Architecture Technology, 41*(4), 301–304.
115. Feng, W., Cao, Y., & Ren, H. (2003). Research civil engineering cost estimation method based on case fuzzy reasoning. *Journal of Civil Engineering, 36*(3), 51–56.
116. Hu, Y., Wu, X., Guo, S., et al. (2006). Analyze the new institutional economics of investment control in the Three Gorges Project. *People's Yangtze River, 37*(2), 14–16.
117. Lu, M., Han, X., Kong, X., et al. (2011). Research on pre-control of engineering cost based on BP neural network and TOC. *Journal of Xi'an University of Architecture and Technology (Natural Science Edition), 43*(1), 106–112.

118. Liu, J., & Han, W. (1994). Fuzzy influence diagram method for risk analysis of engineering projects. *Journal of Systems Engineering, 2*, 81–88.
119. He, S., & Fu, H. Economic explanation and risk prevention of engineering quality risk. *Civil Engineering and Environmental Engineering, 28*(6), 106–110.
120. Tang, W., Qiang, M., Lu, Y., et al. (2006). Research on project risk management based on partnership. *Hydropower, 32*(7), 1–4.
121. Wang, M., & Zhang, C. (2008). Fault analysis and control countermeasures for urban underground engineering construction. *Journal of Architecture and Civil Engineering, 25*(2), 1–6.
122. Wang, F., Hu, Q., Huang, H., et al. (2010). Dynamic risk management of the construction of foundation pits in multi-station rail transit. *Journal of Underground Space and Engineering, 06*(5), 1027–1032.
123. Zhao, H., & Yin, Y. (2011). Analysis of factors affecting rational risk sharing of ISM based engineering projects. *Journal of Beijing Institute of Technology (Social Science Edition), 13*(6), 15–19.
124. Yao, B. (2003). On the information construction of civil engineering. *Journal of Civil Engineering, 36*(9), 88–90.
125. Zhang, J., Han, B., Li, J., et al. (2006). 4D visualization management of construction site. *Construction Technology, 35*(10), 36–38.
126. Zhu, G. (2008). Some understanding of engineering management informatization. *China Engineering Science, 10*(12), 32–35.
127. Liu, R. & Sun, K. (2010). Discussion on the connotation and extension of engineering management informatization. In *China engineering management forum*.
128. Wang, M., Zhang, S., & Cheng, Q. (2010). Research on standardized management technology of major projects. *Science and Technology Progress and Countermeasures, 27*(6), 23–26.
129. Wang, Y., & Wu, Y. (2012). Research the theory of wisdom construction and key technical issues. *Science and Technology Progress and Countermeasures, 29*(18), 13–16.
130. Liu, Y. (2009). Analysis of engineering management theory system. *Science and Technology Progress and Countermeasures, 26*(21), 56–58.
131. Wang, Z., Yang, Z., & Ding, J. Modern engineering management theory and knowledge system framework (II). *Journal of Engineering Management, 25*(3), 256–259.
132. Wang, Q., & Wang, M. (2012). Reflections on the framework of China's engineering management theory system. *Science and Technology Progress and Countermeasures, 29*(18), 6–8.
133. He, J. (2013). On the core of engineering management theory. *China Engineering Science, 11*, 4–11.
134. Wang, Q., Wang, M., et al. (2013). Construction of a conceptual model of engineering management theory system. *China Engineering Science, 15*, 103–107.

Chapter 3
Engineering Management Methodology

The method is the sum of the ways, strategies, tools, means, and procedures for humans to understand and adapt to the world. The methodology is a theoretical system of how humans understand the world and adapt to the world. Advances in engineering technology depend on methodological innovations, from carriage to space shuttle, from beacon tower to microwave communications, all are the same [1]. The resolution of a series of complex scientific issues in engineering management also requires methodological guidance. The engineering management methodology is a methodological system for studying engineering management issues. It is dedicated to the active handling of various qualitative or quantitative problems encountered in engineering management. It is comprehensive, knowledge-based, practical, and innovative. It is an integration and development of modern scientific methods within the scope of engineering management.

Strategic major projects are characterized by difficult decision-making, high investment, complex technology, and wide-ranging influence. Their construction and operation also face severe challenges of various complex environments, including nature, society, humanities, technology, which need comprehensive integration of multiple disciplines and technologies. Modern engineering management is a comprehensive and systematic practice process that must be creatively carried out engineering management practices under the guidance of methodology and in combination with engineering practice.

According to the general framework of the methodology, this chapter divides engineering management methodology into three levels: engineering management philosophy methodology, engineering management general scientific methodology, and engineering management specific scientific methodology. To construct a complete engineering management methodology system, it is necessary to systematically sort out and summarize the specific method sets applied in engineering management theory and practice and explore the relevance and common problems between the methods. A perfect engineering management methodology system is a booster to promote innovation in engineering management methods. In the engineering

J. He, *Principles of Engineering Management*,
https://doi.org/10.1007/978-981-99-1168-4_3

management methodology in three levels, the systems science and project management methods are self-contained systems. The level of outstanding contributions of the two is different. The systems science method is more remarkable because of its significant value as a general scientific methodology. At the same time, the contribution of the project management method lies in its relatively complete specific method system. Therefore, in terms of the level of general scientific methodology, this chapter focuses on systems science methods; in terms of the level of specific scientific methods, it focuses on project management methods.

3.1 Engineering Management Philosophy Methodology

Philosophy is about the world view. Philosophy has the highest generality; it is the abstraction and generalization of the general nature and regularity of the whole world, including nature, society, and human thinking. For this reason, philosophy has the widest permeability. All fields of human production and life have infiltrated philosophy. All natural sciences, social sciences, and thinking sciences are influenced by philosophy. Engineering management theory and practice are no exception.

The methodology of engineering management philosophy is the most abstract and highest level of thinking in engineering management. From the perspectives of systems view, dialectical view, and harmony view, through the philosophical speculation of the engineering management process, mode, and law, we can get the general principles and thinking mode to guide engineering management research and practice [2, 3], that is management philosophy methodology.

The engineering management theory formed in engineering management practice is an interdisciplinary subject of "engineering" and "management" in terms of disciplines. The direct theoretical source of engineering management theory is engineering and management, and it is the combination and integration of the two. However, engineering is a complex system in the natural environment, social environment, and human environment, involving many complicated factors of nature, society, and culture, so economics, law, sociology, psychology, etc., are also the theoretical basis of engineering management theory. The engineering management theory is not perfect without embedding these disciplines. What is the theory that runs through management, engineering, economics, law, sociology, and psychology? That is philosophy, so engineering management theory is supported by philosophy as the most profound theory, which shows that engineering management methodology needs to be speculated from the philosophical level.

The study of engineering management theory needs to be supported by philosophical theory. Marxist philosophy should become the "dominant" philosophy of studying engineering management problems and constructing engineering management system theory. At the same time, we must also consider the valuable ideas in the Chinese traditional philosophy and Western philosophy that are compatible with Marxist philosophy.

Marxist philosophy's guidance to scientific methodology is mainly embodied in the principles of scientific practice, materialist dialectics, and historical materialism principles. Therefore, standing at the height of philosophy, the philosophy of engineering management should include seeking truth from facts, contradiction analysis, unity of knowledge and conduct, dialectical thinking, and unity of truth scale and value scale, which together constitute the essence of engineering management philosophy methodology.

3.1.1 Seeking Truth from Facts

Seeking truth from facts is an ancient proposition in China, including naive materialism and dialectical thinking. Mao Zedong borrowed this proposition to solve the Party's ideological line. To this end, he explained it from dialectical materialism: "facts" are everything that exists objectively, 'truth' is the internal connection of objective things, that is, regularity, 'seeking' is research [4]. It can be seen that seeking truth from facts embodies the unity of materialism and dialectics, the unity of objective regularity and subjective initiative, and becomes a proposition of dialectical materialism.

Seeking truth from facts is a worldview of dialectical materialism and a methodology of science. As Mao Zedong's metaphor, the method is important, like the boat and bridge crossing the river. To engage in revolution, construction, and reform, we must seek truth from facts. The method of seeking truth from facts should be implemented in all work. Of course, engineering management must also adhere to seeking truth from facts to achieve the goal and results.

All aspects of project management should follow the principle of seeking truth from facts, starting from objective reality and combining subjective initiative with objective law. In this way, engineering planning and design are scientific and reasonable, engineering implementation is effective, and operation is healthy.

Seeking truth from facts is a basic method of engineering management and a basic position and attitude that engineering managers and engineering management theory researchers should have. Investors, decision-makers, managers, engineers, workers in the engineering community, and government officials related to the project, if there is no realistic attitude, the engineering activities will deviate from the correct track, leading to errors or even failures. The core value of engineering management theory is "people-oriented, collaborative innovation, harmony between nature and human, and building harmony." If we want to clarify the "core value," we must first understand the attitude of seeking truth from facts. This "core value" itself is a study that seeks truth from facts. It is realistic to extract and summarize from engineering practice rather than subjective imagination and judgment. Not only the "core value" of engineering management theory, but all other research on engineering management issues should follow the method of seeking truth from facts and have a realistic attitude.

Seeking truth from facts is the foundation of engineering management innovation, and it is also a guide to ensure the correct direction of innovation. The development and progress of engineering and its engineering management must also rely on innovation. Engineering innovation is at the core of the national innovation system. Without engineering innovation, there is no industry formation and development, and there is no economic progress and improvement of the entire country. Of course, this also includes the role of engineering management innovation. However, engineering management innovation must be realistic, proceed from the actual needs of engineering management development, select the direction, topics, and tasks of engineering management innovation, strive to explore the inherent internality of engineering management, and carry out "targeted" collaborative innovation. Only in this way can we succeed, and only in this way can it be truly valuable engineering management innovations.

Seeking truth from facts is the highest standard of engineering management innovation. Engineering management innovation cannot be "untargeted" and cannot be a talk on paper. Therefore, engineering management researchers must go deep into the project construction site, do an in-depth and meticulous investigation and research work, and strive to find out the regularity of engineering management. With various scientific and technological methods, we can refine and summarize the conceptions, principles, and methods of general rules of engineering management. Engineering management innovation should also focus on the existing experience that previous people summed up. However, as the project is constantly changing and developing, we must focus on the practical engineering management practice and conduct a practical investigation. "Theory is gray, and the tree of life evergreen." Suppose engineering management researchers hold books based on existing experience and ignore the extremely lively engineering management practices. In that case, it is impossible to understand the actual "facts" of engineering management, and it is impossible to obtain the "truth" of engineering management. It is, of course, impossible to make real innovations in engineering management theory. For the actual workers of engineering management, the most important thing is to be good at summarizing the continuous development of engineering management practice experience and timely propose new ideas, new principles, and new methods of engineering management. For high-level engineering managers, it is also necessary to go deep into the grassroots and the engineering construction site for investigation and research. Otherwise, engineering management innovation will be impossible to start. Investigation and research are the essential requirements of seeking truth from facts. Without investigation and research, it is impossible to seek truth from facts. Without seeking truth from facts, there is no effective engineering management innovation.

Engineering management innovation should adhere to the "moderation" principle. Too "advanced" and too "complex" innovation will certainly encounter "barriers" or even "traps" in actual work due to the lack of applicable conditions or deviation of engineering practice, which is not only useless but also wastes resources. "Moderation" is a philosophical category, that is, to maintain the limits of matter and quantity, neither "overdone" nor "undone." Confucius said that "Too much is as bad as too little" means "moderation." There were many deviations in the engineering management innovation practice; the most prominent one was the "advanced" research divorcing the actual project management conditions. This so-called "advanced" is not necessarily a scientific foresight, and many of them are cranky things or are copied from the imported items. Therefore, engineering management innovation's "moderation" problem is like shooting the arrow at the target. Too "complex" engineering management innovation is not "moderation" either. Countless formulas and rules will make the actual engineering manager feel confused. Engineering management is as complex and simple as anything else. Engineering management innovation research summarizes and abstracts the complex engineering management and expresses in a series of simplistic "paradigms" and concepts, principles, and methods. The "moderate" engineering management innovation seeks truth from facts, depending upon the objective conditions, practical needs, and application suitability.

What kind of experience does engineering management innovation need to learn from abroad? What kind of innovation do we need by ourselves? These need to be combined with seeking truth from facts to find the answer. Some researchers in engineering management theory believe that only the things from western are good. Therefore, a lot of Western views have been mechanically forced into the engineering management theory of China. As a result, Chinese engineers cannot understand these theories, which is useless for severe separation from Chinese engineering management. Of course, foreign good things should be used for reference, but more importantly, the valuable experience accumulated by Chinese engineers in long-term engineering practice should be valued. In fact, many useful management ideas are included in ancient China's philosophy, economics, military science, sociology, and political science, such as Confucius' thoughts of "Lizhi Ritual" and "Renai Kindness"; Sun Wu's idea of making correct war decisions based on "five fundamental factors, namely, doctrine, weather, terrain, commander and politics; Laozi's leadership of "people is the core"; Han Fei's control thoughts of the combination of "law," "managing strategy," and "power," etc. Although these ideas are at the national level or in the military, they also penetrate ancient engineering management activities and are useful references for today's Chinese engineering managers. Modern engineering management practices in China have created many successful engineering management experiences. The "san lao si yan" style formed by the development of Daqing Oilfield is an engineering spirit and a management theory and method created by Chinese oil workers. "Two bombs and one satellite" project, manned spaceflight project, Three Gorges Project, Qinghai-Tibet Railway project, etc., China's engineering people have many successful creations in engineering management. Therefore, China's engineering management innovation should be based in China because research and applied engineering management theory cannot be separated from China; this is seeking truth from facts. Learning from foreign things must also be combined with the reality of China. For example, the "flattening" organization, the "project system," and the bidding system, which are introduced from the West, are combined with the actual engineering management in China, incorporating many "Chinese elements" and thus enabling them to be widely implemented in the field of engineering management in China. This is the embodiment of introduction, absorption, innovation, and the manifestation of seeking truth from facts.

How does engineering management innovation carry out "collaborative innovation"? Linking up decision making, planning, organization, command, coordination, and control; making appropriate arrangements for the early planning, design, implementation, and operation; achieving quality, cost, schedule, occupational health and safety, environmental protection goals, and integrating resources, contracts, risks, technology, information, culture, etc., these need collaborative innovation. To achieve collaborative innovation, we need to seek truth from facts. For example, the "system engineering theory" founded by the famous Chinese scientist Qian Xuesen is based on the reality of the engineering system. It reveals the internal relations and regularity of many parts, aspects, and elements of the engineering system. "System Engineering Theory" has not only been successfully applied in China's rocket engineering and "two bombs and one satellite" project, but also has affected all subsequent aerospace and military engineering practices in the long run, and has also affected all areas of engineering construction in China. The world recognizes System Engineering Theory as both a theory of engineering management innovation and a method of engineering management innovation. Similarly, the "overall method" and "optimization method" proposed by China's famous mathematician Hua Luogeng is also based on the reality of China's economic construction, and they are management theories based on investigation and deep thinking. They have not only been applied in the management of industrial and agricultural production but also successfully applied in China's engineering management practice. Whether it is system engineering theory, overall method, or optimization method, they all emphasize "participating items" and "related parties" in economic construction and engineering activities; all parts, aspects, links, and elements must be considered and the key points be highlighted. Therefore, they are all theories of "collaborative innovation." Qian Xuesen and Hua Luogeng are models of seeking truth from facts and collaborative innovation in the scientific and engineering fields.

Therefore, whether or not we can adhere to seeking truth from facts is fundamentally related to the success or failure of the project. It is the fundamental ideological method that engineering management must adhere to, and it is also the basic principle that engineering management theory research and practice must conform to [1].

3.1.2 Contradictory Analysis

The contradiction is universal. Engineering, as a transformation activity of changing society and the construction of artificial nature, its contradictions with people, nature, and society are inevitable. Based on the particularity of contradiction, the contradiction between engineering and people, nature, and society manifests itself as the specific contradiction in the stages of planning, design, construction, and operation with the development of engineering activities. Therefore, engineering management must face up to the contradictions and be good at solving contradictions to promote the healthy development of engineering management. As an objective existence, a contradiction exists in the development of all things. Engineering management with engineering activities should not be afraid of contradictions, nor can it avoid contradictions. Instead, we should start from the contradiction of identity and struggle characteristics and find the correct way to resolve contradictions. In engineering management, it is crucial to distinguish between internal and external contradictions, subjective and objective contradictions, major contradictions and secondary contradictions, and imbalances in many aspects of contradictions, correspondingly identifying methods for coping and resolving and resolving conflicts in order of priority [5]. Mao Zedong put forward the famous "playing the piano" working method in the article *The Working Methods of the Party Committee*, which is about the art of prioritizing and solving problems. In the book *On Contradictions*, Mao Zedong emphasized the importance of analyzing and resolving contradictions. He used a lot of space to talk about the particularity of contradictions. This is based on the consideration of the particularity of the Chinese revolution and is to solve the problem of the Chinese revolution. The contradictions in engineering management are complex and diverse. Not only the contradictions between engineering and nature, engineering and society, engineering and people are different, but the contradictions in each engineering management stage are also unique. For example, the contradiction between subjective initiative and objective regularity is more prominent in the decision-making stage. The contradiction between internal functionality and external aesthetics is more pronounced in the design stage; The contradiction between progress and quality is more pronounced in the implementation stage; The contradiction between safety guarantee and operational efficiency is more pronounced in the operation phase, etc. Although there is universality in the particularity of contradiction, the focus of resolving contradictions must be placed on particularities to find specific ways to resolve specific contradictions. So-called "one key opens one lock" is what this means. The specific analysis of specific problems is the living soul of Marxism. Engineering management work will not really succeed if the engineering manager does not understand the "soul" of this dialectic. There is no doubt that the study of engineering management theory must also get this "soul."

"Harmony" is one of today's most popular propositions of multi-faceted significance. It is also of vital importance to engineering management. One of the core values of engineering management is "building harmony." As a philosophical proposition, "harmony" is based on the recognition of contradictions and differences, and it is the embodiment of the unity or identity of contradiction. The contradiction under the connotation of "harmony" is not the absolute and rigid identity, but the identity between multiple opposites under certain differences, the "harmony in diversity" spoken by Confucius. Harmony does not mean completely eliminating and obliterating contradictions but following the principle of universality of contradictions to achieve the "identity" of contradictions. Engineering and engineering management are full of various contradictions. Engineering management focuses on adjusting and resolving these contradictions to achieve harmonious, orderly, and healthy implementation and operation of the project. The contradiction in engineering management is the premise. It is precise because of contradictions that it is necessary to adjust and solve and pursue harmony. At the same time, harmony is also relative. When the engineering activities reach harmony at a certain stage, it is only basically or in general. The contradiction is still there, but at this time, the opposite side of the various stakeholders in the engineering activities does not occupy a dominant position, and the dominant position is the identity. However, at another stage of the project, the contradictions may be highlighted. At this time, the engineering management needs to adjust and resolve the new contradictions so that the various stakeholders in the project can achieve new harmony. In the same way, once a project is completed satisfactorily, and a harmonious goal is achieved, another new project begins, and the pursuit of a new harmonious engineering goal begins. Contradiction, harmony, contradiction, harmony ..., this process is cyclical and never-ending.

Therefore, contradiction is the driving force for the development of all things and the source of engineering management's progress. In industrial management practice, the contradiction analysis method is fundamental to understanding engineering management and building a harmonious project.

3.1.3 The Unity of Inner Knowledge and Action

As early as the pre-Qin period, China put forward the philosophical proposition of the unity of knowledge and action and gradually enriched and developed it in the later period. Knowledge is understanding; action is practice, the ancients called Jianlv. The relationship between knowledge and action is the relationship between understanding and practice. For this reason, the subtitle of Mao Zedong's famous book *Practice Theory* is *On the Unity of Understanding and Practice, knowledge and action.* In the relationship between understanding and practice, Marxism pays more attention to practice. The practical point of view is the primary basic point of Marxist epistemology. This is also a remarkable sign that Marxist philosophy is different from all previous old philosophies. Philosophers only explain the world differently, but the problem is changing the world [6]. Of course, the unity of knowledge and action in ancient China cannot be equated with Marxist epistemology, but it also has similarities and fits. Wang Fuzhi in the late Ming and early Qing dynasties said: "the action is prior to knowledge, and the action can lead to knowledge; knowledge means understanding, action because knowledge; knowledge is verified by action, knowledge is proved by action; doing knowledge and action together, mutual dependence and promotion." There are many such arguments in ancient times, all emphasizing the view of practice first, objectively showing the relationship between understanding and practice. Practice is the basis of understanding; conversely, understanding has a guiding role in practice. Therefore, as long as we give the meaning of Marxist philosophy to the unity of knowledge and action, it can become a proposition of dialectical materialism epistemology and become a proposition of practice theory.

The unified view of knowledge and action, or the Marxist view of practice, adheres to epistemological materialism and epistemological dialectics; it is the proposition of dialectical materialism epistemology. The concept of the unity of knowledge and action is a Marxist world view and a Marxist epistemology and methodology.

Engineering is a practical activity that adapts to nature, uses nature, and creates artificial nature. The function of engineering management is the decision-making, design, organization, command, coordination, and control of these kinds of practical activities. The concept, principles, means, and methods of engineering management are all derived from practice, which is a summary and generalization of practice, rather than subjective self-production. Obviously, practicality is a distinctive feature of engineering management activities. If engineering management theory separates the practice, it will inevitably fail and be useless. Therefore, the study of engineering management theory must follow the point of view of practice first, pay attention to the investigation and summary of the practice of engineering management activities. Only the engineering management theory condensed from practice can effectively guide the practical activities. In the process of guidance, the correctness and rationality of the theory are tested, revised based on the practice, and then supplemented and enriched to promote the development of the theory. Only in the process of engineering management practice to engineering management theory to engineering management practice can the theory and practice of engineering management be continuously upgraded. This is Marxist epistemology, and it is also the method of unity of knowledge and action. In the actual engineering management practice, some engineering managers and decision-makers always think they are "smart," like "Captain's call" decision or taking action for granted. They are unwilling to be deeply involved in reality and do a thorough and meticulous investigation and research on the natural, social, economic, political, cultural, scientific, and technological factors related to engineering. As a result, they make decision-making mistakes, and the actions deviate from the objectives of the project and cause harm to the country and people. From an epistemological point of view, these decision-makers and managers lack the practical point of view and deviate from the unified view of knowledge and action requirements. They are "lazy" and "voluntarists," let them guide engineering management practices and will fail.

It is a basic viewpoint of historical materialism to believe that the masses of the people are the creators of history and the decisive force for promoting social development. Adhering the leadership method of "doing everything for the people, relying on them in every task," carrying out the principle of "from the masses, to the masses," that is, the methods of the mass line. Engineering management must also adhere to the practices of the mass line. The method of the mass line is the concrete manifestation of the unity of knowledge and action. In the article *Several Issues on Leadership Methods*, Mao Zedong pointed out that "in all practical work of our party, all correct leaders must come from the masses and go to the masses. That is to say, the opinions of the masses (the scattered and nonsystematic opinions) are collected (through research, turned into a collective system of opinions), sent to the masses for propaganda and interpretation, and then are turned into opinions of the masses, so that the masses can insist them and reflect in action, then test whether these opinions are correct in the mass action. Then they will be concentrated again from the masses and be insisted by masses. This infinite loop is more accurate, more vivid, and more prosperous than before. This is the epistemology of Marxism [7]. This classic expression raises the mass line to epistemology and rises to the concept of the unity of knowledge and action.

Engineering management is oriented to engineering practice. The main body of engineering practice is the people of the engineering community, the people engaged in engineering planning, design, implementation, operation, and decommissioning. The engineering management theory comes from practice, that is, from the practice of the masses of the engineering community. It then returns to the practice of the masses of the engineering community. The mass line runs through every stage of the engineering activity. Taking the mass line in the decision-making stage of engineering planning is important. It is necessary to emphasize that it is not possible to rely on the individual or minority of the leadership to make a decision, but to listen to the experts' opinions fully, and to broaden the public opinion and conduct feasibility demonstration of democratization. More designers or design organizations need to be mobilized to participate in the competition in the engineering design stage. For various programs, a variety of experts with insights should be considered, their opinions are compared and evaluated, and the best design plan to be finalized. The implementation and operation phase of the project may involve the construction task and the application task shared by thousands of engineering people. It is more necessary to take the mass line and give full play to the wisdom and strength of the owners of the engineering community to achieve the determined engineering objectives.

Therefore, the method of unity of knowledge and action appropriately expresses the relationship between understanding and practice in engineering management. In engineering management activities, using the unity of knowledge and action, and adhering to the mass line is a crucial method to achieve engineering management objectives and promote the development of engineering management.

3.1.4 Dialectical Thinking

Engels said that if a nation wants to stand at the highest peak of science, it cannot be without theoretical thinking for a moment. Thinking is an abstraction of the essence of objective things. It is composed of a series of concepts and categories and is in the form of logic. With the development of human cognition, thinking, as the highest spiritual achievement of humans, cognition presents a diversified form and constitutes a colorful human "network of thoughts." Each of the proven concepts and categories is a "knot" on the network of thoughts. Thinking can be divided into scientific thinking, technological thinking, artistic thinking, engineering thinking, etc., and philosophical thinking, dialectical thinking, is the highest abstraction, and thus the most profound reflection of the most general nature of objective things. Dialectical thinking must be integrated into all human thinking and among all human activities. Engineering management activities and engineering management theory research, of course, also need to have dialectical thinking.

The content and form of dialectical thinking are rich and varied. The series of laws and categories of dialectics is the embodiment of dialectical thinking. The dialectical thinking methods of inductive and deductive, analytical and comprehensive, abstract and concrete in the long evolution are the forms of dialectical thinking, which are the tools and methods to guide people to know the world and engage in scientific research. Engineering management also needs to use these methods that are proven to be effective.

"Induction" is a way of thinking that generalizes general principles from individual facts. It is a form of reasoning that goes from individual premise to general conclusion. "Deductive" is a thinking method that draws individual conclusions from the general principle and is a form of reasoning that derives individual conclusions from the general principle. In general, the former is from individual to general thinking, and the latter is from general to individual thinking. Engineering management is a very rich and practical activity. It requires researchers of engineering management theory to choose to "analyze" various engineering projects, engineering projects in various industries, and representative engineering projects, engineering objects, or engineering phases in various fields, summarize the general significance and raise it to theory. This is the "anatomy of the sparrow" method, the meaning of "typical experiment." What is generalized from individual projects, because of its universal theoretical significance, can be used to guide all engineering activities of the same category, homogeneous, and process, so that different individual conclusions can be obtained. Then, once again, through induction and deduction, again and again, the relatively stable principles, categories, and methods in engineering management theory can be formed in such a cycle.

It should be noted that this kind of induction and deduction should not only be directed to large representative projects, such as the Qinghai-Tibet Railway, the Three Gorges Project, the Daqing Oilfield Project, and the manned spaceflight project. It can also be applied to projects that have a small scale, but have a "sample" meaning, such as one or several of the railway stations along the high-speed railway, one or several of the small hydropower stations, and one or several of many countless residential areas in the city. It should also be noted that this kind of induction and deduction should be for successful projects and failed projects. In fact, failed projects tend to be more inductive and deductive.

"Analysis" is a way of thinking in which the cognitive object is decomposed into various parts, various aspects, and various elements. An analysis is essentially the analysis of the contradiction of things. "Integration" is a way of thinking that combines various parts, aspects, and elements of an object into a whole in the process of thinking. The integration is essentially based on the holistic nature to determine the internal relations of various parts, aspects, and elements. Any project is a complex system composed of many complicated factors such as science, technology, culture, economy, law, politics, morality, etc. It is also composed of several links such as planning, design, implementation, and operation. It involves a variety of stakeholders such as owners, designers, builders, supervisors, users, etc., and is also closely related to government agencies. Therefore, when engaging in engineering management activities and researching engineering management theory, we must pay attention to the way of "divergence thinking." Firstly, we scientifically decompose complex projects and then study them in different categories to determine the characteristics and laws of each part, aspect, and element. Finally, we use the "convergence thinking" method to integrate the various components, aspects, and elements that have been decomposed and studied, summarize the characteristics and laws of the entire engineering system, and form the specific strategies, methods, and measures for solving engineering problems and managing projects.

The "abstract" and "concrete" thinking method is important area of Hegel's logic. Marx applies it successfully to the study of *Capital*. "Concreteness" is divided into perceptual concreteness and rational concreteness. Perceptual concreteness is a kind of "chaotic representation" that people form through the senses to the whole thing. It is a vivid, rich, but rather general perceptual knowledge. The categorization of rich and vivid sensibility is abstracted one by one to reach the understanding of the essence of many parts, aspects, and elements of things and then further moves to rationality. Rational or concrete thinking refers to the complete and concrete reproduction of the organic whole of a certain abstraction of an object in accordance with its intrinsic connection. As Marx said, concreteness is concrete because it is a synthesis of a series of abstract rules. Obviously, the process from abstraction to concreteness is a process of analysis and synthesis, a process from perceptual cognition to rational cognition.

Engineering management is a practical activity and intellectual activity. The theoretical research of engineering management is creative mental work, and both follow the law of thinking from perceptual to rational. In the 1970s, when engineers and technicians began to survey and design the Qinghai-Tibet Railway, the three major problems of the Qinghai-Tibet Railway: perennial permafrost, ecological fragility, and plateau hypoxia are still at the level of perceptual cognition. Russia, Canada, and other foreign countries have limited experience. After all, it is the first time to build the world's highest plateau railway. After several generations of researchers and engineers repeatedly explored, researched, and experimented on the snow-covered plateau, they gradually reached a rational and specific thinking level, understood its nature and laws, and formed an overall understanding of the special geography, climate, and climate ecology of the Qinghai-Tibet Plateau. On this basis, the Qinghai-Tibet Railway has creatively proposed a series of solutions to solve the three major problems, ensuring the construction and operation safety of the Qinghai-Tibet Railway, such as slate gas cooling, gravel slope protection, and other active cooling measures, establishing a high altitude disease prevention and treatment system, designing environmental programs such as wildlife passages based on animal habits. In this process, the builders of the Qinghai-Tibet Railway also follow the law of abstraction to concrete thinking and gradually explore the specific implementation plan under rational cognition on the basis of perceptual knowledge. In the operational stage, this thinking process is still going on.

Engels said: "Where history begins, where the thought process should start, and the further development of the ideological process is merely a reflection of the historical process in an abstract, theoretically consistent form [8]." The unity of history and logic means that the structure and evolution of logic are consistent with the objective development history of the object, and the history of the understanding of the object. But this does not mean that the logical category has a one-to-one correspondence with historical events, but it has a certain deviation. Because the logical category is an abstraction of the essence and laws in the historical process, it can reflect history more deeply and correctly.

Engineering activities and their engineering management have existed since ancient times, accompanied by the historical process of the entire human society. It can be said that the history of engineering and its history of engineering management is part of human history. The long-standing, evolving engineering and its engineering management activities can be described as rich and varied and present a development process from low to high, from simple to complex. The theory of engineering management should reflect the development history of ancient and modern domestic and foreign projects. The researchers of engineering management theory should follow the historical context of engineering and engineering management. The research of engineering management theory should begin when the history of engineering and engineering management starts. However, the study of engineering management theory is not necessary, and possible to make a detailed description of the details of engineering and its engineering management (if necessary, it is the task of engineering history and engineering management history research). As research on engineering management theory, the focus should be on the refinement of engineering management concepts, the argumentation of category, the revealing of laws, and the understanding of essential. Although this kind of understanding must be based on the history of engineering and engineering management, it must exclude many accidental, secondary, non-essential factors in the history of engineering and engineering management and focus on inevitable, dominant, and essential research. In this way, the history of engineering and engineering management can be reflected more profoundly, correctly, and comprehensively. In the study of engineering management theory, it is necessary to cite typical examples in the history of engineering and engineering management. Still, the purpose of the examples is to lead to some rational engineering management theoretical conclusions. For example, speaking of historical architecture, the ancient Chinese garden architecture or Siheyuan is to prove one of the core concepts of the engineering management of "Harmony between nature and human." Of course, there can be other extensions of management concepts.

3.1.5 Unity of Truth Scale and Value Scale

Due to the problem of diversified engineering management value, economic interests are no longer the only target. Adhering to unifying the scale of truth and the scale of value can effectively avoid the limitations of "Covering the Eye with a Leaf," truly realize the "harmony between nature and human" and build a harmonious project management purpose.

The ideas of seeking truth from facts, contradiction analysis, the unity of knowledge and action, dialectical thinking, and the unity of truth scale and value scale have become the essence of engineering management philosophy methodology. In engineering management practice, if they are properly applied in engineering management practice and engineering management theory research, the significance will be unquestionable.

3.2 General Scientific Methodology of Engineering Management

The general scientific methodology of engineering management is the principle and method generally applicable to engineering management. Compared with philosophical thinking methods, it is more operational and has available procedures, steps, or rules. Compared with the engineering management philosophy methodology, it solves the methods and paradigm problems generally applicable in engineering management practice and research fields and belongs to the comprehensive horizontal method.

As a typical comprehensive and applied interdisciplinary subject, engineering management spans the fields of management, economics, sociology, law, engineering, and art. Therefore, the principles and methods of generally thinking applicable in engineering management must also be the endogenous products of engineering management and its adjacent disciplines. General science methodology such as systems science, information science, and mathematics has been widely used in the engineering management field, plays an important guiding role, and will continue to have far-reaching effects.

The universal connection between the material world and its holistic thinking is systems thinking. The world-famous ancient Chinese water conservancy project Dujiangyan, which has a history of more than 2200 years. It has formed a complete diversion irrigation system of "water diversion for irrigation, flood diversion for disaster reduction" through the fish mouth water diversion, the Bottle-Neck Channel water guiding, and the Feishayan flood diversion to achieve harmony among humans, engineering, and nature. The overall concept is the outstanding application of systems thinking. The "Ding Wei made a palace" project, which was praised by the ancients as "one stroke and three benefits," regards the soil-burning bricks, transportation building materials, and garbage backfill as a series of coherent links and is organically linked with the palace construction project. The idea of problem solving is a typical naive systems thinking. China's manned spaceflight project is also a national key project with a large scale, complex system, great technical difficulty, high reliability, safety requirements, and high risk. Throughout the implementation of the project, it is necessary to integrate a variety of theories, techniques, methods, tools, and equipment to coordinate and optimize the relationship between people and nature, people and people, people and organizations, organizations and organizations to achieve both technical requirements and rational allocation of facilities, funding, and human resources to reduce costs and ensure quality goals [9].

Information science is a theory and knowledge about the nature of information and the law of information transmission. It is an emerging subject that studies measurement, communication, exchange, reception, and information storage. As a methodology, the information science method refers to the use of information to examine the behavior and function of the object. Understanding the motion law of an object can be achieved through the acquisition and processing of information. The engineering information has a large volume, complex types, wide sources, scattered storage, and a complex application environment. It has the characteristics of non-consumptive, systematic, and inconsistency in time and space, which is always in dynamic changes. In today's big data era, people gradually realize that the engineering implementation management process is a process of information management. The development of information science and advanced information technology provides the basis of methods and means for scientific management of engineering management [1].

Mathematical thinking is about the concept of mathematics, theoretical methods, and the law of the generation and development of morphology. It is the understanding of the nature of mathematical knowledge and techniques. The mathematical method expresses the state, relationship, and process of things in mathematical language and derives, calculates, and analyzes to form the method of interpretation, judgment, and prophecy of the problem. The engineering management research method first paid attention to the empirical research method or the experimental research method, which was determined by people's understanding of the subject attributes. With the deepening of people's understanding of engineering management issues and the development and introduction of normative research methods, mathematical methods are gradually being widely used in engineering management, typically in operations research, mathematical statistics, fuzzy mathematics, etc. [1].

In the above general scientific methodology, the breadth and depth of the application of systems science methods has not only stayed at the level of general scientific methodology, but it has formed a relatively complete systems science methodology system, infiltrating into philosophical thinking and concrete methods, with the meaning, function, and roles at three different levels of the methodology of engineering management. In system science, the thinking mode (entirety, complex, level, optimization, development, etc.) and research methods that are based on the whole and adequately handle the relationship between the whole and the part are the most general methodologies for understanding the system, transforming the system and constructing the system. They belong to the level of philosophical methods. The basic methods of proceduralization of research systems and the procedural work steps for solving system engineering problems are methodologies applicable to many related fields and have certain universal significance. They belong to the level of general scientific methods. The scientific method proposed or applied in combination with specific system engineering problems is the methodology of system engineering research specific reality system, which belongs to the level of specific scientific methods. The systematic scientific method embodies the basic ideas and methods of the above three different levels of methodology. The three are interdependent, mutual influencing, and complementing each other, constitute a unique set of thinking methods, theoretical foundations, basic procedures, and method steps.

However, in terms of the application of engineering management, the guiding significance of the general methodology level is much greater than the other two methodological levels. Therefore, the following focuses on systems science methods, including the concepts of systems science and systems engineering, four typical systems science methods, and emerging systems science methods, namely the large system decomposition coordination method.

3.2.1 Systems Science and Systems Engineering

System science is a new comprehensive scientific category formed by the system as the research and application object, centering on systems thinking and integrating multiple subject contents. According to its development and current situation, systems science can be divided into narrow and generalized. The narrow systems science refers to a discipline that includes theoretical basis and practical application. The theoretical basis refers to the systematic theory that clarifies the characteristics and laws of the system. Its practical application refers to system engineering. Generalized systems science refers to the subject group composed of the system's basic theory and application development as the research object. It focuses on the relationship and attributes of various systems, reveals the law of its activities, and explores various theories and methods of the system. Systems science has the characteristics of cross-discipline, comprehensive, holistic, and transversal, which is also a distinctive feature of systems science that is different from other scientific theories.

Modern science and technology show a trend of mutual dependence and mutual promotion: the technicalization of science and the scientificization of technology, making science and technology increasingly integrated. Qian Xuesen regards modern science and technology as a whole system, vertically divides it into departmental systems such as natural science, social science, mathematical science, systems science, thinking science, and human science, and horizontally divided it into three levels: basic science, technical science, and engineering technology, thus creating the architecture of modern science and technology. Among them, systems science includes system theory, information theory, cybernetics, synergy, and many disciplines such as operations research, systems engineering, and information technology. These disciplines are all independent scientific theories, but they are closely related to each other and infiltrate each other. They tend to be integrated and unified in development and form a comprehensive science with a strict theoretical system [10–12].

System engineering was born in the 1950s. It is based on general systems theory, cybernetics, information theory, operations research, computer science, and self-organization theory, dissipative structure theory, synergy theory, and catastrophe theory generated by its development. It is both an organization management technology and a universal scientific method to all systems [13].

When using systems engineering theory and method to study and solve real system problems, it is necessary to proceed from the whole, fully consider the relationship between the whole and the part, and carry out an overall design, rational development, scientific management, and control coordination according to specific system objectives, to achieve the best overall effect or significantly improve system performance. Compared with general engineering techniques and management methods, system engineering has the following characteristics [14].

3.2.1.1 The Integrity of Research Thoughts

When using systems engineering theory and method to study system problems, we must adhere to the thinking method of integrating holistic approach and the analysis method of reductionism, that is, based on a detailed understanding of the relationship between the various elements of the system, starting from the whole, study the relationship between system and elements [15], thus understand the overall emergence of the system, reveal the inherent characteristics of the system and the laws of motion, and scientifically understand the overall situation.

The South-to-North Water Transfer Project is the basis for solving the problems of sustainable development in the Beijing-Tianjin North China region in the twenty-first century. Systems engineering should be applied in all stages of planning, construction, and operation management. The decision-makers placed the South-to-North Water Transfer Project in the economic, social, and environmental contexts and considered the water source and water-deficient areas as a whole. The three lines of the Middle, East, and West were designed to solve the issue of water shortage in the region of Jiangsu, Shandong, and Anhui, Northwest, and Beijing, Tianjin, and Hebei, respectively.

3.2.1.2 Diversity of Research Methods

When studying system engineering problems, we must flexibly choose scientific methods based on the needs of actual issues. The methods describing system engineering problems generally combine qualitative description with quantitative description, global description and local description, and qualitative description and uncertainty description. Methods for analyzing and researching system problems are typically model analysis and simulation experiments, system analysis and integration, system prediction and control, system evaluation, and system design.

The South-to-North Water Diversion Project should consider the current economic conditions, the adjustable water volume of the water supply area and the actual water shortage in the water receiving area, and the Danjiangkou Reservoir dam heightening and the water supply after joint dispatching with the Three Gorges Project. It is necessary to consider the adjustable water volume for the high water supply rate (95%) of the city, and also to consider the adjustable water volume of the agriculture and ecological environment in the water receiving area in the Yangtze River flood year and the drought year in North China. It is necessary to consider the temporary increase in water volume during the Han River floods and the drought and less rain in North China and consider the need for water replenishment of the Yellow River cutoff problem [16]. The engineering research has gone through half a century; the planning scheme has been changed several times. Each scheme has comprehensively applied the diversified research method combining qualitative and quantitative, model analysis and simulation experiment, system prediction and control, system evaluation and design to demonstrate the scheme's technical feasibility and economic rationality. The scheme also needs social support finally.

3.2.1.3 Use the Comprehensiveness of Knowledge

The research object of systems engineering is mainly the large complex system dominated by people or involved by people, so the problem of handling system engineering must be both scientific and artistic [17]. As a discipline, system engineering is an interdisciplinary subject formed by the intersection of natural science and social science. Therefore, when studying systems engineering problems, it is necessary to use several natural sciences and technical sciences such as mathematics, physics, chemistry, biology, information, and technology, as well as humanities and social sciences such as economics, sociology, psychology, and behavioral science.

The South-to-North Water Diversion Project runs through the Chinese mainland and spans the four major river basins of Jiang, Huai, Huang, and Hai. It is a huge system of engineering covering multi-disciplinary and multi-industry in politics, economy, humanities, law, science, and technology. Therefore, in the work of the various stages of the South-to-North Water Diversion Project, the comprehensive application of knowledge should be fully reflected. For water transfer and water supply, basic industries and infrastructure, economic benefits and social benefits, we should use the knowledge in the fields such as water conservancy engineering, civil engineering, environmental engineering, economics, and social science, study the relationship between water transfer and the economy, and seek the best intersection of water transfer and economic development.

3.2.1.4 The Universality of the Application Field

The subject property of systems engineering determines that it has an extensive range of applications, including technology systems engineering, industrial systems engineering, agricultural systems engineering, transportation systems engineering, construction systems engineering, military systems engineering, ecological environment systems engineering, resource systems engineering, economic systems engineering, social systems engineering, management systems engineering, etc.

In China, systems science and systems engineering were emerging and developing with developing "two bombs and one satellite." The first scientific and technological plan in the history of China, "The Outline of the Science and Technology Development Vision for 1956–1967 (Draft)", puts the development of cutting-edge technology represented by missiles, atomic bombs, and hydrogen bombs in a prominent position. In April 1958, Mao Zedong said at the second meeting of the Eighth National Congress of the Communist Party of China that we would also engage in artificial satellites. Since then, the vigorous development of "two bombs and one satellite" has begun, and in 1960, 1964, 1967 and 1970, missiles, atomic bombs, hydrogen bombs and "Oriental Red One" satellites were successfully launched. On the cornerstone of the strategic development of "two bombs and one satellite," China's aerospace industry has formed a complete supporting system of research, design, production and test, and aerospace industrial system. It has a series of aerospace models such as launch vehicles, satellites, manned spacecraft, and missile weapon systems, and successfully implemented major engineering tasks represented by manned spaceflight and lunar exploration projects, achieving a historic leap from nothing, from small to large, from weak to strong, and laying the international status of China's space power. After more than 50 years of development and construction, systems science and systems engineering have become the core competitiveness and significant soft power of aerospace science and technology. They promote the normalization and scientific track of China's aerospace scientific research management, guide the entire national defense scientific research work, and provide effective guidance for large-scale projects involved in socialist modernization.

3.2.2 *Typical System Science Methodology*

In recent decades, systems science has formed a variety of scientific methodologies with specific influence. In engineering management practice, for the particular engineering problems, it is often necessary to creatively and comprehensively apply and develop the basic ideas and methods of these methodologies, and use relevant systems engineering analysis techniques to carry out system modeling, system analysis, system prediction, system design, system integration, system evaluation, and system decision making. The main systems science methodologies include Hall, Checkland, integrated integration, and Wuli-Shili-Renli methods.

3.2.2.1 Hall Methodology

Hall methodology [18] was proposed by the famous American communications engineer and systems engineering expert A. D. Hall in the 1960s. It provides a method for solving the planning, organization, and management problems of large and complex systems with a manageable structure and has been widely used. According to the Hall methodology, the entire task of system engineering can be decomposed into seven phases and seven steps that are closely connected before and after. At the same time, various knowledge is required to complete the activities in each phase and each step considered. This forms a three-dimensional structural model consisting of time, logic, and knowledge dimensions, as shown in Fig. 3.1 [19].

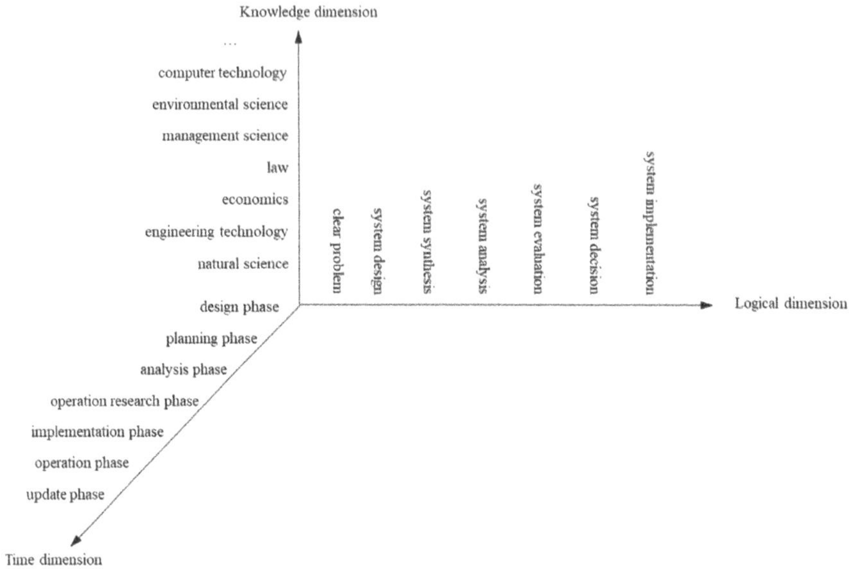

Fig. 3.1 Hall three-dimensional structural model

The time dimension expresses the work required in each phase of the system engineering from the start to the final completion, including seven stages such as the planning phase, the design phase, the analysis phase, the operation research phase, the implementation phase, the operation phase, and the update phase. The logical dimension refers to the logical sequence and working steps that should be followed in each stage of system engineering. It is generally divided into seven steps: understanding the problem, system design, system synthesis, system analysis, system evaluation, system decision, and system implementation. Knowledge dimension refers to the various professional knowledge and management knowledge required to complete the engineering activities in the above stages and steps, including natural science, engineering technology, law, economics, management science, environmental science, computer technology, and other knowledge.

Hall three-dimensional structure mainly focuses on engineering systems and pays more attention to quantitative analysis methods. Engineering systems often lack samples and information and influence system construction, program evaluation, and factor impact analysis using systems engineering methods. Therefore, models and simulation methods occupy a critical position. Only in this way can we have a deeper understanding of the problem, thus helping to inspire ideas and accelerate the process of systems engineering research [20]. Hall methodology emphasizes the goal, and its core content is optimization. It believes that the real problems can be summarized into engineering problems so that the quantitative analysis method can be used to obtain the optimal solution. The methodology has salient features, such as the integrity of research methods (three-dimensional), the comprehensiveness of technology application (knowledge dimension), the scientificity of organizational management (time dimension and logic dimension), and the problem-oriented (logic dimension) of system engineering work.

3.2.2.2 Checkland Methodology

With the continuous deepening of systems science research, the application fields are also expanding. The systems science theories and methods are increasingly applied to economic and social development strategies and organizational management research. The behavior of these systems is difficult to describe with mathematical models and can often be judged by human intuition, using semi-quantitative and semi-qualitative methods. Since the mid-1970s, many scholars have proposed a variety of soft systems science methodology based on Hall methodology for such complex large systems with ill-structures. The most representative of these is the Checkland methodology proposed by Professor Checkland of the University of Lanchester, UK[21]. The core is not to seek the "optimization" of the system but to "investigate, compare" or "learn" from the current situation investigation and model comparison, and learn ways to improve the existing system. The problem processing process of the Checkland methodology is shown in Fig. 3.2 [22].

The Checkland methodology is more suitable for studying "soft" system problems such as social economy and management. The core is the comparison and learning, that is, learning the way to improve the status quo from the comparison between model and reality. More emphasis is placed on the combination of the basic qualitative and quantitative methods.

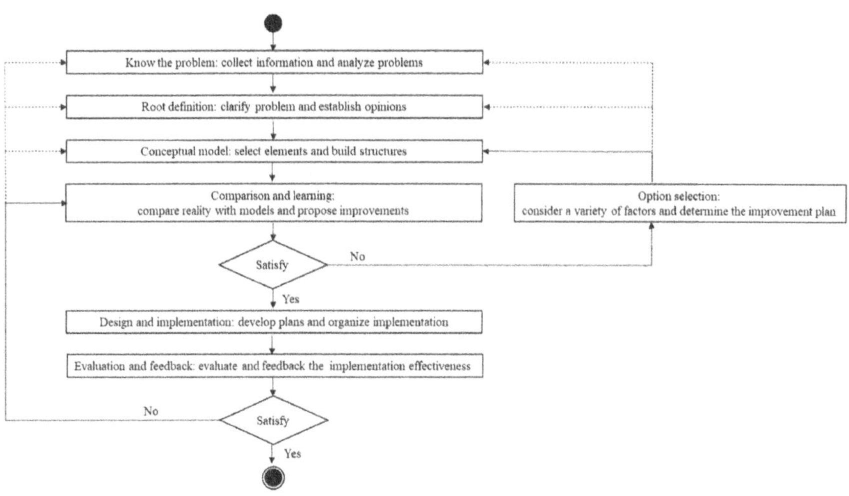

Fig. 3.2 The problem handling process of Checkland methodology

3.2.2.3 Integrated Integration Methodology

While creating China's space industry, Qian Xuesen has concisely generalized some creative and forward-looking important academics and suggestions of great value and formed Qian Xuesen's systems thinking with systems science and comprehensive integration system as the core concretized the system methodology. In the early 1980s, the "comprehensive integration from qualitative to quantitative methods" and its practical form "from qualitative to quantitative integrated integration research system" were put forward. The two were collectively referred to as integrated integration methods. The overall design department uses integration methods and applies systems engineering technology [23], forming a set of methodologies and practical approaches that can be effectively operated.

The "two bombs and one satellite" project involves a large number of industries and personnel and the professional and technical complexity. The organization of the project itself is extremely challenging. Faced with such large-scale scientific and technological engineering, scientific organization management methods and techniques are inevitable to realize the effective organization of personnel and complete the research and development of highly reliable products quickly. Qian Xuesen uniquely integrated the idea of engineering cybernetics into the aerospace science and technology industrial system. After continuous development and improvement, China Aerospace has formed a total design department, model command system, and model designer system, three-step model development route of "pre-researching by one generation, developing by one generation, and producing by one generation," and four technical state control stages of "plan, initial sample, sample, and equipment stereotype." Each model in the aerospace system is an engineering system with an overall design department for each model. The overall design department is composed of professionals familiar with the various aspects of the engineering system and is led by a broad-based expert (called the chief designer). On the one hand, the overall design department puts the system under the large system it belongs to and develops technology from the realization of the large system. At the same time, the overall design department regards the system as an organic whole composed of several subsystems and considers the technical requirements of the subsystems from the perspective of ensuring system technology coordination. The contradiction between sub-systems, the contradiction between sub-systems and system, and the contradiction between system and large system should be coordinated according to the principle of overall service objectives. The practice of the overall design department is the embodiment of the systems science methodology. It has problems in how to rationalize and optimize the allocation of resources and involves institutional mechanisms, development strategies, planning plans, policy measures, and decision-making and management issues. Whether pre-study, model development, or various management tasks, system ideas, systems, and methods such as overall optimization, system coordination, environmental adaptation, innovation development, and risk management are constantly running through.

The basic idea of the comprehensive integration methodology is: when studying the system engineering problem, firstly we form empirical assumptions of system problems from the system as a whole, through the organic combination of scientific theory, practical knowledge, and expert judgment, for example, judgments, guesses, ideas, plans, etc. Then we use modern information science and technology to establish a highly intelligent human–machine combination and a people-oriented decision analysis system. Through the joint discussion of relevant multi-disciplinary experts and human–computer interaction, based on repeated comparisons, systematic analysis, and system integration, we reach the stage of understanding problems from qualitative to quantitative and gradually be able to make clear scientific conclusions on empirical hypotheses. As a scientific methodology, the integrated approach is based on the theoretical foundations of thinking science, the methodological foundations of systems science and mathematics, the technical foundations of information technology, and the philosophical foundations of dialectical materialism, realizing the comprehensive integration of information, knowledge, experience, wisdom, and creativity, and providing a standardized and structured effective method for studying complex giant systems such as societal systems, human systems, and geographic systems [10].

The workflow of the integrated integration methodology is: (1) Collect information on actual problems and provide basic data for building models; (2) The expert group analyzes and studies the problem, clarifies the system characteristics, and determines the modeling ideas; (3) Combine theoretical knowledge with empirical assumptions and gradually form a system model from qualitative to quantitative; (4) Computer simulation research on system behavior; (5) The expert group analyzes and evaluates the simulation results, validates the results, and proposes the revisions; (6) Adjust the system model according to the above results, and then run and evaluate the system until satisfactory results are obtained. The workflow of the integrated integration method is shown in Fig. 3.3 [24].

The comprehensive integration methodology is achieved through the integration of qualitative integration into combining qualitative and quantitative integration, and then from qualitative to quantitative integration. This method system has comprehensive advantages, overall advantages, and intelligent advantages. It is an effective method for dealing with complex systems and complex giant systems and societal systems. It has been successfully applied in the fields of social-economic systems engineering.

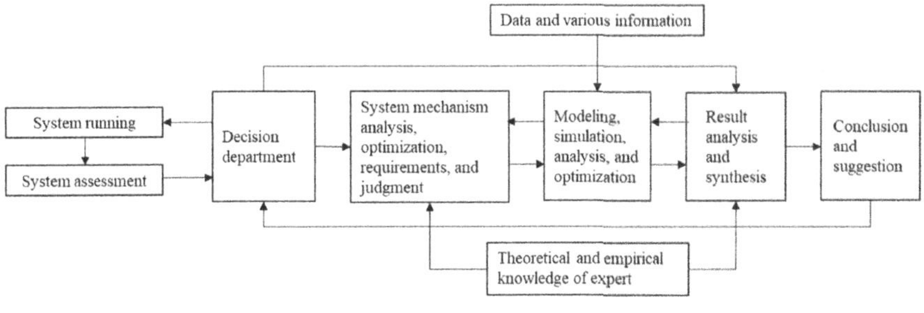

Fig. 3.3 Workflow of the integrated integration method

3.2.2.4 Wuli-Shili-Renli Methodology

The Wuli-Shili-Renli methodology [25] was proposed by Professor Gu Jifa and others of the Chinese Academy of Sciences in 1995. The Wuli is mainly involved in the mechanism of material movement, usually using natural science knowledge. The Shili is the reason for doing things, mainly to solve how to arrange these things. The Renli is the truth of being a human being. The core idea of Wuli-Shili-Renli methodology is: system engineering workers must not only understand Wuli, the natural sciences, and understand what the world is like, but also know Shili, be familiar with scientific methodology, and be good at choosing scientific and rational methods to deal with things. They should also understand Renli, master the art of interpersonal communication, fully understand the value orientation of various departments within the system, and coordinate all parties' interests. Combining these three aspects, using the logic of human rational thinking and the comprehensiveness of image thinking to organize practical activities, makes it possible to produce maximum benefits and efficiency and achieve creative results. The main contents of the Wuli-Shili-Renli methodology are shown in Table 3.1 [19].

The Wuli-Shili-Renli methodology emphasizes the cross-infiltration and integration of natural sciences, engineering technology, and social sciences. The work steps of the Wuli-Shili-Renli methodology include: understanding intentions, investigating and analyzing, forming goals, building models, making recommendations, implementing programs, and coordinating relationships. The coordination relationship is at the core of the Wuli-Shili-Renli methodology, reflecting the comprehensiveness and sociality of systems science. Its workflow is shown in Fig. 3.4 [26].

(1) Understand intention—identify the problem to be solved and understand the intention of the decision-maker. In most cases, the decision maker's desire to solve the problem, or the system's intentions may be clear or ambiguous, which requires communication and coordination, because the decision-makers stand

Table 3.1 The main contents of the Wuli-Shili-Renli methodology

	Wuli	Shili	Renli
Reason	Material world rule	Theory of management and work	Theory of human, discipline, and norms
Object	Objective material world	Organization, system	People, groups, relationships, wisdom
Focus	What is it (functional analysis)	How to do it (logical analysis)	What should be done (humanistic analysis)
Principle	Honest, pursue the truth, be as correct as possible	Coordinated, efficient, smooth as possible	Humanity, effective, flexible as possible
Required knowledge	Natural science	Management science, systems science	Humanistic knowledge, behavioral science

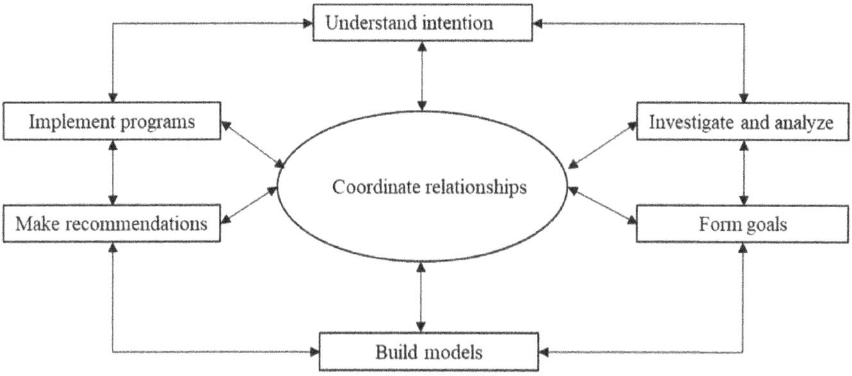

Fig. 3.4 The workflow of the Wuli-Shili-Renli methodology

at different angles with different understandings about the issue and intentions, which requires analysts to understand their intentions, but also to understand the intentions of the people involved.

(2) Investigate and analyze—investigation and analysis is a physical analysis process that can only be concluded after intensive and careful investigation and analysis. To carry out investigation and analysis, it is necessary to coordinate the relationship with the respondents, strive for the active cooperation of the respondents (experts, the masses), and deal with the necessary materials and information.
(3) Form goals—after comprehending and understanding the intentions of decision-makers, conducting in-depth investigation and analysis, and obtaining relevant information, it is necessary to determine the system goals to form targets. These goals may not be exactly the same as those of the original decision-makers, and they may be changed after extensive analysis and further consideration. Therefore, to reach a consensus on the goals formed, continuous coordination is needed throughout the process.
(4) Build models—the model is more general. In addition to the mathematical model can also be a physical model, a conceptual model, an operational program, a running rule, etc. At this stage, the design work is mainly carried out using Wuli and Shili; that is, the corresponding methods, models, steps, and rules are selected to analyze and process the target. In the field of engineering management, when applying the Wuli-Shili-Renli methodology, it is necessary not only to follow the principles of comprehensive integration, human-oriented human–machine integration, iteration, and learning, but also to apply human rational thinking and image thinking, then propose proposals, coordinate the relationship between the parties, effective implementation, and achieve engineering management objectives, based on natural sciences, humanities, social science, and behavioral science.

3.2.3 Large System Decomposition Coordination Method

Large systems generally refer to systems with numerous influencing factors, diverse mission objectives, large system scales, complex structures, and randomness. Conventional modeling methods and optimization methods are difficult to use for the analysis and design of such systems. In 1960, Dantzig-Wolfe proposed a large-scale system decomposition coordination method when studying the decomposition algorithm of large-scale mathematical programming. The basic idea is: firstly, the large complex system is decomposed into several simple subsystems to achieve the correct control of the subsystems. Then, the coordination strategy between the subsystems is proposed according to the large system's total tasks and overall goals to achieve global optimization.

There are many research results on the decomposition and coordination methods of large systems, including large system decomposition and coordination methods based on hierarchical control structures and large system decomposition and coordination methods based on probabilistic networks and agent technologies.

3.2.3.1 Large-Scale System Decomposition and Coordination Method Based on the Hierarchical Control Structure

The hierarchical control structure refers to a hierarchical structure formed by hierarchically arranging the subsystems and their controllers constituting the large system. Large-scale system decomposition and coordination method based on hierarchical control structure is a kind of iterative optimization method [27]· It mainly includes decomposition and coordination methods based on association prediction principle and decomposition and coordination method based on association balance principle. The basic idea of the decomposition and coordination method based on the principle of association prediction is:

(1) Firstly, the high-order complex large-scale system is decomposed into several low-order subsystems, and an associated parameter vector is set between the large system and each subsystem, that is, the coordination parameter vector α, and according to the prediction result, for each subsystem initially set the coordination parameter vector value;
(2) According to the overall optimization principle of the large system, set up a coordinator to coordinate the optimization process of each subsystem to guide the optimization activities of each subsystem;
(3) Each subsystem is independently optimized according to the initially given coordination parameter vector α, and the decoupled parameter vector s obtained after optimization is reported to the coordinator of the large system;
(4) The coordinator corrects the coordination parameter vector α according to the overall target of the large system and the current effect of the system, and communicates it to each subsystem;

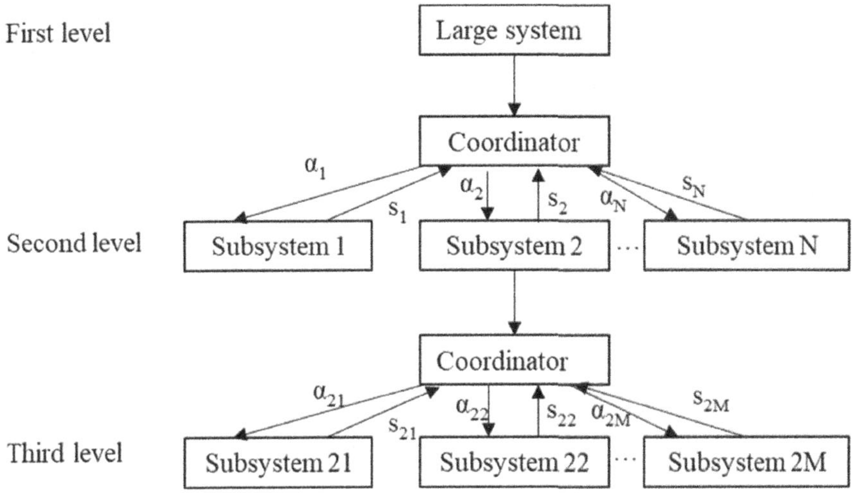

Fig. 3.5 Schematic diagram of a three-level hierarchical control structure

(5) Repeat iterations and coordinate optimization until the solution of each subsystem converges to a satisfactory solution to the overall problem of the large system.

According to the decomposition and coordination strategy, the complex large system can be decomposed step by step to form a multi-level hierarchical control structure. Figure 3.5 is a schematic diagram of a three-level hierarchical control structure.

The decomposition and coordination method based on the principle of correlation balance is different from the decomposition and coordination method based on the principle of association prediction. According to the requirements of the decomposition and coordination method based on the principle of correlation balance, when each subsystem performs local optimization, the association constraint is not considered, and the subsystem treats the associated parameter vector as an independent optimization variable. The coordinator gradually corrects the optimization objectives of each subsystem through the intervention signal to ensure that the association constraints are satisfied, thereby achieving the overall system optimization.

3.2.3.2 Large System Decomposition and Coordination Method Based on Agent Technology and Probabilistic Network

An agent is in an environment and, as part of this environment, capable of perceiving this environment and taking appropriate actions. It can establish its own behavioral norms and can influence environmental changes. An agent has characteristics of autonomy, responsiveness, pre-action, and sociality, enabling reasoning and planning. Multi-Agent technology can effectively simplify the design, implementation, and maintenance of large complex systems. Multi-agent-based collaboration and inference solving can effectively reduce internal conflicts of the system and reduce the difficulty of solving problems and solving complexity. Multi-Agent technology has always been an important technology for the dynamic modeling of large complex systems.

The probabilistic network model is one of the graph models. It can describe the system structure in a flexible and intuitive form. It can perform efficient reasoning computing in an uncertain environment to make planning and decision-making, realizing the functions of the system.

A combination of multi-agent technology and probabilistic network can be adopted for a large complex system with randomness. Through the analysis of the system, the system functions are gradually refined according to the process from top to bottom and from shallow to deep and decomposed into a collection of several subsystems (the subsystem can continue to decompose). Probabilistic network models are established for the subsystems, and each model is managed by an intelligent agent to rationally arrange the subsystem objectives, functions, and resources [28]. Through the interaction and negotiation between agents, the joint and coordination between subsystems is achieved, thus forming the system's overall model and realizing the system's overall function. The decomposition and coordination process is shown in Fig. 3.6. This kind of decomposition-combined complex large-system modeling method can effectively reduce the complexity of large-scale system modeling and reasonably solve the problem of system decomposition and coordination.

The basic idea of large system decomposition and coordination based on agent technology and probabilistic network is: to comprehensively use the theoretical knowledge and practical experience, as well as appropriate system decomposition methods, layer-by-layer decomposing the large complex systems, establishing a probabilistic network model and management agent organization of the subsystem, then establishing contact and collaboration modes between the agents, that is, between subsystems; and finally optimizing the overall system through effective collaboration between the mutually cooperating agent organizations, as shown in Fig. 3.7.

This kind of large complex system based on a "decomposition-joining" strategy, as a general scientific methodology of engineering management, can effectively reduce the complexity of large systems and reasonably solve the coordination problem

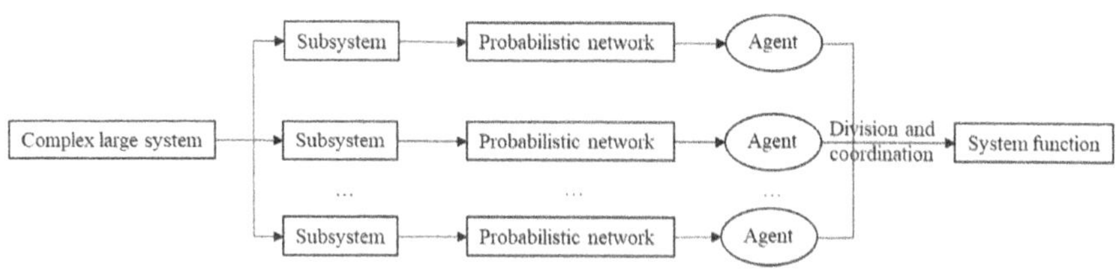

Fig. 3.6 The decomposition and coordination of large complex system

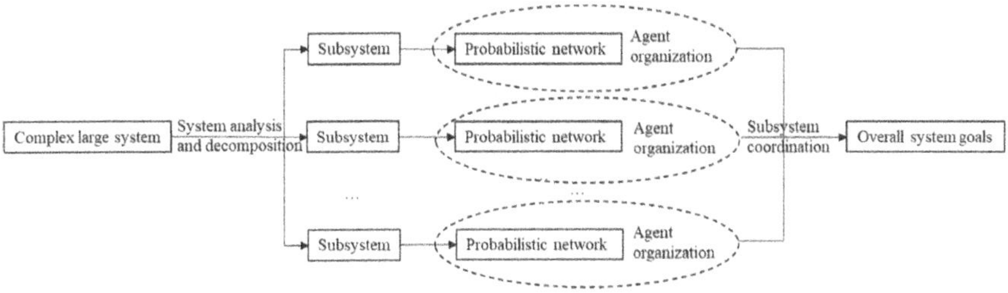

Fig. 3.7 System model based on agent technology and probabilistic network

between system decomposition and subsystems, which has important significance and value.

3.3 Engineering Management Specific Scientific Methodology

The specific scientific method of engineering management is a more operational method and means for specific problems and objects within a specific research scope. It is an important tool to promote the refinement of engineering research. It is also an important tool to promote the scientific management of engineering under the guidance of engineering management philosophy methodology and general scientific methodology. The specific scientific methodology of engineering management needs to systematically sort out and summarize the specific method sets applied in engineering management theory and practice and explore the relevance and optimization of each method. This section combines engineering management theory and practice to explain the specific scientific methods commonly used in engineering management and then focuses on project management methods.

3.3.1 Common Research Methods

Currently, research methods commonly used in the engineering management field include questionnaire surveys, case analyses, simulation, statistical analysis, investigation research, literature research, comparative analysis, data mining, experimental research, scenario analysis, etc.

3.3.1.1 Case Study Method

The case study is a research strategy whose focus is on understanding the dynamic processes in a single context. The scientific benefit of case studies is that it opens the way to discover problems. It can also provide insight into further research or lay the groundwork for hypotheses in terms of the unique meaning of its methodology.

The case study method provides an effective way to achieve the localization of engineering management theory and promote the innovation of engineering management theory in the Chinese context. The engineering management theory is derived from engineering management practice. Through the comparison, analysis, induction, and deduction of engineering management practice, the Western theory can be revised to a certain extent so that the engineering management theory with Chinese characteristics can be truly formed.

3.3.1.2 Simulation Method

At present, system dynamics and computational, experimental methods are commonly used simulation research methods in engineering management. In the 1950s, Professor Forrest first proposed a system dynamics method based on computer simulation technology. Subsequently, the method was widely applied to the quantitative research of social, economic systems. System dynamics is based on systems theory and fully uses the principles of cybernetics and information theory. Through the model, the system's internal structure is characterized by components such as flow, product, rate, and auxiliary variables to explore the inherent causality of system problems and help solve problems.

Computational experiments simulate the real system through a computer to summarize the behavioral characteristics and interrelationships of the microscopic subjects within the system and reveal the system's evolution law and the mechanism of action between the system and the environment. A scenario modeling method of computational laboratories comprehensively applies multiple disciplines and theories such as computer technology and complex systems theory.

At present, people's expectations for computer simulation technology are getting higher and higher. The simulation technology is gradually used to describe complex systems, even those composed of many different systems, and is applied to solve engineering problems. Most design, manufacturing, and property management in product development need to be described, inspected, and verified by computer simulation, including assembly process simulation, machining process simulation, and production planning and scheduling simulation. The solution to these problems is urgently needed to deepen the research on simulation technology.

3.3.1.3 Statistical Analysis Methods

The statistical analysis method refers to understanding and revealing the interrelationships, changing laws, and development trends of things to achieve a correct interpretation and prediction through the analysis and study of the quantitative relationship between the scale, speed, scope, and extent of the research object. Statistical analysis methods have received more and more attention and are widely used because of their scientificity, objectivity, and accuracy. It is simple in methodology and small workload but requires high data integrity and accuracy. Commonly used statistical analysis methods include structural equation modeling (SEM), AHP method, scientometrics, etc.

Structural equation modeling method. Structural Equation Modeling (SEM) is a combination of factor analysis and path analysis. It also tests the relationship between observed variables, latent variables, interference or error variables contained in the model and then obtains the direct effect, indirect effect, or total effect of the independent variables on the dependent variables.

Analytic Hierarchy Process (AHP). When using AHP to analyze a problem, first, the analyzed problem is layered. According to the nature of the problem and the overall goal, the problem is decomposed into different factors. These factors' interrelated influences and subordination are combined and assembled at different levels to form a multi-level analytical structure model. Finally, the problem is transformed into the ranking problem of comparison of advantages and disadvantages of the lowest level relative to the highest level (total target). With these rankings, we finally evaluate and make decisions about the analyzed problem.

Scientometrics is a branch of science that uses mathematical methods such as mathematical statistics and computational techniques to quantitatively analyze the investment in scientific activities such as scientific research personnel, research funding, output such as number of papers, number of citations, and processes such as information dissemination, communication network formation, and determining the regularity of scientific activities. Scientometrics attempts to find intrinsic or quasi-regularities of scientific activities through quantitative methods and provides guidance for more efficient research activities. Typical scientometric issues include the productivity issue of scientific research, the optimization of research funding, the prediction of discipline development trends, and the identification of funding priorities through scientometric methods and indicators. It can be a tentative exploration of the mainstream of engineering management to introducing methods such as scientometrics and information visualization technology into the academic research field of engineering management [29].

Word frequency analysis and co-citation analysis are two basic statistical analysis methods, as follows:

(1) Word frequency analysis mainly through the statistical analysis of the frequency of keywords appearing, finding research hotspots in research fields or disciplines, and observing the development or transfer trend of research hotspots brought about by the change of keywords [30].
(2) Co-citation analysis judges the degree of correlation between the two documents by simultaneously cited frequency statistics. The higher the frequency of co-citation, the more similar the subject background of the two, the closer the relationship between the two. The clustering group of key node literature can reflect the relevance theory and joining time points to reveal the evolution path of the theory.

Data plays an increasingly important role in today's economic development. How to obtain effective information from massive data has long been a research topic in various disciplines, and the role of statistical analysis methods has become increasingly prominent. Combining engineering issues, exploring how to effectively collect, arrange, and analyze data with randomness to make inferences or predictions about the engineering problems under investigation until the research for recommendations for specific decisions and actions have become an urgent need in the engineering management research field.

3.3.1.4 Investigative Research Method

The investigative research method refers to a research method that directly acquires relevant materials by analyzing the objective conditions and analyzing the materials. The investigative research method provides a detailed understanding of the survey subjects through questionnaires, interview surveys, sample surveys, field observations, etc., researches phenomena and problems. It is widely used in descriptive, explanatory, and exploratory research.

Questionnaire method. A questionnaire survey is a method of collecting information in a written (or electronically written) manner. The method assumes that the researcher has identified the questions to be asked. These questions are printed on the questionnaire, compiled into a written question form, submitted to the respondent, and then retrieved and analyzed to conclude. The key to applying this method is to prepare a questionnaire, select the subjects and analyze the results.

Interview method. It refers to the basic psychological research method of understanding the psychology and behavior of the interviewee through interviews with interviewees and interviewers face-to-face. The interview method can be developed in different forms according to the characteristics of the research object, the purpose of the interview, etc., and quickly collect a large amount of analytical data. Therefore, this method is widely used. Although the interview method is widely used and the collected data is relatively authentic and reliable, it also requires specialized interviewing techniques, taking time and effort, and high cost.

Mao Zedong once pointed out that it "should be based on actual things that exist objectively and draw rules from them as a guide for our actions [31]." To this end, it is necessary to have detailed information, scientifically analyzed, and comprehensively studied, as Marx said [32]. This is especially true for engineering management scientific research in the Chinese context. For example, for the particularity, diversity, and complexity of China's major infrastructure projects, researchers must grasp the first-hand information of the project through investigation and research methods and gain an in-depth understanding of the operating characteristics and laws of the engineering system to provide a reliable basis for scientific decision-making.

3.3.1.5 Literature Research Methods

The literature research method is a scientific method to search for, compare, analyze, judge, and sort the related documents for the research object to find out the essential attributes or internal laws of the research object and prove the research object.

Building your own research based on other people's research is a prominent feature of modern academic research. Through literature research, we can obtain the historical context and main path of the development of a certain research field, and at the same time, discover the problems that the academic community faces and needs to solve. The engineering management discipline has formed a vast subject knowledge system, and new theories, techniques, and methods are emerging. Through the organization of the existing literature in engineering management, the visualization research on the development law and the mainline of engineering management with the help of scientific measurement software has quietly emerged.

3.3.1.6 Comparative Analysis Method

Comparative analysis is the comparison of objective things in order to achieve the purpose of understanding the nature and laws of things and making correct evaluations. In scientific research, the comparison is one of the most commonly used research tools. It originated in the comparative study of Japan and Europe in the 1970s. Several significant results have been achieved in its application in management, such as the 7S model of McKinsey and Porter's diamond model. Comparative research methods are also indispensable in the field of engineering management research, such as in urban rail transit topics. In the National Natural Science Foundation of China project entitled "Research on Investment Control of BT Projects by the Government from the Perspective of Repurchase Contracts," through the comparative analysis of three BT project cases, Du Yaling concludes that BT mode can be divided into weak BT, strong BT, and standard BT, according to the control right. It provides a valuable reference for the rational allocation of project control rights under BT mode.

3.3.1.7 Data Mining Method

Data mining is to search for hidden information in a large amount of data through an algorithm. It is a computing method based on Internet technology. In this way, shared hardware and software resources and information can be provided to computers and other devices on demand.

As we all know, sufficient and valuable information is a prerequisite for scientific decision-making. Engineering decision-making is a comprehensive, integrated system work involving many aspects such as socioeconomic factors and natural factors. The support of a large amount of data is the premise and basis of scientific engineering decision-making. Based on "big data," the use of "cloud computing" and other advanced information processing tools and computer-aided decision-making means to study decision-making management is a scientific and forward-looking research field.

3.3.1.8 Experimental Research Method

The experimental research method is that the researcher designs the corresponding experiment according to the nature of the research problem, simplifies the real environment, controls certain variables, repeats the experimental phenomenon, and observes the results to find out the law.

The experimental research method was first applied in the natural sciences and gradually developed into the main research method of the natural sciences. From the beginning of the Renaissance, natural science established the connection between theory and empirical facts; it promoted the rapid development of natural science because of the adoption of experimental methods. In recent decades, researchers in various disciplines have increasingly recognized the importance of experimental methods for developing disciplines. They have begun to work on applying experimental methods to their respective disciplines. The same is true in the field of engineering management. The experimental methods of psychological behavior that emphasize the role of human behavior in solving various scientific problems are gradually applied to engineering decision-making, engineering management implementation and control, and social issues involved in engineering management.

3.3.1.9 Scenario Analysis Method

Scenario analysis, also known as scripting or foreground description, is a method of predicting what might happen or the consequences of predicting an object, assuming that a certain phenomenon or a certain trend will continue. It is often used to make assumptions or predictions about the future development of predictive objects. It is an intuitive qualitative forecasting method.

Zhu Yuezhong proposed that before the scenario analysis, people need to analyze the past history retrospectively and then make a series of reasonable assumptions about future trends or establish certain goals that they hope to achieve in the future; that is, to conceive the blueprint for the future or the development prospects, and then analyze the various feasible ways to achieve this goal and the measures that need to be taken [33].

In traffic engineering, Zong Beihua applied the scenario analysis technology to the enterprise strategy in China's transportation industry in the Ministry of Communications' project entitled "Study on the application of scenario analysis in port and shipping enterprises." Zhong then applied the technology to "Hainan's 30-year development strategy related research", which provides a scientific decision-making basis for Hainan's 2020 transportation strategy [34].

3.3.2 *Project Management Method*

The project management method originated from the engineering practice of human creation activities is a unique method of project management formed in engineering practice. And it is also a method derived from other areas but is often and effectively used in project management. Engineering is the most common and typical type of project, so project management methods are widely used in engineering management.

Project management refers to project-oriented, system management methods through a temporary and flexible organization, efficient planning, organization, guidance, and control of the project to achieve dynamic management and comprehensive coordination of objectives throughout the project and optimized management activities [35]. Modern project management begins with large-scale national defense engineering projects, from the traditional emphasis on planning to focus on change. The standard of success of the project is to meet or exceed the stakeholders' expectations and thus adopt a management method system that integrates the idea of centralized change, system thinking, and balance [36].

3.3.2.1 Project Argumentation and Decision Method

Argumentation and decision-making are the first phase of the project life cycle, including project opportunity research, program planning, feasibility study, assessment, and decision making. Project argumentation is an important part of the project start-up phase. Through the argumentation of the necessity, feasibility, and rationality of the project, it is concluded whether the project is feasible. Project decision-making is the scientific process of selecting the optimal plan based on the feasibility study and project evaluation to achieve the project's expected objectives. Commonly used methods for project argumentation and decision making include factor stratification method, scheme comparison method, SWOT analysis method, project financial evaluation, project national economic evaluation, project environmental assessment. Among them, financial evaluation, national economic evaluation, and environmental assessment are discussed in detail in the relevant chapters of this book.

1. **Factor stratification method**

The factor stratification method is commonly used in engineering project opportunity research. It stratifies many factors involved in engineering projects according to project opportunities, project problems, and project implementers' advantages and disadvantages. Through factor stratification analysis and subjective scoring, opportunities and issues identified, and strengths and weaknesses, judgment and decisions can be made.

The factor stratification method is a combination of qualitative and quantitative methods. First, the factors affecting the project are layered by class through qualitative analysis; then, the quantitative method is used to score subjectively according to the factor's influence. After stratified evaluation, the favorable and unfavorable factors affecting the project will be displayed visually to facilitate engineering decision-making.

2. **Scheme comparison method**

The scheme comparison method is a method of selecting the best scheme by comparison. It is a general optimization of the schemes by using the indicators of multi-program evaluation and the comprehensive evaluation method. The scheme comparison method can compare and analyze many proposals put forward in the project opportunity research and feasibility study and select the advanced and feasible technology and economic and social benefits plan, which is the basis of detailed argumentation.

The scheme comparison method should abide by the principle of comparability, that is, the engineering objectives are comparable (satisfying demand), the consumption cost is comparable, the price is comparable (the price system must be consistent), and the time is comparable.

3. **Evaluation index system**

The evaluation index system refers to the method of evaluating the project in many aspects through multiple indicators. The basic idea is to select multiple indicators for the relevant objectives of the project and conduct a comprehensive evaluation according to the different weights of each indicator. The content includes:

(1) Determining the evaluation index system, that is, determine each indicator item and the weight of the indicator items;
(2) Collecting data and performing the same measurement processing on the indicator items of different measurements to determine the standard value;
(3) According to the criteria specified above, combined with the weights, the processed indicators are aggregated and calculated;
(4) According to the change of the evaluation index, summarize the law of change and conclude.

4. **Project portfolio optimization method**

Project portfolio optimization refers to the investment selection and resource optimization and configuration of multiple projects or project groups under available resources and corporate strategies to achieve organizational strategic goals. Project portfolio optimization emphasizes the organization's strategic goals, portfolio investment returns, and multi-project optimization and resource allocation under resource constraints. Project portfolio optimization must follow the following principles [36, 37]: (1) Strategic: corporate strategic goals guide project portfolio optimization; (2) Dynamic: project portfolio optimization focuses on dynamic changes of every project and between projects, a timely configuration of project resources, and resource conflict resolution is obtained; (3) Emphasize the integrity of the organization, systematically select each project, evaluate the status of each project in the portfolio and its compliance with organizational goals; (4) Emphasize the importance of project selection. The key to portfolio management is to choose the right project.

Through project portfolio optimization, ensure that the project portfolio and resource allocation align with the company's strategic goals, thereby maximizing corporate profits. The portfolio optimization process is a dynamic continuous execution and cyclical process. As the environment changes, the portfolio and optimization will change. Therefore, there is a need to track and optimize portfolios dynamically.

3.3.2.2 Project Planning Method

A project plan is a set of task sequences and relationships defined to achieve a specific project goal and is a formal document that guides project execution and control. The project plan generally includes overall project introduction, organization description, management procedures and technical procedures, schedule, budget, quality standards and control methods, risk planning strategy, project stakeholders analysis, multi-objective optimization, etc. Project planning is the premise of project implementation. Throughout the project activities, it is the guarantee that the project objectives can be effectively realized. Commonly used project planning methods include work breakdown structure, network planning technology, resource cost curve, responsibility matrix, etc.

1. **Work breakdown structure (WBS)**

The work breakdown structure is the main method for project scope planning. By creating WBS, project deliverables are decomposed into smaller and easy-to-manage units to achieve project-wide quantitative management. WBS is a hierarchical decomposition structure for deliverables. It decomposes project tasks according to its own rules in a systematic, interrelated, and coordinated manner, resulting in a tree-like hierarchical structure. The more the structure hierarchy goes down, the more detailed the definition of the project components, and the bottom layer of the project, leaves, is called the project work unit. WBS is also a functional document that provides the basis for project quality, schedule, cost, and risk management. The project team follows the WBS implementing project. The project stakeholders verify the completion of the project through measurement, inspection, and testing methods to decide whether to accept the project.

The Qinghai-Tibet Railway Golmud-Lhasa section started construction in 2001. During the construction process, the electric load increases, and the power load along the railway area is also growing faster. Therefore, the Anduo-Naqu-Dangxiong-Liuwu 110 kV substation is planned to construct [38]. In order to clarify the scope of the project and refine the management, the project team worked on the work breakdown structure of the Qinghai-Tibet Railway power transmission project, as shown in Fig. 3.8.

2. **Network planning technology**

Network planning technology is the management technique for project planning and control. Its core is to express the working relationship and project time in a network and reveal the key links in project progress control. The basic network planning

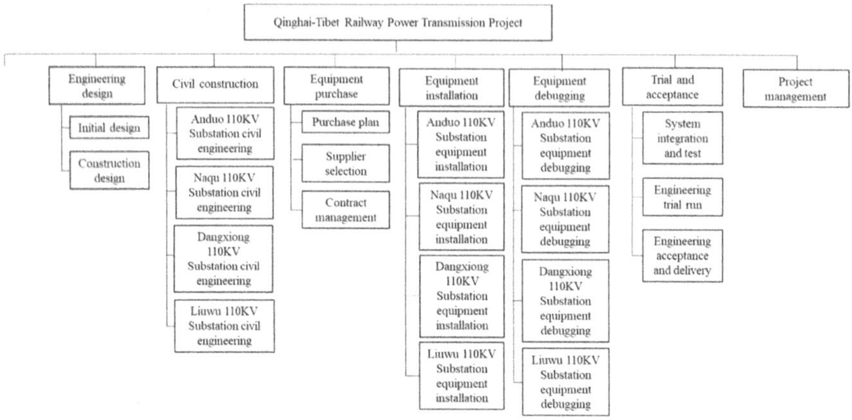

Fig. 3.8 Decomposition diagram of the working structure of the Qinghai-Tibet railway power transmission project

methods include Program Evaluation and Review Technique (PERT) and Critical Path Method (CPM).

(1) PERT. PERT is more focused on evaluating and reviewing work arrangements and is more suitable for research and development projects with uncertain working hours. (2) CPM. CPM finds the key route in the work plan and controls it by analyzing the relationship between the cost and the construction period and analyzing the earliest start time, the earliest end time, the latest start time, and the latest end time of each work.

The Three Gorges Project is huge in scale and complex in technology, and it is necessary to prepare a network plan [40]. Using the network planning technology to compile the Three Gorges Project schedule plan can clearly express the logical relationship between the work, effectively control the project duration, and facilitate the schedule's optimization. In the planning and design stage of the Three Gorges Project, the progress network plan review of different plans was carried out, and the probability of completion on schedule was analyzed to lay the foundation for the successful completion of the project.

3. **Resource cost curve**

The resource cost curve is a common method and tool for resource planning and control. It is a curve based on the relationship between the total amount of project resources or human resource expenditures and the project progress time. The resource curve takes time as the abscissa and the accumulated resource cost as the ordinate to form an S-shaped curve, as shown in Fig. 3.9. If the planned and actual resource cost cumulative input is reflected in the same graph, it is called the double S curve.

The resource cost curve mainly shows the project resource cost plan and consumption through a two-dimensional plane curve. The drawing method is simple and easy to interpret, so it is widely used. The resource cost curve can be used to succinctly reflect the plan and actual situation of the project throughout its life. At the same

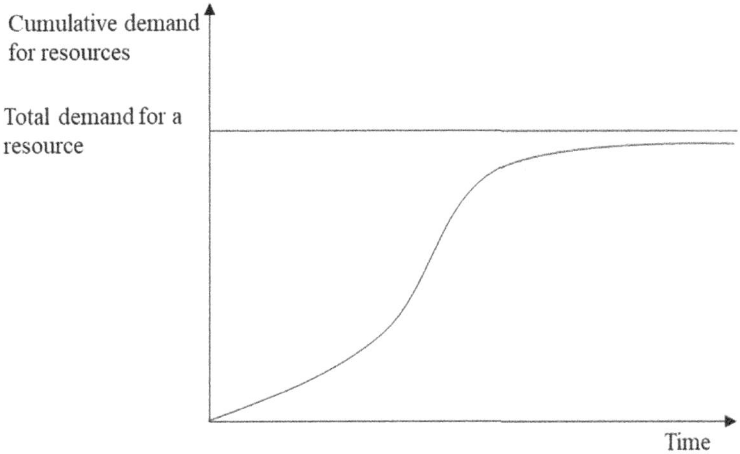

Fig. 3.9 Resource cost curve

time, during the implementation of the plan, through the actual situation reflected by the resource cost curve, the actual situation is compared with the plan, and it can be concluded whether the project is over-expenditure to obtain the deviation between the actual progress of the project and the plan, providing relevant evidence for the project cost and progress control.

4. **Responsibility matrix**

The responsibility matrix is a matrix structure that establishes the relationship between the work breakdown structure and the project organization structure to clarify the roles and responsibilities of the project team members. The "row" of the responsibility matrix represents the work element, the "column" represents the organizational unit, and the symbols in the matrix represent the roles and responsibilities of the project members in each work unit. The responsibility matrix has the characteristics of simple production and straightforward interpretation. The responsibility matrix can reflect the work responsibilities and mutual relations between the various work departments or project members.

Table 3.2 is the functional responsibility matrix of the China Railway Engineering Corporation Malaysia Railway Project. The responsibility matrix can clarify the responsibility relationship of the project's overall work, and the various departments and members of the project team can understand the organization's work goals and individuals, thus achieving employee self-management. At the same time, the responsibility matrix can also promote consensus among team members on various work responsibilities and achieve seamless integration.

Table 3.2 The functional responsibility matrix of the China railway engineering corporation Malaysia railway project

Management task	Management level						
	Commerce department				Finance department		
	Project manager	Business manager	Budgetor	Business engineer	Finance minister	Project accountant	Project cashier
Preparation of project qualification management and promotion plan	C	I					
Project related registration and annual inspection	C	I					
Professional certification and promotion		A					
Project public relations promotion plan	C	A D			A D		
Establish and update contact guides		A D			A D		
Etiquette letter, piece, greeting card preparation	C	R			R		
Etiquette gift preparation	C	R			R		

(continued)

Table 3.2 (continued)

Management task	Management level						
	Commerce department				Finance department		
	Project manager	Business manager	Budgetor	Business engineer	Finance minister	Project accountant	Project cashier
Multimedia promotional materials		A			A		
Charity event	C						
Social group activities	C						
Market promotion	C	A			A		
Public entertainment	C	A			A		

(continued)

Table 3.2 (continued)

Management task	Management level											
	Engineering department									General department		
	Construction manager	Planning engineer	Design e engineer	Equipment engineer	Bridge culvert engineer	Track engineer	Civil engineer	System engineer	Safety engineer	General minister	Secretary	Per sonnel
Preparation of project qualification management and promotion plan	A											
Project related registration and annual inspection	A											
Professional certification and promotion	I											A
Project public relations promotion plan	A D									I E		
Establish and update contact guides	A D									I E		

(continued)

Table 3.2 (continued)

Management task	Management level											
	Engineering department									General department		
	Construction manager	Planning engineer	Design e engineer	Equipment engineer	Bridge culvert engineer	Track engineer	Civil engineer	System engineer	Safety engineer	General minister	Secretary	Per sonnel
Etiquette letter, piece, greeting card preparation	R										H	
Etiquette gift preparation	R										H	
Multimedia promotional materials	A									I E		
Charity event										A H		
Social group activities										A H		
Market promotion	A									A H		
Public entertainment	A									A H	H	

Note A-assist; C-control; P-propose; E-exchange; D-provide data; R-review; I-initiate; H-handle

3.3.2.3 Project Implementation Control Method

After the project planning work is completed, it enters the organization implementation stage. Using organized management to arrange various human and material resources and then carry out organized activities to achieve the set goals and plans. The main task at this stage is to generate business and management work for project outputs. The project implementation organization includes two processes of project planning execution and control. During the execution of the project plan, the project team must manage various technical and organizational interfaces of the project, that is, coordinate various relationships inside and outside the project. After the project plan is implemented, the project manager and team need to control the project, find the deviation between the project implementation and the plan, and take corrective measures. Project implementation control methods include stakeholder management, production factor management, quality control methods, earned value analysis, and comprehensive change control.

1. **Stakeholder management**

A project often involves the interests of many organizations and individuals. In particular, the benefits involved in engineering projects are extensive. These direct and indirect individuals and organizations are project stakeholders. Because of the target disagreement, project stakeholders may influence the project, its deliverables, and project team members, thus affecting the normal implementation of the project. The project management team should clarify the scope of internal and external stakeholders involved in the project, understand the interests of all stakeholders, and manage the influence of stakeholders on the project.

The enlightenment of stakeholder theory on project management is reflected in four aspects: First, when formulating the project implementation plan, it is necessary to analyze and compare technical and economic factors and examine the potential impact of stakeholders. The second is to evaluate positive and negative views when choosing a program and take appropriate measures to reduce the obstacles to the project. Third, the attitudes and influences of stakeholders are dynamic. At different stages of project implementation, stakeholders' analysis and evaluation are dynamic and continuous. Fourth, before the project changes occur or major decisions are implemented, it is necessary to analyze the impact of the decision on all stakeholders and their possible responses and develop a coping strategy.

Manage the project stakeholders; on the one hand, it is easy to get support from the stakeholders and win more resources for the project. On the other hand, through communication and coordination with project stakeholders, stakeholders can be informed of the project, thereby reducing the adverse impact of project stakeholders on project implementation. Therefore, it is necessary to manage project stakeholders.

Cleland [41] proposed a project stakeholders management phase model [41], which is shown in Fig. 3.10.

According to Cleland's model, the first is to identify the stakeholders, then collect their information, identify the tasks or roles undertaken in the project, analyze their strengths and weaknesses, predict the strategies and behaviors they intend to take, and

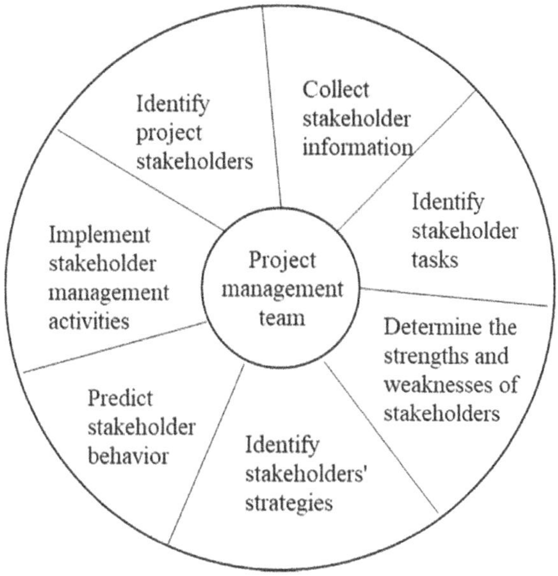

Fig. 3.10 Cleland project stakeholder management model

determine the stakeholder management strategy. Stakeholder management strategies include maximizing stakeholder support and minimizing project disruption, taking action to turn neutrals into supporters, and opponents becoming supporters.

2. **Production factor management**

The factor of production refers to the various elements of productivity that affect the project. These elements are simply referred to as "5 M", that is, manpower, machine, material, money, and management. Production factor management refers to the management of the allocation and use of elements, the purpose of which is to save living and materialized labor. By managing the production factors of the project, it is possible to optimize the allocation of production factors, optimize the combination, and realize the dynamic management of production factors and the rational and economical use of resources during the operation of the project.

The main contents of production factor management include:

(1) Optimized configuration of production factors. The optimized configuration of production factors refers to the rational allocation of various elements, which meets the production needs of the project and maximizes the value of the essential resources.
(2) Optimized combination of production factors. The optimal combination of production factors has two aspects: one is the optimization of the elements themselves, that is, the process of improving the quality of various elements. The second is to optimize the combination based on the combination, seek the joint force of the overall structure, and effectively form productivity.

(3) Dynamic management of production factors. The dynamic management of production factors refers to the project construction organization method generated according to the dynamic process of the project itself. Since the project implementation process is a dynamic process that is constantly changing, the basic content of dynamic management is to effectively plan, organize, coordinate, and control the various factors according to the inherent laws of the project, so that it can flow reasonably in the project and realize dynamic configuration and overall optimization [42]. Commonly used dynamic management includes dynamic balance method, daily scheduling, accounting, production factor management evaluation, on-site management and supervision, ABC classification, storage theory, and value engineering.

3. **Quality control method**

Quality control is the process of supervising the project's implementation status and implementation results, comparing the actual situation of the project implementation with the quality planning standards, identifying the deviations, and analyzing the causes. Quality control should also identify ways to avoid quality problems and identify solutions to improve quality. Quality control runs through the entire process of project implementation. To put it simply, quality control is an activity that continuously checks, measures, and adjusts the operations and activities of the project implementation process to meet the organization's quality objectives. The content of quality control is to ensure the quality of the project and to carry out quality planning, process control, inspection, acceptance control, abnormal factor analysis and elimination on the production factors, processes, plans, acceptance, decision-making, and other factors of the project. The main methods of project quality control include the histogram method, control chart method, arrangement chart method, flow chart method, causal relationship diagram, and trend analysis.

4. **Earned value analysis**

Earned value analysis analyzes the deviation between the actual execution and the plan of the project objectives (such as schedule and cost), so it is also called the deviation analysis method. The earned value analysis method obtains the deviation between the progress and the expense by measuring and comparing the budgeted cost of the completed work, the actual cost of the completed work, and the budgeted cost of the planned work, thereby judging the implementation of the project budget and progress. It is called the earned value method because it introduces a key parameter—earned value, which is the completed work budget.

The earned value analysis method is comprehensive for summarizing work (task), cost and schedule planning, and execution deviation. Under the premise of actually completing the same task, compare the difference between the budgeted cost and the actual cost, and get the cost difference. Under the premise of spending the same cost, the tasks completed by the plan are compared with the tasks actually completed and obtain the difference in progress.

The earned value analysis method involves three main parameters, namely the budgeted cost of the work scheduled (BCWS), the actual cost of work performed (ACWP), and the budgeted cost of work performed (BCWP).

The main evaluation indicators of the earned value analysis method include two deviation indicators (CV and SV) and two performance indicators (CPI and SPI):

(1) Cost deviation (CV) = BCWP − ACWP. When CV < 0 indicates that the project implementation is in an over-expenditure state; when CV > 0, the project is in a state of savings; when CV = 0, the actual cost is consistent with the plan.
(2) Progress deviation (SV) = BCWP − BCWS. When SV < 0 indicates that the project implementation progress is delayed; when SV > 0, the project progress is ahead; when SV = 0, the actual progress is consistent with the plan.
(3) Cost Performance Index (CPI) = BCWP/ACWP. When CPI < 1 indicates that the project implementation is in an over-expenditure state; when CPI > 1, the project is in a state of savings; when CPI = 1, the actual cost is consistent with the plan.
(4) Progress Performance Index (SPI) = BCWP/BCWS. When SPI < 1 indicates that the project implementation progress is delayed; when SPI > 1, the project progress is ahead; when SPI = 1, the actual progress is consistent with the plan.

5. **Comprehensive change control**

The one-time and unique nature of the project makes the project face many uncertain factors in the planning stage. These uncertainties will inevitably lead to inconsistencies in the project's scope, progress, cost, and quality in the actual implementation process, and various changes occur, and such changes are often interrelated. For example, changes in scope will inevitably lead to changes in costs or schedules. Therefore, it is necessary to manage project changes comprehensively.

Change is absolute and normal, while stabilization is relative and abnormal. For project managers, many uncertainties are unpredictable, and it is difficult to predict and prevent all changes. Therefore, the purpose of the project's comprehensive change control is not to control the occurrence of changes but to manage the changes to ensure that the changes are carried out in an orderly manner.

The project integrated change control system realizes project comprehensive change control. The integrated change control system defines the procedures, methods, and management practices for integrated change control. It includes the process of change management, responsibility division, authorization level required for authorization and authorization change, document management of change, tracking supervision of change, etc. The content of the comprehensive change control includes the effect on the cause of the change to ensure all the parties agree to the change. And confirm that changes have occurred and are managed at the same time as actual changes occur. In the process of comprehensive change control of the project, it is necessary to maintain the soundness of the performance metrics, ensure that product range changes are reflected in the project scope definition, and coordinate cross-domain changes. The process of the project's comprehensive change control system is shown in Fig. 3.11.

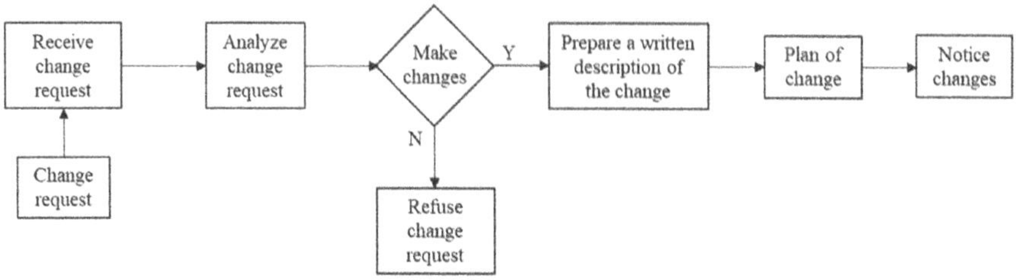

Fig. 3.11 Project change control system

The comprehensive change control will result in updating the content of the project plan and related supporting materials, correct some of the factors that interfere with the project objectives in the process of comprehensive change control to ensure the effectiveness of project management and identify deviations, learn lessons, and document them into components of the project history database within the organization.

3.3.3 *Project Closing and Evaluation Method*

Closing is the final stage of the whole process of project implementation. When the project is completed according to the requirements of the target or if the project is terminated for some reason, the project needs to be closed. In the final stage of the project, it is necessary to summarize and evaluate the final results of the project completion, implement the data compilation and archiving, end the project implementation activities and processes, and complete the project management. The closing of the project mainly includes two parts: contract closing and management closing. The closing of the contract is to check the contract, check whether all the requirements of the contract have been completed, and whether the contract is effectively executed. Management closing includes archiving various project documents, asset transformation, resource disband, and a summary of lessons learned from the project. Common methods for project closure include the checklist method and expert scoring method. Post-project evaluation methods mainly include the logical framework method and comparative analysis method.

1. **Checklist method**

The checklist method is a unique structured method. The checklist method is the inspection work at the end of the project, the plan is compared with the actual situation, and the check of completion of the project that is being inspected. The closing checklist varies from project to project, mainly reflecting the progress of the project plan, the definition of the remaining work, the self-inspection by the project team and check by other parties, self-evaluation, and the ending meeting minutes.

2. **Expert scoring method**

When making project termination decisions, the project investors, owners and builders generally invite relevant experts to judge the key factors affecting project implementation, whether changes have been occurred including the resolution of technical difficulties, changes in positioning markets, policy adjustments, and changes in material prices.

A project termination decision will be made if a key factor changes, causing the project to not continue or continue but fails to achieve the project objectives. If there are no major changes to these key factors, these factors or impact indicators (such as progress, cost and quality indicators) are further determined and decided.

3. **Logical framework method**

The logical framework method is a method for project design, planning, and evaluation. The logical framework method combines several dynamic factors related to the project and must be considered simultaneously and evaluates an engineering project by analyzing the relationship between them and their objectives and actual results. The core is to reveal the logical causal relationship between the levels of things, that is, if "a certain condition is provided", "then" it will produce some kind of result; these conditions include the factors inherent in the things and the external conditions required.

The logical framework method reflects the changes in the project development process by constructing unified evaluation indicators, such as project benefits, costs, and progress indicators for pre-assessment, implementation assessment, and post-assessment of the project, respectively.

4. **Comparative analysis method**

Comparative analysis is a method of comparing the results of the same indicator at different times or in different scenarios to find the difference. Comparative analysis can be divided into "before and after comparison" and "with or without comparison." "Before and after comparison" is to compare the situation before and after the implementation of the project to determine its role and benefits. "With or without comparison" is to compare the actual existence of the project and the state that may occur when there is no such project, to measure the true benefits and effects of the project. The comparative analysis method can also be divided into two types: absolute number comparison and relative number comparison according to analysis needs.

References

1. He, J., Xu, C., Wang, Q., et al. (2013). Engineering management methodology. *Chinese Engineering Science, 16*(10), 4–9.
2. Yin, R., Wang, Y., & Li, B. (2013). *Engineering philosophy* (2nd ed.). Higher Education Press.
3. He, J., & Wang, M. (2008). Philosophical thinking on engineering and engineering management. *China Engineering Science, 10*(3), 9–1.

4. Mao, Z. (1991). *Selected works of Mao Zedong* (Vol. 3, p. 801).
5. Mao, Z. (1991). *Selected works of Mao Zedong* (Vol. 1, p. 306). People's Publishing House.
6. *Selected works of Marx and Engels* (Vol. 1, p. 61, 1995). People's Publishing House.
7. *Selected works of Mao Zedong* (Vol. 3, pp. 899–900, 1991). People's Publishing House.
8. *Selected works of Marx and Engels* (Vol. 2, p. 43, 1995), People's Publishing House.
9. Hu, S., & Zhang, Q. (2004). China's manned spaceflight project—A model for the successful practice of systems engineering. *China Aerospace, 10*, 3–6.
10. Glansdorff, P., & Prigogine, I. (1971). *Thermodynamic theory of structure, stability and fluctuations*. Wiley-Interscience.
11. Haken, H. (1983). Synergetics; An introduction. Springer.
12. Zeeman, E. C. (1976). Catastrophe theory. *Scientific American, 234*(4), 65–83.
13. Qian, X. On system engineering. Hunan Science and Technology Press.
14. Wang, R. (2009). Review of the three Gorges project demonstration. *Journal of China Three Gorges University (Natural Science Edition), 36*(6), 1–6.
15. Cheng, H., & Han, Y. (2012). Engineering management systems thinking and engineering life cycle management. *Journal of Southeast University (Philosophy and Social Sciences Edition), 14*(2), 36–40.
16. Li, Z. (2002). Planning, construction, and management of south-to-north water transfer project with system engineering thinking. *Hebei Water Resources, 4*, 16–17.
17. Von Foerster, H. (1984). *Principles of self-organization—In a socio-managerial context*. Springer.
18. Hall, A. D. (1962). *A methodology for systems engineering*. Princeton.
19. Liu, J., Zhang, W., & Zhu, J. (2011). *Systems engineering*. Tsinghua University Press.
20. Wu, M. (2010). Project management innovation and practice based on system engineering. *Engineering Construction, 42*(4), 54–56.
21. Checkland, P., Scholes, J. Soft systems methodology in action. Wiley.
22. Yu, B. (2009). *Systems engineering theory*. University of Science and Technology of China Press.
23. Yu, J., & Zhou, X. (2005). From integrating thoughts to integrating practice—Methods, theory, technology, engineering. *Journal of Management, 2*(1), 4–10.
24. Wang, Y. (2011). *Systems engineering* (4th ed.). Mechanical Industry Press.
25. Gu, J. (2000). Wuli-Shili-Renli (WSR) methodology. In *Systems science and engineering research*. Shanghai Science and Technology Education Press.
26. Chen, H. (2006). *Introduction to systems engineering*. Higher Education Press.
27. Jamshidi, M. (1986). *Large system: Modeling and control*. Science Press.
28. Yang, S., & Hu, X. (2007). *Modeling and solving methods for complex decision tasks*. Science Press.
29. Wang, M., et al. (2013). A visualization analysis of engineering management theory research. *Scientific and Technological Progress and Countermeasures, 30*(23), 1–5.
30. Chen, Y., Wang, X., & Zheng, G. (2007). Analysis of management theory frontiers based on knowledge mapping. *Science Research, 25*(sl), 22–28.
31. Mao, Z. (1991). *Selected works of Mao Zedong* (Vol. 3, 2nd ed., pp. 799, 801). People's Publishing House.
32. Mao, Z. (1991). *Selected works of Mao Zedong* (Vol. 3, pp. 795–803). People's Publishing House.
33. Zhu, Y. (2000). The enlightenment of the medium and long-term energy development scenario analysis method to China's future energy conservation planning. *China Energy, 5*, 5–6.
34. Zong, B. (1992). Application of scenario analysis in port development strategy. *Journal of Shanghai Maritime University, 4*, 28–35.
35. Wang, C., Li, J., Ji, J., et al. (2008). *Introduction to modern project management*. Mechanical Industry Press.
36. China (Shuangfa) Project Management Research Committee, China Project Management Knowledge System (Revised Edition). Publishing House of Electronics Industry.

37. Jiang, C., & Ma, K. (2008). Research on project strategy compliance evaluation based on evidence theory. *Journal of Hefei University of Technology (Natural Science Edition), 31*(10), 1660–1663.
38. Project management case and WBS template—A case study of Qinghai-Tibet railway power transmission project. http://www.docin.com/p-49522759.html.2010.4.13
39. Cai, Z. (2012). Large-scale international project management—Marketing and implementation of large-scale international engineering projects. http://wenku.baidu.com/view/af6126d726fH705cc170a4c.html.2012.6.2
40. Wu, C. (1990). Research on the network planning of the construction progress of the three Gorges project. *Water Resources and Hydropower Technology, 6*, 28–30.
41. Cleland, D. I. (1998). *Project management handbook*. Jossey-Bass New York.
42. Liu, W. (2010). Construction management of construction enterprises. *China New Technology and New Products, 23*, 179.

Chapter 4
Engineering Management Decision Theory

4.1 Overview

Decision-making refers to making a decision on whether to conduct an action, which is the core of modern management and runs through the whole process of management activities. Simon, a famous American management scholar and Nobel laureate in economics, even said: "Management is decision-making." With the comprehensive application of systems theory, operations research, computer science, behavioral science, etc., modern decision-making has formed a relatively complete theoretical system, including the type, criteria, processes, and methods of decision making. The engineering decision was made in the Song Dynasty. In the Political Policy of the Dynasty-the Yellow River, Zeng Gong said: "since it is difficult to understand water's trace in detail unless you investigate thoroughly, it is hard to make decisions without being fully informed." It says that for the governance of the Yellow River, if you do not conduct in-depth investigation and study and understand the law of water flow, it is difficult to make correct decisions and govern.

In fact, the ancient Chinese government had strict regulations on engineering design and construction. In the Tang Dynasty, the standard working hour quota (i.e., middle term work) was stipulated, and the seasonal working hour quota was increased or decreased by 10% (i.e., long-term work and short-term work). This continued until the Qing Dynasty, and the government issued a decree of "Ying Shan Ling," which stipulates the scale and form of various types of buildings [1]. With the rapid development of modern engineering, a large number of large-scale projects like the Three Gorges Project, the Qinghai-Tibet Railway, and other large-scale projects have emerged with complex decision-making processes involving a wide variety of technologies, large organizational structures, long duration, and numerous participants. In the decision-making process of these projects, it is necessary to collect a large amount of information, including technology, personnel, organization, environment, and ecology, etc. And using various empirical and rational decision-making methods to make a reasonable evaluation and judgment of the decision-making plan with certain evaluation criteria or selection mechanisms.

J. He, *Principles of Engineering Management*,
https://doi.org/10.1007/978-981-99-1168-4_4

As described in Chap. 1, engineering management's function is to make decisions, plan, organize, direct, coordinate, and control the project to achieve the expected objectives [2]. Generally speaking, engineering management is characterized by systems, comprehensiveness, and complexity [3]. Engineering decision-making is the core part of engineering management. The basic decision-making process and criteria are consistent under similar situations regardless of the object and stage of engineering decision-making. However, the process and standards of engineering decision-making will change accordingly over time, following changes in the social environment and social patterns. This change reflects the development and evolution of engineering decision-making. It is the sublimation process of engineering management along with the "people-oriented, harmony between man and nature, collaborative innovation and harmonious construction." Major project decision-making involves many complex engineering factors such as site selection, technology application, construction safety, ecological protection, etc. It must go through a long period of careful and meticulous forensic and demonstration work [4]. This chapter discusses the essence and regularity of engineering decision-making. It systematically elaborates the connotation and characteristics of engineering decision-making, the objectives and tasks of engineering decision-making, the procedures, the modes and methods of engineering decision-making, and the engineering decision-making system with practical cases. It focuses on the connotation, tasks, and methods of engineering decision-making. Each level shows different aspects of engineering decision-making. The connotation of engineering decision-making is abstraction and sublimation to the philosophical level of "harmony between nature and human." The task of engineering decision-making is to achieve the philosophical level. The method of engineering decision-making is the means of achieving the task of engineering decision-making.

4.2 The Connotation and Characteristics of Engineering Decision-Making

4.2.1 The Connotation of Engineering Decision

4.2.1.1 Decision-Making and Decision-Making Theory

Decision-making is similar to but different from judgment and choice made in people's life; it is more rational, scientific, and complex. Selecting a technology through scientific experiments and simulation is a decision-making process. It is also a decision-making process by collecting information, repeatedly investigating, making feasibility studies, and finally deciding to start the project. The former is a micro process, and the latter is a macro process. Decision-making can be defined as the process of analyzing, comparing, and selecting alternatives by adopting scientific means and methods and collecting relevant information guided by specific objectives.

After the Second World War, the theoretical system of the decision-making process, criteria, types, and methods began to form and establish. Herbert A. Simon (1916–2001), the Nobel Laureate of economics, put forward the continuous finite comparative decision-making theory. He carefully analyzed and studied the impact of human behavior on decision-making results and proposed that the concept of bounded rationality has an important impact on decision-making. His view that "management is decision-making" played a decisive role in promoting the status of decision-making theory and the development of decision-making theory. Another representative is James G. March (1916–), who is recognized as one of the most contributing scholars in organizational decision-making. He has made outstanding achievements in the fields of organization, decision-making, and leadership. He worked with Herbert Simon to develop and perfect the school of decision-making theory. His main works are on *How Decision-Making Comes into Being* and *Organization*, which he co-authored with Simon, and *Theory of Corporate Behavior*, which he co-authored with Celtic. In addition, other theories such as complete rational decision-making theory, irrational decision-making theory, rational organizational decision-making theory, and realistic progressive decision-making theory have also discussed and studied the criteria, processes, and methods of decision-making from different perspectives. Modern decision-making theories and methods mainly include modeling and analyzing uncertain decision-making problems, decision-making theories and practices under the network environment, group decision-making theories and methods, multi-objective decision-making theories and methods, network data fusion, and online decision-making theory, etc. The research results formed have been widely used in social and economic fields. With the rapid change of the social and economic environment and the emergence of big data, decision-making is becoming increasingly complex and uncertain. Service-oriented decision-making in infrastructure construction and environmental protection has become increasingly prominent. At the same time, the complexity of decision-making issues has increased, which has led to a significant increase in the complexity and scale of decision-making groups. Decision-making gradually relies on data, especially big data analysis. Decision-making based on big data analysis is a deep decision-making problem and a basic problem of decision-making science in a large data environment.

The main points of decision-making theory are as follows: taking decision-making as guidance, highlighting the position of decision-making in management, and adjusting the role of decision-makers. The formation, development, and application of decision-making theory indicate that decision-making has been established as an independent science and plays a vital role in guiding decision-making problems in various fields. With the development of decision-making theory, decision-making science is becoming more and more rich, refined, and integrated. The deep integration of decision-making science and other scientific fields will restructure and reshape the theory, principles, and decision-making methods, forming a unique set of decision-making disciplines. However, there is no particularly complete or applicable decision-making scientific system, including most theories. This is related to the short formation period of decision-making theory and the wide application scope of decision-making.

4.2.1.2 Understanding of Engineering Decision-Making

Engineering decision-making is a cognitive process of decision-making, so it is a cognitive activity. The method of engineering decision-making begins with the activity of cognition, that is, in the stage of putting forward and analyzing problems in engineering decision-making, the method of engineering decision-making is also the method of recognizing decision-making problems. Decision-making is a process in which the subject realizes a specific goal, relying on an understanding of the law of development of things and objective reality, developing subjective initiative, applying the cognition based on objective practice to the real world through judging and choosing decision-making alternatives [5]. The cognitive process of engineering decision-making is particularly complex. For example, the critical decision-making problem of deciding whether a project will be constructed or implemented involves various potential indicators such as science and technology, value factors, and human factors. The quantification process of these indicators and determining their interaction requires a lot of time, effort, and elaboration. Many engineering decision-making errors are caused by insufficient depth and breadth of understanding of decision-making issues, incomplete collection of information on various indicators affecting decision-making results, and inadequate quality and authenticity of the information. Especially in the decision-making process of public works, some policymakers excessively pursue political achievements, exaggerate, conceal or tamper with crucial information, which affects the quality of understanding of decision-making issues and directly leads to significant errors in the results. On the other hand, the pursuit of fast decision-making and quick start-up makes many public works neglect the in-depth study of decision-making issues, deciding to invest in construction before a complete understanding of the necessary decision-making issues is available; it causes many problems in the later stage of the project, such as high cost and inadequate funds.

Engineering is the process of adapting and utilizing nature. The knowledge of the objective world acquired from practice can help people understand the essence of engineering, understand and study decision-making problems, and choose the best decision-making. It is a decision-making subject that needs to give full play to its subjective initiative, recognize the essence of the problem from a large number of phenomena and influencing factors, determine the goal of decision-making, ascend from irrational perceptual knowledge to rational knowledge, quantify the influencing factors, and achieve the final goal of decision-making by choosing and optimizing feasible alternatives.

Modern society is in an era of massive information, and the Internet model has replaced many traditional industrial and commercial models in the past. There are many factors affecting engineering decision-making, but these factors can also be represented in different forms of "information," so the decision-making based on data information will become the primary mode of engineering decision-making in the future. From the beginning to the end, a project is also an active process with ample data information. How to judge the validity and practicability of the data is also a complex problem for decision-makers. In some cases, only a small part of the data may be found in the mass data to form small samples. In other cases, a large amount of data may be related to engineering decision-making problems, resulting in huge sample data. Therefore, engineering big data will be a new feature of engineering decision-making.

4.2.1.3 Ideas of Engineering Decision-Making

Engineering decision-making is essentially decision-making but only applied in the engineering field, so it has characteristics different from general. Engineering ultimately serves people, so "people-oriented" and "harmony between nature and human" are the highest realm of engineering decision-making activities. The realization of the essence of engineering activities depends on the reasonable introduction of an "abnormal state" into a "normal state." The reasonable criterion of "normality" is whether the "normal" environment embodies "people-oriented" and rises to the philosophical level of engineering decision-making. It is the abstraction and sublimation of engineering decision-making at the ideological level on the basis of the essence of engineering decision-making, that is, whether it can meet human needs and improve the quality of life of the public. At the same time, we should protect the natural environment and realize that human beings are "in the center of the heaven and the earth," thus becoming the guiding ideology of engineering decision-making. The idea of "harmony between nature and human" is one of the fundamental concepts of Chinese classical philosophy and one of the most significant differences between Chinese philosophy and Western philosophy. Chinese traditional philosophy differs from western philosophy in terms of the living environment, social condition, historical background, and cultural heritage. Chinese traditional philosophy is dominated by the unity of humans and nature, while western philosophy is based on the division of humans and nature. Chinese traditional philosophy takes philosophy of life as its core, morality and art as its spirit, intuition and comprehension as its method, full of poetic realm, highlighting the value function of the best and the most beautiful, that is, the so-called "conform to the heaven and belong to human." Lao Tzu advocated that "Tao follows nature" and that heaven, nature, and human are all connected. The purpose of life is not to recognize and conquer nature but to love all things. Therefore, Chinese philosophy is full of human feelings, giving affection to all things and seeking communication between humans and nature. When Chuang Tzu wrote *The Theory of Equality of Things*, he said, "Heaven and nature are born with me, and all things are with me."

From the relationship between nature and artificial nature, the essence of "harmony between human and nature" is the unity of nature and artificial nature. The evolution of engineering activities shows that the evolution of nature and the evolution of engineering activities on the earth's surface are consistent to a certain extent. Engineering activities make the changes on the ground and in the sky generally the same; they embody the unity of humans and nature.

Engineering decision-making is an important part of engineering management activities, a series of intelligent activities that are carried out around the solution of decision-making problems. Engineering decision-making comes with engineering and goes with engineering, which often determines the success or failure of engineering activities. The theory of engineering decision-making should be a systematic knowledge system that examines the intrinsic links among the subjects, tasks, forms, and decision-making methods from the perspective of engineering itself. Therefore, the idea of engineering decision-making should have the following characteristics:

The idea of engineering decision-making should embody the core idea of engineering management. As mentioned earlier, engineering is often referred to as "one-time" activities because of its special characteristics of time, scope, and processes. However, these seemingly "one-time" projects still have many commonalities, which can help people better understand the project. The theory of engineering decision-making must be established from practice, from the actual project to find the commonness, and from the real experience to refine and sublimate. The task of engineering decision-making is to make the decision based on engineering service. From a micro perspective, engineering decision-making is a single decision-making problem such as the Three Gorges Project. From a macro perspective, it is a set of similar water level problems. Based on engineering requirements, the task requirement of the project will change with the different stages of the project. At the same time, the decision criteria and methods of the project will also vary accordingly. These decision-making tasks not only retain their uniqueness but also reflect some commonalities of decision-making problems. To sum up, the idea of engineering decision-making should be an in-depth study and analysis of all subjects, tasks, forms, and methods of universal decision-making under the whole engineering system, reflecting the theory of subjects, tasks, forms, and methods, and discovering the extension and connotation characteristics of engineering decision-making from the two-dimensional span of time and space.

The thought of engineering decision-making has its particularity. It is the theory of the close integration of humans and nature, which is determined by the essence of engineering. At present, there is no perfect engineering decision-making theory. Still, long time engineering construction practice has formed a preliminary engineering decision-making idea: the particularity of engineering decision-making problems, the rationality standard of engineering decision-making, "people-oriented," and the goal of engineering decision-making pointing to the organic unity of "nature and human."

4.2.1.4 Rationality Criteria for Engineering Decision-Making

The connotation of engineering decision-making determines that its rationality should be judged through the criterion of "human-oriented." Human-oriented is first related to people's sense of mission. Man is the rational agent of the co-evolution between humans and nature, but the earth is not only ready for human beings. The purpose of the earth's survival is not necessarily consistent with that of human beings. However, human development facts show that it is unrealistic that human beings separated from nature or that nature separated from human beings. In his book *Sustainable Development in the Perspective of Philosophy*, Chen Changshu holds that: it is in a long historical process that human beings gradually emerge from natural selection and survival competition, gradually get rid of the passive position dominated by natural forces, and gradually learn to imitate nature, and then become the main body of utilizing nature. The history of the human being as the subject is continuing. Perhaps we will see that human beings will also become the subject of adjusting and integrating nature. Chen Changshu expects human beings to "act as the subject of relying upon nature, learn from nature, conform to nature, protect and adjust nature" to achieve "human-oriented" [6].

4.2.1.5 Engineering Decision-Making from Phenomenon to Essence

The process of engineering decision-making is the process of finding hidden objective laws from a large number of phenomenal data, and it is the process of peeling off phenomena to see the essence. Philosophically speaking, this essence can be regarded as the origin of engineering decision-making. In terms of practical activities, it can be regarded as the goal of engineering decision-making. Another statement of the decision-making goal is the realistic reflection of the origin of engineering decision-making. For the decision-making problem that takes cost as the primary consideration, stripping off the appearance is the decision-making problem that takes the economic value as the main influence index. The decision-making problem with the environment as the main consideration object is the decision-making problem with the ecological value as the primary influence index. By analogy, the process of establishing the decision-making goal of the decision-making problem represents a sublimation process of the whole decision-making problem from phenomenon to essence. By further combing these sublimations, we have realized the recognition of the origin of engineering decision-making.

4.2.1.6 Engineering Decision-Making from Irrationality to Rationality

One of the most important steps in the process of engineering decision-making is the collation and processing of information related to decision-making. It needs to identify the authenticity and reliability of information and rank and select according to its relevance. A good understanding of decision-making is a comprehensive understanding of the importance, but it is not equal to the information flooding. In the process of collecting and sorting information, decision-makers have a perceptual understanding of decision-making problems in their minds and conceive the goal or alternatives of decision-making in advance. This perceptual knowledge is helpful to the final decision-making but not scientific. Understanding and sorting the information regarding decision-making problems can be beneficial when it reaches a rational knowledge level. That is to say, analyzing information using scientific methods, making up for the critical problems that may be neglected in perceptual knowledge, and playing a positive role in selecting decision-making methods or alternatives. The complexity of the project determines this.

4.2.1.7 Interaction Between Practice and Cognition in Engineering Decision Making

The ultimate goal of engineering decision-making is the smooth implementation of the project, so engineering decision-making must be based on practice; that is, all feasible alternatives of decision-making should be screened and selected on the premise of implementation. Understanding engineering decision-making must proceed from history and practice. Although historical practice experience may not provide perfect information, accumulation and reference of engineering practice is the cognitive basis of engineering decision-making. From numerous decision-making practices, people have summed up the general procedures, methods, and models of decision-making, resulting from sublimation from practice to understanding. Conversely, these cognitive results will guide and help analyze and solve other decision-making problems in the future. Many engineering decision-makers in China give more importance to practice than knowledge and lack the process of sublimation from practice to knowledge. Therefore, they often deviate from rational knowledge and blindly believe in experience and subjective judgment.

4.2.2 Characteristics of Engineering Decision-Making

As a subset of general decision-making, engineering decision-making has the commonness of general decision-making and its own characteristics. Engineering decision-making mainly includes the evaluation methods of technology, economy, ecology, environment, risk analysis procedures and methods, as well as decision-making criteria. If we regard "business decision-making" and "internal enterprise decision-making" as micro-decision-making, engineering decision-making is mid-aspect decision-making. Its position in a country's economic construction is significant. Some mid-aspect decision-making scale is also very large, such as the decision-making of the Three Gorges Project; because of the huge scale of the project, its impact is also enormous. Engineering decision-making has the following characteristics.

4.2.2.1 Initiative of Engineering Decision-Making

Engineering decision-making serves engineering, so engineering decision-making has the attribute of "engineering." The difference between engineering decision-making and ordinary decision-making is that it is a creative decision-making process. In addition to the elements of general decision-making, it must add the special elements brought by "engineering." Engineering is a creative process from scratch, and its decision-making scheme is the exercise and simulation of this creative process. Therefore, in a certain sense, engineering decision-making is transformed from a passive selection of alternatives to subjective initiative creation of alternatives, based on the abandonment of the passive and negative nature of ordinary decision-making, moving toward more active and innovative decision-making. The "one-time" characteristics of the project may result in a severe lack of reference experience for similar projects. Practical activities may not provide decision-makers with suitable engineering decision-making plans and methods. Engineering decision-makers must have subjective initiative and create plans or methods to meet the decision-making objectives under the existing knowledge and technology systems. This is based on the understanding and summary that decision-makers made on the historical engineering practice. Therefore, engineering decision-making should have initiative, aiming at solving engineering tasks and engineering problems using science and technology and the collection and accumulation of information, carrying out the process of design, evaluation, and selection about engineering plans.

4.2.2.2 Whole Process of Engineering Decision-Making

In a narrow sense, engineering decision-making only refers to the planning and feasibility study in the early stage of the project, that is, whether the project is invested or not. However, with the continuous understanding of the inherent meaning of the project and the gradual improvement of the understanding of the concept of decision-making, the essence of engineering decision-making has been extended and expanded within the scope of the project. The selection of design alternatives and implementation plans have also been included in engineering decision-making. However, the extension of concept scope does not mean the generalization of engineering decision-making; all selection activities are regarded as engineering decision-making activities. Therefore, in this book, only the decision-making that significantly impacts the direction, procedures, and results of engineering activities in the stage of engineering activities is defined as engineering decision-making.

4.2.2.3 Stage and Hierarchy of Engineering Decision-Making

The extension of the scope of engineering decision-making has multiplied the types and quantities of engineering decision-making problems, but its characteristics are becoming more and more apparent. With the continuous development and renewal of engineering technology, new decision-making problems are constantly emerging, and engineering decision-making problems and concepts must be constantly expanded and upgraded. Therefore, the concept of engineering decision-making proposed in this book is different from the previous definition of decision-making in engineering concepts. It is broad and dynamic which has stages and levels.

Firstly, the phase characteristics of engineering activities determine that engineering decision-making has phase characteristics; that is, decision-making tasks are closely related to phase tasks of engineering activities. Engineering decision-making has the characteristics of continuity; that is, the decision-making results of the previous stage affect the decision-making input of the next stage, and the decision-making results of the next stage are feedback and support to the decision-making results of the previous stage.

Secondly, engineering decision-making also has hierarchical characteristics; that is, in each stage of the project, the main decision-making sequence is clear; in each stage of the project, some engineering decision-making is dominant, and some are supplementary. At present, the project objectives are the control of quality, time limit, and cost, and relate to fundraising, risk analysis, use and maintenance, and local economy and environment. Therefore, the project objectives and decision-making should be considered in a broad sense. Firstly, the project strategic decision-making level is determined, and the appropriate project management mode is reasonably determined to effectively control the risks arising from the project manpower allocation and the project organizational behavior. Secondly, the decision-making level of engineering financing determines the quantitative investment decision-making plans of engineering cost estimation, which meets the financing requirements. Thirdly, the subcontracting planning layer calculates the expected benefits and risks and analyses the influencing factors of implementation. Then it is the project operation layer, which mainly includes the planning and management of the supply chain, the planning and allocation of operational resources, the allocation and flow of key personnel, the subcontract strategy and laws and regulations, supervision and inspection, etc. Finally, the ecological environment layer, engineering construction should not destroy the local ecological environment, including engineering ecological environment assessment, ecological restoration, and other decisions.

4.2.2.4 Comprehensive Engineering Decision Making

Engineering decision-making involves many factors such as technology, economy, and management. Firstly, engineering involves the fields of science and technology. To manage engineering well, it needs the comprehensive application of various sciences, technologies, and knowledge. Secondly, the emerging management sciences, such as systems theory, operations research, information theory, behavioral science, and computer systems as tools, have increasingly become effective means of comprehensive engineering management. Effective management means should be used in the whole process of the project from establishment to delivery and use. At the same time, only scientific management can do a good job on comprehensive engineering management. Thirdly, project management aims to improve economic benefits, so project management should also serve this goal. Engineering is to create the best economic benefits with the most economical life cycle. On the one hand, comprehensive management should be carried out throughout the project's life cycle to reduce costs; on the other hand, efforts should be made to improve the utilization rate and operational efficiency of the project. In short, engineering technology is the foundation, the economic benefit is the purpose, and management is the means. The three are linked together as a whole; only by combining each other can we achieve the project management objectives.

4.3 Objectives and Tasks of Engineering Decision-Making

4.3.1 Objectives of Engineering Decision-Making

The connotation of engineering decision-making determines that the goal should be "harmony between nature and human." "Harmony between nature and humans" means that humans and nature should adapt to each other and coexist peacefully. We should not talk about conquering and being conquered. The contemporary engineering activities form the consistency between natural and artificial evolution and achieve the "harmony between nature and humans" practically. This is the first meaning of "harmony between nature and humans." From the relationship between natural and artificial nature, the other meaning of "harmony between nature and human" is the correspondence between natural and artificial nature, that is, humans should live in harmony with nature and make the natural environment and social environment unified. Human acts according to the law of nature, and nature develops to benefit human society. Within the allowable range of natural regeneration ability and natural coordination ability, human uses science and technology to develop and utilize nature, and through human practice, to establish a new balance for nature that is beneficial to mankind, and to establish a virtuous circle of material exchange, energy circulation, and information transmission [7].

Because of the diversity and complexity of constraints, the ultimate consideration of engineering decision-making is actually the objectives reflected by various feasible plans, including economic, technological, social, ecological, etc. These objectives reflect the influence coefficients of feasible plans under specific decision criteria. For example, the assessment of the occupational hazards must be included in the engineering decision-making in petrochemical construction projects. Through defining the construction projects that may cause occupational hazards, the hazard factors should be analyzed, the degree of hazards should be evaluated, and the protective initiatives should be proposed finally. In this process, the decision-making object has shifted from the decision-making itself to the various impact objectives that may affect the decision-making.

The variability and phase characteristics of engineering decision-making tasks reflect the complexity of various factors affecting the decision-making results, that is, engineering decision-making contains a number of impact objectives, which are the perspectives of observing engineering decision-making tasks, and some of them are primary, some of them are secondary, and all of them are interrelated. These sets of objectives that affect the decision results form the various elements in decision criteria. Decision-making needs to find one or more primary objectives in these impact objectives or the balance among these objectives. These objectives include economic objectives, technical objectives, environmental objectives, safety objectives, ecological subjects, etc. Decision-making may vary due to specific projects or specific issues. However, if individual characteristics and categories are pulled away, the selection of the main impact objectives of decision-making will be consistent over a long time. The objectives of engineering decision-making are described in the following.

4.3.1.1 Technical Objectives

At the beginning of construction, Qinghai-Tibet Railway was facing three global engineering problems: permafrost, alpine anoxia, and ecological fragility, which was a challenge in the history of engineering. Since the existing railway engineering technology could not be directly applied to the plateau, the Qinghai-Tibet Railway had made many breakthroughs and contributions in engineering technology. For example, in permafrost technology, the technical design idea of "active cooling, cooling foundation, and protecting permafrost" had been established, which enriched the theory of permafrost engineering and improved the construction level of plateau permafrost railway.

From this, the technical objective is an important objective affecting engineering decision-making. Large-scale projects such as Qinghai-Tibet Railway and High-speed Railway can be implemented only when the technological breakthroughs can meet the project's needs. Therefore, technological objectives occupy a dominant position. Large-scale projects are often accompanied by a new generation of the engineering technology revolution. Engineering has become the carrier and promoter of new technologies. Satisfaction with technical objectives is often a necessary condition.

4.3.1.2 Economic Objectives

Economic objectives are often mentioned in areas such as national or regional development planning, reflecting the items and numerical values of certain socio-economic phenomena. Economic objectives in engineering decision-making have their own characteristics, which are often expressed as maximization of engineering benefits. For example, the economic objectives of a construction project are reflected in the cost data of various engineering materials and activities, such as walls, floors, doors and windows, hydropower, etc. The data directly reflects the cost of construction projects. The maximum benefit of the project can be calculated. Therefore, based on ensuring safety, the choice of materials and engineering technology to maximize economic benefits is the first consideration in project decision-making, with economic objectives as the leading factor.

4.3.1.3 Social Objectives

The social objective of engineering can be called the social evaluation objective of engineering. Usually, social evaluation objectives are most widely used in public works, such as highway construction, water conservancy construction, etc. The object of public works is society, which is to meet the development needs of the country or region and improve the social welfare of the country or region as the project's ultimate goal. Therefore, the impact of public works on the economic development, life construction, and other aspects of the surrounding areas must become the primary decision-making factors. However, with the enhancement of public awareness, many non-public projects also put social goals in the important position of project decision-making. Social goals have also expanded from simply considering economic development and construction to social groups' acceptance and recognition of engineering projects. This is also one of the reasons why more and more projects are moving away from urban areas and residential areas.

4.3.1.4 Ecological Objectives

The Three Gorges Project had a huge scale and many influencing factors. The possible ecological impact had always been the top priority of the decision-makers, and it is the focus of the Three Gorges Project in the demonstration and decision-making. In 1985, the State Council requested the State Planning Commission and the State Science and Technology Commission to set up the Expert Group on Ecology and Environment Demonstration of the Three Gorges Project. They conducted a thematic review on the normal water level, the environmental capacity of the reservoir area immigrants and the impact on the ecological environment of the plain, lake and estuary areas in the middle reaches of the Yangtze River, especially on the ecological environment. This item, once the focus of the Three Gorges Project, is an important goal affecting decision-making. This reflects the improvement of people's awareness of ecological environment protection and is also the core embodiment of harmony and unity between engineering and nature in the process of adapting to and utilizing nature. At present, considering ecological objectives is a prerequisite for every project and is also an important issue to be considered in the construction and operation of the project.

4.3.1.5 Security Objectives

Security objectives play a decisive role in engineering decision-making in chemical, mining, and nuclear power industries. Because of the particularity of the technology or products of these projects, once a security incident occurs, it will cause huge loss of personnel and property. Therefore, in the process of risk assessment, technical plans selection, and operation management, safety must be the primary decision-making goal of these projects. Sometimes, some new technologies and materials are preferred not to be used to ensure the reliability and stability of the project. With the gradual improvement of safety awareness, the proportion of safety objectives in the decision-making of all engineering projects is being gradually emphasized.

4.3.2 Tasks of Engineering Decision-Making

4.3.2.1 Characteristics of Engineering Decision Tasks

The characteristics of engineering decision-making tasks can be reflected and demonstrated from 14 special research reports formed at the end of the research stage of the Three Gorges Project, which mainly have the characteristics of variability and stage. This kind of decision-making problem determining the success or failure of the project occupies a large proportion in the whole set of project decision-making problems. But once the project is put into construction, there are still various decision-making problems of different sizes in each process stage. These decision-making problems will show different characteristics because of the different stages of the project, the different objectives, the different stakeholders involved. We will find different characteristics of decision-making tasks for the same decision-making problem from different perspectives. The above-mentioned phenomenon is determined by the complexity of the project and is an inevitable problem in the decision-making process. Therefore, the project decision-making task has the following characteristics.

1. **Polymorphism characteristics of engineering decision-making tasks**

The decision-making problem of the same project can be analyzed from different perspectives and has different characteristics, so it has polymorphism characteristics. For example, the decision-making of the Three Gorges Reservoir Level is a decision-making problem of which technology can be used to satisfy the rated water storage from the perspective of technical requirements. Considering safety demand, this is the decision-making problem of whether the downstream basin can meet its safety needs in flood season after the depth of water level is stipulated. From the point of view of ecological demand, it is a decision-making problem whether the determination of water level depth will change the ecological environment of the surrounding and downstream areas.

Different decision-making problems may have similar characteristics if they are analyzed from the same perspective. For example, in the process of project investment, the decision-making of how much to invest in the earlier stage and whether to build a factory building in the project construction process are different in the stages of the project process and the representation of the decision-making objectives. But, they are all financial decision-making problems with the lowest cost and the greatest benefit. From this point of view, if we look at these engineering decision-making problems from different perspectives, we will find the different characteristics of engineering decision-making in multiple spaces or surfaces. The decision-making characteristics in each aspect present different categories to meet certain objectives. From the functional point of view, engineering decision-making tasks can be divided into financial decision-making, safety decision-making, quality decision-making, and progress decision-making. From the perspective of scope, it can be divided into strategic decision-making and tactical decision-making. From the goal point of view, it can be divided into optimal decision-making and satisficing decision-making. From the perspective of subject decision-making mode, it can be divided into individual decision-making and group decision-making. Decided by the polymorphism characteristics of engineering decision-making tasks, no matter from which perspective to understand the task of engineering decision-making, the characteristics of engineering decision-making tasks cannot be exhibited completely.

2. **Phase characteristics of engineering decision-making tasks**

Engineering activities generally include engineering planning, design, implementation, and operations. Because the characteristics of engineering decision-making are closely related to the characteristics of engineering, engineering decision-making tasks have phase characteristics; that is, the decision-making objectives of engineering decision-making tasks are different at different stages of engineering activities, or the main decision-making tasks are various at different stages of engineering activities.

4.3.2.2 Contents of Engineering Decision-Making Tasks

1. **Phase characteristics of engineering decision-making tasks**

Engineering planning is the process of planning future engineering tasks, engineering processes, engineering effects, environmental requirements for engineering activities, and the procedures and steps for engineering implementation. The decision-making task in the stage of engineering planning is to determine whether the project's purpose is reasonable and whether various technical and non-technical elements can be effectively integrated. At the same time, on the basis of the analysis of the organizational and social environment of the engineering system, the strategic plan and arrangement of the target project are formulated according to the analysis results, and the reasonable arrangement of the time, sequence, and direction of each step is made.

Planning and feasibility study refers to the systematic demonstration and analysis of the proposed project from the perspective of the needs of national economic development. Then the task of decision-making is to obtain whether the project is to be constructed or not, that is, investment decision-making in the usual sense. The time required for decision-making tasks in the planning stage is mainly determined by the size of the project and the number of factors involved. For large-scale projects involving the national economy, development, and environment, detailed planning and decision-making are needed. For example, the time span of decision-making analysis of the Three Gorges Project, South-to-North Water Diversion Project, and High-speed Railway Project are very long, from a few years to a few decades. For projects involving national security and competitiveness, such as the Two Bombs and One Satellite project, manned space engineering, large aircraft project, Baosteel construction project, and Daqing Oilfield construction, the necessity of the project does not need too much analysis. It needs to decide which parts should be built immediately under the existing conditions and which parts should be pre-prepared and research design. It is found that the decision-making task in the stage of engineering planning is to do feasibility analysis of project construction through extensive preliminary investigation, survey, and investigation of the project alternatives, and pre-evaluating the possible risks in the project construction and budgeting. The possible damage to the ecological environment caused by engineering construction is also analyzed.

From the perspective of the theory of engineering evolution, the decision-making in the stage of engineering planning is a pre-planning activity that indirectly introduces an "abnormal state," which embodies the unity of causality and purpose in the process of engineering decision-making. The existence of an "abnormal state" is a natural process, that is, a causal process and the engineering decision-making process is an entry to the "abnormal state" process, that is, a purposeful process. By employing the function of engineering decision system, the evolutionary essence of engineering can be achieved; that is, the purpose of engineering activities is consistent with the causality of natural process [8].

2. **Decision-making tasks in the engineering design phase**

Engineering design is a key link between the overall planning and the implementation of the project. Engineering design is essentially a leading process of transforming knowledge into real productive forces. In a sense, it can also be said that "design is a process of pre-virtualization of the construction and operation process of the project". In engineering activities, design work is of special importance. Successful design is the premise, foundation and important guarantee for the smooth construction and successful operation of the project. The mediocre design indicates the mediocre project, while the poor and wrong design will inevitably lead to the failure of the project [9]. Perfect planning and design can improve the feasibility and safety of the project, establish a sound project implementation plan, and reduce the risks that may exist in the project construction. For example, the choice of the project site directly affects the success of the project. The infrastructure projects such as the Three Gorges Project, the South-to-North Water Diversion Project, the railway project and the bridge construction all need to choose the appropriate construction sites. On the one hand, it can ensure the efficient construction of the project, on the other hand, it should consider the impact of the project on the social environment. The construction sites of national high-tech safety projects, such as the Two Bombs and One Satellite, Manned Space Engineering, and Large Aircraft project, need to consider safety, technical environment, and other factors. For the construction of Baosteel, Daqing Oilfield and other industrial projects, we also need to consider many factors, such as technology, market, future development, and human resources.

The decision-making task in the engineering design stage is mainly to further demonstrate the projects that have been identified and to determine the construction site, scale, technical scheme and resource selection scheme of the project. In engineering design activities, the project has the characteristics of regularity and purposiveness by exerting the function of engineering decision-making system. By incorporating the "abnormal state" into regularity, the "normal state" is constructed through purposiveness, and the "abnormal state" indirectly changes to the "normal state" through the combination of regularity and purposiveness, which embodies the evolutionary nature of engineering activities.

3. **Decision-making tasks in the implementation stage of engineering**

The implementation and construction plan of the project will change with time and technological development, and there may be some inadequate considerations in the early stage of the project, which needs to be further improved in the construction process. From a systematic point of view, special management is a problem that every project needs to consider. For example, in the process of engineering management of Qinghai-Tibet Railway, five control objectives (engineering quality, environmental protection, health and safety, construction period and investment scale) have been set up to meet the special requirements of the project. In the process of Baosteel Construction, coordinated efforts, unified dispatch, and unified arrangement of construction have ensured the progress and quality of the project, and budgetary estimates and cost-saving sharing have been adopted to control and save funds. In the stage of project implementation, construction and production raw material supply are becoming increasingly tense and prices are rising, resulting in high cost of many projects. The traditional way to reduce project cost is mainly to reduce the construction cost by improving the process and manufacturing initiatives, but when the construction cost is reduced to a certain extent, it is difficult to improve. Therefore, after detailed planning and design decision-making, the main task of the project implementation stage is to determine how to organize the construction of the project, and it also needs to continue to implement construction decision-making, mainly for the progress of the project, quality, cost, safety, technology and other aspects of decision-making. Among them, project schedule decision-making mainly includes schedule planning, plan execution, conflict coordination optimization and schedule risk decision-making. Engineering quality decision includes quality inspection and quality evaluation. Project cost decision-making mainly includes cost accounting and control, cost analysis and prediction. Engineering safety decision-making mainly includes safety identification, safety assessment, safety early warning and risk resolution.

In the process of engineering implementation, the level of technology and engineering has changed. Firstly, the result of engineering integration innovation leads to the integration of many original technologies, which is constantly integrated into new technologies, and the engineering of new technology integration also contains more and more powerful functions of introducing "abnormal state". Secondly, the emergence of technological inventions is suitable for engineering needs. Engineering decision-making needs not only integration of mature technologies, but also integration of technologies that are suitable for engineering needs found in the process of engineering construction activities. In short, in the process of engineering implementation, the compatibility of technology and engineering in engineering technology system promotes the rational introduction of "abnormal state." It is necessary to introduce "abnormal state" through technology, isolate and protect "abnormal state" through engineering construction, so that "abnormal state" can change to "normal state" indirectly to achieve the essence of engineering.

In addition, in the stage of project implementation, it is necessary to make decision on project organization and control, especially to collect engineering information systematically, process and process engineering information, obtain the judgment of project operation status, and adjust engineering management behavior accordingly, in order to achieve the goal of project management. Therefore, in the stage of project implementation, the function of engineering decision-making system can prevent and eliminate the negative effects of engineering activities and prevent the deviation of the goal of introducing an "abnormal state."

4. **Decision-making tasks in operation stage of engineering**

The decision-making of engineering operation management is an effective way to maximize the efficiency and benefit of the project by adopting advanced management technology based on existing construction achievements.

After the implementation of the project, a new artificial nature has been constructed, and it has entered the stage of operation. Engineering assessment will be involved before and during operation. The engineering operation process is the key stage to embody the engineering target group, and it is also the real proof to evaluate whether the engineering concept is correct, whether the engineering decision-making is appropriate, whether the engineering design is advanced, and whether the engineering construction is excellent [3]. Engineering assessment is the last stage of engineering activities which is the comprehensive, objective, scientific and fair inspection and evaluation of the project's objectives, implementation process, input–output benefits, etc., determine whether the expected objectives of the project have been achieved and whether the main benefits of the project have been achieved. It is targeted to meet the needs of owners and other engineering communities to know the progress of the project and confirm its degree of achievement.

Operational management decisions exist in all projects, such as the routine operations scheduling and maintenance in the Three Gorges Project, South-to-North Water Diversion Project, Qinghai-Tibet Railway, High-speed Railway, and Sutong Bridge; the relocation and resettlement of in the Three Gorges Project and the South-to-North Water Diversion Project; the operation and maintenance of Two Bombs and One Satellite and Manned Space Engineering; the production scheduling and maintenance of Baosteel and Daqing Oilfields, etc. Therefore, after completing the project construction, it is necessary to make operational management decisions for the project.

From the perspective of engineering evolution theory, in the engineering operation and evaluation stage, the indirect introduction of "abnormal state" effect through engineering activities is tested by the function of the engineering decision-making system, and the essence of engineering evolution is confirmed.

The above four stages of engineering decision-making are carried out, which does not mean that the problem of project decision-making exists independently in each stage. On the contrary, they are connected with each other. But the importance of decision-making in each stage is different. If project decision-making is the key to the success or failure of the whole project, then the decision-making in the planning stage is the most crucial stage of the whole project decision-making.

4.4 Procedures, Models, and Methods of Engineering Decision

4.4.1 Engineering Decision Procedure

The procedure of engineering decision-making refers to all the logically related linkages to each other in completing the whole project decision-making. The procedure or process of engineering decision-making is unified, standardized, and reasonable. It is necessary to ensure that the decision-making can be carried out smoothly and reasonably. The engineering decision-making process does not specify or guide the decision-maker to choose any specific decision-making method. Still, it will remind the decision-maker to do the related activities in the necessary time and stage, which may be a compelling choice of decision-making methods. Normalized procedure flow is an essential way of scientific management. For example, if Zhuhai Airport had a reasonable decision-making procedure during the decision-making period, no necessary feasibility study would be missing, and major decision-making mistakes would be avoided.

Understanding the project and establishing a perfect decision-making mechanism are important sessions to ensure reasonable and adequate decision-making. These linkages need standardized procedures for implementation and supervision. Through continuous practice, people have put forward standardized decision-making procedures in the decision-making process. These procedures have sorted out and summarized the various linkages in the decision-making process macroscopically. By quantifying the various factors in the decision-making problem, the complexity of the decision-making problem and the impact of irrational factors is minimized. Essentially, the decision-making process is a complete dynamic process of asking, analyzing, and solving questions. The ordinary decision-making process generally includes five basic steps: asking questions, analyzing questions, formulating plans, optimizing alternatives, and implementing feedback. Because of the diversity of technology, personnel, and resources involved in engineering, and the complexity of decision-making environment, more in-depth analysis and understanding of decision-making issues must be done before feasible plans are prepared. Therefore, the engineering decision-making process generally consists of five steps, as shown in Fig. 4.1.

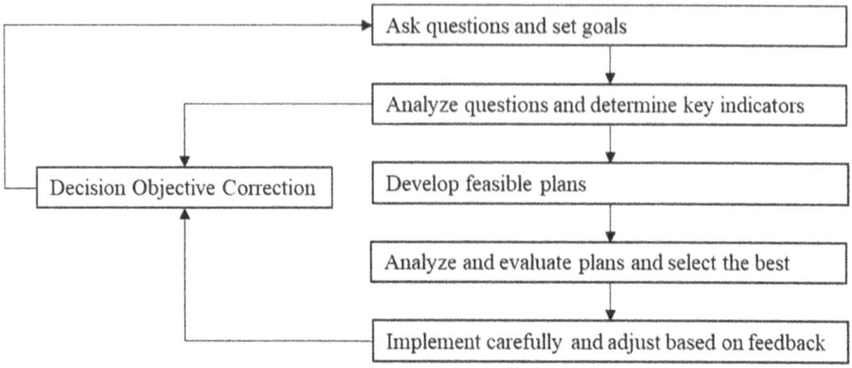

Fig. 4.1 The procedure of engineering decision-making

4.4.1.1 Ask Questions and Set Goals

The goal of decision-making is the key to engineering decision-making and the direction of decision-making. To avoid practical difficulties in implementing follow-up activities, decision-making questions and target positioning play a significant role. Excessive high decision-making objectives will increase the difficulty of decision-making and affect the work in the follow-up stages. And lower decision-making objectives will waste engineering resources. In many past large-scale project decision-making cases, there were often "over-fulfillment" and "ahead of schedule" and other decision-making objectives of engineering construction. A substantial reduction in the construction period will lead to a significant decline in the project's quality. This is a substantial reason for the occurrence of many jerry-built projects. The correct decision-making goal should be a scientific goal based on many investigations and researches.

4.4.1.2 Analyze Questions and Determine the Key Indicators Affecting Decision-Making

Once the objectives of decision-making problems are determined, it is essential to analyze and study the key indicators affecting decision-making. The analysis has two functions: one is to clarify the complexity of decision-making problems and the intrinsic relationship between critical indicators; the other is to verify the correctness of decision-making objectives further. Many practical engineering decision-making problems are very complex. For example, the mechanism of engineering decision-making tasks is usually unclear, and it is difficult to describe them by conventional methods, which is called the "black box" problem. The main decision-making modes of engineering decision-making tasks are scattered, and the decision-making subjects are complex and diverse. The uncertainties of engineering decision-making tasks are considerable, and there are too many decision-making factors. Most engineering decision-making tasks are multi-objective decision-making problems. Therefore, in the analysis stage of decision-making, it is often necessary to analyze and solve them by establishing mathematical models. If the analysis results do not find a feasible plan, we need to go back to the first step and reexamine the correctness of the decision-making objectives.

4.4.1.3 Developing Feasible Plans

The feasible alternatives of decision-making are sorted out or drawn up based on the previous analysis. Usually, the feasibility plans of engineering decision-making are not single, especially in the process of multi-objective or multi-decision-maker, it is easy to have multiple "feasible solutions." The selection and formulation of these feasible solutions is the output of the analysis stage, which must meet the implementation conditions and decision objectives. The reservoir level of the Three Gorges Project is a typical example of selecting several feasible plans. In formulating feasible plans, domestic and foreign experts had considered and calculated various factors such as water storage, sediment discharge, migration, and so on, and finally selected the executable plans according to the conditions of production and construction technology at that time in China.

4.4.1.4 Analyzing and Evaluating Plans and Selecting the Best

A large number of feasible plans does not mean that the decision-making is easy. On the contrary, too many feasible plans will bring more difficulty and intensity to subsequent evaluation and judgment, especially in large-scale engineering projects with economic, social, and technical indicators and other factors. It is important for decision-makers and evaluation decision-making methods to choose one plan that can balance the indicators and achieve the situation of "multi-win." In modern engineering projects, people have begun to find the best feasible plan by establishing mathematical models to analyze the proportion of various indicators. However, in many large-scale engineering projects in China, quantifying the decision-making indicators by mathematical methods has not been popularized and promoted. Judgemental decision-making methods occupy most of the "decision-making market." In many engineering projects, although in the absence of a viable plan, the implementation decision of the construction are still being made.

4.4.1.5 Implementing Carefully and Adjusting Based on Feedback

All precautions and delicacies in the early stages are for the smooth development of the follow-up work of decision-making. The construction period of large-scale engineering projects is long; it usually cannot find problems or errors of directional or guiding decisions in the short term. Once a significant mistake in decision-making is found during construction, it will bring huge losses to the project, and in many cases, such losses are irreparable. Without adequate market research or risk prediction, many domestic engineering projects rushed to start. As a result, projects were set aside because of the lack of funds and the contradictions between contractors and developers. The frequent occurrence of "uncompleted projects" may be caused by the mistake of engineering decision-making or by the error of decision-making implementation. Careful implementation of decision-making results is a necessary condition to ensure the smooth progress of the project. The re-verification in the implementation process is also feedback to the original decision-making objectives. On the other hand, it also guarantees the correctness of decision-making objectives in the next stage.

The procedure of engineering decision-making is different from the method. It is a process and means to solve the problem of engineering decision-making. It may be necessary to use a unique and applicable scientific method in each process or stage. The complexity of the engineering decision-making process increases the difficulty of engineering decision-making, and the selection of decision-making methods needs to be more careful. Different methods may be applied at different stages of the procedure for the same decision-making problem, and different methods may be applied at the same stage for horizontal comparison. Because of the different stage objectives, engineering decision-making methods have many characteristics. Therefore, decision-makers need to choose different decision-making methods according to the different characteristics of decision-making tasks. The research and exploration of decision-making methods must also follow the characteristics of engineering decision-making tasks. The process of analysis, judgment, and decision-making is the cognitive process of engineering problems itself. Through the cognitive results, people can follow certain decision-making procedures, adopt corresponding methods, and ultimately achieve objective, rational, and fair decision-making and judgment. Although the process of engineering decision-making can be divided into several stages, the objectives of each stage and the decision-making methods used may be different; from a broader perspective, they mainly solve two major problems: understanding the engineering decision-making problems and looking for the engineering decision-making mechanism. Deviations usually cause the major mistakes in engineering decisions in the cognition of decision problems, or problems in the decision mechanism.

For example, the Three Gorges Project of Yangtze River has been significant in the past 20–30 years, and its decision-making has undergone a long process. As early as 1919, Dr. Sun Yat-sen put forward the idea of the Three Gorges Project. Shortly after founding New China, in 1954, the Yangtze River Basin suffered from floods. The State Council began planning for the Yangtze River Basin. In February 1958, Premier Zhou Enlai led relevant leaders and experts from China and the Soviet Union to investigate and discuss. In March, *the Opinions on the Planning of the Three Gorges Water Conservancy Project and the Yangtze River Basin* was passed. At the same time, the Yangtze River Basin Planning Office was established. Because of the enormous scale, long construction period, and numerous immigration problems of the Three Gorges Project, which was not compatible with China's national strength,

Gezhouba Project, the transitional project of the Three Gorges Project, took the lead in the early 1970s. The successful completion of Gezhouba Project had solved the power problem in central China, and the implementation and construction of the Three Gorges Project had been put on the agenda. Due to the project's complexity, the feasibility study of the Three Gorges Project began in the early 1980s, and it took nearly nine years before it was finally completed. At the fifth session of the Seventh National People's Congress in 1992, *the Resolution on the Construction of the Three Gorges Project of the Yangtze River* [10] was formally adopted. The last feasibility study took only a few years, but the decision-making for the construction of the Three Gorges Project began in the early 1950s and ended in the early 1990s. After nearly 40 years, a large number of demonstration work was overturned, restarted, overturned, and restarted one after another, and finally, 14 special demonstration reports and 1 feasibility study report were formed.

The fourteen thematic demonstration reports include geology and earthquakes, key buildings, hydrology, flood control, sediment, shipping, power systems, electromechanical equipment, resettlement, ecology and environment, comprehensive planning and water level, construction, investment estimation, comprehensive economic evaluation, and many other important decisions. Each topic had set up a special group, which invited 412 experts from 17 departments and units under the State Council, 12 institutes of the Chinese Academy of Sciences, 29 colleges and universities, and 28 provincial and municipal professional departments to conduct detailed argumentation and research. Among them, the water level of the Three Gorges Dam is the most time-consuming and controversial decision-making problem in the Three Gorges Project. Various viewpoints remain deadlocked from the initial 150 m to 175 m and 180 m. In the early 1980s, Gezhouba project began to generate electricity. At that time, if the water level was too high, the burden of migrants was too high, and the sediment problem was more complex, so the water level was set at 150 m. But many experts believe that the water level is set too low. In April 1984, the State Council, taking into account experts' opinions, raised the dam crest to 175 m, leaving room for flood control. In September 1984, the Chongqing Municipal People's Government filed an application with the State Council to raise the water level of the Three Gorges Dam to 180 m so that the 10,000-ton fleet could reach Chongqing directly. Faced with various voices, in May 1986, the panel invited the World Bank and Canadian consulting firms to carry out feasibility studies of the Three Gorges Project in parallel with domestic institutions. After many comparisons of alternatives by experts at home and abroad, the height of 156 m [11] of the project's first phase was finally established. This seemingly simple numerical problem took two years and eight months. The decision-making process is cautious and meticulous because the project's decision-making is related to the lifeline of the project and is the key to the success or failure of the project. Once the decision-making plan is determined, many human and material resources are invested, the project can no longer be turned back. This is one of the reasons for the long and vital investigation and research work in the early stage of the project. That is also the main reason why people spend a lot of time and money in the engineering decision-making phase.

4.4.2 Engineering Decision Model

4.4.2.1 The Change of Engineering Decision-Making Model

The goals of influencing decision-making results are never single and independent, especially projects involving complex multi-technology, multi-resources, and multi-organizations. Decision-making is necessarily a process of multiple constraints and games that affect goals. However, among many influential goals, there will always be a goal that accounts for a more significant proportion of the decision-making process than other goals. We can call it the "dominant decision-making goal." The dominant decision-making objective may be economic, technical, or social. It will adopt different dominant decision-making objectives according to the characteristics of decision-making tasks or constraints of engineering resources. Ideally, the decision-making objectives of engineering decision-making tasks are determined only by the characteristics of the project itself. Still, in some specific cases, the dominant decision-making objectives will be affected by the economic and social environment of the country and region, resulting in variability.

Engineering cannot be separated from the impact of regional development on its own, but at the same time, the implementation and construction of many projects have also brought breakthrough changes to the local economic and social development. Since the founding of New China, China's economic and social development has undergone many transformations. Economic transformations and industrial restructuring have affected the decision-making criteria orientation in many large-scale projects. The decision-making criteria orientation determines the choice of the leading decision-making objectives. As a result, with the development of China's economy and society, the dominant decision-making goal of China's engineering decision-making task has formed a series of evolutionary processes, which is a miniature and direct reflection of China's economic and social development.

China's economic and social development has roughly gone through four major stages.

The first stage: planned economy. In the first 30 years after the founding of the People's Republic of China, the State Planning Commission planned and formulated goals in various fields of economic development. The factories produce products according to the national plan, the countryside grows crops according to the national plan, and the commercial departments purchase and sell the products according to the national plan. The planning departments formulate all varieties, quantities, and prices. This system enables China's economy to develop steadily in a planned and targeted manner, but it also severely restricts its vitality and speed of development [12].

The second stage is the transition from a planned economy to a market economy. As a symbol, the Third Plenary Session of the Eleventh Central Committee was held in December 1978. The reform of the rural economic system took the lead. The household contract responsibility system and the dual management system of unification and decentralization replaced the "three-level ownership, team-based" of people's commune system and implemented in rural areas throughout the country. In October 1987, the Thirteenth National Congress of the CPC put forward that the socialist planned commodity economy system should be a unified system with the plan and market, a new mode of economic operation. Generally speaking, it is a mode in which the state regulates the market and guides enterprises [13].

The third stage is the preliminary establishment of a socialist market economy. *The Decision on Several Issues Concerning the Establishment of a Socialist Market Economy System*, adopted at the Fourteenth and Third Plenary Sessions of the Fourteenth Central Committee in October 1992, puts forward the basic framework of a socialist market economy system. Many large and medium-sized state-owned enterprises have been transformed into wholly state-owned companies, limited liability companies, or limited companies. Many national industry headquarters have been reorganized into holding companies, and a number of large-scale enterprise groups have been developed using capital as a tie across regions and industries. Many small state-owned enterprises have carried out reforms through restructuring, joint venture, merger, leasing, contractual operation, joint-stock cooperative system, and sale [13].

The fourth stage is to improve the socialist market economy. *The Decision on Several Questions Concerning the Improvement of the Socialist Market Economy System*, adopted at the Third Plenary Session of the Sixteenth Central Committee, made a comprehensive plan to establish a sound socialist market economy system. Following the requirements of coordinating urban and rural development, regional development, economic and social development, harmonious development of human and nature, domestic development and opening-up, the government of China actively promoted reforms in various fields [13]. Among them, particular emphasis was placed on creating and maintaining a fair, competitive environment, promoting sustainable economic and social development, and improving scientific and democratic decision-making [14].

Different economic systems affect the social value system to a certain extent, the criteria of resource utilization and allocation, and thus directly affect the choice of the dominant decision-making objectives of engineering decision-making tasks, which are embodied in the following aspects:

1. **The Leading Decision-Making Objectives Under the Planned Economy System**

Under the planned economy system, resources are regulated by the state in a unified way, and the macro-control plan replaces the market feedback. In this case, resources are assumed to be impossible to waste, and both supply and consumption are planned to achieve absolute unity. Many engineering projects paid less attention to resource allocation and cost calculation. Economic objectives are not the dominant decision-making objectives for engineering decision-making under this system.

In China's early stage of development, the most important goal was to collect all resources to rapidly restore agricultural and industrial productivity, promote all-around rapid development in all fields, and increase influence and deterrence worldwide. During this period, China introduced a large number of advanced technologies in the industrial area from the Soviet Union and rapidly formed a "Great Leap

Forward" type of production activities throughout the country, which affected the decision makers' dominance of the project decision-making task. The Sanmenxia Project, which began planning simultaneously as the Three Gorges Project, was a typical large-scale project during that period. It broke ground in April 1957 and completed the main part of the dam in only about four years. During the construction of the project, there was constant controversy. Many experts had asked for re-investigation and feasibility analysis on the design plans of "water storage and sediment control" or "flood control and sediment discharge." But the decision-makers who were eager to make achievements blindly believed in the greatness of the technological force and had not passed rigorous scientific demonstration. They decided in a hurry. The main technology of Sanmenxia Dam was provided by the Leningrad Hydropower Design Institute of the Soviet Union, but the design institute had no experience in conducting water conservancy projects on such sandy Yellow River, so the sediment problem with serious consequences was neglected at that time. The blindness of the decision-making stage and the insufficiency of planning and design made the Sanmenxia Dam have to be rebuilt twice after its initial operation. The operation mode changed from "water storage and sediment retention" to "flood control and sediment discharge." The repair work continued until the 1990s. The Sanmenxia Dam can be said to be "in ruins." Premier Zhou Enlai admitted in his conversation with the Vietnamese Water Conservancy Delegation in June 1964 that "we have fought unprepared battles in the Sanmenxia Project and our scientific attitude is not enough." Of course, the Sanmenxia Project has made significant contributions to flood control in the Yellow River Basin in the past 40 years. Suppose safety, ecological, and economic objectives can be taken into account more during the decision-making period, the Sanmenxia Project may be more beneficial to the lives of the people in the Yellow River Basin than it is now.

The dominant decision-making goal under the planned economy system was biased towards technical goals due to the time background. The security and reliability problems brought about by new technologies had been neglected in this period, and the economic, ecological, and safety decision-making objectives had also been paid less attention under the flawed decision-making theory.

2. **The Leading Decision-Making Objectives Under the Transformation of Market Economy System**

China's economic system in the second and third stages of development is a microcosm of the dominance of the market economic system. Although the planned economy still exists, the rise of the market economic system once suppressed the macro-control of the planned economic system, presenting a "one-sided" situation. During this period, many state-owned enterprises began to be privatized, the strength of private enterprises began to be strong, and the stimulation of market mechanism made the social and economic development present a prosperous picture. Behind this situation, the social value had changed, and economic benefits had been put in the first place for most business operators to pursue their goals. This change of values also directly affected the selection of the dominant decision-making objectives of the project decision-making task. Technological goal orientation had begun to turn into an outstanding economic goal, and the pursuit of maximum benefits had become an important criterion for engineering decision-making.

Excessive pursuit of economic benefits will lead to ignorance of technical and secure objectives. With the rapid development of the market economy system in China, the "incomplete project" and "jerry-built project" have appeared. In the process of project cost management, project decision-makers cut corners and sacrifice safety for low cost, coupled with inadequate supervision by regulatory authorities, resulting in low-quality projects. In constructing public projects, such as highways, bridges, housing, and other projects, excessive pursuit of economic benefits was more serious. On January 4, 1999, the Qijiang Rainbow Bridge in Chongqing, which was completed for only three years, suddenly collapsed in its entirety, killing 40 people in the accident. Xinsan Highways in Yunnan Province collapsed on the second day of opening. The Tuojiang Bridge in Hunan province collapsed before completion. The Kunlu Highway in Yunnan Province, which cost 380 million yuan, was only opened for 18 days, resulting in subgrade subsidence and pavement cracking. At 1 a.m. on November 25, 2007, the two-story waiting hall of the West Passenger Station in Houma City, Shanxi Province, which was under construction, collapsed, causing at least two deaths. Ten hours before the collapse, the capping ceremony was held for the waiting hall. There are many forms of subcontracting cooperation in engineering projects. The consequence of focusing only on economic objectives is to focus only on the low quotation of the contractors, but not to examine and inspect the construction qualification of contractors related to safety issues.

Under the market economy system, the dominant decision-making goal of engineering decision-making tasks directly reflects maximizing benefits. It reflects the social value change and affects the internal structure and decisive factors of engineering decision-making criteria. After so many failed engineering cases, people began to reflect on the importance of the leading decision-making objectives in engineering decision-making.

3. **The Leading Decision-Making Objectives Under the Socialist Market Economy System**

After 2000, China began to enter a period of perfection of the socialist market economy. However, China's socialist market economy is different from the traditional planned economy. It is not a pure market economy, it is the combination of planned economy and market economy, and it is an effective unity of macro and micro. Depending entirely on planned economy, there will be neglect of economic benefits, safety benefits, over-confidence in the effective and rational allocation of resources, resulting in waste of resources. However, totally relying on the market economy will lead to the blind pursuit of economic benefits, neglect of social benefits and safety benefits, and over-confidence that the automatic regulation of the market can make up for the deficiencies in all aspects. In recent years, the subprime economic crisis in European and North-American countries is the result of the excessive free market economy. With the continuous improvement of China's socialist market economy and the gradual transformation of social values, people pay more and more attention to social, ecological, and safety benefits beyond economic objectives.

This change in values is also directly reflected in the dominant decision-making objectives of engineering decision-making tasks. In recent years, in the decision-making of engineering projects, ecological objectives have become one of the items that must be examined in the feasibility study. The feasibility study report must give the project an important decision-making goal to determine its impact on the surrounding ecological environment, whether it is good or bad, whether it has advantages and disadvantages, or whether the advantages outweigh the disadvantages in the project decision-making. On the other hand, with the improvement of the quality requirement of the project and the enhancement of the human-oriented consciousness of the project itself, the construction safety consideration and personnel safety consideration in the project are very important considerations in the decision-making process. This is a necessary means to eliminate the "incomplete project" and "jerry-built project" project and an effective way to protect the safety of construction personnel. Therefore, at this stage, the project decision-making objectives are the aggregation of multiple factors, a multi-constrained game problem, and a process in which the dominant decision-making objectives and the secondary decision-making objectives restrict each other.

With the increase in the number of engineering projects, people begin to distinguish the types of engineering projects and decide the decision-making objectives of engineering decision-making tasks according to different projects. In mining industry, due to the lack of strict state supervision, the increase of private mining operations and the excessive pursuit of economic benefits under the market economic system, the operation is not standardized, the construction site jerry-building and material-reducing landslides occur frequently, and the personal safety of construction personnel cannot be effectively guaranteed. Therefore, the safety goal must be the dominant decision-making goal in mining engineering, which is the basis of economic benefits and even more important than the technical and economic goals. Other ecological engineering projects, such as biogas and environmental protection projects, have gradually come into people's vision in recent years. These projects usually put the economic goal at the end, mainly investigating the impact on project decision-making from the ecological goal to create a good natural atmosphere for people's living environment and living conditions. With the improvement of humanistic thought and consciousness in Chinese society, attention to social value has increased. Many projects begin to consider social effects more in decision-making, such as choosing construction sites and taking into account the feedback of people and society. For example, for those projects that will cause pollution problems to a certain degree, the hazards caused by the tasks need to be publicly and appropriately explained when selecting the site, and feedback from residents in the surrounding areas will be collected and considered to achieve the high degree of social decision-making goals.

4.4.2.2 Decision-Making Model

In addition to inadequate understanding of engineering decision-making could lead to decision-making errors, the wrong choice of decision-making mode will also lead to significant deviations in decision-making results. The decision-making mode generally includes establishing the decision-making subject, the division of decision-making power, decision-making organization and mode, etc. Perfect decision-making mode can ensure the full analysis of decision-making problems, ensure the smooth progress of each stage in the project decision-making process, and ensure the quality of decision-making results. Through repeated deliberation and demonstration, a perfect decision-making mode needs to form a standardized decision-making system and relevant laws and regulations. Fundamentally speaking, cognitive errors in decision-making problems are also manifestations of imperfect decision-making models. The decision-making errors of many large-scale projects in China are largely caused by the unclear and imperfect decision-making mode, which will lead to the ambiguous cognition of the subject, procedure, and means of decision-making. At this time, the result of decision-making depends on the experience or subjective consciousness of decision-makers rather than scientific decision-making methods. For example, Fuzhou Changle International Airport has suffered heavy losses since its opening on June 23, 1997, with a 4-year debt of more than 3 billion yuan. The State Audit Office disclosed that the decision-making errors of the former leadership of the State Power Company caused significant losses; in which 3.28 billion yuan (42%) was lost or potentially lost due to individual leaders' violation of decision-making procedures or unauthorized decision-making. Because of the lack of required decision-making methods, decision-makers tend to place their subjective will too high, resulting in that subjective consciousness being more emphasized than scientific decision-making means. On the other hand, even if they have a decision-making mode but cannot implement it strictly, they may override the decision-making model. Although necessary decision analysis and feasibility studies have been carried out for this kind of project, to cater to decision-maker's preferences, the data of feasibility report is manipulated to make the mode of decision-making into form and display. Therefore, a reasonable decision-making model must have sufficient assumptions and norms for mutual supervision and review among decision-makers. Since most of China's current large-scale projects are initiated by the government, decision-making power is usually in the hands of a small number of officials. With the continuous development of China's social democratization, the participation of the masses in the decision-making of public engineering projects is getting higher and higher. The decision-making opinions of the third party have gradually entered the decision-making horizon of large-scale engineering projects. Establishing a reasonable decision-making model to help the government make a good judgment on the decision-making of large-scale projects and avoid low-level and unnecessary decision-making errors is an important issue in China's engineering decision-making management.

The complexity of projects determines the diversity of decision-making modes. Different economic types and legal structures will affect decision-making modes' establishment and selection. Even within the same project, due to the change of decision-making problem and the change of decision-making body structure, the original decision-making mode will be changed. The decision-making mode presents a form of decision-making in the process of decision-making. Engineering decision-making models mainly include the following:

1. **Centralized Model**

The centralized mode is the highly centralized decision-making, which attributes all decision-making power to one or several people. In the specific historical development period, centralized decision-making mode will bring rapid economic benefits to engineering construction. Because of its fast speed and simple process, it does not need repeated deliberation and demonstration. In some events requiring quick response and judgment, centralized decision-making mode is necessary and effective. It has won the initiative in time, avoiding further losses or hazards. For example, when unexpected engineering accidents such as mine collapse occurs, centralized decision-making mode will make the fastest judgment and provide instructions, concentrate manpower and material resources in the shortest time, make full efforts to fix unexpected, and avoid further expansion of casualties. But the centralized decision-making mode depends too much on the decision-maker's quality, experience, and decision-making ability. Especially in the decision-making of the start-up and development of major projects, decision-making affects the development and planning of an industry. If the decision-maker cannot do the comprehensive feasibility study, it will lead to the investment decision-making error of the project and bring substantial economic losses.

2. **Expert Demonstration Model**

For a long time, the major engineering projects in China have adopted the expert demonstration mode. Expert demonstration model is an effective combination of scientific decision-making and government will. Before starting the project, the decision-maker will select experts in relevant fields and set up an expert demonstration committee according to the related environment, technology, and personnel involved. The expert committee members should come from many disciplines, such as engineering, technology, natural science, sociology, ecological environment, and system engineering. The selection of argumentation experts should follow the principle of combining qualitative and quantitative analysis and construct a corresponding evaluation index system and mathematical evaluation model. The selected experts should have a profound academic foundation, rich engineering experience, and a high sense of responsibility. The responsibility of experts is to make a detailed analysis and data acquisition of engineering decision-making problems by scientific means. For complex decision-making problems, the corresponding mathematical model should be established by mathematical methods to carry out effective analysis and put forward reasonably feasible schemes. The final feasibility report presented by the expert committee is examined by government departments, discussed at the expert meeting, and selected the feasibility plan. Because the experts come from different academic backgrounds and have different engineering experiences, the different feasible studies will be available for the same engineering problems. This is the essence of the expert demonstration mode. Only by multiple examinations at different levels can we reasonably and effectively demonstrate engineering problems and avoid major mistakes in decision-making. The expert demonstration model plays a vital role in many large-scale engineering projects in China.

However, under the stimulation of the rapid economic environment, weakening the expert demonstration model began to appear. Some tend to formalize the mode of expert argumentation, while others even ignore the expert process. Once the experts cannot fulfill their duties and seek truth from facts, the model of expert argumentation will lose its effectiveness, and the decision-making process will become a direct reflection of the subjective consciousness of the decision-makers. A great deal of practical evidence shows that the probability of deviation and error is very high in this kind of decision-making which lacks actual and objective demonstration and judgment. The disadvantage of expert demonstration mode in a specific environment is prominent in China's political climate. Improving the expert demonstration model and enhancing the effectiveness of the decision-making model is a thorny problem facing policymakers.

3. **Expert Demonstration + Public Participation Model**

Expert argumentation mode is a powerful combination of science and decision-making. Expert argumentation and decision-maker wills are indispensable dialectical unity in significant engineering projects. However, with the evolution of the value system in engineering tasks and the transformation from economic indicators to social indicators, the third-party decision-makers in the existing decision-making model gradually begin to appear, that is, the public will. With people's increasing attention to environmental issues, the start-up and implementation of many engineering projects are related to the development of industry and the safety of life and production of the surrounding residents. With the rapid development of the network information age, the related information of engineering projects can be widely disseminated in different media. People begin to care about the intrinsic relationship of their interests in large-scale engineering projects. In the past, the will of decision-makers could take economic development as the main decision-making goal without considering the masses' will. However, in the continuous progress of China's democratization, the attempt to conceal the drawbacks of engineering projects has been more and more widely concerned and complained about.

From Dalian to Xiamen, from Shina to Qidong, from Ningbo to Kunming, the major chemical projects that emerged in recent years were opposed by residents and public opinion and triggered mass incidents, which is worth thinking deeply about by policymakers. The emergence of the value system of engineering projects' social and environmental value will inevitably lead to corresponding decision-making objectives. As a new force, the public will appear in the decision-making bodies of many projects. The decision-makers of projects can no longer ignore the participation of the masses' will and must accept the masses' will as the important factors in the decision-making. Therefore, the effective combination of expert demonstration and public participation is the most reasonable decision-making mode for major engineering projects in China at present. The results and feasibility reports of expert demonstration need to be open to the public. The decision-makers need to explain the advantages and disadvantages of the project to the public. This is an important key to reflect the project's social value and a benign trend of the development of modern society.

4.4.3 Engineering Decision-Making Method

The method of engineering decision-making is essentially the means to achieve the task of engineering decision-making based on the engineering decision-making path. People know nature through experience and utilize nature through action. Engineering is the process of adapting and using nature, and engineering decision-making are the necessary means to achieve the state of "harmony between nature and humans." A clear, organized, and systematic understanding of nature can form a theory, and a method can be formed by standardized, logical, and effective utilization of nature. Implementing the engineering decision-making process and mechanism needs the support of engineering decision-making methods; otherwise, their effectiveness will be difficult to guarantee.

4.4.3.1 Connotation of Engineering Decision-Making Method

The engineering decision-making method refers to the sum of the categories, principles, theories, and means universally applicable to engineering decision-making activities and play a guiding role. There are many objectives affecting engineering decision-making. Some of the decision-making objectives involved in engineering decision-making are not explicit, and they have internal relationships and interactions. The specific connotation of engineering decision-making methods is how to recognize the proportion of these impact objectives in decision-making, accord with the project decision-making objectives, and ultimately calculate or estimate the proportion through effective ways. The essence of the engineering process should be rational behavior, but irrational factors are interpenetrated in rational behavior because of their complexity. The increase of irrational factors and uncertainty makes engineering decision-making more difficult [15]. It is important for engineering decision-makers to find and use scientific decision-making methods scientifically.

Scientific engineering decision-making methods should have a comprehensive understanding and in-depth exploration of all the scientific technologies and value and human factors involved in the scope of space and time of the project. Since the reform and opening-up, China's economic development has grown rapidly, and various large-scale projects have been launched and implemented. When the level of practice and engineering management theory has not reached the level of objective demand, decision-makers have begun to use unscientific decision-making means to plan and construct engineering projects. Many decision-makers do not realize the systems and complexity of engineering decision-making and lack the necessary study and research. On the other hand, with the rapid economic development, many decision-making theories remain in the financial field, and policymakers tend to pay more attention to the economic value of projects. This phenomenon makes decision-makers lack a clear understanding of engineering decision-making, unable to grasp scientific and effective decision-making methods, resulting in decision-making errors. Some industries and local governments have made a blind investment and repeated construction at a low level. Some local governments have not carried out scientific and comprehensive decision-making research at all and have not carried out necessary feasibility studies on the project, so they have rushed to invest in construction, which violates the core idea of scientific development. China's "airport construction phenomena" in the mid and late 1990s brought some cases of decision-making mistakes. For example, Zhuhai City in Guangdong Province planned to build "the most advanced airport in the country" around 1995 and decided to invest 4 billion yuan (but the total cost was 6.9 billion yuan) to build Zhuhai Airport. But in fact, within a radius of less than 100 km, there are many international airports such as Shenzhen, Guangzhou, Hong Kong, and Macao. This makes the passenger flow of Zhuhai Airport difficult to meet the expected requirements for a long time, so it directly affects the repayment capacity. Policymakers in Zhuhai had hoped that the airport would repay bank loans and arrears by operating income, but they were unexpectedly plunged into huge losses.

Similarly, Mianyang Airport in Sichuan Province was completed and opened in 2001 and lost more than 38 million yuan that year. During the decision-making period, these projects did not carry out a scientific investigation on the airport passenger handling, geographical location, traffic demand, and the follow-up issues of airport construction. They lacked a significant engineering feasibility study in engineering decision-making [16]. The main reason for these engineering decision-making errors is that too many irrational factors interfere with the normal implementation of decision-making. There is no scientific engineering decision-making method as effective support for decision-making. Scientific engineering decision-making method is not only a simple choice of A or B mathematical method; it should include practical issues such as feasibility study of the project, avoiding investment risks, coordinating the comprehensive benefits of various stakeholders, etc.

4.4.3.2 Quantitative Method for Engineering Decision

There are many engineering decision-making methods, including traditional methods and some new methods. Most of them are quantitative analysis methods. There are different decision-making methods for different decision-making problems. Among them, the typical decision-making methods are analytic hierarchy process, multi-objective decision-making method, fuzzy decision-making method, grey decision-making method, uncertain multi-attribute complex group decision-making method, dynamic evolution, and simulation method of engineering decision-making process, complex engineering risk decision-making method, dynamic equilibrium decision-making method of engineering project ecosystem evaluation, prediction method of resource and environment factors based on big data analysis.

1. **Analytic Hierarchy Process**

Analytic Hierarchy Process (AHP) is a multi-criteria decision-making method that combines qualitative analysis with quantitative analysis put forward by T. L. Satty et al. in the 1970s. The method of AHP makes people's thinking process in solving complex, large-scale systems hierarchical, organized and quantitative. Its basic principle is to evaluate the alternatives according to the objective, sub-objective (criterion), constraints and department with hierarchical structure, to determine the judgment matrix using pairwise comparison method, and then to take the component of the eigenvector corresponding to the maximum eigenvalue of the judgment matrix as the corresponding coefficient. Finally, the weights of each alternative are obtained, which can provide a quantitative basis for analysis, decision-making, and control [17]. This method is simple and easy to implement. Because the evaluator can compare the given factors with pairwise comparison, it has high reliability and small error. The disadvantage of this method is that for large-scale problems, for example, if the number of factors in a subset of some factors is large (e.g., more than 9), it tends to have problems, such as the difficulty of judgment matrix to meet the consistency requirements. This method has been widely used in the overall design of launch vehicle systems, comprehensive evaluation of space propulsion systems, cost–benefit decision-making, equipment development demonstration, and planning, etc.

2. **Multi-objective Decision-Making Method**

The basic idea of the multi-objective decision-making (MODM) method is to synthesize multiple objectives of the system into a single goal to measure the system's overall quality to select and rank. After more than 40 years' development, MODM has many methods. However, the technique for order performance by similarity to ideal solution (TOPSIS), a multi-criteria decision analysis method, is widely used in complex engineering systems. The elements of the decision matrix in TOPSIS are composed of system indexes of alternatives. The main idea is to define the positive ideal solution and the negative ideal solution of the decision problem and then find an answer in the feasible solutions set closest to the positive ideal solution and the farthest from the negative ideal solution. Among them, the ideal point is a hypothetical best alternative, which can select the best value from the different attributes of various alternatives. In contrast, the negative ideal solution is the hypothetical worst alternative, which can select the worst value from the different attributes of various alternatives. Positive ideals and negative ideals often do not appear in practice. Still, they represent the extreme situations that we strive to pursue and avoid in decision-making and make a comprehensive trade-off between multiple technical alternatives [16].

The main advantage of TOPSIS method is its simplicity and directness, but its disadvantage is that its result does not have a definite meaning but only represents a relative level. In addition, when determining the elements in the decision matrix and ranking the alternatives, the method uses a deterministic method to give the values. There will be some difficulties in practice, especially in conceptual design decision-making.

This method is generally used to make a preliminary selection among multiple alternatives. The inferior alternatives can be quickly discarded by using this method, and the workload can be reduced by further adopting other decision-making methods. For example, this method has been applied to evaluating and selecting the overall technical alternatives of aircraft manufacturing engineering [18]; this method has also been applied to selecting the overall technical alternative of aircraft propulsion systems [19].

3. **Fuzzy Decision-Making Method**

Fuzzy sets theory methods mainly study and deal with fuzzy phenomena quantitatively, and the most direct application in decision-making is fuzzy comprehensive evaluation (FCE). FCE refers to the comprehensive evaluation of things affected by various factors using fuzzy mathematics. The traditional decision-making method of fuzzy theory has the advantages of simplicity, convenience, and clear concept, so it has specific practical value. In addition, the fuzzy theory can be combined with other decision-making methods to produce some new decision-making methods, such as combining AHP method to produce the fuzzy analytic hierarchy process [20]. The appearance of these new methods makes them more suitable for the decision-making environment of practical complex engineering systems.

The FCE method can better solve the ambiguity in comprehensive evaluation (such as the ambiguity of object attributes in complex engineering system design, evaluation experts' knowledge, etc.). Because some attributes in the design of complex engineering systems are often vague concepts, they are widely used.

4. **Grey Theory Decision-Making Method**

The grey system refers to a system where part of the information is known and part of the information is unknown. Grey system theory is a theory to study and solve grey system analysis, modeling, prediction, decision-making, and control. It was put forward and developed by Professor Deng Julong of the Department of Automatic Control and Computer of the Huazhong University of Science and Technology in the early 1980s. Because many indicators describing the attributes of engineering systems are qualitative. In contrast, the evaluation criteria for quantitative indicators are uncertain; their evaluation practice is often based on the evaluators' knowledge level, cognitive ability, and personal preference. This makes the information available to decision-makers inaccurate and incomplete, that is, gray. Therefore, it is appropriate to use the grey theory decision-making method for this kind of engineering problem, and a grey hierarchical comprehensive evaluation model [21] is formed.

5. **Uncertain Multi-attribute Complex Group Decision Making Method**

Because engineering involves multiple uncertainties, which involve both the uncertainty of the external environment and the uncertainty of subjective preferences of decision-makers, it is necessary to deal with a large number of uncertain information or multiple uncertain information. Traditional uncertainty decision-making methods (such as fuzzy decision-making method, linguistic decision-making method, Bayesian decision-making method, multiple uncertainties of random fuzzy and interval fuzzy decision methods) can basically deal with uncertain information or multiple uncertainties. However, for complex and large-scale projects, many experts in different fields are required to provide wisdom to participate in decision-making. At this time, experts involved in project decision-making will show complexity characteristics. There may be various explicit or implicit relationships and conflicts between experts' decision preferences. This brings a high degree of complexity and uncertainty to the aggregation of decision preferences. These characteristics determine the need for a new engineering decision-making method different from the traditional group decision-making method: the uncertain information multi-attribute complex group decision-making method.

6. **Dynamic Evolution and Simulation Method of Engineering Decision Process**

The decision-making process of engineering argumentation is a complex system engineering. It is a scientific and technological research activity for decision-makers to understand, utilize nature, and live in harmony with nature and a process of reflecting strategic will, judging, choosing, and coordinating multiple values. Therefore, the decision-making process of engineering is a dynamic process of eliminating preference conflicts and evolving from chaos to order. It is the scientific requirement of engineering decision-making to explore its dynamic evolution law and predict its evolution trend. Therefore, an evolutionary game and dynamic simulation system are needed to simulate and analyze the decision-making process and effect of the project.

7. **Engineering Complex Risk Decision-Making Method**

In addition to the project's technical risks, the risks of the project are more embodied in project decision-making, management and credit, project investment and construction implementation, construction management and supervision. They are mainly located in the process of decision-making, implementation and supervision. It has the characteristics of high concealment, derivation, and conductivity. Therefore, its risk control becomes highly complex and challenging because of incomplete and uncertain information. It is necessary to monitor, warn, and resolve the risks in the whole process of the project, formulate management countermeasures, guide the project's complex risk control, and improve the accuracy and efficiency of risk control. The above new environment requires the support of complex risk decision-making methods.

8. **Dynamic Equilibrium Decision-Making Method of Engineering Project Ecosystem Evaluation**

Essentially, engineering is a socio-economic project and an ecological project. There are close and complex interrelationships between them. Large-scale projects such as the Three Gorges Project, the West–East Gas Pipeline Project, and the South-to-North

Water Diversion Project have more or less unforeseen negative effects during the implementation of the project (for example, additional costs of migration and higher costs of environmental protection) or positive effects (such as accelerated regional economic and social development). Therefore, to improve the quality of project evaluation, a dynamic equilibrium decision-making method based on the ecosystem is needed. According to the combination (economic impact, social stability, environmental ecology, etc.) decision-making index system, according to the degree of ecological impact (the possibility of ecological risk occurrence and its impact on project value), the method designs index weight and ecological dynamic correction coefficient (specific index assignment and relative dynamic change of weight at different time points), and multi-angle embedding project-related subjects (investors, contractors, managers, the public, etc.), using the dynamic evaluation model of portfolio equilibrium to judge the degree of ecological damage. The decision path is identifying key evaluation factors—the possibility estimation of ecological damage occurrence—cost and time impact assessment—the equal value impact assessment—the combined dynamic equilibrium decision of ecological damage degree. This method synthesizes the related subjects of engineering projects and considers the balance between ecological indicators. It is a composite dynamic equilibrium decision-making method.

9. **Prediction Method of Resource and Environment Factors Based on Big Data Analysis**

The purpose of forecasting is to provide the decision-making system with the necessary future information for decision-making. In engineering management, because the complexity of external factors such as the environment is constantly increasing, it isn't easy to envisage the success of engineering construction without predicting. Therefore, it is no exaggeration to say that it is difficult to make scientific decisions without scientific predictions. New forecasting methods are needed in the big data environment, such as real-time dynamic forecasting methods, missing information forecasting methods, massive data mining and analysis methods, etc. The forecasting results are used to assist decision-making.

4.5 Engineering Decision System

The engineering decision-making system is a system formed with the support of decision-making methods to solve the decision-making problems at each stage of the project as the core to achieve and complete the corresponding decision-making tasks in each stage.

4.5.1 Characteristics of Engineering Decision System

Engineering decision-making is systematic, and the system theory is its basic guiding ideology. It requires the analysis of the system from the point of view of the system as a whole, focusing on the overall effect of the system, regardless of the merits and demerits of individual factors, and sometimes even sacrificing local and short-term interests for the overall and long-term interests of the system [22]. Because of the particularity of the project itself, compared with the general decision-making system, the characteristics of the engineering decision-making system also have particularity, with the following characteristics [23].

4.5.1.1 Complexity of Decision-Making

Decision objectives of simple decision-making systems are generally easy to quantify, can be pursued by a subjective judgment for the best goal. The objectives of complex decision-making systems are relatively vague and difficult to quantify. They always appear in the form of system objectives, and there are often no fixed patterns or operational procedures to follow. Simple decision-making systems often do not consider the impact of the external environment. In contrast, complex decision-making systems must fully consider the impact of the system's external environment, including the economic environment, political environment, social environment, and so on. Therefore, a complex decision-making system must be put into the economic, political, and social system for decision-making.

4.5.1.2 Diversification of Decision Subject

The decision-making of engineering (especially complex large-scale engineering) involves a wide range of fields, so the decision-making subject has hierarchy and complexity. According to the different scopes of decision-making, engineering decision-making can be divided into micro-decision-making, middle-level, and macro-decision-making. From the perspective of discipline classification, micro-decision-making is a kind of decision-making at the level of engineering technology; middle-level decision-making is decision-making in the sense of single discipline or a single field, such as ecology, economics, and sociology; macro-decision-making is decision-making that reflects how science, technology, economy, and society achieve coordinated development. Therefore, engineering decision-making is divided into several decision-making units of different levels and disciplines. For different decision-making units, it is necessary to select appropriate decision-making methods, experts in corresponding fields, and departments with different responsibilities. In addition, engineering decision-making involves multiple interests, and representatives of each party participate in engineering decision-making based on their different interests and values. Each decision-making subject's power and degree are not the same, thus forming a decision-making group of different levels and influences.

4.5.1.3 Diversity of Decision-Making Objectives

For general conventional engineering, long-term economic benefits can be regarded as the fundamental goal to pursue, while other benefits can be achieved as limited conditions or subsidiary objectives. Major engineering decisions have not only one objective value but also multiple objectives. There are different levels among the objective values. The evaluation criteria are not single but multi-criteria. Therefore, it is multi-objective decision-making. The overall goal of the decision-making should be the system's overall optimization. In the decision-making process, the lower-level goals are subordinate to the upper-level goals and are interrelated and conflicted with each other among different levels or at the same level.

4.5.1.4 Dynamic of the Decision-Making Process

Engineering decision-making is a complex and cyclical process, not just a moment of choosing a project. It is based on the judgment and prediction of a series of uncertainties in the future, and the future situation is constantly changing. Therefore, the engineering decision-making system should be dynamic. Firstly, at the macro decision-making level, such as project establishment, scientific prediction, and moderately advanced decision-making (such as project size and site selection) should be made from the perspective of sustainable development, which is determined by the contradiction between the long-term nature of project construction, the time lag of investment effect, and the demand of the rapidly changing external environment for the project. Secondly, at the micro-level decision-making, such as the construction and implementation of the project, the adjustability of the decision-making should be maintained to prepare contingency measures to adapt to possible changes.

4.5.2 The Role of Engineering Decision System

Engineering decision-making systems should promote the integration of engineering natural resources, the development of the engineering industry system, the creation of social value, and the formation of engineering artificial nature.

4.5.2.1 Promoting the Integration of Engineering Natural Resources

Engineering activities are always carried out under certain natural conditions, so the important content of engineering decision-making activities is to integrate natural resources, that is, to make decisions on the integration of resources in engineering activities.

First, it includes the decision-making of coordinating the relationship between the environment and engineering. Engineering activities should be developed in the direction of "green" through decision-making efforts. When carrying out the project, we should pay attention to the fact that the construction project cannot exceed the environmental affordability, especially the ecological affordability of the environment. Second, it includes making the best use of society's limited resources and putting the idea of ecological cost compensation into the construction activities to promote the sustainable development of green engineering construction.

4.5.2.2 Promoting the Development of Engineering Industry

Engineering is the material basis of industrial development. After the transformation of a certain type of engineering activity, it shows the corresponding industrial form. Therefore, it is necessary to make decisions on transforming engineering into corresponding industrial problems. First, it includes the decision-making that can transform engineering into industrial activities, especially the characteristics of scale, profitability, transformation, and structure. Second, it includes whether the economic benefit is reasonable decision-making. In each stage of engineering decision-making, it is necessary to consider whether the economic benefit is a reasonable "cost–benefit" ratio. Therefore, it is necessary to carry out detailed technical and economic benefit analysis and create more feasible alternatives with the reference of economic values. The economic value is considered the center of the evaluation scale, judges the satisfaction degree of various alternatives, then carries on the choice analysis, etc.

4.5.2.3 Promoting the Creation of Social Value

Because engineering construction is to create artificial nature through creation, engineering and society are close. The production of engineering activities will inevitably lead to changes in social relations, so engineering activities are always carried out in a specific institutional environment. Therefore, it is necessary to make decisions on the innovation and integration of institutional elements related to engineering activities. Firstly, it includes the social value of engineering decision-making, emphasizing the interests of social subjects, but it does not deny the diversity of value evaluation. The social value of engineering decision-making is neither harmful to the fundamental interests of the social subjects nor against the achievement of the law of social development. Secondly, it includes making decisions based on changes in social policies, laws, rules, and regulations, and various interest groups, to promote enterprises to establish a social responsibility concept, to use appropriate methods and skills to resolve conflicts and contradictions in the process of environmental management, and to make the general public from suspecting projects and rejecting projects to understanding projects, to ensure the smooth progress of project activities.

4.5.2.4 Promoting the Formation of Engineering Artificial Nature

Because the most essential feature of engineering is to form engineering artificial nature through integration, the function of the engineering decision system should be to create an artificial nature.

As a means of changing nature and regulating society, engineering must serve the purpose of human beings and meet the needs of society. The exertion of regularity and purposiveness of engineering can promote the achievement of the artificial nature of engineering. Through regularity, it can accommodate "abnormality"; through purposiveness, the "normality" can be constructed; through the combination of regularity and purposiveness, the "abnormality" indirectly enters the "normality" and constructs the engineering artificial nature [24].

The contradiction between "normal" and "abnormal" is an eternal fundamental contradiction, which requires the introduction of "abnormal" through technical decision-making and the isolation and protection of "abnormal" through engineering construction to prevent the destruction or damage of "abnormal" to "normal." Hence, the process of creating artificial nature is a process in which technology and engineering cooperate reasonably, and the degree of cooperation deepens continuously, thus changing "abnormality" to "normality" indirectly and constructing engineering artificial nature.

It is necessary to enlarge production scale through engineering and technological transformation decision-making, provide the possibility for the corresponding industry's economic effects and the achievement of profitable goals, and then promote the upgrading of the regional industrial structure so that the direction of engineering investment is consistent with national industrial policies to promote reasonable allocation of various resources elements and construction funds, to provide a structural foundation for the integration of industrial transformative elements, thereby changing "abnormality" indirectly to "normality" and creating engineering artificial nature.

4.5.3 Functional System of Engineering Decision System

Based on the idea of systems, any decision can be regarded as a system. The project decision-making is different from general decision-making because the project has the characteristics of huge investment, complex structure, long construction cycle, and many influencing factors, so improper decision-making leads to huge risks. At present, there are many cases of engineering decision-making errors and severe

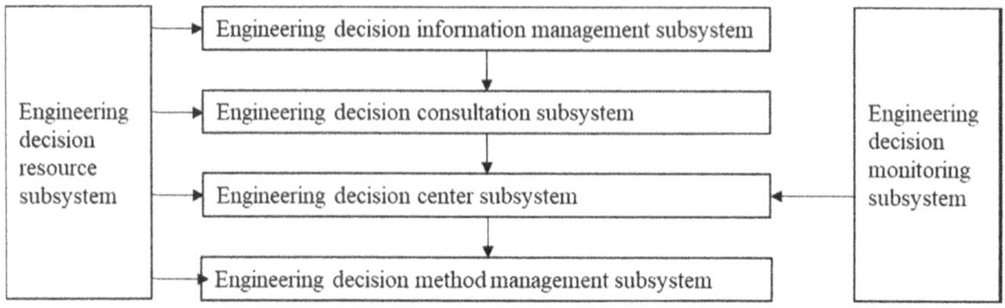

Fig. 4.2 The architecture of engineering decision system

consequences in China. The reason is the lack of an effective system mechanism to effectively organize, manage, and coordinate the complex influencing factors in the decision-making process. Engineering itself is mostly a complete and complex huge system. From the viewpoint of systems theory, it is of great practical significance to adopt systems science methods for analysis to ensure the correctness and effectiveness of engineering decision-making.

The engineering decision-making system is divided into six subsystems: engineering decision information management subsystem, engineering decision consultation subsystem, engineering decision method management subsystem, engineering decision center subsystem, engineering decision monitoring subsystem, and engineering decision resources subsystem. In the process of engineering decision-making, subsystems of different levels, different types, and different functions should be organized. Meanwhile, various methods and technologies should be used to form a smooth flow of information, communication and coordination, and a convenient system structure. The logical relationship of each subsystem (i.e., the architecture of the engineering decision system) is shown in Fig. 4.2.

4.5.3.1 Engineering Decision Information Management Subsystem

In an engineering decision-making system, because all kinds of information restrict or even conflict with each other, the function of the decision-making information management subsystem is first to organize and collect all related information to ensure the completeness and integrity of information. Second, the complex information is filtered and refined to control the appropriate amount of information. Then, the decision-making information subsystem is needed to identify the information to ensure the correctness and accuracy.

4.5.3.2 Engineering Decision Consultation Subsystem

Decision-making consulting subsystem refers to a system that entrusts relevant professional departments and experts to comprehensively process and analyze the information output from the above information subsystem and then formulates a feasible plan after full demonstration and research. The function of the decision-making consultation subsystem is to use engineering knowledge to form different forms of uncertain decision-making information, pay attention to the study on a plan that is not feasible while presenting the feasibility report, stand on a relatively independent position, and put forward opinions and plans for engineering decision-making to avoid decision-making errors.

4.5.3.3 Engineering Decision Method Management Subsystem

The management subsystem of the engineering decision-making method uses various decision-making models and methods in the process of engineering decision-making. Its main means are various qualitative investigation and quantitative mathematical models and methods. According to different decisions, corresponding decision-making models and methods are constructed to form a database of engineering decision-making methods and their management systems. In this way, it can be used by other subsystems. For example, the analytic hierarchy process and the fuzzy comprehensive evaluation method are selected to divide the objectives at all levels of engineering decision-making and establish an evaluation index system for quantitative evaluation.

4.5.3.4 Engineering Decision Center Subsystem

The main task of the project decision-making central subsystem is guided by the goal of decision-making and based on the mastery and processing of the information related to decision-making. Making full use of the long-term accumulated experience and knowledge of decision-makers, adopting the means of comparison, analysis and balance, the most reasonable plan is selected from various alternatives in the decision-making consultation system. The appropriate decision-making method is selected from the subsystem of the engineering decision-making method for evaluation and decision-making, and the best decision-making alternative is obtained.

4.5.3.5 Engineering Decision Monitoring Subsystem

Because of the uncertainty of decision information and decision environment, the process of engineering decision-making is also uncertain. Therefore, the main task of the project decision-making monitoring subsystem is to monitor the whole process of project decision-making, including the implementation of decision-making procedures, the achievement of decision-making objectives, and the completion of decision-making tasks. Once problems occur, timely corrections are made to avoid large mistakes in decision-making, minimize the cost of decision errors, and reduce the risk of decision errors.

4.5.3.6 Engineering Decision Resource Subsystem

In the process of engineering decision-making, to effectively solve the problem of engineering decision-making, abundant resources of engineering decision-making need to be supplemented, including engineering decision-making knowledge base and its management system, technology base and management system, case base, and management system etc. Engineering decision-making knowledge base is a knowledge system composed of all kinds of knowledge related to engineering decision-making. In the whole knowledge repository of human beings, engineering knowledge is an enormous quantity and the richest content. A considerable part of it is related to engineering decision-making knowledge. First, engineering decision-making knowledge includes a variety of natural science knowledge, technological knowledge invention, technical know-how, and so on as its basic knowledge. Second, the knowledge of engineering decision-making includes management, economics, sociology, and other humanities and social sciences. Third, engineering decision-making knowledge includes all kinds of experience related to engineering decision-making. More importantly, this kind of knowledge is not a simple stacking of all types of knowledge mentioned above, but a combination of them and transformed into practical, feasible, and operable knowledge following the actual engineering situation, and to make it into practical engineering knowledge that guides engineering decision-making practices, and the management of this knowledge forms an engineering decision-making knowledge system.

When a new technology is invented, it should be connected with the existing technology to form a new system and then applied in production. There are complementary mechanisms within the engineering technology system and among all kinds of technologies. One form of technological change may affect or influence other forms of technological change. Different technologies have mutual synergy and form a technology library. The management of these technologies forms an engineering decision-making technology system. As far as a specific project is concerned, many kinds of technologies can be applied. The comparison and selection of different technical plans, the optimization of combination, and the performance-price ratio after implementation are important contents to consider in engineering decision-making. The important function of engineering decision-making technology systems is determining whether technology and engineering have the best matching. In engineering activities, there are special technological inventions and creations. These technological inventions and creations are an integral part of engineering activities and serve the overall engineering goal. Engineering innovation decisions must integrate mature technologies and new technologies suitable for engineering needs that were invented in the process of engineering activities. Therefore, it is necessary to use engineering decision-making technology systems to promote the development of engineering-required technologies.

References

1. Xia, Z. (2012). Analysis of management thinking of large-scale construction projects in ancient China. *China Market, 23*, 34–35.
2. He, J. (2013). A preliminary study on the core of engineering management. In *China engineering science and technology forum*.
3. He, J., Chen, X., & Hong, K. (2005). On engineering management. *China Engineering Science, 10*, 5–10.
4. Yin, R. (2011). *Theory of engineering evolution*. Higher Education Press.
5. Mu, J., & Xu, J. (2009). The cognitive roots of decision-making errors. *Journal of Harbin University, 11*, 41–46.
6. Chen, C. (2000). *Sustainable development in philosophical perspective* (pp. 107–113). China Social Sciences Press.
7. Zhang, M. (2008). Ecological nature view: A new nature view facing ecological crisis. In *Proceedings of 2008 forum on disaster prevention and reduction of China association for science and technology*, 9, pp. 824–828.
8. Li, B. (2002). *Introduction to engineering philosophy*. Elephant Publishing House.
9. Yin, R., Wang, Y., & Li, B. (2011). *Theory of engineering evolution*. Higher Education Press.
10. Zou, J. (1992). Resolution on the construction of the three Gorges project. *People's Yangtze River, 4*, 1.
11. Qian, Z. (1992). Demonstration of the three Gorges project. *Wenweipo*, 1992.1.12.
12. China's economic system. http://news.xinhuane
13. The historical evolution and basic direction of China's economic system reform. http://www.cser.org
14. Fan, H. (2006). The historical evolution and basic direction of China's economic system reform. *Reference for Economic Research, 48*, 8–14.
15. Zhu, W., Wang, M., & Zheng, J. (2015). Philosophical thinking on project management. *Theoretic Monthly, 8*, 47–51.
16. Liu, X., & Song, B. (2004). Decision-making theory and method for conceptual design of complex engineering systems. *Systems Engineering Theory and Practice, 12*, 72–77.
17. Xiao, H. (2006). Accountability "decision-making errors." *Procuratorial View, 23*, 24.
18. Kirby, M., & Mavris, D. (2000). A method for technology selection based on benefit, available schedule and budget resources. In *2000 World aviation conference* (p. 5563).
19. Cartagena, M., Rosario, J., & Mavris, D. (2000). A method for technology identification, evaluation, and selection of aircraft propulsion systems. In *36th AIAA/ASME/SAE/ASEE joint propulsion conference and exhibit* (p. 3712).
20. Che, A., Lin, Z., & Fang, Y. (1998). Application of fuzzy set theory in QFD. *Journal of Systems Management, 2*, 55–57.
21. Zhou, Q., & Zhang, D. (1999). Multi-objective grey relational degree decision-making model for engineering system design and application. *Systems Engineering and Electronic Technology, 1*, 1–3.
22. Li, Y. (2010). *Research on the decision-making subject of public investment and construction project* [Master's Thesis]. Chongqing University.
23. Li, Y. (2010). On the structure, function and characteristics of decision-making system for major engineering projects. *Manager Journal, 10*, 186–186.
24. Zhai, M. (2013). *Research on the essence of scientific and technological production relations and industry-university-research cooperation* [Doctoral Thesis]. Northeast University.
25. Wang, L. (1996). Comparison of Chinese and Western philosophical thoughts on the unity of nature and humans and the separation of nature and humans. *Journal of Guangxi Normal University (Philosophy and Social Sciences Edition), 2*, 66–67.
26. Cai, Y., & Huang, Z. (2012). Moral education reform from the perspective of cultural ecology—A changing balance. *Agricultural Education in China, 4*, 5–8.
27. Chen, Y. (2009). Understanding of ecological and environmental impact assessment of the three Gorges Project. *pydropower, 12*, 31–33.

28. Deng, J., & Pan, Y. (2005). On the supervision of the power of the "leader." *Journal of Guangzhou Socialist College, 2*, 9–12.
29. Lu, G., Fu, C., Wu, J., & Liu, Y. (2008). Research on the decision-making process and characteristics of major projects: A case study of the three Gorges project. *China Science and Technology Forum, 8*, 20–24.

Chapter 5
Organization Theory of Engineering Management

The organization is the "main promoter" of the progress of human civilization. The earliest organizational form was the primitive tribes with biota as the main body. They gathered strength and emerged wisdom, which enriched the human ability to fight against nature. Afterward, family organization and clan organization linked by consanguinity shifted human social life into social life. The resulting private ownership promotes state organization with hierarchical relation as the axis, marking the beginning of the human practice of socialized organizational production mode. Since modern times, the organizational form has become more and more diverse. Social organizations, enterprise organizations, and public welfare have sprung up like bamboo shoots after a spring rain, creating a diverse and organic social ecology. Nowadays, people are always committed to one or another, large or small organizational form; individuals become the cornerstone and cells of the organization, born, grown, and finally ended.

The purpose of the organization is to make ordinary people do extraordinary things. Organizations have been accompanied by human practice since their inception. By integrating and regulating design ideas, human resources, materials, and technical methods, engineering organizations have realized the transformation of engineering from conception to the entity. Engineering organization is a collective state of people in engineering activities and reflects the cooperative relationship among different interests and different levels of people in engineering management activities. It is also a process of resource transformation. Under the guidance of engineering objectives, raw materials, equipment, capital, and personnel are allocated and coordinated in a planned way. The efficiency of engineering organizations plays an essential role in achieving engineering objectives, so the theory of organization is called the mother science of project management.

Tracing the source, people cannot help asking: What is an engineering organization? Why does an engineering organization exist? How did the engineering organization evolve? What are its characteristics? What is the future of it?

Around these questions and reflections, this chapter first clarifies the concept of engineering organization, that is, what is engineering organization and why it exists.

J. He, *Principles of Engineering Management*,
https://doi.org/10.1007/978-981-99-1168-4_5

From the perspective of etymology and semantics, the original meaning of organization is analyzed. The connotation of the organization is expounded from a narrow and broad perspective. The different meanings of organization in different theoretical schools are also analyzed. Furthermore, it elaborates the definition, form, constituent elements, tasks, the core of the engineering organization, and the dialectical logical relationship between engineering and management.

Second, it explores the evolution process of engineering organizations in the entire engineering management activities. From the process of its establishment, existence to change and disappearance, it analyzes the changes caused by the expansion of the scale of the engineering ontology and the increase of the participating parties, studies the evolution caused by the external environment of politics, law, economy, society, culture and technology, analyses the evolutionary path of engineering organization form from the early time "sea of people strategy" to the middle period "human–machine combination" to the modern "human–machine-network combination," and explores the development law of its symbiotic interaction with the engineering ontology and external environment.

Third, an engineering organization is an agglomeration with essential aggregation, efficiency, and adaptation characteristics. Why do they agglomerate? How does the periodicity of agglomeration come from? Where will the evolution of engineering organization cohesion go? After answering these questions, this chapter tries to find the expression of the efficiency of an engineering organization, analyze its formation and development, existence and fading away, as well as the modeling effort, and then analyze the organization's adaptability from the power source and operation mode of organization.

Finally, aiming at the development trend of engineering management theory and practice in the future, the structure of future engineering organizations is discussed from the perspectives of networking, organizational flexibility, and organizational boundaries. And the theoretical paradigm of a future engineering organization is forecasted from the traditional organizational theory paradigm to the new self-organization theory paradigm.

5.1 Overview of Engineering Organization

People have different opinions on the understanding level and concept of "organization," especially with the development of productivity and science. The socialized division of labor becomes more and more elaborate. The concept of organization has been given different connotations in various disciplines and types of work. Harold Koontz, an American management scientist, once said that perhaps the most confusing semantics in management is the word "organization" [1]. The interpretation of the word "organization" can be divided into verbs and nouns, static and dynamic. It can also be understood from the aspects of organization and environment adaptation, organization form, or management. Due to the different research perspectives, there are many definitions. In engineering construction, it is difficult to give a simple definition of project organization in engineering construction because of the numerous participants and the complexity of managing technology. Therefore, it is necessary to gradually deepen the interpretation of engineering organizations from the understanding of the general organization.

5.1.1 General Organization

Before answering the question "What is an engineering organization" it is necessary to clarify the basic concept of organization.

5.1.1.1 Interpretation from Etymology and Semantics

In Chinese, the word "organization" is "Zuzhi." The original meaning of "Zu" refers to a wide and thin ribbon, while the original meaning of "Zhi" refers to cloth and silk. For example, *Shuowen Jiezi-Mi Bu* said: "Zu, Shou genus. The narrow one can be used as the ribbon of a hat. "Zhi is the general name of cloth." The original meaning of the words "Zu" and "Zhi" is to make cloth from silk and linen. For example, the note of Gaoyou from The Annals of Lü Buwei—*Xianji* said: "Zuzhi means artisan weaves by hand." It also refers to the formation of sentences and words in poetry or speech. For example, Liu Hsieh said in his book The Literary Mind and the Carving of Dragons (one)- Yuandao: "carving sentiment, organizing rhetoric." It also means arranging and rectification. Qiu Fengjia in Qing Dynasty said in his poem *Dream*: "Drove the sun and the moon, alternated day and night, and never stopped. Time passed quickly, but reorganizing (Zuzhi) the native land has not yet been achieved". The ancients defined the term "Zuzhi" as the meaning of some elements forming a whole.

In English, the word "Zuzhi" is "organization." The word has undergone a process of evolution from the combined state of an organism, namely from organ to its integrated organism, to the coordinated action (organize) between organs and the result of the action called an organization. In 1873, Herbert Spencer, a British philosopher, first extended the connotation of an organization to the field of social science, regarding the organization as a system or society of combinatorial integration [2].

Nowadays, the organization mainly includes two meanings: one is the effective working collective formed by association, which corresponds to the "organization" with noun attribute. According to the definition of James D. Mooney, a classical organization theorist, an organization is a form that connects all kinds of people and achieves common goals. Chester I. Barnard, an American management scientist, points out that an organization is a collaborative system that can coordinate the activities of two or more people. E. Gross and A. W. Etzioni argue that organization is a social unit that people intentionally construct to achieve certain common goals. Typical organizations are enterprises, troops, schools, churches, and prisons. Although these definitions have different emphases on the understanding of the organization, they all emphasize that organization is a system of individuals or groups.

The second meaning is to organize a large number of people and coordinate their actions to achieve a common goal, that is, to correspond to the "organization" with the verb attribute. According to Henri Fayol, "an organization is to provide all the raw materials, equipment, capital and personnel needed for the operation of an enterprise [3]." The organization with verb attribute is one of the functions of management. They are subordinate to the plan. After the organizational goal is determined, the tasks needed to achieve the goal are allocated scientifically and reasonably, and the most suitable persons coordinate all the tasks. Organizing is a process of creating, maintaining, and transforming an organization to achieve its goals so that the members can play their respective roles. As the environment changes, the organizational structure needs to be constantly adjusted to ensure efficient operation.

5.1.1.2 Interpretation in a Narrow and Broad Sense

In a narrow sense, the organization refers to a collective formed by different people to achieve common goals and cooperate with each other. The definition focuses on the collectives formed by human beings with specific forms, for example, associations, enterprises, troops, etc. Today, organizations are generally believed to be social groups formed by people assembled in a certain form to accomplish certain goals and objectives. The organization is the basic unit of the whole society and the cornerstone of the continuous and orderly operation of the whole society.

In a broad sense, an organization is a system that links many elements in a certain way. Modern society is composed of many and various organizations. In contrast, the broad sense of an organization contains a wider range of objects. Many experts and scholars have studied the organization from different perspectives using systems theory, cybernetics, information theory, dissipative structure theory, and synergy theory. The broad sense of organization is similar to the concept of a system to some extent.

5.1.1.3 Interpretation of Different Theories

The traditional school of organization theory regards the organization as a tool to achieve goals and attaches great importance to its "instrumental value." It holds that organization is a power-responsibility structure or personnel arrangement constituted by certain rules and procedures to achieve common goals. As Max Weber put forward, the hierarchical organization has strict rules and hierarchical systems. Personnel is assigned to different posts and given different powers according to their ranks. Individuals pursue maximum benefits and grade promotion in their posts and achieve the organization's common goals by forming a joint force [4]. In that background, the organization constructed according to the traditional organization theory has more substantial productivity and competitive advantages than the family workshop production unit.

The School of behavioral science holds that an organization's individuals, goals, and information are related. It is a system of interpersonal interaction. Organizations cooperate with each other. It is necessary to explain the behavior of individuals and groups in different environments and put forward feasible management strategies. In traditional theory, the concept of "economic man" puts forward the concept of "social man," which pays attention to social and psychological factors. Organizations are divided into formal organizations and informal organizations. Informal organizations can have a positive or negative impact on formal organizations. Organizational communication is regarded as an important research content.

The modern school of organization theory holds that organization is a comprehensive system including structure subsystem, technology subsystem, psychosocial subsystem, goal (value) subsystem, and management subsystem. It is in three environments, namely physical, cultural, and technological, which determine how people act and interact in the system. Organizations are man-made open systems that exchange materials with and maintain close ties with the external environment. An organization can survive only by adapting to the changing environment and constantly making internal adjustments to balance the relationships among various complex roles.

To sum up, an organization is defined as a collaborative system with flexible boundaries composed of people for a common goal. Explained as follows:

(1) Organization has common goals. It is the foundation of the existence of an organization. They can generate strong attraction and enable members of the organization to gather and carry out activities together. For example, the purpose of an aerospace engineering organization is to complete a space launch plan, and the pursuit of an enterprise organization is to complete production and sales tasks.
(2) Organization is made up of people, who are the basic elements of an organization. Many participants connect and coordinate with each other through various forms, gathering everyone's intelligence together, supplemented by an effective management model, thus forming an efficient group to achieve the effect of "the more, the merrier."
(3) Organization has flexible boundaries. An organization is an open system that exists in a certain environment. Their boundaries distinguish the organization from the surrounding environment, determine the organization's scope of activities, and play a filtering role. In the manufacturing process of Boeing 787 aircraft, after multiple comparisons, Chengfei Civil Aircraft Company was selected as the sole supplier of the rudder, while other competing parts suppliers were filtered. Another important criterion for Apple to choose suppliers is confidentiality. If a supplier violates this criterion, it will be sanctioned by Apple and filtered out in the organizational system of Apple's production.
(4) Organization is a collaborative system. There are complex relationships among individuals, individuals and groups, as well as groups and groups. It relies on

good institutional norms and effective leadership so that members of the organization can perform their duties, coordinate to ensure the efficient operation of the whole organization, and achieve organizational goals. The Beijing Olympic Organizing Committee has 13 departments: secretarial administration department, general planning department, international liaison department, etc. It organizes multi-department settings within the organization, cooperates closely with each other, and ensures operational efficiency.

5.1.2 Engineering Organization

As its name implies, an engineering organization is the extension and development of the organization in the field of engineering in a general sense. However, engineering organizations are different from other organizations because of their unique characteristics.

5.1.2.1 Definition of Engineering Organization

From the point of view of human history, every exploitation and utilization of resources with historical significance will change people's thinking, production, and lifestyle and play an irreplaceable role in promoting human development. The use of fire has separated humankind from the era of eating animals and birds raw. The discovery and use of stone, bronze, and iron have brought about a qualitative leap in the history of human agriculture. The invention of the steam engine has brought humankind into an unprecedented period of rapid development. Today, the application of electricity, nuclear energy, wind energy, solar energy and information network make human activities more colorful and have achieved unprecedented modern human civilization.

From historical sites to modern large-scale projects, where are the main drivers behind these projects? Is it engineering? Is it an individual? In fact, human activity is ultimately about creation and use. Creation is literally defined as making things. Broadly speaking, as long as a product is produced through certain labor, even simple duplication or improvement, it can be called creation, and this kind of "thing" can be either a product with physical form or a product with the virtual state (such as technology, patents). In a narrow sense, creation is to use all kinds of resources to produce objects with physical form through labor. Human engineering activities used labor, material, machinery, and other resources organized technology integration and innovation, creating housing, bridges, roads, water facilities, and other physical objects, creating a new "artificial nature." However, engineering is a kind of creative activity with more division of labor and more participants than workshop-like creative activity with individual or collective units. It will produce more diverse ideas and technologies. It is necessary to build a more appropriate organization, namely an engineering organization, to coordinate the activities of all parties. Creation in this chapter refers to creation in a narrow sense.

The development of science and technology has provided enormous space for engineering management. After World War II, project management emerged mainly due to military engineering. Along with the significant development of the economy, project management has penetrated water conservancy, electric power, medicine, chemical industry, IT, and other industries. At the same time, the mega-scale, complex and multi-subject characteristics of the project require more and more organizations. The project's strong planning, procedural and legal characteristics greatly impact the economy, society, and environment; its construction rule is different from the production rule of general commodities. This determines that the engineering organization should take on the management and control of the whole project system. In order to achieve the goal of the project, the engineering organizers should make rational and effective use of various engineering management theories and methods in engineering activities to promote the smooth progress of engineering construction.

In addition, engineering management is a collective social action; the engineering community is another form of the engineering organization. It is difficult for any project to be completed only by the engineer group. The actual project activities need different members, such as owners, contractors, supervision units, financial institutions, government departments, and the public, to perform their duties and responsibilities and work together. Consequently, the engineering community is a hierarchical, multi-role, division of labor and cooperation, multi-interest, and complex primary body system of engineering activities assembled under specific engineering activities to achieve specific engineering objectives. It is a complex system consisting of owners, contractors, supervisors, engineers, financial institutions, government departments, the public, and other stakeholders. Given the multi-complex system of the construction engineering community, building an organizational structure that can be improved continuously is imminent. The engineering community is an active group formed based on a common engineering paradigm. It has the same engineering value to achieve the goal, making the construction of engineering organizations with natural rationality and allocating organizational resources reasonable and efficient.

Through the above description and analysis, according to the definition of organization in organizational behavior and the requirements of engineering science for organization, an engineering organization is defined as a flexible group that focuses on creation goals, pools forces from all parties to allocate and use tangible resources and intangible elements efficiently.

First, an engineering organization has very clear goals of creation. The goal reflects the direction and potential results of the organization's efforts. The objectives of the engineering organization include general objectives and operational objectives. The overall goal is the fundamental reason for the existence of engineering organizations and the ultimate benchmark and guidance of engineering organization behavior. Operational objectives are the main tasks to be accomplished in the management activities of an engineering organization. The specific objectives of each major task are to provide guidance for the organization's daily decision-making and activities, including performance, resources, market, employee development, innovation, and other goals. The China Lunar Exploration Project is divided into circling, landing, and returning. The main objective of the circling phase is to launch a satellite ("Chang'e-1") flying around the moon to carry out global, integrated, and comprehensive exploration of the moon. The main goal of the "landing" phase is to achieve a soft landing on the lunar surface and a lunar patrol detection. The main objective of the "returning" phase is to launch a lunar soft lander, obtain the lunar samples automatically and return safely to Earth. The three stages of implementation are operational objectives (main tasks in stages). The achievement of operational objectives is the basis for achieving the overall objectives of the lunar exploration project.

Second, engineering organization has the function of adhesion. Modern engineering is becoming more and more large and complex. It is difficult for a single subject to accomplish the project objectives. It requires the participation of multiple subjects. However, this is not simply a superposition, but the integration of resources, coordination of various engineering activities, unification of direction, stimulation of cohesion, and achievement of the effect of "$1 + 1 > 2$". Taking China manned spaceflight engineering as an example, the project involves more than 110 research and development units of 13 systems and more than 3000 cooperation and support units. To ensure the achievement of the three-step strategic goal, the central government has set up the China Manned Space Engineering Office, which is responsible for the special management of space engineering. The government has established the China Manned Space Engineering Office to implement the special management of large-scale systems engineering, give full play to the cohesion of engineering organization, and coordinate the related work of the units mentioned above.

Third, engineering organization plays a catalytic role. The investment of human, financial, material, information, experience, knowledge, and other resources is the premise to ensure the achievement of the goals. However, the simple investment of resources is not enough. Engineering organization integrates resources of all parties, puts resources into the suitable activities at the right time, makes them undergo "chemical reaction," and produces new materialized productivity. The catalytic role of engineering organizations highly depends on modern technology. For example, with the help of the Three Gorges Project Management System (TGPMS), the creation activities are real-time monitored and feedback, ensuring the effectiveness of resource input and improving the quality of creation activities [5]. In addition, according to relevant statistical data, more than 280 million cubic meters of concrete, 460,000 tons of reinforcing steel bars, and 260,000 tons of metal structures were used in the Three Gorges Project. These creation resources were reasonably and orderly invested in the construction of the Three Gorges Project following the creation plan, ensuring the smooth achievement of the construction goals.

In addition, engineering organization also plays a lubricating role. The project involves many participants who use contractual relationships to achieve mutual benefits. However, the requirements and expectations of each entity are different, and conflicts and frictions are unavoidable. By establishing communication channels, coordinating interpersonal relationships, and emphasizing common goals, engineering organizations can minimize conflicts and frictions among participants, improve overall work efficiency and achieve a win–win situation.

Finally, an engineering organization is a combination of group organization and flexible organization. Engineering organization includes traditional organizations with specific organizational structure, organizational size, and organizational habits. They gather to form a "group organization" for the common purpose of creation." For example, Apple published *Supplier Social Responsibility Progress Report* in 2012, announced its 156 suppliers and partners, covering 14 categories. This group organization not only relies on power distribution and affiliation to ensure the healthy operation of each small group but also depends on the contractual relationship to maintain the interaction between the groups, which makes the participants more flexible and accessible in entering, joining and exiting, to ensure the coordinated operation of this large engineering organization. Therefore, the engineering organization is a kind of flexible group organization, after all.

5.1.2.2 Form of Engineering Organization

Engineering organization is holistic, and many elements and members need to be combined in an orderly way to achieve the most effective connection. The connection among the components of an engineering organization includes the vertical link of different levels, the horizontal cooperation connection of the same level, and the communication relationship between each other. These relationships interact to form a diversified form of the engineering organization. Because the scale, shape, technical requirements, environment, and objectives of the project are not equal, there is no fixed organizational form suitable for all projects, and an appropriate organizational form should be adopted according to the project's situation.

Engineering organization has different forms, but the reason lies in the evolution and combination of several primary forms of organization. Just as human beings differ in the diversity of gene expression, in the final analysis, this diversity is due to the different combinations of DNA segments. Therefore, this chapter mainly discusses the following primary forms of engineering organizations. They are effective "DNA fragments" of various engineering organizations, which deduce the diversity of engineering organizations together.

1. **Linear organization structure**

Linear organization structure is the earliest and simplest form of organization in an engineering organization. Linear organizations carry out vertical leadership strictly according to a hierarchical division from top to bottom. Subordinates only carry out orders from superiors; no special functional departments are set up, all management activities and functions are carried out by the superiors themselves, as shown in Fig. 5.1.

The advantage of linear organizational structure is that it is more concise than other organizational structures. It has a clear superior-subordinate relationship and division of labor, and it can make decisions quickly. The disadvantage is that it requires the high comprehensive ability of leaders at all levels and requires them to be proficient in various knowledge and skills. It is good at the functional division of labor, so it

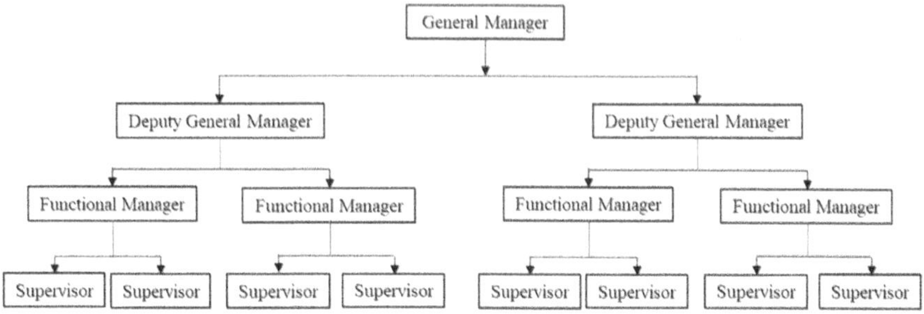

Fig. 5.1 Linear organization structure

isn't easy to use in large and complex projects. Therefore, the linear organizational structure is only suitable for those projects with small-scale construction, simple production technology, fewer participants in the project, or a specific sub-project after project decomposition.

2. **Functional organization**

The functional organization structure is to set up functional departments of human, financial, material and production, supply, and marketing management separately in the organization. Each functional department can issue orders to lower-level departments within their respective business scope. The functional structure requires superiors to be fully authorized to relevant functional departments. The content work of engineering is often divided according to the functions of relevant departments. Various professional and technical departments charge technical work, financial analysis is responsible by the financial department, personnel management is responsible by the human resources department, etc. Under the leadership of senior managers, the heads of functional departments form a coordination layer and personally arrange and implement the work of relevant personnel in their respective departments. This structure is shown in Fig. 5.2.

The advantage of the functional organization is that it can adapt to the engineering projects with complex engineering technology and detailed management and give full play to the professional role of the functional organizations. The disadvantage is that it often leads to multiple orders, which is not conducive to the necessary centralized leadership and unified command. It easily makes the subordinates not know what to do, affecting the normal operation of the management mechanism. Functional organization is usually used in some industrial projects with high repeatability of products, but it is challenging to apply to variable projects.

3. **Linear-functional organization**

The linear-functional organization has the characteristics of both linear and functional organizations. As shown in Fig. 5.3, this form of engineering management organization can be generally divided into two categories. That is, according to the unified command principle, the linear organization exercises command to the organization at all levels, and according to the principle of specialization, the functional

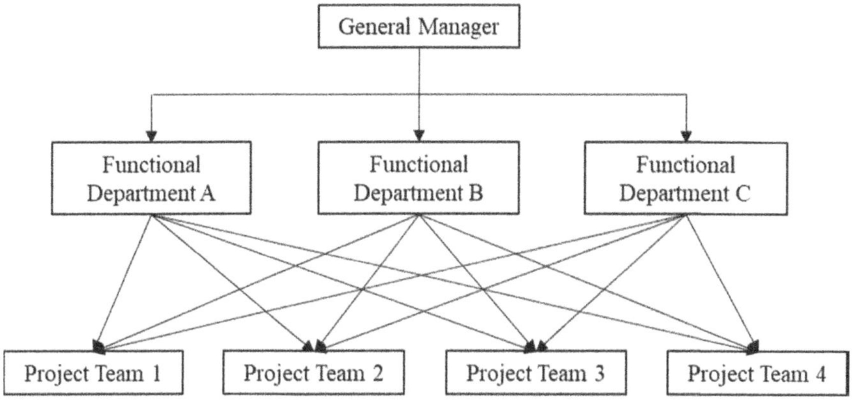

Fig. 5.2 Functional organization structure

organization exercises the functions of the organization. The managers of linear organizations exercise command according to their responsibilities and powers and make corresponding decisions to be responsible for the work of their departments. The functional organization guides according to the business. It does not directly give orders to the departments but assists the linear management in making decisions. This organization is a form in which the higher-level unified leadership command and functional departments help decision-making, playing a management role. Subordinate departments must both obey the orders of higher-level leaders and receive guidance and supervision from other functional departments at the same level.

The advantage of the linear-functional organization is that it can fully play the professional advantages of various functional departments under unified management. The disadvantage is that the functional department only has the right of consultation and must ask for instructions from the leaders, resulting in inefficient work.

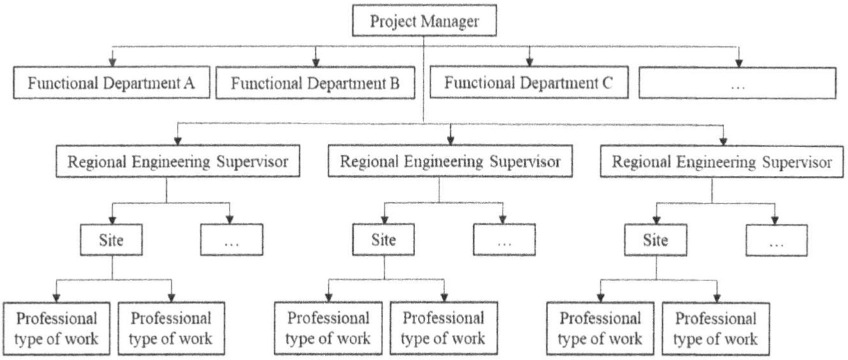

Fig. 5.3 Linear-functional organization structure

Therefore, it is not often used in construction projects but mostly in industrial projects such as manufacturing.

4. **Multidivisional system**

The division system originated from General Motors Corporation in the United States, a highly decentralized management system. It is suitable for enterprise groups and joint companies with large-scale, multi-product, broad customer groups, and large regional spans. The multidivisional system is usually the upper-level headquarters that controls personnel decision-making, budget control, and supervision. The lower-level departments are divided into different divisions according to the project's location, the characteristics of the targeted customer group, and the functional grade of the product. They are separately responsible for the research and development, design, procurement, production, and sales of the related products. Each division has independent accounting and independent operation and has substantial autonomy. The headquarters control the performance of each division through metrics such as profit; that is, each division is a relatively independent organization. The structure is shown in Fig. 5.4.

There are many advantages of the multidivisional system. Firstly, it can better adapt to market changes, and each department chooses the most suitable production mode according to its environment. Secondly, as a highly decentralized organizational form, the enthusiasm and creativity of the participants have been better played. Thirdly, the decentralization of headquarters can reduce the burden on senior managers by evaluating the performance metrics of each division. Fourthly, competition can be formed among various divisions to promote the efficient operation of the whole organization jointly. The shortcomings are also apparent such as the vicious competition among multiple divisions and the high requirements for headquarters management. Otherwise, it is easy to get out of control. The organizational form of division system is often used in large-scale projects across regions.

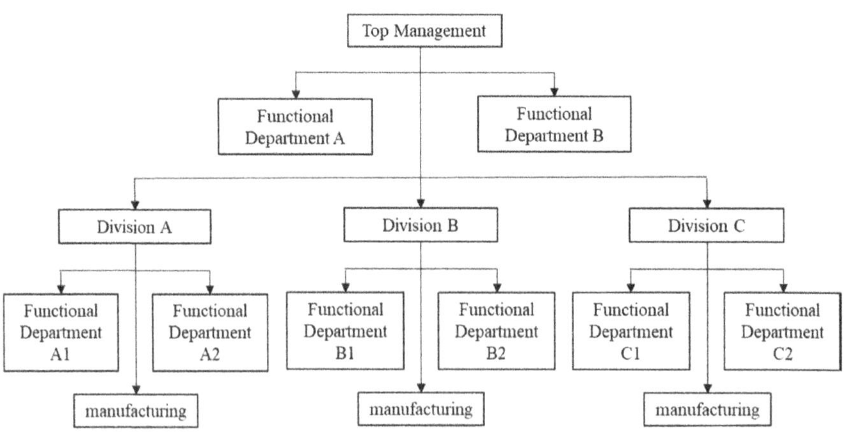

Fig. 5.4 Multidivisional system organizational structure

5. Simulated decentralization system

Simulated decentralization is a kind of organizational structure between linear functional systems and multidivisional systems. Based on the linear functional system, the organization simulates the model of separate accounting and independent operation of the multidivisional system. It artificially divides the organization into several production units, but they are not really "divisions." Due to the limitation of product categories and technical conditions, it isn't easy to classify them independently strictly. For example, in a large manufacturing enterprise, there are two kinds of products, A and B, divided into production units according to their product categories. However, because product A is an essential material for manufacturing product B, it is difficult for the two production units to be treated as divisions. Faced with this situation, large-scale organizations can adopt the form of simulated decentralization so that each production unit has an independent functional organization, as far as possible has greater autonomy, and assumes "simulated" responsibility for profits and losses in order to mobilize their enthusiasm for production and improve production efficiency, as shown in Fig. 5.5. The advantage of a simulated decentralized system is that it can reasonably solve the problem of decentralization caused by large scale enterprises. At the same time, it can reduce the burden of senior managers. The disadvantage is that there is no independent external market in the simulated decentralization system, and it is difficult to assess the performance of each production unit. The tasks of each production unit are closely related, and it is difficult to define their respective rights and responsibilities. At the same time, there are also shortcomings in information communication between management of production units. The organizational form of simulated decentralization is often used in projects with continuous production.

6. Matrix system

The linear functional system has shortcomings, such as less horizontal communication and less flexibility. The matrix system is an organizational form to make up for these shortcomings. It has two different departments, horizontal and vertical, under

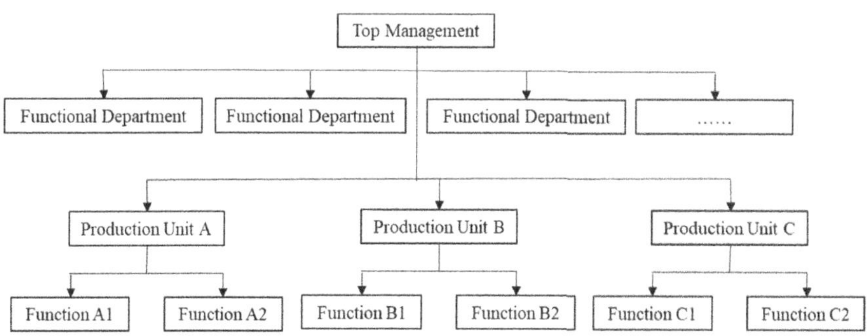

Fig. 5.5 Simulated decentralization system organizational structure

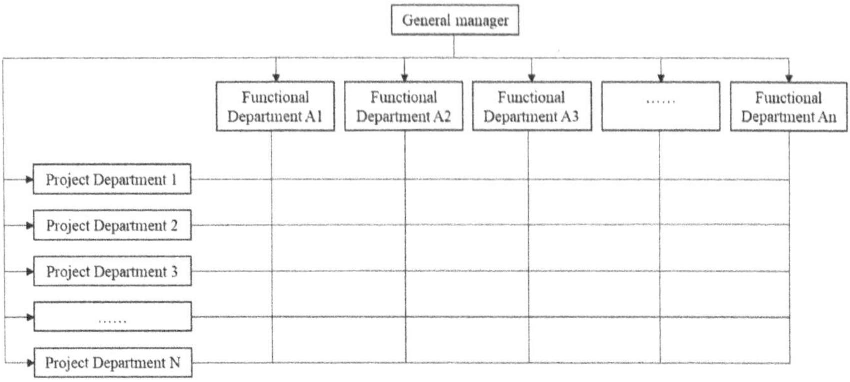

Fig. 5.6 Matrix system organizational structure

the top management. Horizontal departments are the functional management departments, while vertical departments are the project departments for specific tasks, as shown in Fig. 5.6.

The advantage of a matrix system is that it can set up an independent cross-functional department for a project, so when a new project is started, a special project team can be set up to complete the work of the whole life cycle of the project. Personnel from the corresponding functional departments can be drawn to participate in the work according to their specialty at the appropriate stage. When there is a conflict between the vertical and horizontal departments, the highest commander can coordinate the task to ensure the completion of the job. Under this organizational structure mode, personnel are mobile and can be allocated flexibly according to the situation of the task. The project team and the person in charge are also temporarily organized and appointed in this organizational structure. The personnel will be disbanded after completing the task and return to their original units. Therefore, the organization form is suitable for horizontal collaboration and key projects.

The advantages of a matrix system are as follows: first, it has substantial flexibility. It is developed based on the project. It has straightforward tasks, clear objectives, and can gather the professional talents of different functional departments. Plus, teamwork fully stimulates the creativity of project team members and overcomes the difficulties in the project; Second, it can effectively overcome the disconnection between different departments in the linear functional structure and strengthen the cooperation and information exchange between various departments. The biggest disadvantage is that it will lead to double leadership. As the project participants are from different functional departments, they mainly listen to the administration of functional departments. It is difficult for someone to temporarily assume responsibility for effective management, resulting in too big responsibility and unequal responsibilities and powers. Project team members are also prone to temporary ideas, which have a certain impact on the work. The matrix structure is a common organizational form in an engineering organization, especially suitable for all kinds of major projects, complex projects, and projects requiring technology innovation.

5.1.2.3 Elements of Engineering Organization

The constituent elements of engineering organization can be summarized as follows:

1. **Have a clear goal of creation**

Chester I. Barnard believes that goal is a necessary condition for the existence of an organization and argues that every intent of an organization should be understood and accepted by its members. Coordination and cooperation within the organization are based on a clear goal, which can provide a basis for the organization's activities. Barnard's organizational theory holds that organizational goals should also be distinguished from those of the members, emphasizing the variability of objectives [6].

The engineering organization is centered on the creation and guided by defining the goal of creation. The project should understand the goal of creation. It is phased and dynamic. It can comprehensively take the engineering organization's ontology and environment into consideration to condense all forces (such as resources, technology, information, etc.) at all stages. Without the goal of creation, it cannot be called an engineering organization. With the corresponding goal of creation, the engineering organizer can determine the direction, play the role of adhesives, mobilize resources of all parties, pool forces of all parties, and form a working group with the ability to create. For large-scale creation activities, the goal of creation is generally achieved by decomposition in a phased and orderly manner. For example, the "three-step" strategy of China's manned space engineering (including the achievement of space-to-Earth commuting, the establishment of space laboratories, and space stations) gradually promote the development of China's manned space engineering.

2. **Own resources**

Resources mainly include five categories: human, financial, material, information, and knowledge.

The greatest resource of an engineering organization is humans, which is one of the essential elements of an engineering organization and the source of its creational ability. For example, in general engineering projects, technicians, managers, and logistical support personnel perform their duties and work together to serve the creative activities and ensure the smooth development of the creative activities.

Finance mainly refers to capital. Without the support of funds, engineering organizations cannot achieve development. Because of the support of funds, all kinds of activities of engineering organizations can be carried out in an orderly manner. Especially, modern engineering has the characteristics of large-scale and complex, which results in the increasing demand for funds. Before the creation activities are carried out, it is necessary to open the access to funds, reduce financing costs and prevent potential funds risks. In the development of creative activities, we should attach importance to the arrangement of funds and meet the creation requirements in a timely and adequate manner.

Materials mainly refer to finished products, semi-finished products, raw materials, and equipment. "Supply goes before troops." As far as engineering organizations are concerned, only when they have material resources such as raw materials and machinery can they play the role of "catalyst." Then follow the plan of creation, allocate and invest resources in a timely and reasonable manner, and produce high-efficiency productivity to meet the specific needs of the development of engineering organizations and accomplish the purpose of creation.

Information resources mainly refer to information (such as industrial development documents), instructions, data (such as price level), and symbols. There are many participants in modern engineering, and their relationship is complex. In addition, the complexity and large-scale of the project will produce a considerable amount of information. The engineering organization needs to make effective use of this resource, maintain creation ability, play the role of "lubricant," and maintain a good cooperative relationship between all parties. For example, the engineering organization should keep track of the market information such as the price, supply, and demand level of all kinds of production materials promptly, which is very helpful to the cost target control.

Knowledge includes tacit knowledge (such as experience and know-how) and explicit knowledge (such as norms and industry standards). The high efficiency of engineering organizations in the process of creation, to a large extent, benefits from the collective efforts of the members of the organization. Every member's background, technology, and experience are a kind of knowledge and an important magic weapon to solve various problems in creation. Engineering organization business process itself is a process of knowledge reengineering. Attention should be paid to the use of knowledge management in the organization, such as knowledge acquisition, sharing, innovation, and application, to enhance the knowledge proportion of the results of creative activities.

3. **Maintain a specific contractual relationship**

The classical organization theory represented by Weber and Fayol emphasizes the organization's rights relationship. The change of the rights relationship within the formal organization will cause the change of the organizational structure and then affect the achievement of organizational goals. Based on classical organization theory, behavioral science theory covers the attention to people. It emphasizes that organization is an economic system and a social system. It should consider the social and psychological needs of people. The organization should go beyond the power and responsibility system, consider the organization's interpersonal relationship, and pay attention to the communication between members and its influence. Modern organization theory considers environmental change based on the former views. Herbert A. Simon believes that the importance of social functions is gradually increasing, while the importance of hierarchical status is gradually decreasing [7]. Peter F. Drucker argues that information technology will constantly change the organization's work and organizational structure [8]. That is to say that the organization should consider the internal and external relations, which is the basis of writing a contract.

Contracts are documents, terms signed by both parties in order to achieve the purpose of buying, selling, and leasing. *Shuowen Jiezi* explains, "contracts are important agreements." There are relatively many types of contracts, such as unilateral and bilateral contracts, real and consensual contracts, named contracts and contracts without names. Ronald H. Coase clearly put forward the contract theory in *the Nature of Enterprises*. He believed that enterprises and markets are two alternative means for coordinating production. The market completes transactions through contracts, while the enterprise organization completes transactions through internal authority [9].

An engineering organization is a group of organizations. The organization's rights and responsibilities are clearly divided, so the tasks have specific executors, and the rights, responsibilities, and interests are equal. At the same time, the management range of the organization is appropriate, the organizational level is clear, and the organizational structure is reasonable. In addition, there is a clear contractual relationship among the organizations, which stipulates the rights and obligations of the participants, guarantees the communication between the parties, maintains the good interaction between the parties, and guarantees the normative and flexible nature of the engineering organizations, to form an orderly, logical and efficient collaboration system. Creative activities cannot be carried out smoothly without various multi-party participants, and the responsibility and rights of participants are highly dependent on contract-based bonds. For example, the components of a Boeing aircraft come from about 5000 suppliers worldwide, and the basis for proper and effective management of so many suppliers is a contractual relationship.

4. **Composition and structure of elements**

Whether in the natural or social field, the structure of things determines their function to a certain extent. In nature, two substances with the same element but different structures are allotropic, such as diamond and graphite (Fig. 5.7). Graphite is a crystalline mineral of carbonaceous elements with a layered hexagonal structure and ample distance between layers, one of the softest materials. Diamond is an octahedron, colorless and transparent, but it is the hardest substance in nature. The reason is that each carbon atom is tightly bonded within the diamond crystal and supports each other to form a dense three-dimensional structure. Because of this dense structure, diamond has the most extraordinary hardness.

Similar phenomena also exist in engineering organizations. The elements of engineering organization mentioned above include human, financial, material, information, knowledge, etc. The constituent structure of these elements is also an important aspect of the engineering organization. The constituent structure of these elements determines the function of an engineering organization. The function and efficiency of an engineering organization with the same resources may be vastly different if its internal composition and structure are different. Suppose the elements within the engineering organization can maintain close contact for the same purpose, support and depend on each other in their work, and maintain integrity with the outside

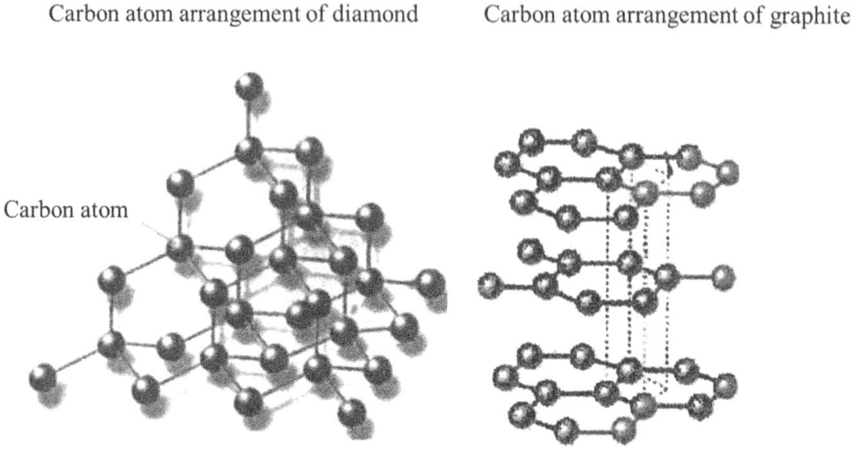

Fig. 5.7 Diamond and graphite atomic composition diagram

world. In that case, there is no doubt that such a close and mutually supportive organization team will develop rapidly and become the hardest "diamond organization." However, if the elements in the engineering organization cannot be closely linked, the joints between them are loose and weak, and there is no common goal orientation, it is bound to be as weak as graphite. The composition of internal elements and the construction of the organizational structure of an engineering organization play a decisive role in the function of external organization, just like diamond and graphite.

5.1.2.4 Characteristics of Engineering Organization

The constituent elements of an organization have an important influence on its characteristics, and the constituent elements of an engineering organization also reflect its characteristics. In addition to the characteristics of general organizations such as purposiveness, holisticness, and openness, engineering organizations have their characteristics including stage, dynamics, and virtuality.

1. **Purposiveness**

Engineering organizations are created by engineering tasks and have a very clear goal orientation. For example, the goal of a manned space project in China is to build a permanent space station. The overall goal of the South-to-North Water Diversion Project is to improve and restore the ecological environment of water-deficient areas in the north. It can be said that because of the large number of stakeholders and interests, the project is a contradiction under the multi-objective.

2. **Holisticness**

Throughout history, mankind's understanding of holisticness has gone through a long time. Many ancient philosophers have their own unique views on holisticness.

In ancient China, the simple concept of holisticness emphasized wholeness, harmony, and coordination, which is strongly supported by "the unity of heaven and earth, and the unity of all things" in *Huainanzi Spirit*. Many ancient Greek philosophers also discussed the integrity of things. G. W. F. Hegel's dialectical holistic view holds that they are all connected in many ways, from the natural world to the spiritual world. Aristotle proposes that the whole is greater than the sum of its parts.

The holisticness of engineering organization includes two meanings. Firstly, the engineering organization comprises several elements, such as the goal of creation, resources, and their corresponding relations, inseparable from each other. When the activity of creation needs, it should be put into the activity of creation in a reasonable and orderly manner according to the pre-determined plan. For example, aerospace engineering needs multi-professionals and the support of capital, technology, materials, and other elements. Secondly, an engineering organization is not simply a combination of the elements, but an organic gathering of the elements, and its overall function cannot be obtained through the simple addition of the functions of the elements. The catalytic effect of an engineering organization makes the whole organization have new functions and attributes different from each constituent element.

3. **Openness**

In his book *General Systems Theory: Basis, Development and Application*, Ludwig von Bertalanffy first proposes that "Life system is essentially an open system, which is defined as a system of exchanging substances with the environment" [10]. The engineering organization is also an open system. It constantly interrelates and interacts with the external environment, chooses, exchanges, and transforms the elements of matter, energy, and information, which are applied in creative activities and display its holisticness.

The development of engineering organizations cannot be separated from this kind of openness, and opening to the outside world is also a prerequisite for the stable existence of organizations. For example, in the process of high-speed rail construction, it is necessary to obtain lightweight materials such as aluminum alloy continuously, fiber-reinforced epoxy resin, and other composite materials from the external environment to manufacture high-speed rail vehicle bodies. High-quality carbon steel, low-alloy, low-carbon high-strength steel, polymer composites are obtained for bogie manufacturing, polyurethane, carbon fiber composites, silicone rubber, ethylene-propylene terpolymer, and other polymer materials are obtained for ballastless track system construction. Because of the open interaction with the external environment, the engineering organization of high-speed rail construction can continuously obtain all kinds of production factors for the selection of creation activities.

4. **Stage**

The life cycle of an engineering organization refers to the time process of the organization from birth to death, which is a natural process. Richard L. Daft divides the life cycle of an organization into four stages: entrepreneurship, collectivization, standardization, and refinement, according to his book *Organization Theory and*

Design [11]. Growth is not easy. Every time an organization enters a new stage of its life cycle, it will be accompanied by a set of corresponding new regulations. However, as a flexible group organization, most engineering organizations have a strong temporary nature. The focus of their work in the life cycle is more related to the engineering attributes, which is different from the general enterprise organizations. In the early stage of the project (decision-making period), the entrepreneurship stage of the project organization mainly focuses on the functions of decision-making and planning. In the mid-term (implementation period), the collectivization stage of the engineering organization focuses on control, organization, leadership, and management functions. In the later stage of the project (implementation period), the standardization stage of the project organization is mainly to ensure delivery and continued operation. In the refinement stage of general organization, for engineering organization, it is at the end of the old project and the beginning of the new project. The old organizational relationship will be dissolved, and the new organization needs to be formed. Engineering organizations in different stages have different working nature, scope, and objectives. Therefore, the managers of engineering organizations should use correct management methods to improve the organization's operational efficiency according to the characteristics of different stages of the organization.

5. **Dynamic**

Karl Heinrich Marx's dialectical materialism holds that matter moves and movement are absolute. Systems theory emphasizes that dynamics is the basic characteristic of a system. The dynamic nature of engineering organization is embodied in three aspects: firstly, the engineering organization can always keep the choice and exchange of material and ability with the environment in which the engineering organization is located; secondly, the creative activity has a life cycle, and the whole life cycle of the generation, development, and extinction of the engineering organization is dynamic change; thirdly, at different stages of the project life cycle, the engineering organization will adopt timely operation mode and project implementation plan to maintain moderate productivity, in order to adapt to the characteristics of different stages of creative activities.

6. **Virtuality**

A virtual organization is the continuation of global collaborative production in the era of an industrial economy and a new thing of enterprise organization in the information age. Specifically, with the help of information technology, organizations break through the space constraints and form dynamic alliances within a certain period to accumulate resources, information, and other resources. For engineering organizations, with the development of information technology, intensified competition, and economic globalization, it is inevitable to develop from traditional organizational form to virtual organization. Virtual organizations can help organizations share core resources and gain strategic advantages. It can be predicted that with the development of science and technology and the intensification of globalization, the improvement of specialization, and the increase of cooperation among organizations, engineering organizations will turn more to virtualization. For example, Airbus A320 aircraft manufacturing process involves many parts from outside the European mainland, China's Hafei Aviation Industry Co., Ltd. provides carbon fiber reinforced composite body parts to it. Undoubtedly, virtual organizations play a vital role in this case.

5.1.2.5 Tasks of the Engineering Organization

Today, the old axiom of the economist Milton Friedman says, "business is business," is no longer entirely accepted by people. Society and people have higher requirements for engineering constructions. They emphasize that engineering projects achieve commercial value and take certain social responsibilities. The focus of project management research has extended from traditional "efficiency" to "society." Therefore, to meet social and public needs, the tasks of engineering organizations have been improved qualitatively, reflecting the special value of engineering activities. At the same time, the cross-domain characteristics of engineering activities and the diversification of stakeholders determine that the project contains multiple social values.

Modern engineering activities are a part of market economy activities. The engineering community members participating in modern engineering activities are naturally "economic men." They pursue the maximization of personal interests without considering social interests. The environmental and social losses caused by various projects are transferred to others and future human beings, resulting in negative externalities. It affects the social value of the project directly. In this case, engineering community members need to fully realize that they have two identities of "economic man" and "social man." They cannot only pay attention to personal interests but ignore environmental factors. They should correctly establish the concept of environmental protection in order to maximize the social value of engineering.

As a "balance valve" for the community players, the engineering organization needs to take into account the internal benefits of the organization and the external benefits of society. On the one hand, from the ontological point of view, engineering activities are multi-dimensional, covering political, economic, social, cultural, and other factors. It is a comprehensive practical activity. Engineering projects need active collaboration and cooperation among various stakeholders. Then the organization members must be able to obtain the corresponding benefits in the construction activities. On the other hand, as a kind of social existence, engineering activities cover a series of links such as project decision-making, feasibility study, engineering design, engineering construction, etc. They all involve social factors. Therefore, any engineering project exists in a certain social structure and social relationship and is limited and restricted by a certain time and space. The social value evaluation of engineering projects should not be neglected in the analysis of engineering activities, especially the systematic analysis of the social risks and possible social impacts of engineering projects.

5.1.3 The Core of Engineering Organization: Engineering Community

Suppose the engineering organization is regarded as a people, finance, material, and knowledge system; in that case, the core of the engineering organization is the collection of the participants, which is the engineering community.

5.1.3.1 Components and Organizational Forms of Engineering Community

"Engineering Community" occupies an important position in the related research of engineering organization theory because engineering activities, as the closest approach for human existence, need to form certain relationships and carry out purposeful and organized social actions in the process of adapting to and utilizing nature [12].

1. **Community**

Aristotle first proposed the concept of community. The first sentence of his article *Politics* is, "Every state is a community of some kind, and every community is established for some good reason" [13]. Aristotle believes that the first community was the family, then the village, and then the city-state community (family, village, and city-state). In 1887, Ferdinand Tonnies, a German sociologist, pointed out in *Community and Society* that the community is earlier than society in the history of human development. Still, people's understanding of community is later than society. The community mainly grows naturally with the ties of kinship, friendship, and ethical harmony. Its primary forms are kinship (blood community), neighborhood (geo-community), and friendship (spiritual community) [14]. In 1917, R. M. Maclver, a British sociologist, distinguished the connotations of society, community, and association in *Community: A Sociological Study*, which further expanded the understanding of community. Although scholars have different definitions of community, the concept of "community is a group, organization or team that lives together for a specific purpose" has been generally accepted by sociologists [15]. Since the 1980s, "Communitarianism" has become an important research content in sociology and management. In communitarianism, "community" includes two aspects; one is the "big community" like a country, the other is the "intermediate community" like churches, associations, professional associations, classes, races, and so on [16]. Since the twenty-first century, with the development of network information technology, all kinds of forums, virtual communities, and communities of practice (CoPs) based on network platforms without space and time constraints have become the research hotspots of many scholars. For example, in knowledge management, communities of practice, as an effective organizational form to promote knowledge diffusion, is considered to be one of the most effective tools for knowledge sharing and innovation. The characteristic of a practice community is to break the boundaries of traditional organizational departments, gather people interested in a particular field of knowledge to form an informal group, and provide a conversational communication platform for employees, thus transforming their knowledge into organizational knowledge [17].

2. **Engineering Community and Its Components**

The process of human development relies on nature, recognizing nature, adapting to nature, and using nature reasonably and appropriately. Engineering activities build a bridge between scientific discovery, invention, and industrial development [18]. Engineering activities are purposeful and organized collective social actions, and the engineering community is the basic organizational form of engineering activities. For a long time, there has been a misunderstanding about the engineering community, which regards engineering as the activity of engineers, so the engineering community is the group of engineers. However, it is difficult for any project to be completed only by the engineer group. Actual engineering activities require the owners, contractors, supervision units, financial institutions, government departments, and public members to perform their duties, responsibilities and cooperate. Consequently, the engineering community is the main body system of engineering activities with hierarchical classification, diverse roles, clear division of labor, mutual cooperation, multiple interests, and complex contents, which is formed in order to achieve the predetermined project objectives under specific engineering activities. It is a complex system consisting of owners, contractors, supervisors, engineers, financial institutions, government departments, the public, and other stakeholders.

(1) **Owner**. In the construction process, investors (government investment, enterprise investment, individual investment, etc.) set up special organizations or appoint special persons responsible for the management of projects as owners. The owner is the owner of the project. In the process of project implementation, the owner can be divided into fund raisers, managers and controllers in the whole process (investment control, progress control, quality control, and safety control, etc.).
(2) **Contractor**. It refers to an enterprise with a certain economic foundation, construction, production, and operation capabilities, a team of technicians and managers, and a business license capable of undertaking the corresponding business, which can meet the owner's requirements to undertake the specific implementation of the project.
(3) **Supervision unit**. Provide professional service activities, including construction progress, investment, quality control, contract management, safety management, and organization coordination to the owner. Engineering supervision is a kind of paid engineering consulting service.
(4) **Engineer**. Engineers have a special position in the engineering community and play multiple roles. Firstly, engineers have the professional technical knowledge and practical experience, and they are the "technical authority" in the engineering community. Secondly, to achieve the project's expected objectives, engineers need to assist project managers in formulating effective engineering

implementation standards and technical management systems. Thirdly, due to the uncertainty of engineering activities, engineers need to choose the best alternative according to the specific environment and conditions in engineering implementation and put it into practice. Engineers run through engineering activities and appropriately integrate production factors such as labor, technology, and management.

(5) **Financial institution**. Modern large-scale projects need a large number of funds, which require the support of financial institutions. Financial institutions include a lender to provide funds for engineering construction, a guarantor of engineering guarantee to provide credit support for the project, etc.
(6) **Government department**. According to the state's laws and regulations related to construction projects, the whole project is examined, supervised, managed, and controlled by national interests.
(7) **The public**. It refers to the public groups, including residents, community organizations, and users, affected by the construction and operation of the project.
(8) **Other stakeholders**. Other stakeholders refer to other participants in engineering activities besides the seven categories mentioned above—for example, NGOs, NPOs, and news media related to engineering projects.

3. **Organizational form of the engineering community**

According to different organizational structures, the engineering community can be divided into "engineering activity community" and "professional engineering community." The former is more essential than the latter. Without the engineering activity community, there would be no professional engineering community. In this sense, compared with the engineering activity community, the professional engineering community is a derived sub-community. At the same time, as a secondary sub-community, it has its necessity to exist. It serves the community of engineering activities, guarantees the rights and interests of the members of the professional engineering community, establishes the industry norms of the engineering community, improves members' professional quality, and enhances the sense of industry identity. A systematic comparison of two different forms of organizational engineering community is shown in Table 5.1.

5.1.3.2 Maintenance Mechanism of Engineering Community

In order to explain the relationship between the elements of the engineering community and its mechanism of action, it is necessary to focus on the analysis of the reasons why different professions, heterogeneous individuals, can unite and that they must unite to become "engineering activity community." That is the necessity and feasibility of establishing the engineering community (Fig. 5.8).

(1) The common goal is the premise of the generation of the engineering community. The engineering community is a division of labor and collaboration system

Table 5.1 The differences between two types of engineering communities in different organizational forms

Type	Main manifestation	Goal	Task	Member characteristics	Interest retention	Duration
Engineering professional community	Association of Engineers, Union, Employers' Association, etc.	Members have basically the same common goals	Not responsible for specific engineering projects	Member homogeneity	Mainly safeguard the legitimate interests of various professional groups	Long duration
Engineering activity community	Enterprise, Company, Project Department	Members have their own goals	Responsible for specific engineering projects	Member heterogeneity	Consider and coordinate the interests of different groups	Short duration, disintegration as project ends

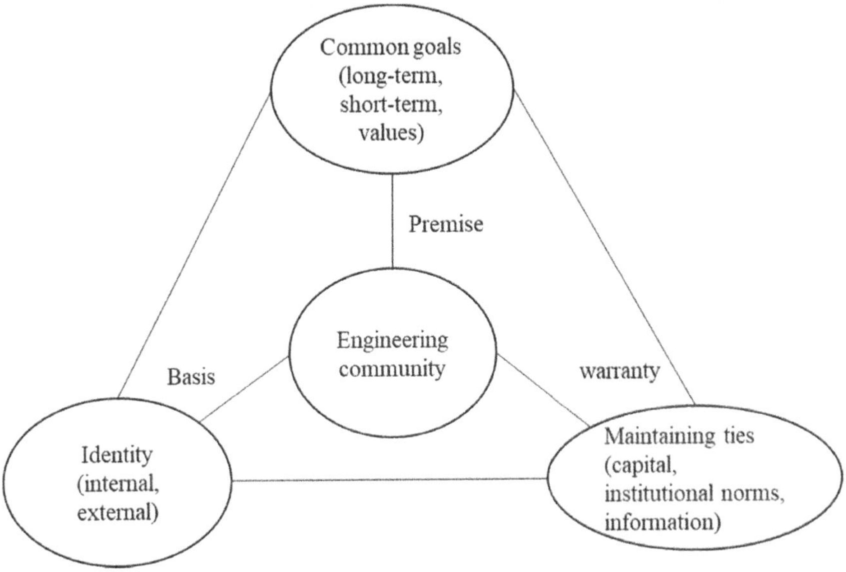

Fig. 5.8 Formation mechanism of engineering community

composed of many "heterogeneous members," and substantive collaboration is generated to pursue common goals by individuals. Therefore, the emergence of the engineering community must first have a common goal. Such goals may be "common short-term goals" or "long-term common goals" among community members or even common values.

(2) Identity is the basis of the generation of the engineering community. From the cognitive and psychological point of view, the "internal identity" and "external identity" of individuals and society to the engineering community are the foundation of the existence of the engineering community. Heterogeneous individuals constitute the engineering community. If the individual does not form a basic sense of identity, the community cannot create and exist. This is the "internal identity" of the individual to the engineering community. In addition, as a part of the social system, without the "external identity" of "society" (specifically expressed as law, social habits, other social groups), the engineering community cannot be established in society.
(3) All kinds of maintaining ties are the warranty of the operation of the engineering community. In order to ensure the healthy and orderly operation of the engineering community, it is necessary to build a link to maintain the operation of the engineering community from the aspects of capital investment and benefit-sharing, institutional norms, information communication, and knowledge sharing. Otherwise the engineering community may face disintegration. Specifically, it covers three aspects [19]: ① Capital bond. Capital refers to monetary capital (financial capital) and physical capital (especially machine equipment and other means of production), human capital, and social capital. Related to this is the interest demand of capital investors, including the acquisition and distribution of economic and other interests. ② Institutional bond, including division of labor and cooperation within the engineering community, various institutional arrangements, management methods, post settings, behavior habits, communication relations, etc. ③ Information bond, including all kinds of professional knowledge and skills necessary for engineering construction and the normal operation, engineers can impart relevant knowledge and information to workers (explicit knowledge sharing) in the form of technical specifications through operation rules, operation details and so on. At the same time, engineering activities are imparted by apprenticeship. In the mode of tacit knowledge sharing, skilled workers (teachers) can impart their operational experience to new employees (tacit knowledge sharing) through words and deeds, thereby improving their operational skills and engineering quality.

5.1.3.3 Relations Among Main Elements of Engineering Community

Each participant in the engineering community is a cognitive actor. According to their different roles, each participant has their interest demands; each participant can achieve their interest demands through engineering activities under rules and regulations, thus resulting in social interaction among the participants in the engineering community.

In this interactive relationship, investors (government, enterprises, individuals, etc.) set up special organizations or appointed special persons as owners to carry out fundraising and the whole process of control. The owner chooses the contractor to carry out the construction through bidding and other means, and the contractor will

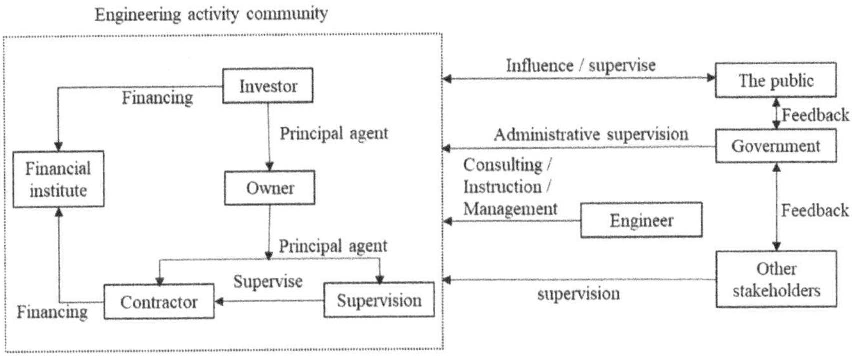

Fig. 5.9 Interactive relationship among the participating entities in the engineering community

also be supervised by the supervisory unit appointed by the owner. In the project implementation process, due to the continuous investment of funds, investors or contractors need to establish a corresponding financing relationship with financial institutions. At the same time, the government departments will carry out administrative supervision on the project according to the relevant provisions of the state and take into account the public interest. As technical, consulting and management experts employed by the engineering community, engineers are responsible for engineering design, technical consulting, technical guidance, and management. The residents, community organizations, users, and other social public affected by the construction and operation of the project are one of the key subjects to supervise the implementation process, output, and social impact of the project. They will also participate in the decision-making work of the project through the demonstration and hearing of the project. In addition, non-governmental organizations, non-profit organizations, news media, and other stakeholders are also a critical component of the whole project supervision, which can build effective feedback channels with government departments. The relationship between the main elements of the engineering community is shown in Fig. 5.9.

5.1.3.4 Interaction in the Life Cycle of the Engineering Activity Community

The engineering activity community arises from human beings' purposeful use of nature. If engineering activities are compared to drama, the community of engineering activities constitutes the "actor" of the theater in engineering activities, and its appearance must follow the corresponding sequence according to the change of the plot. The life cycle of the engineering activity community can be roughly divided into four stages: pre-construction preparation stage, construction preparation stage, construction stage, completion and acceptance stage [20].

Pre-construction preparation stage of the project. Since engineering activities are purposeful social activities of human beings, it is necessary to select an appropriate objective of engineering activities in the "possible world" first. The project's initiator is the person who first proposes the aim of engineering activities. For example, in October 1952, Chairman Mao Zedong proposed the great idea of the South-to-North Water Diversion Project. At the beginning of the nineteenth century, the British-French Submarine Tunnel Project was first proposed by a French engineer Albert. After putting forward the project objectives, government departments, investors, consultancy, and the public need to evaluate and demonstrate the scientificity and feasibility of the project. For example, in the South-to-North Water Diversion Project , the relevant departments, provinces and municipalities, and units of the state have done a lot of planning, survey, design, and demonstration work. The planning and research staff are involved in many disciplines, such as economy, society, environment, agriculture, water conservancy, etc. Nearly 100 expert consultation meetings, symposiums, and review meetings were held during the planning process. Almost 6000 experts attended, including more than 110 academicians from the Chinese Academy of Sciences and the Chinese Academy of Engineering [21].

Construction preparation stage. Enterprise or project department is the most common organization form of engineering activity body in modern society. For the life course of specific engineering activity community, the formal establishment of the project department means the "formal birth" of the engineering activity community. For example, from the beginning of the South-to-North Water Diversion Project, the offices of Wen Jiabao and Li Keqiang as the chairmen of the construction committee have been established successively and the office of administration. At the same time, to carry out the project smoothly, it is necessary to take the legal project person as the leader to coordinate the relationship among contractors, supervision, consultation, design, and other units by controlling the contract.

Construction stage. The whole process of construction activities includes some basic steps such as decision-making, design, implementation, installation, operation, and abandonment. In different stages of engineering construction activities, to complete the corresponding tasks, various interactions are carried out among members of the engineering activity community, such as contractors, suppliers, design units, supervisors, and the public. It is necessary to pay attention to the control of the construction process, solve all kinds of technical problems in time, and pay special attention to the quality of the project. For example, in the South-to-North Water Diversion Project, the participating units take a highly responsible attitude towards the state, the people, and the history, comprehensively strengthening the quality management of the project, take the lead in formulating the detailed rules for the implementation of the quality responsibility lifelong system for national key engineering projects nationwide, and stipulate the retained data during the construction period. It has made the function of "warning for the present, good for the future."

Completion and acceptance stage. After completing the project, government departments, owners, and other organizations check and accept the completion of the project. At the same time, the owner and the contractor, supplier, design unit, etc., settle the project payment. In addition, the completion of the project also means

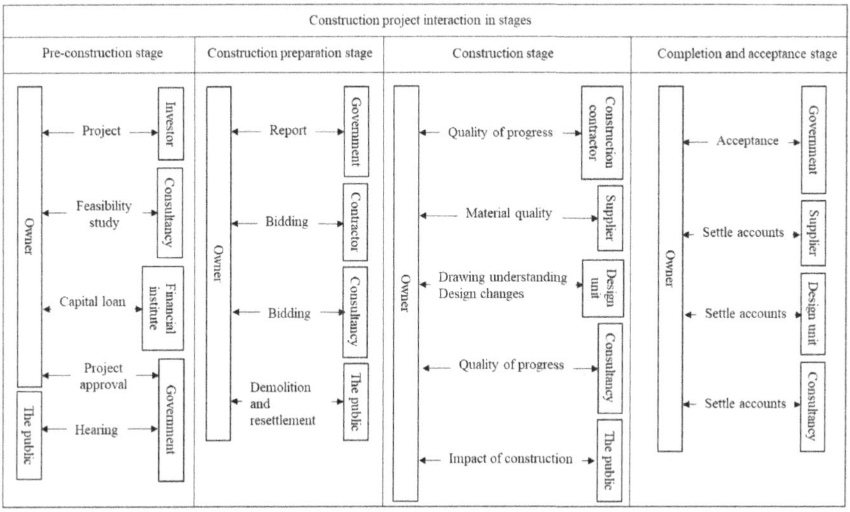

Fig. 5.10 Relations among subjects in the engineering community life cycle

the disintegration of the community of specific engineering activities. For example, the east and middle lines phase I of the South-to-North Water Diversion Project was opened in 2013 and 2014, respectively. After the opening, the relationship between the South-to-North Water Diversion Project Construction Committee and the contractors, survey and design, supervision, construction, consultation, and other participants on the first phase of the east and middle lines' construction ended, the corresponding engineering community disintegrated.

For a construction project, the interaction among the main bodies of the engineering community in its life cycle is shown in Fig. 5.10.

5.1.4 Engineering Organization and Engineering Management

Project management originated from the organization theory [22]. The latest edition of *Project Management Knowledge System Guidelines* issued by the Project Management Institute (PMI) in the US puts the "leadership and organization" at the top of all project management functions, which fully reflects the position of engineering organizations in project management. The organization has the dual role of noun and verb. From the perspective of a noun, an engineering organization is the main body of engineering management and the object of engineering management. From the verb point of view, an engineering organization is a core function of engineering management, which plays an important role in implementing projects.

5.1.4.1 Engineering Organization is the Subject and Object of Engineering Management

An engineering organization is a flexible group that gathers all forces around the goal of creation and efficiently allocates and utilizes all kinds of tangible resources and intangible elements. Engineering organization is the subject of engineering management and the object of engineering management. On the one hand, all stakeholders form engineering organizations according to the rules and carry out engineering management activities as the subject of engineering management. The project life cycle involves owners, design units, survey units, contractors, supervision units, engineers, financial institutions, government departments, the public, and other stakeholders. All stakeholders manage the engineering system's human, financial, material, information, and knowledge following certain contractual relationships to ensure the smooth progress of the project. On the other hand, an engineering organization is the object of engineering management. The main objective of engineering management is to manage the relationship between the main elements of the engineering organization, make all parties of the engineering organization balance, and achieve both internal and external benefits of the organization.

As the subject and object of engineering management, engineering organization decides success or failure. Only through understanding the dual role of engineering organization in engineering management and fully utilizing its dual nature can the project be successfully completed. For example, the Three Gorges Project is huge in the world. The success of its construction will have a far-reaching impact on the direct and indirect interests of the provinces and cities along the Yangtze River. It will significantly change the political, economic, and social life of the whole country. With the keen attention of the Chinese people, the Three Gorges Project organization has completed the basic works of millions of immigrants and dams in 15 years and achieved three major functions of flood control, power generation, and shipping [23]. Constructing the engineering organization to accomplish the three major functional objectives is an urgent problem to be considered in the initial stage of the Three Gorges Project of the Yangtze River.

The primary goal in the Three Gorges Project of the Yangtze River is to alleviate the flood disasters in the middle and lower reaches of the Yangtze River, store and detain the flood water through reservoir regulation, reduce downflow during the flood peaks, reduce the flood control pressure of the Jingjiang Dike, and improve the flood control capacity of the Jingjiang reach and the lower reaches. This is the most important social benefit. Secondly, the abundant hydropower resources of the Yangtze River are utilized to generate substantial electric power, thus repaying huge investment, which is the most direct economic benefit of the Three Gorges Project. At the same time, the construction of the reservoir has fundamentally improved the Chuanjiang channel and will play a significant role in the development of Western China. In the process of engineering organization, multi-level engineering organizations have been constructed based on the three major functional objectives of the Three Gorges Project. With the Three Gorges Project Construction Committee of the State Council as the center, the Office of the Three Gorges Project Construction

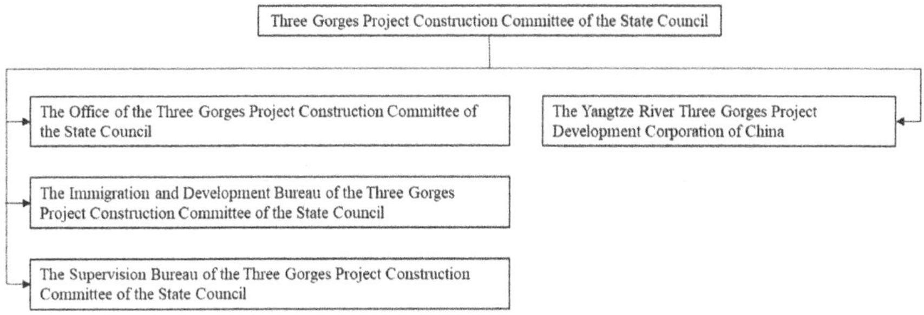

Fig. 5.11 Organization chart of the first-level engineering of Three Gorges Project

Committee of the State Council, the Immigration and Development Bureau of the Three Gorges Project Construction Committee of the State Council, the Supervision Bureau of the Three Gorges Project Construction Committee of the State Council and the Yangtze River Three Gorges Project Development Corporation of China forms the first-level engineering organization. Among them, the Committee office, the Immigration and Development Bureau, the Supervisory Bureau, and the Yangtze River Three Gorges Project Development Corporation of China set up the secondary engineering organization according to their respective functional objectives. For example, the Yangtze River Three Gorges Project Development Corporation of China adopts a matrix organizational structure with the engineering construction department, materials and equipment department, finance department, and other departments. Among them, the engineering construction department consists of project departments and support systems. The organization charts are shown in Figs. 5.11 and 5.12 [24].

5.1.4.2 Engineering Organization is an Important Function of Engineering Management

Besides nouns, the organization also has the connotation of the verb. Fayol, a managerial scientist, once proposes five core elements of management activities: planning, organization, command, coordination, and control. Among them, the organization is an important management function, which mainly refers to organizational behavior or activities. That is, through certain power and influence, to achieve a certain goal, the resources needed are allocated reasonably; in addition, the behavior or activities of dealing with the relationship between people and people, people and things, people and objects.

In the engineering field, the construction process of engineering project products is complex, so it is more necessary to have the organization's role. Starting from the overall construction situation, according to the site's specific conditions, the problem of construction organization should be solved in the best way. Specifically speaking, it is necessary to comprehensively and scientifically deal with personnel, machinery, materials, technology, environment, time frame, and funding to achieve the expected

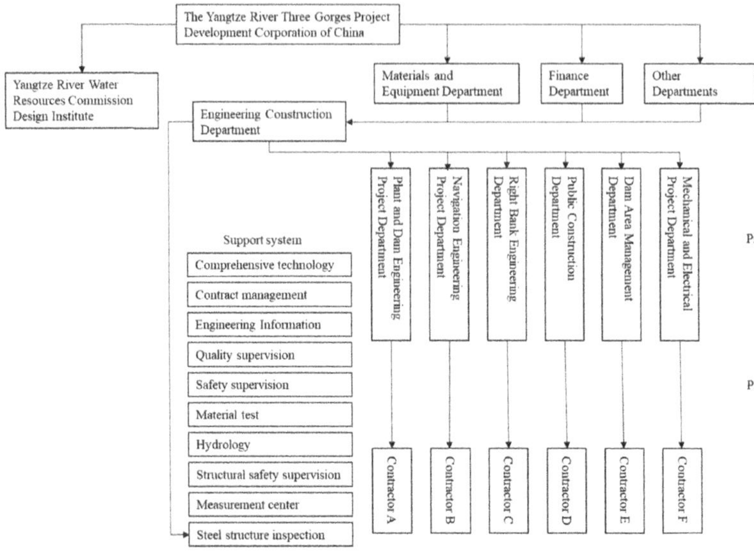

Fig. 5.12 Organization chart of the secondary engineering of Three Gorges Project. *Source* Lu [24]

objectives of quality, progress, investment, and safety to efficiently accomplish the construction task qualitatively. In practical implementation, it is often carried out through the design of construction organization and specific management activities.

For large-scale projects, construction organization design and management activities are often more prominent. Whether they are scientific or reasonable can usually lay the foundation for the whole project's success. For example, the well-known Qinghai-Tibet Railway has encountered many challenging problems in the construction process, such as plateau climate, complex technology, and public pressure. How to build a world-class, high-quality railway on the schedule based on that the plateau ecological environment is not destroyed, ensuring the personal safety of the participants, and withstanding the evaluation of all sectors of society, which is undoubtedly a very arduous task.

Therefore, for the construction organization of Qinghai-Tibet Railway project, on the one hand, the idea of systematic and overall consideration is introduced into the design source of the organization scheme. The whole construction of Qinghai-Tibet Railway is regarded as a systems engineering project. The construction organization design of the whole project is carried out according to system optimization and overall rationality requirements. Based on fully respecting science, the requirements of duration, quality, and investment are met greatly. On the other hand, for the specific operation process, the Qinghai-Tibet Railway is implemented into six key elements to ensure the effective implementation of the organizational plan (Fig. 5.13).

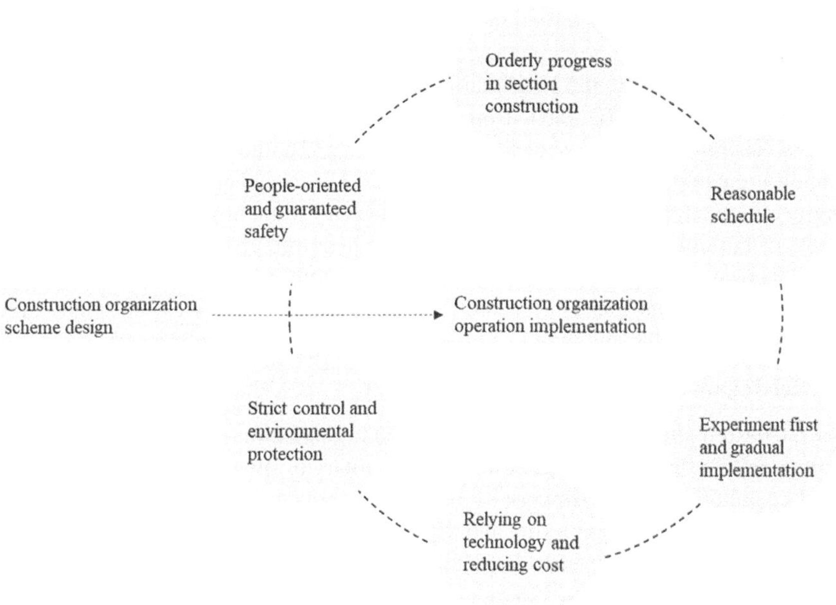

Fig. 5.13 Thoughts on construction organization management of Qinghai-Tibet railway

1. **Orderly progress in section construction**

The whole Qinghai-Tibet Railway construction project is divided into three sections, namely, Golmud to Kunlun Mountains (GK), Kunlun Mountains to Tang-ku-la Mountains (KT), Tang-ku-la Mountains to Lhasa (TL). Construction time in different sections is separately arranged. For example, the section of GK was set to be the first construction in 2001, the section of KT was mainly scheduled to be constructed in 2002 and 2003, and the section of TL was mainly scheduled to be constructed in 2003 and 2004. This avoided too intensive construction and prevented premature investment, ensured the duration of the project, and saved cost.

2. **Reasonable schedule**

The original construction plan is about eight months a year, but the practice of the first year of 2001 shows that the construction season of the Qinghai-Tibet Plateau is only 5–6 months. Since 2002, the headquarters had adjusted the construction time, stipulating May to October as the construction period every year, so that many construction operations could be avoided in winter, thus effectively reducing the health hazards of construction personnel and ensuring the quality of the project.

3. **Experiment first and gradually implementation**

Due to the world-class challenges of permafrost, environmental protection, and cold and anoxic conditions faced by the construction of Qinghai-Tibet Railway, even many valuable explorations have been made by railway builders and scientific researchers at home and abroad, there is still not a set of successful experiences that can be used directly by the railway construction in Qinghai-Tibet Plateau. In order to achieve the goal of a high standard, high quality, and high starting point for the construction of Qinghai-Tibet Railway, the headquarters decided that the design and construction of all projects should be carried out on a large scale only after scientific experiments. Therefore, in Qingshui River, Beilu River, Tuotuo River, and Anduo with four different geological conditions, experiments are carried out to verify whether the engineering materials and measures of marshes and wetlands can meet the requirements of building world-class plateau railway in extremely unstable areas such as permafrost area, slope wetland, and high ice content permafrost area. Facts have proved that this can make the design and construction measures and plans more scientific and reliable and avoid significant quality problems and rework.

4. **Relying on technology and reducing cost**

For the design alternatives of the construction organization, some bridge pile foundations are constructed by the dry method of a rotary drilling rig with high efficiency and less pollution. The earthwork is constructed using machines, concrete mixing is carried out by centralized mixing and filling truck, the whole steel formwork is used for the bridge formwork, and a large number of prefabricated components are used in bridge structures to reduce the local labor. Without these advanced construction organization designs, the number of workers on the Qinghai-Tibet Railway project would at least double.

5. **Strict control and environmental protection**

From the design of construction organization and all kinds of temporary projects, all construction plans need strict approval from the general headquarters before construction can be carried out. The general approval principles include less occupied meadows, environmental protection, and soil and water protection. At the same time, the process of examination and approval is also the process of optimizing the plan. In the original construction plan, part of the permafrost area was designed for road building. However, after field investigation and experimental practice of the headquarters, combined with factors such as global warming, a new plan of modifying the design, adding bridges, and replacing roads with bridges was put forward, which is effective for the permafrost environment. At the same time, the quality of the project is guaranteed. In the nature reserve, the headquarters also specially designed the organization of environmental protection construction. For example, in Kuona Lake in Tibet, a super environmental protection design plan has been formulated. Environmental protection measures have been taken around the lake to protect the beautiful lake from pollution.

6. **People-oriented and guaranteed safety**

In the construction organization design, a health guarantee is the first step; in the bidding, medical equipment, and medical staff should be enough to ensure the safety of construction personnel. To this end, the headquarters has established a series of health management rules and regulations, such as *Plateau Access System, Pre-construction, In-service and Post-construction Medical Examination System, Night*

Checking System, and *New Personnel Training System*. According to the principle of medical security first, before the team of each construction unit enters the construction site, relevant personnel are asked to conduct an on-site hygienic survey of the contracted construction area, in order to provide a scientific basis for the site selection of the construction team and the implementation of health security measures after the construction team enters the construction site.

Therefore, engineering organization design and organizational management activities are an indispensable part of engineering management. Suppose the function of organizational activities is brought into full play, all resources in the process of project implementation are planned and arranged scientifically, making the best use of human resources, materials and machinery and giving full play to production efficiency can better guide the development of project construction work and lay a solid foundation for the high-quality and efficient completion of the project.

5.2 Evolution of Engineering Organizations

The evolution of engineering organization and engineering evolution are concomitant. Before studying the evolution of engineering organizations, it is necessary to know the process of engineering evolution.

Charles Robert Darwin's creation of evolutionism is undoubtedly a landmark scientific event in biology. Since the founding of evolutionism, its influence has extended from biology to engineering, sociology, and other fields. Human engineering activities have undergone a long, tortuous, and complex evolutionary stage. The evolutionary nature of engineering activities originates from human practice's initiative, historicity, and creativity. As a typical form of modern human practice, engineering activities have evolved and developed continuously with the historical changes of human practice form, content, fields, and environment. It can be said that engineering activities are essentially a historical process, changing eternally and endlessly. Engineering evolution refers to applying the ideas, methods, modes of thinking, and research tools of the theory of biological evolution in the study of specific engineering activities. Engineering activities are viewed from a dynamic, historical, and evolutionary point of view, and the connotation, shape, evolutionary process, mechanism, characteristics, and regularity of engineering activities are profoundly revealed.

As the main body of engineering activities, engineering organizations continuously keep evolving in the history of human activities with the changes of engineering. With the engineering activities from simple to complex, from low level to high level, from material engineering to energy engineering to life intelligent engineering, the engineering organization keeps continuous innovation and breakthrough to meet the development requirements of engineering evolution better. Generally speaking, the change of the engineering ontology and the change of the external environment work together to cause the evolution of the engineering organization. There are three types of organizational forms corresponding to this evolution, namely early, middle and modern engineering organizations characterized by "huge-crowd strategy," "human–machine combination," and "human–machine-network combination."

5.2.1 Evolution of Engineering Organization Caused by the Change of Engineering Ontology

Ontology is "Benti" in Chinese. In *Cihai*, "Ben" refers to the origin and origin of things. "Ti" is mainly interpreted as the manifestation of objective things or certain thoughts. Engineering ontology can be understood as the existence of engineering, the activities generated by engineering, the resources needed by engineering, and the thinking and knowledge system generated by engineering activities. From the point of view of Marx's materialist dialectics, engineering ontology refers to a whole composed of the stream of consciousness, material flow, engineering activities, and communication among the related subjects involved in engineering. The change of engineering ontology can refer to the change of engineering itself or related engineering activities, the change of consciousness and material among the related subjects, and the change of engineering information flow. As the scale of the project is getting larger and larger, the number of participants keeps increasing, the complexity of the project is getting higher and higher, the content of engineering science and technology is getting higher and higher, and the project itself is becoming more and more sensitive to the impact of the environment.

Among the various explanations for the change and development of engineering organizations, the most direct one regards the change of engineering organization as the result of the change of engineering ontology. As pointed out by Peter F. Drucker's "goal management" or Alfred D. Chandler's "strategic choice" theory [25], changes in strategy or objectives precede changes in organizational structure and lead to changes in organizational structure. Traditional engineering has low requirements for technology and productivity. It mainly relies on simple labor force accumulation and task allocation to solve the problem. For example, compared with today's, the construction of the Great Wall, an ancient mega-project, relies not on various high-tech and advanced equipment but sufficient human resources and strict execution of orders from superiors in each construction section. In this case, most of the decision-making power of the project is concentrated on a few senior managers. The corresponding engineering organization only needs a simple, linear structure to ensure the effective execution of the orders. For example, the Great Wall of Hexi in the Han Dynasty was built by Wuwei, Zhangye, Jiuquan, and Dunhuang counties. The tasks of the counties were assigned to their subordinate counties in turn and eventually assigned to the construction sections, let the workers of the defensive strongholds finally complete. Due to the low technology content and low productivity level of the project itself, the complexity and regularization of the organization are very low.

With the expansion of the scale of the project itself and the improvement of productivity, in civil engineerings such as buildings, roads, and bridges with high technical maturity, in order to achieve the goal decomposition and allocate resources reasonably, the loose organizational structure of the project is transformed to bureaucracy, to accomplish the professional division of labor among the participants and improve the efficiency of the creation. For example, large international engineering companies such as Bechtel, Kellogg, Fluor, etc., in the United States have adopted a bureaucratic organization of business departments to establish professional branches worldwide by their business areas. Each branch company has a project management department, project control department, quality management department, design department, procurement department, construction department, etc., to meet the needs of professional construction of the project itself.

In the late stage of industrialization, the project is developing at an unprecedented speed and scale. The project participants are interrelated and interdependent at more levels. The importance of information flow in engineering activities is constantly increasing. Knowledge and innovation are gradually becoming the key to leading projects in many engineering fields. It makes the engineering organizational structure flatter and more flexible. For example, as the major national defense science and technology project in the United States, the "Missile Defense Program" [26] requires the support of high-tech engineering entities, the participating units are diversified and dispersed including national research institutions, laboratories, universities that undertake basic research. In addition, many companies responsible for applied research and technological development, such as Boeing, which manufactures missiles, Lockheed Martin, which manufactures boosters, TRW, which manufactures satellite communications equipment, Raytheon, which manufactures interceptors, forming an olive-shaped supply chain network with the main contractor authorized by the federal government as the core, multi-level supplier support, and combination of production, education, and research. Faced with tens of thousands of suppliers, the main contractor is only responsible for the engineering system's overall R&D, design, and assembly. In contrast, the R&D and production management of the engineering subsystem is the responsibility of suppliers at all levels. The units are not subordinate to the administrative subordinates but are connected through contracts to build a good information platform for cooperation. This flexible organization reduces the management complexity of the main contractor and is more conducive to technological innovation.

5.2.2 Evolution of Engineering Organization Caused by External Environment of Engineering

Chandler had a precise description of the relationship among environment, strategy, and organization: "structure (organization) follows strategy, strategy follows environment." The external environment of the engineering organization influences the

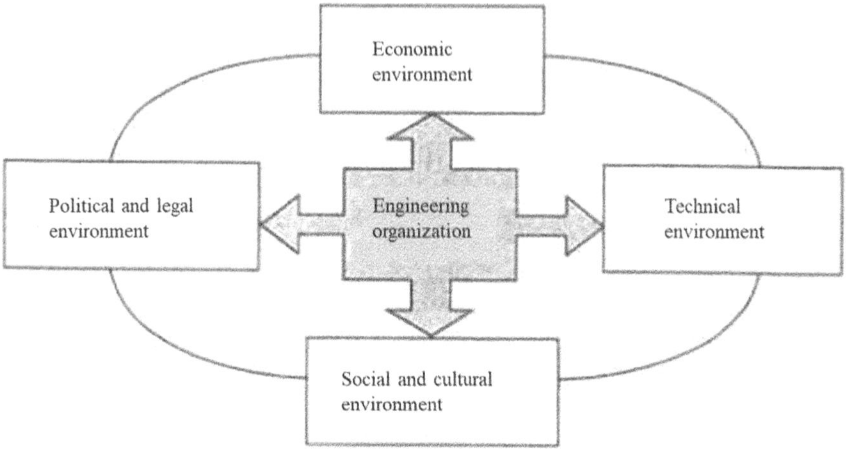

Fig. 5.14 The PEST analysis of the external environment

development of the organization. In turn, the growth of the organization affects the environment. The interaction between the organization and the environment promotes the evolution of the engineering organization. The external environment of an engineering organization can be analyzed by the PEST model, as shown in Fig. 5.14. Kast, a representative of the school of systems theory, points out that the critical step in implementing organizational change is to examine the status quo, that is, to investigate and analyze the status quo of the internal and external environment of the organization. Winter and Taylor believe that dramatic changes in the external environment trigger changes in corporate organizations.

5.2.2.1 Political and Legal Environment

The will of the ruling party is embodied by politics, while the will of the state is embodied by law. The political and legal environment covers the country's social system, the nature of the ruling party, economic system, political system, tax system, environmental protection law, anti-monopoly laws and regulations, and relations with major powers.

Taking the change of economic system as an example, the reform and opening-up drive China to achieve the transformation from planned economy to market economy. Accordingly, the engineering organization is also changing. Judging from China's national defense engineering development history, a purely government-led organization model was implemented before 1992. For example, in the Two Bombs and One Satellite Project, the state leader is responsible for the overall establishment of an organizational model in which an overall design department, a technical command line, and an administrative command line cooperate with each other. Later, it gradually developed into a "government-led + enterprise participation" model. It was under the government departments' overall responsibility and involved large enterprises in research and development. The Manned Space Project was led by the Central Committee of the Party and the State Council, in which the General Equipment Department is responsible for the overall management of the top level of the project. The General Equipment Department, the Ministry of Industry and Information Technology, the Chinese Academy of Sciences, and the China Aerospace Science and Technology Corporation are jointly responsible for constructing various systems of the project. Up to now, in the mode of "government leading + enterprise entity," the role of government in organizational structure has changed from "leader + executor" to "leader + supervisor." For example, China Commercial Aircraft Co., Ltd. is set up by the state to involve Shangfei, CAAC, Baosteel, and Sinochem as shareholders to carry out a market-oriented operation, under the supervision of the government, all enterprises give full play to their advantages in technology and management [27].

5.2.2.2 Economic Environment

The economic environment is divided into macroenvironment and microenvironment. The macro-environment mainly focuses on the population and its changes, national income, gross domestic product, and other indicators to reflect national economic development and development speed. The micro-economic environment mainly focuses on consumers' income, consumption tendency, savings situation and employment level in the area where the enterprise is located or served. The engineering organization's key economic variables include GDP and its growth rate, disposable income, consumer propensity, inflation rate, consumption mode, economies of scale, interest rate, exchange rate, foreign economic situation, import and export factors, and monetary and fiscal policies, etc. For example, a country's high-speed economic environment will attract many domestic and foreign organizations to participate in market competition, inevitably impacting the original engineering organization form and leading to organizational change. After implementing a diversification expansion strategy in 1992, Haier Group faced fierce competition from many well-known enterprises at home and abroad in various fields. The problems of engineering organizations in their production and manufacturing are also increasingly apparent. The traditional hierarchical organization structure affects the efficiency of information transmission and decision-making and produces small group doctrine in departments. The trivial rules and regulations restrict the development of innovation. Therefore, Haier Group has adopted a large number of organizational innovation strategies, separating the financial and procurement functions of various departments and integrating them into independent fund flow promotion department and logistics department to achieve unified procurement and settlement within the group. Haier also merged the human resources, technology quality management, information management, equipment management, and other functional departments into an independent service company. Eventually, a professional process-oriented flat organizational structure was formed, which broke through the original vertical business organizational structure, greatly improved production efficiency, and occupied 40% of the market of small refrigerators in the United States.

With the development of the global economy, international competition intensifies, the form of an engineering organization is changing, and the mergers and acquisitions in the world are endless. The strategic alliance is forming, the boundaries of engineering organizations diluted, and cross-regional virtual organizations have become the main tools for quickly responding to competition and accurately mastering the market.

5.2.2.3 Social and Cultural Environment

Social and cultural environment refers to residents' educational level and cultural level, religious beliefs, customs and habits, aesthetic level and values, in a region or a country. Educational level has an impact on the levels of residents' needs. Certain activities are banned or resisted because of religious beliefs and customs. Residents' recognition of the objectives of engineering organizations, activities, and organizations themselves are affected by values. Aesthetic concept influences people's attitudes towards the contents, forms, and results of activities of the engineering organization.

The social and cultural environment influences the engineering organization in the way of "moistening things in silence." Under the high-power culture in Chinese traditional culture, people obey authority, recognize the existence of inequality, and have strict hierarchical order. Therefore, decision-makers tend to have more centralized power and more power hierarchies (such as a linear type). At the same time, there is a psychological conflict with the development trend of flattening, flexibility, and networking.

5.2.2.4 Technical Environment

The technological environment mainly includes the technological level, technological policy, new product R&D capability, and technological development direction of a country or region. In terms of technological level impact, due to the increase and decentralization of the total number of technical personnel, especially the rapid spread of information technology, countries or organizations have gradually changed their pursuit of a comprehensive set of technologies, focusing on the deep development of technology. Then the division of professional work is gradually refined, with a more significant trend of internationalization, the industry is gradually transformed into a new mode of production globalization. Therefore, it is difficult to accomplish all kinds of projects, whether from the research and development of individual products or even at the national strategic level, by relying solely on their own strength. More and more projects adopt the "centralized + decentralized" organizational model, move to the centralization of engineering decision-making and supervision agencies and the decentralization of implementation agencies in the form of "main manufacturer (contractor) + supplier + subcontractor." For example, although parts of Boeing aircraft come from around 5000 suppliers worldwide, the core technology of aircraft manufacturing is still mastered by a few companies, which is the core part of the engineering organization.

5.2.3 *Evolution of Engineering Organization Morphology*

From the "carrying things by shoulders" of the farming era to the large-scale machine production initiated by the industrial revolution to the Internet and Internet of Things era brought about by modern electronic and information technology, we have repeatedly witnessed the changing social productivity. Correspondingly, the change of productive forces has caused the change of social production mode, which has experienced three major changes, from "family and handicraft workshop production" to "large-scale mechanized production" and then to "networked and globalized production." In this process, the forms of engineering organizations in the agricultural society, the industrialized production society, and the modern science and technology information society have also undergone significant changes. The development of an engineering organization can be divided into three stages.

5.2.3.1 Early Engineering Organizations: Characterized by Huge-Crowd Strategy

In an ancient agricultural society, family and handicraft workshops were the main modes of production. Restricted by the level of productivity development, "creation" activities mainly rely on people's strength, and the use of tools is limited. The man was the most important engineering element at that time. The corresponding engineering organizations emphasized "people-centered": the bond between people, the strength and wisdom of the group, the coordination between people and things, people and people, and the maximization of human's "creation" ability. It is the original model of engineering organization evolution, but also as a basic model of engineering organization, it is still widely used today.

Huge-crowd strategy is the most typical characteristic of early engineering organizations, such as the large-scale gathering of human resources for engineering construction or manufacturing. For example, the Great Wall of China belongs to an ancient defense project with the longest construction period and the largest amount of construction in the world. It is an outstanding representative of early engineering organizations. At that time, there was no machinery, and all the work was done by manpower. According to historical records, nearly one million workers built the Great Wall during the Qin Shihuang period, reaching one-twentieth of the total population of the Qin Dynasty. The early engineering organization has the longest history. Whether it is the Dujiangyan project and the Great Wall project in ancient times, or many enterprises engaged in low-end assembly and manufacturing industry nowadays, all use the huge-crowd strategy as the starting point for constructing an engineering organization.

The early engineering organization is an omnipotent organization covering many functions such as planning, command, coordination, and control. Due to the low level of standardization in production and construction, the engineering organization has a low degree of the engineering division. Due to the low level of information technology, the engineering organization is often restricted to engineering implementation sites. Under low-tech conditions, to ensure the rapid execution of orders and the unity of action, the organizational structure is mainly linear. Taking the Great Wall of the Ming Dynasty as an example, tens of thousands of people participated in the construction process is a typical engineering organization characterized by the huge-crowd strategy. Each defense area is organized by dividing and contracting out by sections. The governor and other officers served as the top leaders in overall planning. There are supervisors and specific construction division personnel during the construction, including the higher-ranking governor and principal military commanders. Meanwhile, the specific construction division personnel include the general managers of the actual organizational engineering construction, the lower-level managers, and the broad masses of people who perform specific construction activities.

5.2.3.2 Medium-Term Engineering Organization Characterized by the Human–Machine Combination

In the 1860s, James Watt improved the steam engine, marking the beginning of the first industrial revolution, promoting human society into the "steam age," and facilitating handicraft workshops to be transformed into large-scale machine industrial production. It has also changed people's living and working environment and has opened up socialized large-scale output. At the same time, the discovery of electricity was a revolution in human history. This period also coincides with the rapid development of industrialization and urbanization in Western countries. Engineering "creations" activities began to use secondary energy extensively. With the increasing complexity of technology, the scale of projects is increasing, and the degree of standardization is getting higher and higher; most of them are carried out around machines.

In line with the industrial revolution, engineering organizations have shown new development trends. The complexity and scale of technology have led to the deepening of the division of labor within the organization. The transformation from manual workshop to machine mass production has led to the beginning of machine-centered engineering organization, showing the characteristics of the human–machine combination.

Based on the complex needs of engineering management, the organizational model also presents diversification characteristics, including linear, functional, matrix, and other organizational models. As a result, engineering organizations have entered the medium-term form, typically represented by General Electric Company of the United States. In the 1970s, GE adopted a typical functional structure with six executive departments under its headquarters to share the work with the top management. These six executive departments are responsible for the coordination activities of its 9 groups, 50 divisions, and 49 strategic business units. There are many business and functional departments under the strategic business units, which are responsible for specific professional affairs. The medium-term engineering organization still takes specific management activities as its main responsibility, and the level of specialization is prominent, which can be called substantive specialization organization.

5.2.3.3 Modern Engineering Organization: Characterized by the Human–machine-Network Combination

In the 1970s, the information technology revolution promoted the change of social and economic structure and made human society enter the era of large-scale networked production. With the help of the Internet platform, human "creation" activities have undergone essential changes. Various market players, supported by information elements, have launched large-scale networked collaborative production based on opening, equality, sharing, and global mutual assistance.

Accordingly, engineering organization presents a new development trend: informatization brings about flat organization structure, globalization division of labor and cooperation promotes contractual organization relationship, network technology leads to virtualization of organization form and ambiguity of organization boundary, which opens the development stage modern engineering organization. Modern engineering organization consists of organization core and periphery. The organization core mainly undertakes the function of "virtual creation." It mainly produces technology, standard, or appearance design. The periphery participants of the engineering organization undertake the substantive production function, as shown in Fig. 5.15.

The firm core of the organizational structure of modern engineering is supported by core competitiveness, which is connected by network and market contract. Its flexible shell is formed based on social resources, and the external boundary of organizational form is ambiguous. Modern core functions are further reduced and more focused on maintaining core competitiveness, so it has a more vital ability to innovate. The more use and aggregation of external resources has the higher output efficiency. The more attention is paid to the use of contracts to connect participants, the more flexible it is to organize and decompose. Thus, it has a more vital ability to adapt to the environment.

The engineering organizations in the three stages mentioned above have successive relationships from their appearance, but there is no substitution relationship between them. They coexist today, and they are related to each other, which constitute the ecosystem of engineering organizations. Modern engineering organizations are located at the high end of the food chain in this ecosystem, and their downstream

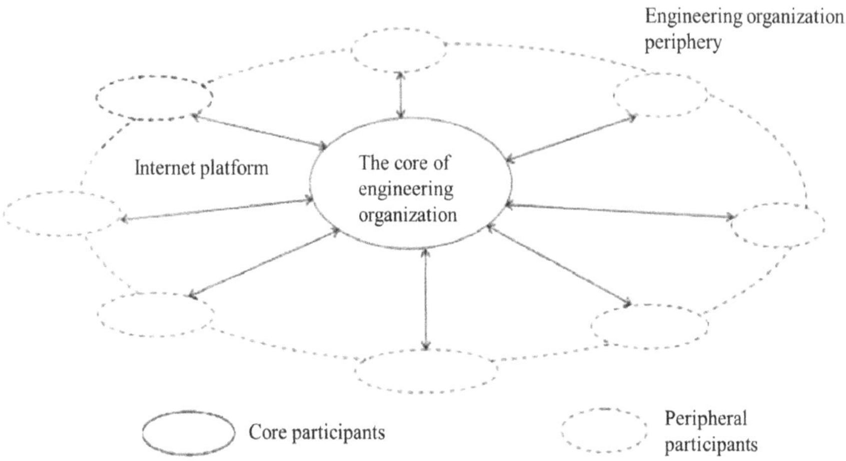

Fig. 5.15 Modern engineering organization

areas include early and medium-term engineering organizations. Taking the production of the iPhone as an example, Apple, as the core of this engineering organization, is a typical modern engineering organization, only responsible for creativity and design. The raw materials and components of the iPhone are supplied or manufactured by more than 150 professional companies around the world, such as processor suppliers AMD, most of which belong to the medium-term engineering organization. All raw materials and parts are finally assembled into Foxconn and other original equipment manufacturers (OEM). The OEMs assemble the final products, mainly labor-intensive enterprises, and belong to the typical early engineering organizations.

5.3 Aggregation, Efficiency, and Adaptation of Engineering Organization

The engineering organization is the same as the shell of the creation. Inside the chassis, the efficient operation of the body is achieved through the effective allocation of resources. Outside the chassis, the high adaptability of the body is achieved by exchanging material, information, and energy with the surrounding environment.

5.3.1 Element Aggregation of Engineering Organization

Various forms and types of engineering organizations have gathered many people, money, and things in a limited time and space background with a fascinating charm

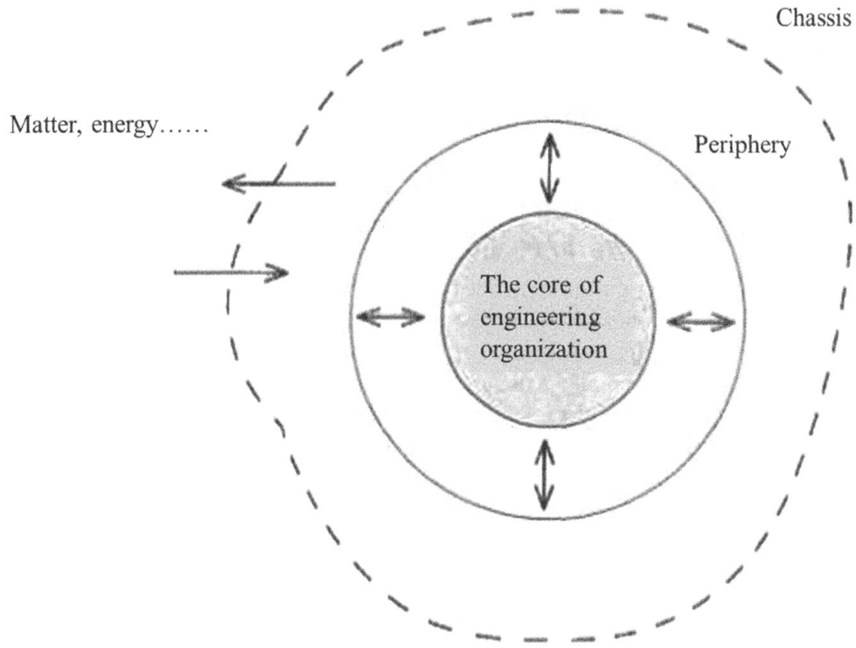

Fig. 5.16 The chassis of engineering organization

and magical power. They have created things from nothing, from small to large, and have benefited thousands of generations.

An engineering organization in any form is a chassis consisting of the core and periphery of the organization (Fig. 5.16). In such a chassis, the collaboration between the core and the periphery of the organization is the basis for maintaining efficient operation. The organizational core attaches great importance to the shaping and maintaining organizational decision-making power and operational power. The organization's periphery emphasizes the overall operation of the creation to concretize the creative activities. The chassis is not a simple superposition of the organization's core and periphery but a system based on the organization's operation mechanism and contractual relationship. The chassis itself is not closed but highly open and flexible. Matter, information, and energy are always selected and exchanged with the external environment of the creative activities. Therefore, the boundary of the engineering organization shell is sometimes blurred or clear.

5.3.1.1 Reasons for Factor Aggregation in Engineering Organizations

The engineering organization is magic because it integrates the crystallization of human wisdom and achieves the cluster effect of "two heads are better than one." Aggregation is the process of selecting the most suitable engineering production factors (such as workforce, technology, materials) in disorder and chaotic state under the guidance and promotion of engineering organization, and achieving the high-intensity aggregation of elements in limited space of creation, and providing the best quality regular products to users. This aggregation process can be visually reflected in Fig. 5.17.

1. **Engineering organizations rely on special forms of association and sharing mechanisms to shape aggregated "chassis"**

From manned space engineering to deep-sea exploration to general civil engineering construction projects, according to the type, scale, nature, and complexity of the project, the engineering organization rationally determines the management level, management span, management department, and management responsibilities based on the principles that are most conducive to decision-making, target control, coordination organization, and information communication, allocates the best resources, capabilities and rules, and forges aggregating entirety.

Engineering organization is the main body of the creative activity. Although the external environment of different creative activities is different, the alternative organization modes are ever-changing, based on the special external environment and internal characteristics of the creative activities, the engineering organization must select the best design mode to ensure that the management level, management span, management department, and management responsibility of the engineering organization are most in line with the requirements of aggregation. For example, under the special background of the rapid development of national defense science and technology, China's manned space engineering is under the special management of the Central Special Commission. According to the scientific and technological procedures and intelligent division of labor of the project, the General Equipment Department, the National Defense Science and Technology Commission, the Chinese Academy of Sciences, the China Aerospace Science and Technology Group Corporation, and other institutions constitute a cross-sectoral and highly centralized organizational management system. The system of the engineering organization ensures that the organization can obtain (gather) the most valuable information and resources from

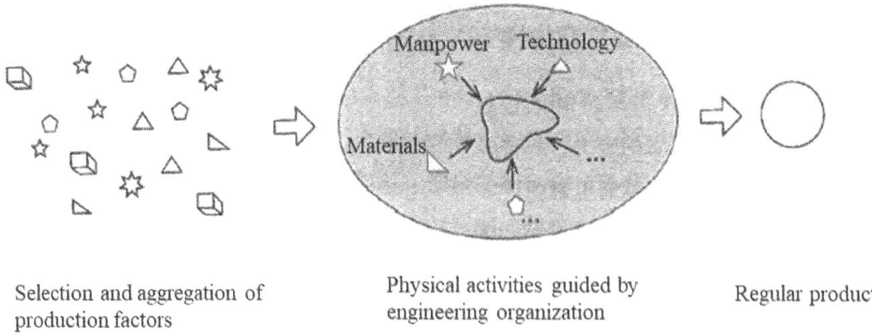

Fig. 5.17 The process of factor aggregation of engineering organization

the rapidly changing external technological and economic environment to judge the changes of the external environment in time and respond scientifically and reasonably.

Perfect engineering organization design is only the first step to ensure the success of factor aggregation, just as the promoters of a great campaign need not only generals but also soldiers. Otherwise, the aggregation can only be on paper. Within the framework of engineering organization, to avoid "make bricks without straw," it is necessary to equip engineering organizations with various factors of engineering products, such as workforce, technology, experience, and knowledge, to ensure the smooth development of the gathering activities, and the construction of "ability" of engineering organizations. Various factors of production are reasonably and orderly invested in the activities of creation according to the plan of creation. In addition, the gathering function of any engineering organization depends on strong engineering rules. For example, because of the special strategic significance of space engineering, participants must abide by a strict confidentiality system, act according to regulations, and distinguish between rewards and penalties to truly achieve the qualitative change from ordinary engineering organization to "creation machine."

2. **The primary task of factor aggregation in an engineering organization is human**

Human is the most active factor of production in an organization, and the mission of an engineering organization is managing people. According to a certain leadership system, department settings, hierarchical division, division of responsibilities, and it constitutes a flexible entirety. An engineering organization is a flexible group composed of people. The external environment and the project itself are complex and changeable, and the creative activity itself is a huge system engineering. It is difficult for a single force to change physical activities qualitatively. It is like ants moving. It is difficult to achieve a large "moving" project with the power of an ant alone. It needs the collective efforts of the "ant colony." Like ants, humans are a part of "ant colony" in creational activities and are the core elements of engineering organizations. The orderly development of creational activities highly depends on human initiative, and the achievement of efficient management of humans becomes the essence of the engineering organization.

Gathering people comes from the need for engineering organizations to develop the "engine" of the core. Nevertheless, no advanced engine can set sail without the full cooperation of wings. In human management in engineering organizations, the whole composed of leadership system, department settings, hierarchical division, and division of responsibilities is particularly critical. They are the "wings" to ensure the starting of an engineering organization. Engineering organization highly depends on the efficient integration of such elements within the organization, which is an important feature in the practice of engineering organizations to manage people.

The management of people in engineering organizations is mainly carried out in two aspects. On the one hand, whether in the core of the engineering organization or as other quasi-organizations in the periphery of the organization, the gathering of diverse manpower is the basis to ensure the smooth work of the organization. For example, the West–East natural gas transmission project needs management talents, power technology talents, economic talents, and legal talents. All kinds of talents work together for the common goal of creation. On the other hand, engineering organizations have the effect of adhering to multi-participant; that is, engineering organizations contain multiple single organizational forms, which retain specific organizational objectives, organizational structure, and organizational size. They are "satellite" organizations around the organization's core, serving the creation activities together. Within the organization, although the power distribution, membership, and organizational objectives of different individual organizations are different, and some of them are even antagonistic, the existence of contractual ties has maintained the development of the organization. Engineering organization is a unity of contradictions based on contractual relationships, and the management of different participants becomes a part of the daily management of engineering organizations.

Multi-stakeholders jointly completed China National Stadium (Bird's Nest) under a contractual bond relationship. The general contractor of Bird's Nest is the Beijing Urban Construction Group. The design is undertaken by Herzog & De Meuron Design Firm in Switzerland, ARUP Engineering Consulting Company in the UK, and China Architectural Design & Research Institute. The financial consultancy services are provided by the National Development Bank, PwC Consortium of America, and HSBC Bank of Hong Kong. In addition, several other relevant organizations offer specialized services for constructing the Bird's Nest. In the project construction process, the effective management of multiple parties is an important guarantee for the smooth implementation of this project.

3. **Forming an efficient, flexible system is the aim of factor aggregation of the engineering organization**

The high openness of engineering organizations is manifested in their strong ability of self-learning and regulation. Even in the case of "domestic trouble and foreign invasion," it can quickly restore its proper shape, structure, layout, and function and promote the creation from disorder to chaos and then to order.

An engineering organization is a platform for exchanging creation activities with the external environment and resources. It has a strong openness and is a "window" for materialized labor. Creation activities are the process of changing the quantity of various engineering production factors into qualitative changes. Engineering production factors (such as materials, equipment, etc.) depend on the efficient command of the engineering organization to participate in the creative activities in a reasonable and orderly manner. However, the activities of creation are not roses all the way. The characteristics of engineering organizations and activities of creation and the complex and changeable external environment in which the engineering organizations are located make the engineering organizations easy to fall into the whirlpool of "domestic trouble and foreign invasion."

On the one hand, an engineering organization is a contradictory and unified group with multi-participation. Each individual organization has its own specific objectives. The contradiction of objectives makes it possible for engineering organizations to have "domestic trouble." The complexity of the creative activities also increases the possibility of "domestic trouble" for the engineering organization. On the other hand, any creative activity's external technological and economic environment is in a state of change. The accumulation of the quantitative changes of such changes will inevitably lead to qualitative changes in the creative activity, and then there may be "foreign invasion." When the creative activity is in the state of "domestic trouble and foreign invasion," the organization is not helpless. Relying on the strong self-learning and regulating ability, the organization can feedback the bad external state, adjust the form, structure, layout, and function of the organization efficiently, meet the needs of the creative activity, and achieve the transformation from disorder to order of the creative activity.

5.3.1.2 Periodicity of Factor Aggregation of Engineering Organization

The aggregation of engineering production factors in the life cycle of an engineering organization is not unchanged. In the initial stage, engineering organizations attach importance to selecting production factors and establishing a contractual relationship. The input of factors is not apparent, and the physical accumulation of production factors is emphasized. The implementation stage is the most concentrated input stage. The supply and demand of production factors are large, and the total input can account for 70%-80% of the total input of factors. This stage pays attention to the "chemical reaction" of different production factors. In the end stage, the demand for production factors is again reduced to a lower level, showing a physical and regular process. During the whole life cycle of the engineering organization, it presents an inverted U-shaped arc (Fig. 5.18).

1. **Initial phase**

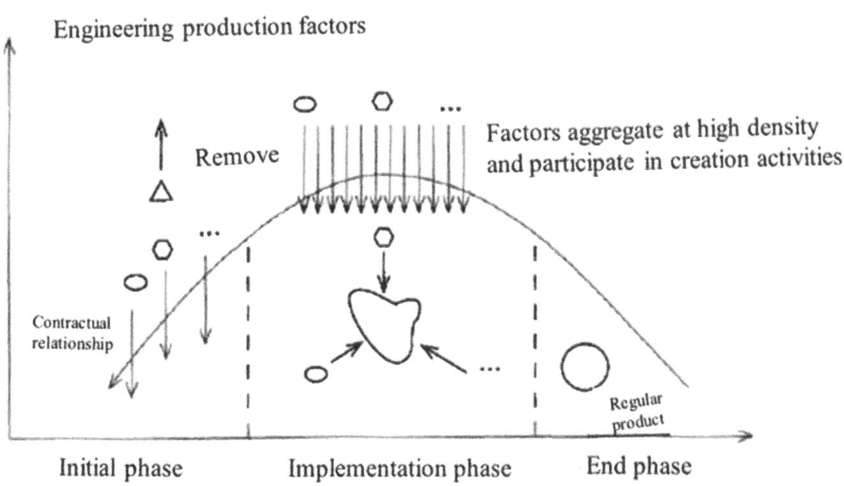

Fig. 5.18 Input and aggregation of factors in the life cycle of the engineering organization

In the initial phase, various factors of engineering production such as workforce, resources, semi-finished products, finished products, rules, regulations are gathered in the forms of far or near, more or less, virtual or actual. Engineering organizations must have recognition, excavation and deployment capabilities beyond imagination. Like the Olympic Games, it organizes athletes scattered worldwide to gather and provide venues and "guides" athletes into the stadiums.

Engineering production factors are the basis for carrying out creation activities, and the spatial dispersion of engineering production factors is the basis for generating engineering organization. In the initial stage, the space of engineering factors of production presents a dotted layout, and its typical characteristics are highly chaotic, disordered, and unbalanced. Besides human resources, engineering factors of production are material rather than conscious and cannot spontaneously carry out the optimal combination according to the needs of creation. Even conscious human factors of production are difficult to achieve efficient convergence due to creative activities without external intervention, which requires an "organization" to intervene in engineering factors of production to achieve factor aggregation. In addition, the activity of creation is a huge systematic project. Diversified input of factors of production becomes necessary, such as human resources, information, experience, and knowledge. It is difficult to deal with complex and changeable activities of factors of production by individual strength alone. The "organization" guiding the creative activity is a pluralistic group organization. For example, in the "Yu Gong Moves Away the Mountains," the endless descendants constitute the group organization in this mountain-moving project.

Engineering organization is the engine of selection, allocation, and collection of engineering production factors. With the rapid development of science and technology, the homogenization of business types of organization groups increases. Thus the selectivity and substitutability of engineering production factors are becoming more expansive in time and space. However, not all engineering production factors are suitable for creative activities. Selecting the most suitable elements of production activities is an important task for the engineering organization at the initial stage. However, even the seemingly "most appropriate" factors of engineering production are not necessarily optimal. The process of creation is unique. Many factors are affecting it, and the factors of production are in dynamic change. There are potential contradictions and conflicts between any factor of engineering production and other factors of production. This requires engineering organizations to have the vision of identifying the optimal engineering production factors in the initial stage, to efficiently allocate the scattered and disordered engineering production factors to form a combined force, to shape the optimal combination of engineering production factors, to achieve the complementary advantages of engineering production factors, and to ensure the maximum productivity.

The initial stage is the basis for any project to proceed smoothly. In the initial stage, engineering organizations need to overcome the obstacles of dispersing engineering production factors in space, identify the most suitable production factors, and efficiently allocate engineering production factors in space to achieve the spatial agglomeration of engineering production factors and complete the creative activities.

Taking the production of Boeing 787 as an example, Boeing adopts a global supply chain strategy. Its suppliers are located in 135 corners of the world. Boeing is only responsible for tail wing production and final assembly (about 10% of the production). The rest of its components are preferred to cooperate with scattered global suppliers. Boeing is responsible for selecting the most suitable production factors for 787 passenger aircraft in the world and guiding the orderly flow and convergence of production factors in the global scope, such as aircraft wing from Japan and landing gear from France. It lays the foundation for the transformation of the creative activities of the 787 passenger aircraft from quantitative to qualitative.

2. **Implementation phase**

In the implementation phase, the engineering organization is the "headquarters" that directs the creative activities, balancing the three contradictions of quality, progress and cost in ideal and reality, blending various material forms, accommodating various contradictions of interests, and promoting qualitative change into quantitative change until it can provide the regulated products with the greatest satisfaction to the users.

In the implementation stage, the engineering organization carries out materialized labor to achieve the specific object of creation. It has a clear goal orientation, and around the object of creation, it needs to make a careful plan beforehand. In the implementation stage, according to the plan, the transformation of engineering production factors from quantitative change to qualitative change is gradually achieved, so the plan guarantees the creative activities. Creation plans need to focus on the quality, progress, and cost of creative activities. However, these "three goals" are not harmonious because high-quality, fast-moving, and low-cost projects do not exist. They can only be the goal system to maximize the balance of creative activities and balance the three major contradictions. In addition, there are many factors affecting the activity of creation in the process of creation, and most of them are unpredictable. The engineering organization should observe and feedback the plan of creation and its implementation in time, especially the implementation and completion of the "three major objectives," and correct the existing or potential deviations in creative activity. According to the actual situation of the creation, the organization and objectives of the project should be adjusted timely, the implementation of the creative activities should be guided reasonably, and the ideal and reality of the creation activities should be balanced. In the Three Gorges Project, the Three Gorges Project Information Management System (TGPMS), organized by the Three Gorges Project Corporation, is an integrated engineering management database system [28], which speeds up the feedback speed of engineering information. Effective management and control of the Three Gorges Project have been achieved. Based on the objectives and control standards of the project plan, the information obtained can be feedback in the implementation process, and the deviations can be corrected and adjusted. The adoption of the information management system of the Three Gorges Project has changed the traditional working model of the project organization. The management and control of the project by different management bodies are more coordinated and efficient. The organization can control the data of all phases of the project management in an all-around way, focusing on the core of the project management, strengthening

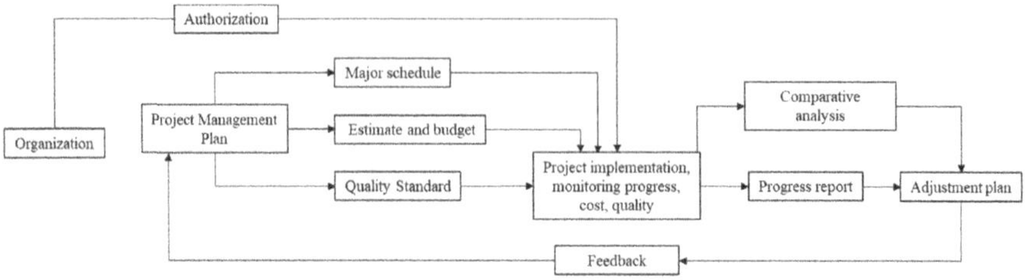

Fig. 5.19 Three Gorges Project information management system. *Source* Lu [24]

the control of information management, organizing more precise control, analyzing and predicting the project, and significantly improving work efficiency. The work efficiency ensures the smooth implementation of the Three Gorges Project (Fig. 5.19).

The implementation stage is the most important period for the implementation of materialized labor. Under the guidance and command of engineering organizations, all kinds of engineering production factors change from quantitative to qualitative. Under the plan's direction, this stage is to combine and match all kinds of engineering production elements most suitable for the creative activities selected in the initial stage of the engineering organization and to arrange the timing of different production factors participating in the creative activities rationally and orderly. It efficiently configures the engineering production elements and ensures the supply schedule and quality of all kinds of engineering production factors before and during creation activities. Taking the "three major objectives" of creative activities as the key points of control, ensuring that all kinds of engineering production factors are reasonably and orderly combined and allocated under engineering organizations' guidance, and promoting the smooth implementation of creative activities.

An engineering organization is a group of organizations with different stakeholders. The implementation stage is the stage when main stakeholders participate in. Although different stakeholders participate in the creative activities by contract, each stakeholder, as an independent organization, participates in the creative activities to achieve the different "wishful thinking" of their respective organizations, which inevitably results in interest friction among different stakeholders. The high frequency and apparent interest contradictions of all parties in the creative activities make the engineering organization undertake a lot of work in coordinating the interests of all parties and play an important role. The organization and management activities are critical and complex and directly determine the success or failure of the creative activities.

3. **End phase**

At the end phase, the engineering products are completed, and the organization will "mail" the regular products to the users, and organizations will disappear.

Qualified engineering products are delivered to the user, reaching the user's satisfaction, then the creative activity is finished. The uniqueness of the creative activities makes the engineering organization show temporary characteristics. After completing the creation activities, the engineering organization will naturally disintegrate and cease to exist. However, the different monomer organizations that make up the engineering organization do not disappear with the end of the creation and the disintegration of the engineering organization. These organizations will be re-involved in other creative activities to provide services. Creation organizations reappear in other spaces in different combinations of monomer organizations, and the cycle runs back and forth. For example, the Architectural Design and Research Institute of China, as a service provider in the field of architecture, provides services in the Olympic Bird's Nest project and involves other organizations to participate in creative activities. This institute also offers architectural services in creative activities such as Beijing Metro Line 4 and Baidu Building.

5.3.1.3 Evolution of Factor Aggregation of Engineering Organization

The evolution of engineering organization's morphology has gone through the initial engineering organization characterized by huge-crowd strategy, the later engineering organization characterized by human–machine combination, and the modern engineering organization characterized by human–machine-network combination. The evolution process of engineering organization's morphology also reflects the aggregation characteristics of different engineering organization factors at different times.

As the evolution of engineering organization, the characteristics of factor aggregation of engineering organization are different in different periods, which is also the result of the changes of engineering ontology and external environment. On the one hand, the changes of engineering ontology, such as the simplicity of the earliest creation activities and the weakness of engineering complexity, are mainly the accumulation of human factors of production. The workforce is the absolute main factor of the creation activities at that time. However, with the increasing complexity of engineering itself, the complexity and professionalism of engineering organizations are enhanced. Engineering organizations gradually move from dispersed to integrated, resulting in enhanced engineering production factors gathering. Various kinds and combinations of engineering production factors are aggregated in the space of creation, especially the wide adoption of knowledge and modern technology. It improves the efficiency of creative activities and gradually becomes the core factor of production participating in creative activities.

On the other hand, changes in the external environment, especially modern technology, have changed people's life and way of thinking. The availability and substitution of all kinds of engineering production factors have been greatly enhanced. The spatial and regional problems of production factors no longer exist. In the process of large-scale production activities, the supply chain of the global industrial chain of engineering production factors has become a universal choice. In addition, the "centralized + decentralized" organizational model and the "manufacturer + supplier + subcontractor" organizational structure are more used in the creative activities, which gives the suppliers of engineering production factors more opportunities to participate in the creative activities. At the same time, to achieve business development, different individual organizations adopt the joint mode of engineering production factors to form a strategic alliance, which further promotes the aggregation of different engineering production factors in the creative space.

1. **Factor Aggregation of Early Engineering Organization**

As mentioned above, the early engineering organization is a typical huge-crowd strategy engineering organization. During the time, due to the lack of resources such as communication and transportation, it was difficult to transfer the creative activities on a large scale. It was difficult for engineering organizations to effectively select and aggregate the elements of engineering production outside the limited space of creation even if they were available, which resulted in weak cross-space selectivity and availability of the elements of engineering production. In addition, limited by the level of productivity development and the limitation of the understanding of objective laws, most of the engineering production factors are not effectively used in creative activities. Even if some of the engineering production factors are adopted, their selection and aggregation show a high degree of proximity to creative activities in terms of space. "Obtaining raw material locally" has become a common practice. For example, the construction of the Great Wall of Qi during the Spring and Autumn Period was limited by the low level of productivity development and the difficulty of information communication and transmission; only relatively few construction materials were available.

The workforce was the absolute subject of engineering production factors in the creation activities at that time. The whole creation process was closely around manpower. The essence of the accumulation of engineering production factors in the creative activities was the accumulation of workforce production factors in the creative space. Before creation, engineering organizations focused on integrating human resources, which was the precondition for gathering other factors of production in terms of space. During the creation, the engineering organization directs the manpower efficiently, pays attention to the cooperation between people, and gives full play to the initiative of manpower production factors in the creative activities to achieve the goal of creation. For example, in the construction of Dujiangyan water conservancy projects, limited by the level of productivity development at that time, only limited productive factors are available. Manpower becomes the most active factor of production in creative activities, and the aggregation of human factors has contributed to the qualitative change of creative activities.

2. **Factor Aggregation of Medium-Term Engineering Organization**

A medium-term engineering organization is a typical engineering organization characterized by the combination of human and machine. Machines are widely used in the creative activities.

In this kind of engineering organization, science and technology gradually play an effective role, and human information exchange is becoming more and more diversified. With the development of transportation technology, communication technology, and graphics technology, transferring engineering production factors in space becomes more convenient. In addition, during the rapid development of urbanization and industrialization in the West, the amount of creative practice activities is large, promoting the efficient flow and convergence of engineering production factors needed for creative activities in a small space. Local material-based acquisition of engineering production factors has gradually withdrawn from the stage. In addition, with the rapid development of science and technology, the level of human knowledge and the deepening of understanding of objective laws, the selectivity and substitution of engineering production factors are greatly enhanced. Many restrictions on creation activities caused by the lack of engineering production factors are reduced, and the production efficiency of creative activities is gradually improved.

In the medium-term engineering organization, the accumulation of engineering production factors is mainly manifested by the accumulation of human and machine elements in the creative space. Human factors in production still play an irreplaceable role in engineering organizations, especially the knowledge, technology, and ability contained in human aspects of production is an important guarantee for organizational management. In addition, the use of all kinds of machine production factors improves the efficiency of the management of the whole engineering organization. It promotes the division of labor within the organization to be more refined, and specialization continues to improve. In addition, with the large-scale mechanization production, the demand for secondary energy production factors is increasing gradually. The standardized flow production method, represented by the Ford Corporation in the US, has dramatically enhanced the efficiency of the creative activities because of the efficient combination and utilization of production factors such as workforce and machines.

3. **Factor Aggregation of Modern Engineering Organization**

A modern engineering organization is an engineering organization characterized by the combination of humans, machines, and networks. In this type of engineering organization, with the rapid development of information and network technology, information exchange and transmission are more diversified and more convenient. The scattered distribution in the space of engineering production factors no longer becomes the restriction factor of creative activities, and the production factors flow and converge in the global network. The globalized supply mode of engineering production factors has been widely adopted, and the whole creation activity shows a strong global cooperation state. In addition, the knowledge of different disciplines is produced and disseminated at the speed of an "exponential explosion." The understanding of objective laws is deepened. The selectivity and substitution of engineering production factors have reached the highest level in history. With the continuous progress of science and technology, the selection and substitution of engineering production factors are pushed to a higher level, and the efficiency of creative activities is constantly refreshed. A large number of global collaborative engineering projects across regions and spaces have emerged.

In this kind of engineering organization, the primary manifestation of engineering production factors is that diversified production factors circulate worldwide and eventually aggregate in the creative space. Organizations tend to be flattened. The motive force for the development of engineering organizations comes from the organizational core, which mainly promotes the global flow and aggregation of virtualized production factors such as information, knowledge, innovation, and network planning. It is mainly because the organizational core pays more attention to the decision-making of the creative activities and the construction of the organization's core competence. In contrast, the outer part of the engineering organization mainly promotes the global selection and procurement of the physical production factors to serve the actual implementation of the creative activities. The basis of global flow and accumulation of engineering production factors is the contractual organization relationship, which makes all kinds of engineering production factors more flexible to participate in or withdraw from the creative activities and improves the adaptability of engineering organizations to the environment. For example, Siemens represents the efficient implementation of global procurement strategies. The procurement committee set up by Siemens is responsible for coordinating global procurement needs and looking for strategic partners worldwide. In addition, a strict supplier selection mechanism for production factor ensures the quality of Siemens' creative activities.

5.3.2 *Operational Efficiency of Engineering Organization*

Engineering organizations are aggregates that operate efficiently through effective allocation of resources and adapt to changes in the external environment by exchanging material, information, and energy with the external environment.

5.3.2.1 The "Efficiency" of the Operation of Engineering Organization

Efficiency is the goal of an engineering organization. The operation efficiency of an engineering organization is reflected in efficiency and effectiveness. The engineering organization can be traced back to the creation, benefits from technology, thought, and natural selection, and constantly evolves to meet the engineering needs of human beings. Its inherent endowment must be efficient to meet the needs of rapid development. On the other hand, the function system and output system constructed by engineering organizations are endowed with clear production function by the production system of human society, which is bounded by the boundary condition of function relationship. Engineering organization pursues the optimal solution of the system in its territory, and their external gains are also bound to be highly efficient to achieve the best satisfactory results for all parties.

China's Two Bombs and One Satellite Project reflects the pursuit of efficiency of engineering organizations [29]. In terms of the institution set up, in 1962, due to difficulty coordinating various engineering systems by the Ministry of Second Machinery that managing the development of the atomic bomb, the Central Committee of the Chinese Communist Party decided to set up a special committee of 15 people. This committee serves as the highest decision-making body for the project. It organizes a large-scale national collaboration for major construction, production, and scientific research experiments in the atomic weapon industry. The establishment of this organization has effectively solved the problem of unclear powers and responsibilities among departments such as the National Defense Science and Technology Commission, the National Defense Industry Commission, and the National Defense Work Office. It has promptly dealt with various technical, material, and organizational management decision-making problems of the project, thus significantly improving the efficiency of the "Two Bombs and One Satellite" development. In terms of organizational structure, the combination of linear structure and matrix structure ensures the efficiency of information dissemination and reduces the occurrence of "multiple leadership." In the project's whole life, the separation of business departments and consultancies is always ensured to maintain the stability of information dissemination.

During the critical period of the development of the atomic bomb, 26 ministries and commissions, 20 provinces and municipalities across the country, including more than 900 factories, scientific research institutions, colleges and universities, and various departments of the People's Liberation Army, participated in the project and solved nearly a thousand issues, which shows the high efficiency brought by the engineering organization. Under the conditions of low economic productivity, the withdrawal of Soviet experts, the outbreak of the "Cultural Revolution," and the scarcity of human, financial and material resources, the "Two Bombs and One Satellite" project could still break through various critical technologies in a relatively short time and achieve remarkable research results. It lays a good foundation for developing China's nuclear industry and satellite industry. It effectively enhances China's national defense science and technology strength and enhances China's international status. The high efficiency brought by its engineering organization is evident.

5.3.2.2 Formation and Development of "High Efficiency" of Engineering Organization Operation

1. **The operational efficiency of engineering organizations comes from the convergence of engineering factors**

The operational efficiency of engineering organizations reflects the group combat effectiveness of engineering organization, which is the combination of engineering production factors and the integration of stakeholders, and the basis of the formation of operational efficiency of the engineering organization.

The combination of engineering production factors shapes the operation efficiency of an engineering organization. Creative activities need the support of different types of engineering production factors, which is the process of "accumulation" of engineering production factors. However, the materiality and non-consciousness of most engineering production factors determine that they cannot spontaneously carry out the optimal combination and matching according to the requirements of creative activities. That is to say, the manufacturing process is not a simple and random mechanical superposition of factors of production, and the guidance of engineering organization is indispensable. The guiding process of engineering organization to engineering factors of production is a process of orderly accumulation of factors of production selection and combination under the action of the engineering organization. It is also a process that promotes the organization to become more efficient step by step. With the appropriate combination of production factors, engineering organizations deepen their understanding of the nature of creative activities and promote the adjustment, standardization, and perfection of engineering organizations' operation mechanisms and development strategies. Exercise the ability to organize and handle creational activities in the combination of production factors (e.g., engineering production). The most direct manifestation of creative activities is the optimal coordination of factors, which promotes the appropriate combination of engineering factors to maximize efficiency under the organizational framework, and the efficient operation of the organization can be achieved.

The integration and unification of stakeholders shape the operation efficiency of engineering organizations. A stakeholder is the core component of an engineering organization. Different stakeholders depend on contractual relationships to participate in creation activities for contradiction but unity. They are in the periphery of the core of the engineering organization, similar to "satellite" around the organization's core, and integrate to form the engineering organization system to maintain the daily operation of the engineering organization. Different stakeholders participate in and withdraw from the engineering organization system in different periods according to the prior creation planning. It makes the engineering organization in a high dynamic balance, occupies and consumes the space and resources in the engineering organization system reasonably, and assists the creative activities in developing reasonable and orderly. In addition, different stakeholders have a clear division of labor in the creative activities and cooperate professionally in the creative activities to maximize their effectiveness in the engineering organization system. It also integrates them into the engineering organization system and ensures the achievement of the high efficiency of the operation of the engineering organization. The operation of an engineering organization is "efficient" does not mean the high-speed operation or evolution of individual factors, but the combination of each factor into a stable "matching pair" or "matching group" under the uncertain constraints of the engineering organization, to maximize the material and energy output of the factors at a certain level. As an example, industrial engineering improves production efficiency depending on the combination and matching of design, technology, production, and process, as shown in Fig. 5.20.

The factors of engineering organizations can be divided into two categories: autonomous factors and non-autonomous factors. The former is purposeful, predictive, conscious, and regular. It usually refers to natural persons or the collection of natural persons who participate in engineering practice. The latter is constancy, passivity, and resources, usually referring to the physical entities, technology, rules, and funds involved in the project.

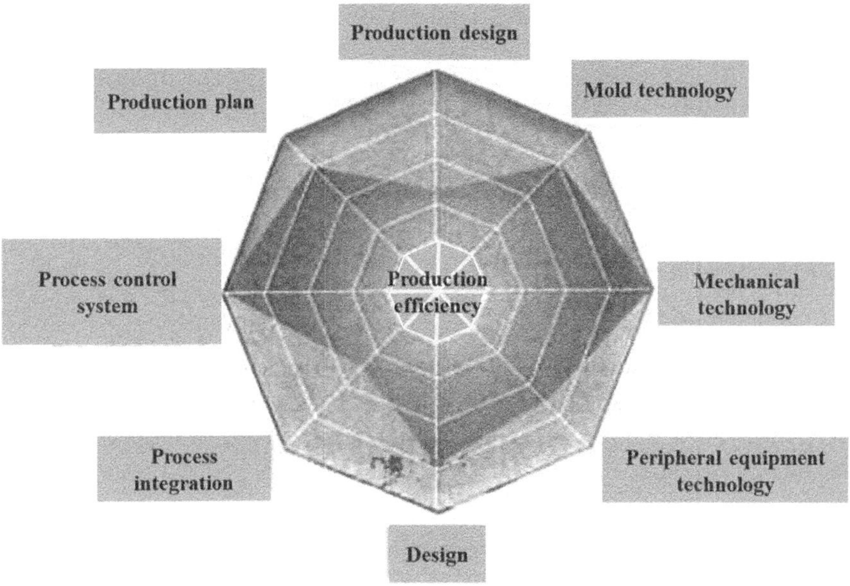

Fig. 5.20 Integrated diagram of production efficiency factors for an industrial project. *Source* http://www.ailab.cn/view/20I6092316601.html

Before implementing the project, each factor is scaled; there is no open form with "embedded interface" based on the corresponding objectives and rules. On the one hand, the formation of an engineering organization brings together certain factors; on the other hand, it makes the above scalar factors "vectorized" (Fig. 5.21). Vectorization is manifested in two forms: one is the linearization of autonomous factors, i.e., the complexity, multi-value, and randomness of autonomous factors' thinking and behavior are regulated at the engineering level as a set of linear behaviors attached to the corresponding engineering construction or management behavior; The two is the "coding" or "rasterization" of non-autonomous factors, that is, all kinds of material entities, technologies and rules, funds and other elements are coded in the time and space coordinates of the engineering objective system, forming the "grid" of the engineering system; then integrate and transform energy according to the coded composition and start the achievement of creation.

2. **The operational efficiency of an engineering organization is based on the multilateral matching of engineering factors**

Factor vectorization aims to establish a "multilateral matching" state among factors. "Multilateral matching" is to determine the matching relationship among the vectorized factors of engineering organizations so that the energy efficiency (weight) of different factors under the constraints of engineering objectives can be maximized as far as possible (for specific projects, such as civil engineering). Or, to form a

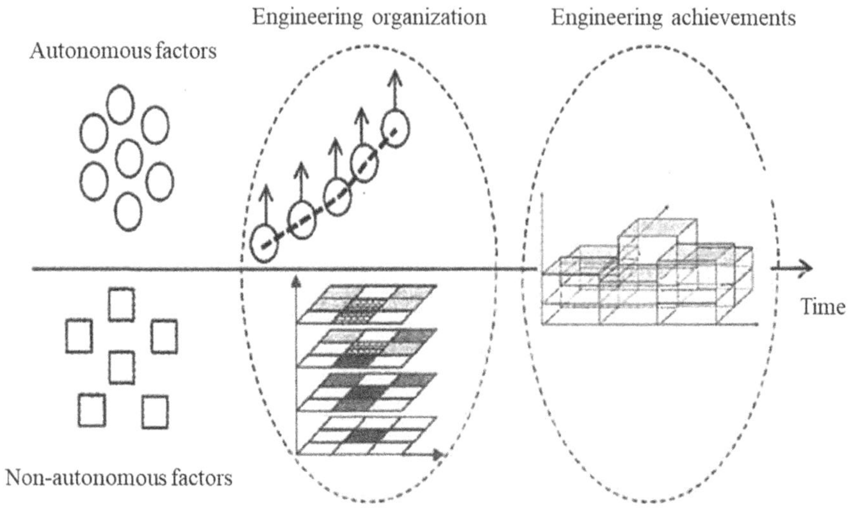

Fig. 5.21 Vectorization process of engineering organization factors

stable matching solution (for stage homogeneous engineering, such as industrial engineering), the above process can be transformed and described as a linear programming problem maximizing the linear objective function. Therefore, the factor "multilateral matching" status established by the engineering organization is essentially the optimal solution set of factors matching. Under the optimal solution, the core of the engineering organization achieves "high efficiency."

Take American space exploration as an example. NASA is just the "director" of the magnificent creation activity. Many commercial technology companies are the protagonists of American space exploration, and these protagonists are the factors of the engineering organization. Under the constraints of space objectives, NASA determines the goals, roles, and relationships of different enterprises in the space program. That is, to determine their mutual matching relationship and make full use of each enterprise's superior technologies and resources to form the optimal solution set of the aerospace plan and achieve "multilateral matching" between the factors.

After selection, the Apollo Manned Moon Landing Program's main equipment subcontracting is provided by different commercial companies, such as command module and service module, which North American Aviation Corporation delivers. The lunar landing craft is designed and manufactured by Grumman Aerospace. These commercial companies play different roles in the plan. They cooperate effectively and form the optimal solution set of the plan, achieve the "multilateral matching" state, and ultimately achieve the efficient operation of the organization. In addition, in the creative activities, the construction of strategic cooperation relations ensures the stability of the "multilateral matching" state to a great extent. The North American Aviation Corporation that has performed well in the Apollo Manned Moon Landing

Program obtained service qualifications based on its previous outstanding performance in the "Mercury Program" in the United States. The formation of similar cooperative relationships can maintain the stability of "multilateral matching."

5.3.2.3 Maintenance and Elimination of the "Efficiency" of Engineering Organization Operation

1. **The "high efficiency" of engineering organizations lies in forming endogenous dissipative structures**

Engineering organizations can construct and maintain their "efficient" state endogenously, mainly because of the formation of their endogenous dissipative structure. An engineering organization is a relatively balanced system in a certain period of time, and the process of vectorization and multilateral matching of the factors of the organization is gradually patterned and regularized. However, the entropy value of this system is getting larger and larger. Once the entropy value reaches the threshold value, the heat death will occur, making the organization unable to adapt to changes in external conditions [30]. The above situation will force the internal configuration of the engineering organization to make endogenous changes, making it transform from equilibrium structure to dissipative structure. The change process of the internal morphology of an engineering organization can be described as a negative entropy introduction process. The core of this process lies in the continuous transmission of new materials, information, and energy to the internal organization (such as innovative management model, the use of new technologies, optimization of management means, etc.), making the organization order increase, disorder decrease, negative entropy greater than positive entropy, and forming a dissipative structure,

The existence of endogenous dissipative structure is an important means for engineering organizations to ensure efficient operation. It promotes the transition of engineering organizations from disorder to dissipative structure to maintain the high efficiency and stability of the engineering organization system. The shaping of internal dissipative structure in engineering is to let engineering organizations maintain a highly open state when the organization is far from its equilibrium. It ensures that the organization can exchange material, energy, and information with the external environment, in which an organization can generate negative entropy flow. The organization can reduce the entropy, achieve the orderly state of the engineering organization in terms of time, space, and function, and ensure the efficiency of its operation. Therefore, in the operation of engineering organizations, especially when there are problems in their operation (such as deviating from the anticipated plans and objectives of engineering organizations), it is necessary to ensure a high degree of openness of the organization, and to ensure the input of factors of production (such as material, energy, and information) in a rational and orderly manner according to the plan, to achieve the shaping of rational and orderly structure of engineering organization, bring engineering organization back to the track of efficient operation again.

The existence of endogenous dissipation makes the engineering organization have better environmental adaptability, which can quickly feedback the changes of the environment and construct a new structural state within the organization, ensuring the organization's transformation from low-level to high-level. It is particularly "precious" in the increasingly complex and changeable creation environment.

However, the activity of creation has the life cycle attribute, and the engineering organization is the constituent of the activity of creation. The engineering organization will certainly disband at some point. The disbanding of an engineering organization results from the increase of the entropy of the engineering organization. In the "later stage" of the life cycle of creation activities, although the exchange of material, information, energy, and other elements always occurs between the engineering organization and the external environment, the intensity of this exchange gradually decreases with the passage of "time." That is to say, the material, energy, and information entering the engineering organization decrease slowly, the organization's negative entropy flow decreases, and the entropy of the engineering organization increases, which leads to the disordered development of the time, space, and function of the engineering organization. The final result of the development is the final disband of the engineering organization.

2. **The threshold of organization entropy is the watershed of the efficiency of engineering organization operation**

Because of a slight random disturbance and amplification, the engineering organization will break the dynamic equilibrium state, enter the unstable stage and achieve a new dynamic equilibrium through self-organization after reaching a certain threshold level. Therefore, a vital factor in modern engineering organizations' competitiveness is accurately knowing the threshold value of organizational entropy. Meanwhile, timely absorb a large amount of material, energy, and information from the environment, make corresponding internal configuration changes, and establish a dissipative structure of the organization [17]. One of the important characteristics of the dissipative structure is that virtual attributes (such as technology, management, and thought) have a higher weight in organizational output than entity attributes (such as resources, funds, etc.). The improvement of output efficiency of an engineering organization lies in the optimization of its dissipative structure.

The basis for the small disturbances inside and outside the engineering organization to develop into great fluctuations is that the highly open engineering organization has a "coupling" function. Coupling is the behavior of many complex factors to strengthen or weaken a certain factor under open conditions. However, this is not simply a linear superposition but the result of many factors acting together. The "coupling" function of engineering organizations makes it possible to strengthen small fluctuations jointly and eventually grow into huge fluctuations, resulting in the sudden change of engineering organizations into unstable states.

The threshold of organizational entropy is the watershed of the efficiency of the operation of engineering organizations. Efficient planning management helps determine the threshold, but this determination is not accurate. The engineering plan is a process of creative activity determined beforehand according to the optimal solution of the established factors of engineering production. It is based on the full consideration of the internal and external environment, and the project factors of production put in according to the plan at the right time ensure to a large extent that the organization can "cross" the threshold of the entropy. However, it is difficult to locate the threshold of the organization entropy accurately. The threshold of organization entropy can be regarded as a signal state. After all, creative activities' internal and external environment is complex and changeable, so emergency management measures are an important supplement to engineering organizations' smooth implementation of planning management. In addition, the rapid development of science and technology is profoundly changing the activity of creation. The dependence on the traditional attributes of the entity in the traditional creation activities has been weakened in the current creation activities, instead of the production factors with virtual features such as management, ideas, decision-making, and knowledge. The input of such virtual production factors is an important means to ensure the efficient operation of the engineering organization. For example, the concept of knowledge management has been widely recognized in different organizations. Knowledge management ensures the efficiency of knowledge acquisition, sharing, and application of the organization and then helps the organization achieve efficiency.

5.3.2.4 Adhering to the People-Oriented Principle to Achieve "Efficient" Operation of Engineering Organizations

1. **People are the most critical resource in an engineering organization**

People are the most active factor in productivity. Only through the play of human subjective initiative can all other resources achieve the process of "creation" of engineering organization and fulfill the purpose of serving humanity. From the microcosmic point of view, tools, technology, time, information, and other resources are static and passive. Only by giving full play to human resources' enthusiasm, initiative, and creativity can it guide other kinds of materials to combine and flow towards the planned goal and form the optimal solution. From the macro point of view, the living soul of the project is also human. The project was inspired by people and impacted the surrounding environment after its completion. The economic benefits of small projects, such as the construction concept of a residential area, directly affect product positioning and later sales, and the comprehensive economic, social, and environmental benefits of massive projects, such as the planned installed capacity of the Three Gorges Project and resettlement of immigrants, and the impact on the ecology of the Yangtze River, which all require to be controlled by humans.

The early engineering organization is the "huge-crowd strategy." The main production factor is people, relying on collective strength and wisdom. It can be said that without a sufficient number of people, there will be no wonders of the world such as the Great Wall and the Pyramid. Medium-term engineering organization's "human–machine" system is also a large machine system created by people to help its production. There is no such era without people. In the "human–machine-network" system of modern engineering organizations, although the relative quantity of human resources has been reduced, the absolute number is still huge when the scale of the project is enormous. The human resources in this organization are reponsible for science and technology, information, and other resources and provide geometrically multiplied production efficiency.

The process of "creation" and development of engineering organizations shows that human resources play an essential role. People-oriented factors are the primary and essential factors in management. The management of engineering organizations is to develop human resources fully.

2. **The subject status of people in project implementation**

People implement the management activity of engineering organizations; whether it is the utilization of resources or the formulation and implementation of systems, it ultimately falls on specific people. People-oriented reveals this essential law and emphasizes the dominant position of human beings. Management objectives can be achieved more efficiently by giving full play to human beings' roles. Only human beings can constantly learn, renew the connotation of human resources, and recombine other kinds of resources to maintain the organization's vitality and promote the organization's restructuring to adapt to the changing external environment. Therefore, human learning is fundamental to ensure the sustainable development of engineering organizations. Innovative results formed in the process of people's subjective initiative integrating tools, technology, information, and other resources, public or private person-to-person contact in engineering organizations, and new spirits created from the interaction and collisions between people in engineering organizations are the driving force for the continuous progress of engineering organizations.

The basic implementation of engineering organization activities and the progress and the development of the evolution of engineering organizations are inseparable from the main role played by human beings. To form an optimized human environment and achieve good management standards, it is necessary to earn "respect for human management." That is to say, it is necessary to respect the personal dignity of members, the value they bring, and their individual potential, and to regard organizational members as the key and fundamental of the development. The goal of human management is to motivate people's enthusiasm. Only when people's enthusiasm is motivated can new ideas and concepts be generated, new inventions and innovations be made in their work, new methods and technological innovations be produced, and the goal of management can be achieved ahead of time.

3. **The direct goal of engineering organization activities is "for people"**

Engineering organization activities are carried out to meet the needs of human beings; their direct purpose is "for people." Suppose the starting point is not based on the guiding ideology of "for people"; in that case, the engineering organization will lose its direction, and its activities will not be efficient or even in the opposite direction, from "high output" to "high destruction." This refers to the construction of engineering organization form from the point of view of serving customers and users and includes other stakeholders affected by the implementation of the project and included in the formation of the engineering organization. From the inside of an engineering organization, the benefits of engineering workers are also important for the smooth implementation of projects. The existence of personnel health and safety institutions within the engineering organization provide a standard-compliant operating environment and perform logistical work such as safety management, labor insurance, and medical insurance.

4. **Humanized engineering organization culture is a catalyst for the efficiency of engineering activities**

In June 2013, Shenzhou 10 successfully launched in Jiuquan and achieved perfect docking with Tiangong-1, successfully fulfilling the task of the project, marking a new achievement for China's manned spaceflight. In the process of achieving this great pioneering work, we have always adhered to the spirit of "innovation, solidarity and cooperation, scientific and realistic, people-oriented, and patriotic dedication." The engineering culture atmosphere that suits the inner demands of human beings has fully catalyzed the positive enterprising spirit of astronauts, and thus successfully completed this time-consuming and arduous project.

The humanized engineering organization culture has a strong incentive function for the engineers in the engineering organization and enables them to apply the humanized idea to the project construction. The astronaut system in the eight systems of China's manned spaceflight project is the first system. Humanized design, development, and production are all centered on the comfort of astronauts. The activities of the whole engineering organization and the products of the engineering organization are highly enthusiastic, orderly, and efficient.

Humanized engineering organization culture requires organizational construction based on respecting human nature. The lofty ideal of organization makes members full of hope and a sense of fullness and guides the good direction for members' hard work. The humanization mechanism of an organization resonates with the needs of its members, gains the general recognition of its members, and maximizes the unity of its effective forces. In the humanized organizational atmosphere, people can satisfy at all levels, so this type of engineering organizational culture can be successfully carried out and continuously develop in the process of practice, form a virtuous circle with members' practical feedback, and maintain vigorous vitality.

In a word, organizational structure and system construction are the necessary conditions to regulate people's behavior. Humanized engineering organizational culture is the soft power to achieve efficient project implementation. Its influence permeates every link and every part of the engineering organization, not only has a positive effect on the members of the organization but also been passed on to the activities of members in the organization, which is the genetic gene to achieve the "efficient" operation of the engineering organization.

5.3.2.5 Construction Model of "Efficient" Operation of Engineering Organizations Under the Background of Information Technology

1. **IT reinterprets the operational "efficiency" of engineering organizations**

In recent years, under the competitive pressure of global economic integration, a tremendous amount of information needs has emerged in both life and work, which has led to the rapid development of the IT market. The scale of the IT system is expanding, and the structure heterogeneity is increasing, ensuring the stability and efficiency of daily life and work and supporting innovation and optimization to help our quality of life.

Faced with differentiated needs from all walks of life, IT services' response speed and service efficiency have become the key to determining the future. Because of this, it has become the consensus of the mainstream IT service providers in the industry to continuously enhance their core competitiveness and ensure their dominant position in the fast-growing market by means of highly efficient R&D.

The complexity of modern engineering is becoming more and more prominent. Some traditional organization modes have been difficult to meet the requirements of large and complex engineering. Modern engineering requires that the organization model should be more flat, reduce the middle level, improve the management scope and efficiency, and the emergence of IT and other high-tech, which has led to various engineering organization models, such as the "engineering cloud organization" model. Compared with the traditional process of factor vectorization and multilateral matching, the internal configuration of engineering cloud organizations is quite different in the temporal and spatial allocation of elements. Although there are also two processes mentioned above, the combination of autonomous factors in engineering organizations has changed greatly due to the direct and flattening of "information setting." The autonomous factors are based on definite and precise objectives (the objective curve comes from the rapid supply and the provision of information technology platform), which directly affect the corresponding non-autonomous factors or autonomous factors that have a coupling relationship with them, so that the "coding" process of non-autonomous factors becomes faster and more accurate. The non-autonomous elements can achieve the "transition" combination and transformation process. The basic principle of the "cloud organization" mode is shown in Fig. 5.22.

The "Engineering Cloud Organization" model establishes the relationship between various flexible related engineering organizations, which makes the efficiency optimization of engineering organizations broader and can adapt to more complex and huge engineering practices.

2. **IT creates unlimited possibilities for the "high efficiency" of the operation of engineering organizations**

Taking IBM's PureScale technology for efficient database transaction processing as an example, in order to take care of customers, investment in hardware and software

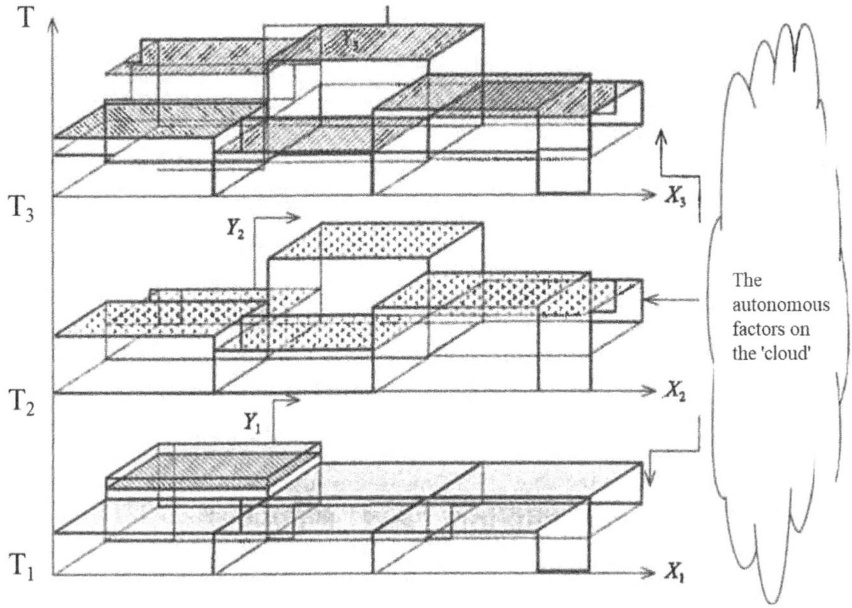

Fig. 5.22 Forms of cloud organization

is increased due to the continuous growth of data volume, choosing economical and efficient plans to expand storage capacity. At the end of 2009, in order to meet customers' growing business needs, Power Systems Labs in Austin, Texas, and IBM Toronto Software Labs jointly developed DB2 to provide customers with continued reliability and almost unlimited computing power. DB2 PureScale can help customers reduce the risk and cost of expanding computing power and ensure no interrupt services. DB2 PureScale can improve the system's horizontal scalability without adjusting applications or databases. This function optimizes the computing power in management and reduces the communication cost in the system.

The complexity of modern engineering is becoming more and more prominent, and some traditional organization modes are difficult to meet the requirements of large and complex engineering. For example, the linear structure lacks the division of functions, the horizontal links between members and organizations are poor, and the managers are required to have high quality, which is only suitable for small organizations. The functional structure emphasizes specialty too much, with poor coordination and adaptability to the environment; it is also inflexible, and members tend to focus on local goals and reduce the overall objectives. The matrix structure used more often also has issues such as lack of stability, double leadership, and large coordination workload. The emerging technology has opened another window to achieve the operational efficiency of engineering organizations.

5.3.3 *Environmental Adaptability of Engineering Organizations*

An engineering organization is an open system that interacts with the external environment. It is always in a certain natural, economic, technological, social, political, and ethical environment, and it keeps in touch with and interacts with the environment. Just like the metabolism of the human body, engineering organizations change their resources and information from the environment into their own essential elements. At the same time, they also release some elements and information into the external environment, react to the environment, and finally achieve self-renewal. Therefore, environmental adaptation is the prerequisite for the upward development of engineering organizations and the carrier for the sustainable development of human society.

Professor Richard L. Daft of the United States points out that "organization refers to a system closely related to the external environment, oriented by clear objectives, with well-designed structure and conscious coordination of activities." [31] Professor Richard Scott thinks that organizations rely on the environment and are conducive to constructing the environment. It is an activity system that makes internal participants rely on each other and relate to each other [32]. Modern management theory holds that there is an open system when an organization interacts with the environment. An engineering organization is an organization that takes engineering as its object of organization. It also has typical organizational characteristics. It is an open, self-organizing, continuous evolution, and spiraling system.

There are a large number of rules and models of non-linear relationships in an engineering organization, which is a complex adaptive system. It can generate adaptive survival and development strategies through "learning" and achieve the creative evolution of the organization. The adaptation of engineering organizations to the environment is the process of active, repetitive, and non-linear interactions between engineering organizations and the environment, and the process of organizational survival and development, and even creative evolution.

5.3.3.1 The Power Source of "Adaptability" of Engineering Organization Environment

Many factors affect the environmental adaptability of engineering organizations. Among them, some factors drive the acquisition or enhancement of the adaptability of engineering organization, and some factors block the acquisition or enhancement of the adaptability of engineering organizations.

1. **Generating adaptability in aggregation**

Birds of a feather flock together. We often categorize things like pine, cypress, birch and then treat them equivalent. This aggregation method facilitates analysis of similar situations; it can be familiar fragments or recombination of classes to generate things that have never been seen before. From this point of view, aggregation is the main means of constructing an engineering organization model. The key to aggregation is to decide which details are irrelevant to the issues of interest and ignore them. This effect is to ignore the differences in detail and categorize things. The modeling process of an engineering organization is an art form, which depends on the experience and taste of modelers. Modelers must decide which features to highlight and which features to eliminate in order to answer questions.

Aggregation also focuses on complex adaptive systems rather than building models only. It explains why the aggregation of simple subjects is bound to produce complex, large-scale behaviors. The ant nest is a good example. A single ant is very weak, and its behavior is stereotyped. Independent individuals have only one dead end in disastrous environments such as human and other species' attacks or floods and droughts. However, when ants gather in groups and build exquisite nests according to the tradition, and work collaboratively around the nest, living and working, breeding offspring, then the group's adaptability is greatly enhanced and can survive in various harsh environments so that their population can reproduce. The same is true of the honeycomb organized by regular hexagons, which is constructed by bees. These honeycombs are extremely delicate, similar to human cities. The bees work and cooperate inside, pollinating plants, extending the intergenerational transmission of plants, and creating delicious honey for humans.

The description of the place where simple organisms gather—ant nest and hive make us understand many more complicated phenomena, such as the intelligence displayed by a large number of interconnected neurons, the wonderful characteristics of the immune system composed of various antibodies, or the construction of the large scale projects by means of organizing the weak individuals to use cold machinery, such as the Pyramid, Great Wall, Dujiangyan, Three Gorges Project, the Qinghai Tibet Railway. It also includes launching spacecraft, building space stations, and exploring tens of millions of kilometers or even billions of kilometers of space. Organisms composed of countless cell types have also demonstrated amazing coordination, and large cities show coordination and persistence. Of course, there is also the coordination and adaptability of the engineering organization to the environment.

2. **Generating adaptability in "flow" systems**

Nodes (subjects) are processors that connect possible interactions, and the adaptation of the subjects determines the appearance or disappearance of connections. Therefore, the "flow" reflects the variability adaptability because of the change of time and the accumulation of experience. The concept of "flow" is by no means limited to the motion of the fluid. For the "flow" of engineering organizations, we can imagine that there are many nodes and the flow of resources connected to a network. For example, nodes are factories, and connections are transportation lines of goods flow between factories. This "flow" forms a three-in-one combination system of nodes (factories), connectors (transportation lines), and resources (materials).

An engineering organization is a typical "flow" system, which links engineering, participants, and production factors to form a complete multi-component system. Its core is engineering, the organization is the link, and the engineering organization is

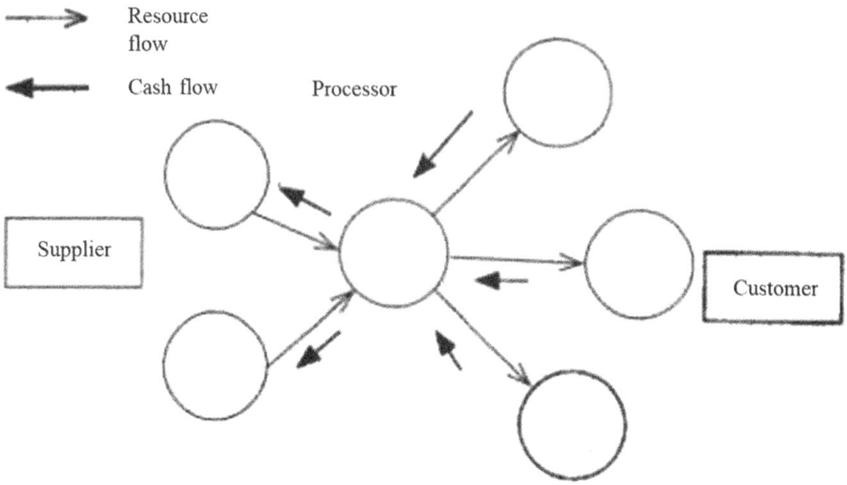

Fig. 5.23 Engineering organization "flow" system

this flow system's node. Contract (contractual relationship) is a connection, similar to a factory's transportation route. The production factors which many participants own are continuously injected into the engineering organization through the contract network. The production factors are integrated into the core of the engineering organization through contracts to achieve specific engineering objectives through the integration of links and connections (Fig. 5.23).

This is true of many major infrastructure projects in China, such as the Qinghai Tibet railway, the Three Gorges Project, and other smaller projects. Taking the Three Gorges Project as an example, the node is the main body of the Three Gorges Project, and tens of thousands of contracts (such as construction contract, material purchase contract, technical consultation contract, survey and design contract) connect the main body of the Three Gorges Project and sharing benefits and risks with many participants are the ties. This kind of "flow" forms node (Three Gorges Project), connector (contract-benefit sharing, risk sharing), resource (material, information, technology, etc.) three-in-one combination system.

Engineering organizations improve their environmental adaptation process by limiting key interactions, i.e., major connections. By selecting those markers that are beneficial to interaction, the negative consequences markers are excluded. The "flow" of engineering organizations has two well-known characteristics: the multiplier effect and the recycling effect.

The first characteristic is the multiplier effect, which refers to the injection of certain resources on some nodes, which will generate more resources through this nonlinear system. The ratio of the magnitude of the chain reaction is a multiplier. The concept of the multiplier effect comes from economics and gradually spreads to other disciplines.

In the field of economics, multiplier benefit is a kind of chain reaction. It is a macroscopic economic effect to measure the change of a variable in economic activity and cause the most general change in the economy. There are many kinds of multiplier effects, such as the expenditure multiplier effect, which means that the change of expenditure on the consumer side will lead to multiple expansion or a fraction contraction of total output. The multiplier effect of increasing or reducing fiscal expenditure is to expand the total economic output or reduce the total economic stiffness; the multiplier effect of increasing or reducing taxes is to shrink the total economic output or enlarge the total economic output.

Regarding regional economics, the multiplier effect refers to the demonstration, organization, and promotion of the surrounding areas through industrial and regional linkages, and this effect is continuously strengthened through circulation and causal accumulation.

Usually, resources transfer from one node to another, resulting in a series of changes. Whether the required resources are information, goods, or money, the network and flow have the characteristics of a multiplier effect, which will be more obvious when the evolution of engineering organizations changes.

The multiplier effect in an engineering organization is manifested in many aspects, such as the change of performance evaluation mechanisms and other systems that may lead to great changes in organizational behavior. The incentive model of the performance appraisal mechanism may spread from publication to execution and gradually evolve into a huge positive or negative effect. The diffusion of new technologies will also show a relatively large multiplier effect.

The second feature is the recycling effect. When resources are recycled, the same raw material input will generate more resources on each node.

In the eco-industry, engineering, and circular economy, in order to reduce the final waste disposal and cost, the waste will be recovered and utilized comprehensively at the output terminal to form reusable resources. Recycling in chemical process refers to the separation of untransformed raw materials after reactions due to the incomplete reactor conversion of materials, which are then sent back to the feed port and added to fresh raw materials for reuse, i.e., the materials are fed back from the downstream of the process to the upstream of the process.

Recycling often occurs in engineering organizations. Recycling keeps rising, from the early and middle stages to the modern stage, and will also develop into some form in the future. For example, some materials and resources can be recycled in the creative activities to achieve the recycling effect, especially in construction projects, such as large-scale recycling of template materials.

3. **Adaptation in diversity**

Each subject has its own limited niche because the subject depends on the environment provided by other subjects. For engineering organizations, diversity is the result that new subjects can produce similar effects and replace the subject. Diversity of engineering organizations is a dynamic model with sustainability and coordination. Once the original state is destroyed, it will quickly return to its original state [33]. However, each remodeling results from continuous adaptation, innovation, and growth. The diversity of engineering organizations is a kind of flexibility compatible with the resources of engineering ontology, engineering environment, and engineering materials. In contrast to rigidity, the diversity of a flexible engineering organization in different periods, different needs of the engineering ontology, and different organizational environments is exactly its flexible manifestation.

Flexible force is infinite, called "the most flexible can tame the most rigid things." Diversity is a form of flexibility, while the core of resource aggregation is its rigidity. The engineering organization must be a complex and adaptive organizational system that coexists with rigidity, flexibility, unity, and diversity.

5.3.3.2 Operating Mechanism of "Adaptability" of the Engineering Organization Environment

The adaptability of an engineering organization is composed of identification mechanism, internal model mechanism, and decomposition and reduction mechanism. These adaptive mechanisms are described in detail below.

1. **Identification mechanism of the engineering organization**

The identification mechanism of an engineering organization is to better gather and define boundaries, just like the same class or school uniforms. Another example is the Flying Tigers team created by American volunteer Chennault. The brave tiger is used as its team logo and flag pattern. The whale teeth painted on the fighter's head are the logo of the Flying Tigers aircraft. Another example is the typical signs of all kinds of flags used by the traditional Chinese army, shop signboards, etc., registered trademarks, corporate logos, and avatars in QQ and WeChat spaces.

Identification provides the basis for selecting indistinguishable targets and subjects, provides the basis for engineering organization selection, cooperation, and specialization, better promotes the selective interaction of organizations, and ensures that the function is not affected by organizational changes.

2. **Internal model mechanism**

An engineering organization is a complex non-linear system, and it is difficult to predict its long-term operation. But to some extent, its short-term situation is predictable. The general prediction behavior is basically based on the simplification of the nonlinear system, that is, linearization, such as the extrapolation prediction model.

Internal models enable engineering organizations to predict certain things. The basic way to build a model is to eliminate the details and emphasize the chosen model. The model we are interested in exists in the internal engineering organization. The main body needs to select the corresponding mode in the input end and transform it into internal change; that is, the model. Must enable the main body to predict that when the mode (or similar model) is encountered again, its consequences will be [34].

3. **Decomposition and reduction mechanism**

Human beings can decompose a complex thing into many parts. Of course, the division of components is by no means arbitrary. Through the accumulation of experience and learning, people can decompose complex things into elements that have been certified and can be reused and then form different combinations.

For example, when encountering the cabin of a flying aircraft that may leak 45 degrees in up front direction, although we have not encountered a similar situation before, we will also adopt a series of methods to deal with it. Human beings are faced with many uncertainties. Regulation is relatively stable. Using the rule of law to deal with the ever-changing reality will achieve the desired results. Of course, it is impossible for us to discover all the potential rules of operation, nor to prepare a complete set of rules for all possible situations. But history will tell us which are more efficient, which are no longer efficient, which can no longer be used, and so on.

Therefore, to adapt to external changes, engineering organizations usually take actions based on historical experience and laws to achieve satisfactory results when encountering new situations. It also reflects the need for engineering organizations to emphasize the accumulation, sharing, and application of technology and experience of previous participating projects.

Decomposition and reduction is a typical linear mode of thinking, a basic way of looking at and understanding the world gradually developed in the primary stage of human beings. Although the world is a non-linear system, it has a certain degree of linear characteristics locally. It is possible to use decomposition and reduction mechanisms within.

With the development of science, the mechanism of decomposition and reduction may become history, and it will continue to exist in the expected future. It will not disappear and may be moved forward.

Engineering organizations improve their environmental adaptability through identification mechanism, internal model mechanism, and decomposition and reduction mechanism. Through the selective interaction of the marking mechanism, engineering organizations can identify the most suitable participants (stakeholders) and the appropriate combination of participants. It is necessary to ensure that the interaction between the various parts of the engineering organization can be maintained when they are constantly changing, which is the basis for improving the adaptability of the organizational environment. In the life cycle of creational activities, when a large number of materials, energy, and information flow into the engineering organization, it depends on the internal model mechanism to predict the changes that may occur in the short term to achieve the most effective control of all kinds of "changes" and improve the environmental adaptability of the organization. Finally, the recurrence of creative activities is changeable, inevitably encountering new situations that are "unexpected." With the assistance of decomposition and reduction mechanisms, engineering organizations rely on previous projects' experience, rules, norms, and rules. Based on the objective analysis of new situations, they adopt appropriate ways to deal with new problems and achieve satisfactory results, and the environmental adaptability of organizations is improved again.

5.4 Future Development of Engineering Organizations

The essence of engineering organization innovation is based on the re-selection and construction of group elements such as people and resources, aiming at new projects. Finally, a new theoretical paradigm is concluded and deduced. The practice of engineering activities proves that the innovation and development of engineering organizations mainly come from the change of engineering ontology, and the transformation of the theoretical paradigm provides a theoretical basis for this innovation.

5.4.1 Development Trends of Future Organizational Structure

An important goal of the existence of an engineering organization is to achieve organization coordination through the design of task structure and power relationships to serve the completion of engineering objectives. Organizational structure is the external formal expression of an engineering organization. How to distribute the work, who is responsible to whom, how to coordinate the internal mechanism, and even all kinds of rules, duties, and rights relations in engineering construction belong to the scope of organizational structure.

Firstly, the change of engineering ontology is one of the reasons for the shift in organizational structure. The change of engineering ontology, including the change of engineering scale, difficulty, resources, and objectives, often starts before the change of its organizational structure and eventually leads to the appearance of a new organizational structure. In addition, the changes of the external environment and engineering purpose of the project, such as politics, economy, society, culture, and technology, also impact its organizational structure. Therefore, the change of engineering organization is the result of the interaction of internal and external engineering (Fig. 5.24). The change of engineering organization is essentially the mapping of the organization's adaptability to the engineering ontology and its external environment. The future evolution of engineering organization structure and the evolution trend of engineering ontology and organization structure have become major topics in engineering management theory.

The adaptability of an engineering organization requires that the organizations adapt to the current environmental requirements and internal conditions within the organization and adapt to the future external environmental requirements and future changes in internal conditions. Therefore, the traditional engineering organization forms still play a huge role today. In some traditional engineering fields, technology

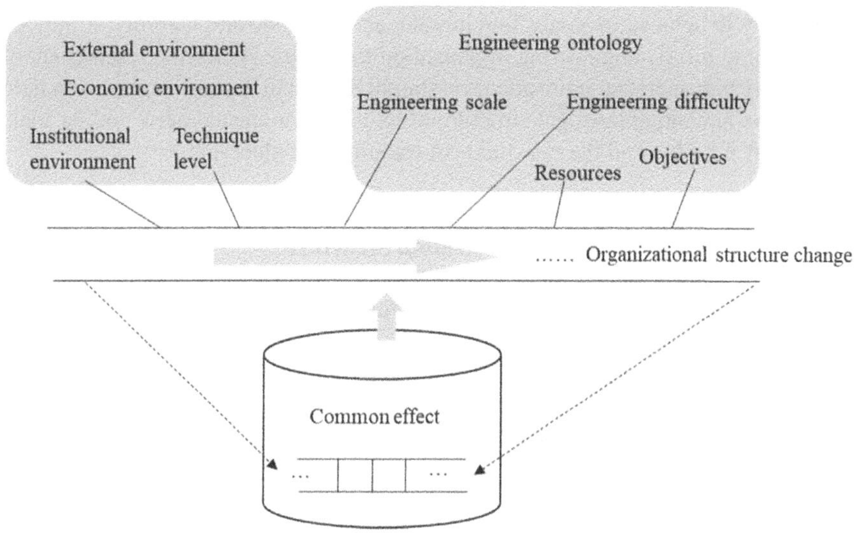

Fig. 5.24 Engineering organization structure change map

updates slowly, knowledge requirements are not high, capital and human investment are the main factors. The inherent organization patterns can fully meet the needs of engineering objectives, and mandatory moving to new organization forms will only be counterproductive. On the other hand, in some new engineering fields, such as aerospace, the speed of change is beyond most people's imagination. If the organizational form does not change, it will inevitably become a rigid giant and be eliminated by competition. It is why a new organizational form has emerged continuously in the past twenty years. At the same time, the change of the political and social environment in which the project is located also forces the organizational form to change accordingly. Like in the transition from a planned economy to a market economy in China, the Ministry of Railways has become a railway corporation. Private capital has been opened to the public in petroleum, telecommunications, and other fields. The organizational form that relies entirely on orders and plans will naturally die out, a new organizational form based on contracts and interests emerges. It is not the reason for the organization itself or the project itself, but the existing soil has changed.

Therefore, it is necessary to have a clear understanding and accurate prediction of the changes of internal and external factors in engineering to judge the future change trend of the engineering organization. It is difficult to predict the future with complete accuracy. Still, certainly, traditional projects will not disappear, and new projects will continue to emerge, and the old and new organizational forms will coexist for a long time. As a project manager, the problem that needs to be solved is whether the organization of the project is compatible with the project, rather than blindly seeking new changes. It is difficult to have an accurate criterion for judging whether the engineering organization needs to change, but when the project shows signs such as difficulty to achieve its goals, low morale, and low innovation ability, it may be the right time for the engineering organization to change. As for the researchers of engineering management, more attention should be paid to the new engineering field because the new organizational structure often has been created here, and its scale and influence far beyond the new fields of traditional engineering.

5.4.1.1 "New Networking" Organization Structure

Under globalization and informatization, engineering organizations have entered a new era of comprehensive innovation of knowledge economy, continuous integration and interactive development of global network technology and modern information. Based on the original networking, the organizational structure will evolve to a new type of network structure with flat organizational form, informatization of organizational level, and virtualization of the organizational platform. Therefore, the future development of the "new networking" organizational structure reflects the following three characteristics.

1. **Flattening**

The flat organization is one of the trends in developing the new network engineering organization structure. In large and complex projects, human resources and other organizational factors increase, where knowledge and information are the main input factors, the importance of labor and capital decreases. The traditional engineering organizations have obstacles in internal information transmission, knowledge sharing, and employee innovation, which lead to the increase of management level, the slowdown of information transmission, the decrease of organizational efficiency, and the rise of management cost.

Due to the increase in the engineering's ability to respond to uncertain factors, with the development of modern information technology, the future network engineering organization will focus on increasing the management scope and transforming from the traditional pyramid structure to the flat structure. Especially the application of high-speed computer management information systems makes applying new networked organizations in engineering possible. The organization achieves common goals through close multilateral links, interaction, and cooperation, rather than relying on hierarchical control as before. Intensive multilateral links and full cooperation are the biggest characteristics of the new network organization, which are the biggest differences from the traditional engineering organization.

Take China's large aircraft project as an example; its organizational structure does not adopt a linear organization form of "overall design + administrative command + technical command," which was fully undertaken by the government during the "two bombs and one satellite" period. Instead, it adopts the network organization with "government guidance with enterprises as the mainstay." The organization includes: the internal network (Shanghai Aircraft Manufacturing Co., Ltd., Commercial Aircraft Corporation of China, China Aluminum Corporation, Baosteel and other design, manufacturing, service providers), vertical networks (design, manufacturing, service, material suppliers), and the general network (it refers to the integrated network of all enterprises in different sectors of the project, such as the integrated management of design, manufacturing and service providers of Commercial Aircraft Corporation). We should achieve the systematic management principle of "taking advantage of each other's strengths, collaborating with each other, performing their duties and achieving unity of objectives" to effectively eliminate the organizational obstacles of previous similar projects and straighten out the relationship between the parties.

2. **Informatization**

With the social progress and economic development, the scale of an engineering organization is expanding, the market competition is intensifying, and the organization's whole social and economic environment is becoming more complex. It requires the degree of informatization of the organization to increase and the network organizational structure and mode of action transforming to informatization. The organizational structure of effective integration of network organization and information technology will be an important way for engineering organizations to flexibly respond to internal and external environmental changes and promote organizational information communication.

The new network organization informatization can achieve hyperlink, multimedia, and all-around dynamic information service. Taking the engineering organization of the 2010 Shanghai World Expo as an example, its organizational structure introduced the concept of information technology, and a standardized engineering organization (Fig. 5.25) was constructed to manage crisis effectively. This organizational structure rationally utilizes the information process of information transmission—information processing—information feedback, establishes the network structure of resource integration and optimization, information data sharing, and interconnection among various departments of the organization, and achieves the goals of "self-stabilization, high efficiency, flexibility, and sharing" of the organization structure [35].

In addition, with the development of information technology in network-based organizations, the parallel and integrated solutions provided can solve the problem that the traditional hierarchical organization structure emphasizes too much on the sequencing of the division of labor. The achievement of project objectives is the result of the joint efforts of owners, design units, construction units, suppliers, and other organizations at all levels. They need task not only arrangement but also close connection. While the sequential division of labor in traditional organizations produces efficiency, it also has the problem of splitting the links between different departments. In the network organization, the participants of each project constantly receive information from other departments through the network, adjust their work arrangements, and send information to other participants, which makes the whole project's promotion manifest as a process of information transmission, collection, and processing. The final engineering product becomes the material embodiment of this data. In the future, network-based organization informatization can change the division of work

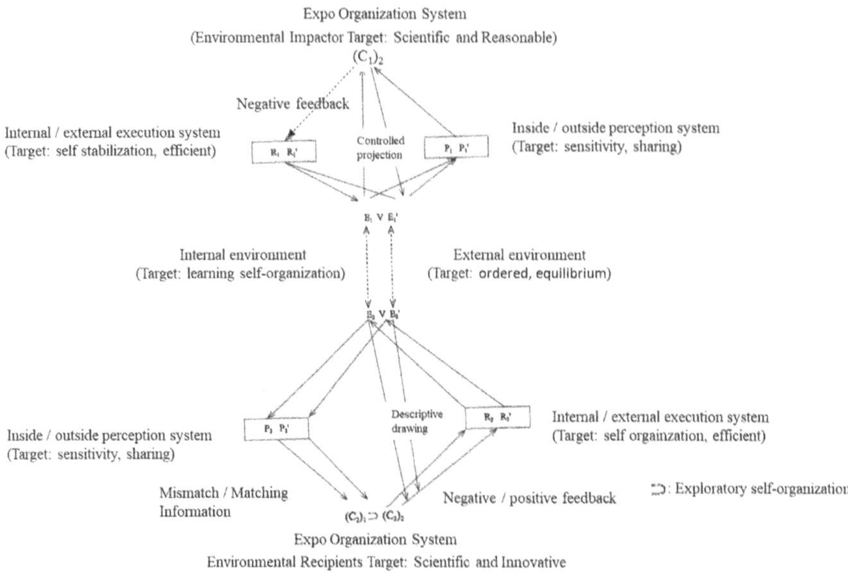

Fig. 5.25 Expo engineering organization. *Source* He et al. [35]

completed in sequence to parallel, shorten the project duration, and optimize the allocation of knowledge resources simultaneously of continuous exchanges and timely adjustments among departments.

Therefore, with the popularization of network information technology, the formation of the information network organizational structure of engineering organizations will be another trend to conform to the direction of the times and improve the level of engineering organizations dynamically.

3. **Virtualization**

For engineering network organizations, there is a certain degree of virtualization. The virtualization of new network organizations has attracted much attention in recent years. By continuously achieving virtualization to participate in the operation of the whole life cycle of network-based organizations, we can identify market trends, quickly gather information resources, and promptly promote technological innovation to meet the development needs of engineering organizations and reduce management risks and save transaction costs. Therefore, enabling the virtualization of network organization is also an important trend of organizational structure innovation [36].

Promoting the virtual operation of engineering network organizations is an efficient and flexible strategy to cope with the intensification of market competition, limited resources within the organization, and the urgent need to broaden information channels. This strategy is to optimize the allocation of resources by retaining the most critical and competitive functions within the organization, virtualizing other functions, and using information network as a platform, thereby maximizing the advantages of the organization's own resources, and to achieve the goal of reducing organizational costs and improving market competitiveness. The virtualization of network organizational structure can integrate resources from different places, and it is an organizational model that transcends the boundaries of space. With the acceleration of economic globalization, a large number of the social labor force will be dispersed outside the existing fixed organizational system. Labor displacement, function outsourcing, and cooperation through network expansion will become a new form of organizational operation. Virtual organizational structure is beneficial for engineering organizations to give full play to their core competitive advantages and promote high-level and high-quality development of engineering organizations.

The virtualization operation of engineering network organization is mainly embodied in two aspects: one is the virtualization of organization form. Through network information technology, the organizational structure will be transformed into an invisible structure without entity form. The organization mainly achieves various organizational transactions through the Internet, thus weakening the role of entity organizational structure. The second is the virtualization of organizational function. With the rapid development of the Internet as the supporting technology, for engineering organization management, design, operation, adjustment, and connection, only retain the organization's core functions, virtualize other functions, and leave them to external organizations.

Therefore, the key to the success of the virtualization of new networked organizations lies in the effective integration of core competencies among member organizations and the core competencies among organizations. For example, the development of e-commerce and the improvement of infrastructure make it possible for engineering organizations to organize, manage and operate inter-and intra- organizationally, handle project and operation, resource sharing, and information queries. It can also help construct a networked virtual work environment using network telephone, video conference, and e-mail, promoting the efficient operation of engineering organizations.

5.4.1.2 Flexible Organizational Structure

Facing the increasingly turbulent and complex organizational environment, the traditional fixed and once-for-all organizational model can no longer meet the needs of modern organizational development. Flexible organizational concept, with its advantages of improving efficiency and decision-making, flexible and highly coordinated communication, has increasingly become the direction of organizational structure reform. Flexible engineering organizational structure refers to the organizational structure that conforms to the current market development needs, has a simple structure, sensitive response, and flexible mechanism. The structure can highly adapt to modern and flexible production technology. Its essence is "people-centered" and achieves humanized management, which is embodied in the following two characteristics [37].

1. **Team organization**

In the future, engineering organizations will be more flexible, that is, more emphasis on the role of people. Since the 1980s, in the emerging fields of information engineering and software engineering, human knowledge and wisdom have had more and more influence on engineering. Innovation has become the key to the success of engineering. People inside the organization consciously and voluntarily dedicate their knowledge and ideas to inspire people's initiative, inherent potential, and creative spirit. Flexible engineering organization is the transformation of engineering objectives into spontaneous actions of personnel in the organization and the internalization of engineering organization norms into conscious awareness of personnel.

The flexible organizational structure is an engineering organization concentrating its members in different functional departments with different professional knowledge and professional skills to a specific dynamic team in order to achieve the same project goal and work together to complete the project task. After completing the project, all members return to their respective jobs. The key to a flexible organizational structure is to meet the diverse needs of various stakeholders in engineering organizations by "team cooperation" to reduce the obstacles faced by engineering organizations and improve the efficiency of organizational decision-making and action. Organizational flexibility will become more important in departments or units with high creative requirements such as engineering design and R&D and difficult to quantify measurement standards. At Google, within the project team, "building as many channels as possible, allowing different people to express different ideas in different ways" is regarded as one of the basic principles of building organizations, empowering employees through Moderator, FixIts, and other internal communication tools, maintaining their innovation channels, inspiring their ingenuity, and allowing free expression of innovative ideas.

2. **Modular organization**

A flexible organizational structure is a modular organization divided by different functions. Modular organizations can quickly recombine with other modular organizational structures to complete specific production tasks with its standardized interface.

The most important feature of flexible organizational structure is that it combines the stability and efficiency of bureaucratic organizational structure with the flexibility and team spirit of temporary organizational structure rather than the negation of traditional rigid organization. In modular organizational structure, rigid organizational management embodied by rules and regulations and objective assessment is the foundation, and flexible organization is the sublimation on this basis. The appropriate combination of rigid and flexible can achieve the project's goal to the greatest extent [38]. A flexible organization can make a non-hierarchical team structure run parallel with a regular hierarchical manager structure. The modular organizational structure achieves interpersonal interaction across functions and departments in the specific environment formed by values and culture. This process also enables knowledge production, dissemination, and accumulation to have progressive and sustained innovation capabilities.

5.4.1.3 Borderless Organizational Structure

The changing trend of engineering organization structure may also occur in the change of organization boundary. Especially in recent years, the rise of borderless organizational structure has attracted wide attention in society. The unbounded organizational structure is a kind of new organizational design composed of an organization whose horizontal, vertical, or external boundaries are not limited by some pre-set [39]. The boundless organization structure is relative to the bounded organization. To fully guarantee the stability and order of the organization, the bounded organization must keep its boundaries. However, no boundary does not mean ignoring the organization's stability or totally denying the necessary control means of the engineering organization. The unbounded organizational structure must also ensure the stability and reasonable presentation of the organizational structure, but it cannot be rigidized.

1. **Modern theory has no boundaries**

The organization of traditional engineering is based on the theory of modern organizations, and a clear and stable organizational boundary is taken for granted. But when the project is big enough, bureaucracy will overflow. Every department only carries out activities following the responsibilities defined by the organizational level. Instead of ignoring the project's overall goal. It does not know the ultimate goal of its responsibility and service, and it will not flexibly adjust the way of fulfilling its responsibilities and fulfilling its contents according to the ultimate goal of service. More and more engineering practices show that the organizational boundaries and scope set by the modern organizational theory are being continuously overstepped.

2. **Organizational activities have no boundaries**

From the internal point of view, the acceleration of information diffusion, the intensification of work tension, and the increasingly close collaborative relationship make the activities of organizations break through the boundaries of traditional organizations. The main manifestations are as follows: firstly, the organizational authority has no boundaries, and the boundaries between superiors and subordinates are blurred; secondly, the organization has no boundaries; that is, the geographical scope of the organization is very broad. Especially in implementing a large-scale project, its internal organization has become a worldwide activity. For example, traffic engineering of highspeed rail, because of the technical requirements of the locomotive, the organization involves many countries such as Germany, France, Japan, etc. We will involve more countries if we consider the parts and components subcontractors. It is more and more difficult to distinguish the organization and the environment from the external boundary of the organization. For example, the development of virtualization has become the core production activity direction of the design organization. The ambiguity of boundary is not an absolute negation of all boundaries. If the engineering organization exists, stability and order are the prerequisites for its existence. The ambiguity of boundary is to break through the boundaries of each other under the premise of guaranteeing such stability and order, to enhance the flexibility and adaptability of the organization.

3. **Organizational ecosystems have no boundaries**

Another changing trend of "borderless" organizational structure is forming an "organizational ecosystem" centered on engineering ontology. The system is formed by the interaction between the community of organizations and the environment, and this group is also composed of enterprises and individuals across sectors and industries. In this system, one or more central enterprises play a central role, leading and influencing the development and movement direction of the ecosystem and establishing and forming a loose, complementary, and mutually beneficial relationship with other organizations and individuals as well as other organizational ecosystems in the process of operation, to achieve common prosperity through cooperation and competition [40]. It should be noted that members and their roles in this system are not static, members enter and exit, members cooperate and compete. Common goals and interdependence are the cornerstones of maintaining system stability.

Take the new US space shuttle project as an example. Because of the characteristics of high investment, high barriers, and high risk, it is difficult for an enterprise to complete independently. The main contractor is selected through a three-level bidding mechanism and then build an olive supply chain organization system with its core, multi-level suppliers and the combination of production, education, and research. The main contractor only undertakes knowledge-intensive R&D, experiment and assembly. In contrast, a large number of primary, secondary, and general suppliers undertake R&D and manufacturing of subsystems, key components, and composite materials according to their specialties. In this organizational system, the main contractor leads the operation of the whole project vertically. At the same time, the suppliers promote the project forward through horizontal contact, competition, and cooperation and achieve the sharing of economic benefits under contract constraints.

4. **Structural complexity has no boundaries**

In addition, the organizational structure presents a trend of complexity due to the emergence of large-scale projects across regions and countries. The complexity here is not only a more detailed division of labor, a wider geographical distribution, more horizontal and vertical hierarchies in engineering organizations, but also a non-linear and immediate linkage mechanism between people in organizations. This mechanism makes the organizational structure more complex and dynamic. At the same time, information communication also strengthens the relationship between organizations. Under the background of globalization, the subsystems within the organization compete and cooperate, which brings about the ambiguity of organizational structure boundary. In this situation, the measurement of organizational structure complexity begins to cross the boundary of organizational structure [41]. It needs a broader perspective to explain this important feature of organizational structure.

In addition to the above trends, supply chain management innovation and integrated management innovation are becoming the future trend of organizational structure. There is room for innovation in the functional system (such as the new decomposition method of engineering objectives), the management structure (such as the adjustment of the functions of engineering management departments, the improvement of internal workflow), the management system (the re-division of responsibilities and authority between departments), and the management behavior (the change of rules and regulations).

5.4.2 *Development Trend of Engineering Organization Theory Paradigm*

In fact, the change and development of organizational structure is only an intuitive representation. As far as its theoretical origin is concerned, it can be regarded as a change of engineering organization theory. The past theory of engineering organization originates from the experience gained in engineering practice and absorbs the knowledge of other management disciplines. The development of engineering organization theory is rooted in various problems encountered in engineering practice and various practical activities. It forms and enriches its own theoretical system by collecting useful theoretical methods for itself and integrating, innovating, and expanding them.

5.4.2.1 Paradigm of Traditional Organizational Theory

From the evolution of classical organization theory, scientific organization theory, to modern organization theory, we can see clearly that the development of organization theory has its inherent law and logic. Traditional organizational foundations emphasize "be organized" or "other organized." "Be organized" refers to the things that are not spontaneously organized but passive, its spatial, temporal, or functional structure passively interfered and driven by the outside world [42]. Both eastern and western engineering management organizations have gone through a long period of being organized. Frederick Taylor's "scientific management" [43] and Peter F. Drucker's "management by objectives" [8], WBS in project management, all of their ideological roots come from "being organized." Under the constraints of the engineering objectives, technical level, and economic environment at that time, "be organized" conforms to the requirements of the objectives. For example, the "centralized + decentralized" organizational model of the US defense science and technology projects, the "one center (general design center), two command lines (technical command line and administrative command line) organizational model implemented by China's early weapons system development, and the military industry committee system of the Soviet Union, belong to this kind. Under this idea, the engineering organization regards the whole engineering organization as a running machine or a precise clock. Everyone is like a gear. Every department is regarded as a gear set. The boundary of each part is clear, and the task is clear. If each part can complete the prescribed task, the project's overall goal can be achieved. Such organizations are structured based on collective and standardized knowledge, characterized by formalization, specialization, standardization, and centralization.

The defect of this kind of organizational form lies in neglecting the differentiation of knowledge caused by specialization. This differentiation has rapidly promoted the growth of organizational knowledge, and the spread and utilization of knowledge have undergone tremendous changes. High-value specialized knowledge has become the main type of knowledge used by social organizations.

5.4.2.2 New Paradigm of Self-organization Theory

For the shortcomings of traditional organization theory, the engineering organization will inevitably transform from the "be organized" paradigm to the "self-organized" paradigm. "Self-organization" refers to the process in which things spontaneously form an organization. A self-organizing system is a process of self-organization, self-creation, self-evolution, self-innovation, and self-development without external instructions. It is a continuous process from disorder to order. It is hard to say that there is a completely self-organizing mode in engineering. But it is undeniable that since the 1980s, due to the entry of human beings into a post-industrial economy characterized by innovation-intensive, the increasing number of technology-intensive and innovative intensive projects has brought opportunities to develop the "self-organizing" mode. Whether it is "learning organization" or "flexible organization theory" or even "cloud organizational structure," the root lies in the emphasis on people in the organization, which is manifested in a flatter organizational structure and fewer levels. To some extent, the "self-organizing" paradigm transforms the organizational structure from a typical pyramid shape into a "central-edge" dual structure that is shared more by the other participants in the project instead of the power of senior managers.

A completely "self-organizing" engineering organization does not exist at present, but this is the trend of future development of engineering organizations; because the intrinsic value of engineering construction itself has been achieved in the process of fulfilling the project objectives. At the same time, the development of each participant who constitutes the engineering organization is influenced by other organizations as well as influencing the whole organization group of the project. In a suitable environment, self-organization may become the basis for the continuous improvement of engineering projects to adapt to the dynamic. Taking an engineering organization as a complex self-organizing system can coordinate the development consciousness of individual organizations and groups and avoid the instability of the whole engineering project caused by the accumulation of local changes. It is also the need to recognize the complexity and dynamics of engineering.

The paradigm of self-organization and the paradigm of being organized is dialectically unified. Organizations need the theoretical support of being organized or other organizations and the nutrients of innovation and growth from the paradigm of self-organization. The principle of self-organization calls for the creation of an organizational structure that stimulates the initiative, continuous learning, and innovation of organizational members and links up independent members through the existence of autonomy.

The application of self-organization theory in engineering management is mainly embodied in management engineering to adapt to the changes of external conditions better and faster. We should reduce the use of the linear management method from top to bottom, use the multi-dimensional structure of the organization, and increase the strength of horizontal process management. However, the application of self-organization in engineering management practice in China is not an easy task because the most important problem in achieving self-organization is the centralization and distribution of power. Influenced by traditional centralized thinking, the idea of centralization began at the beginning of the project and was rooted in the thinking of senior managers in Chinese projects. Also, out of fear of losing control of the project, centralized management has become a safer choice to avoid risks. At the same time, the lack of effective management methods under the decentralization mode makes the decentralization lack effective means. Therefore, the application of self-organization theory mostly appears in IT, network, information, and other market-oriented engineering fields, but few in government-led engineering fields.

In addition, the complex and changeable environment of creation makes the activities of creating more and more complex, which leads to the gradual increase of the openness and systematicness of the "central system" of commanding the activities of creation. On the one hand, the rapid development of information networking makes it possible for the global collaboration of creation activities. Engineering organizations select and adopt materials suitable for creative activities worldwide; that is, engineering organizations absorb suitable global materials to serve creation activities by their strong "aggregation" effect. Openness is the basis for global resources to enter human activities. With the complexity of creative activities, in order to ensure the smooth development of creative activities, the resources needed for creation activities (such as capital, manpower, technology, etc.) become more diverse, engineering organizations become more open. The opportunities for different materials to enter the creation activities will be increased smoothly, and the activities of creation will be more guaranteed. Engineering organization always tends to develop spontaneously and disorderly, that is, to increase the value of entropy. In addition, engineering organization sometimes deviates from the normal track, and the openness of engineering organization exists. It enables the engineering organization to exchange material, capability, and information with the external environment where the creative activities are located at all times, and acquires external resources in time, which leads to negative entropy flow, decreases the organization's entropy, and develops the organization in an appropriate and orderly direction, so as to ensure the efficient operation of the engineering organization. On the other hand, the systematization of engineering organizations is becoming stronger and stronger. The application of new technologies and theories makes the types of engineering production factors involved in the activities of creation varied, and the scope of selection increased. Especially, the development and attention of social responsibility theory makes the scope of human production factors in the activities of creation expand rapidly, that is, the scope of stakeholders. It requires the systematic strengthening of engineering organizations to ensure a clear division of labor among different participants in the activities of creation. To improve the level of specialization, under the guidance of engineering organizations, enhance the understanding of the regularity of creational activities, follow strict and orderly rules, creational activities can be carried out sequentially and systematically in order to achieve the goal of creation.

In short, the continuous development and evolution of engineering organizations come from the uncertainties brought by the changes in the ontology, environment, and project elements. The more uncertainties there are, the more creative points they bring, the more freedom they have in organizational management, and the richer the form of the organization. The result of this change is the emergence of organizational innovation. From the point of view of engineering organization innovation, it can be manifested in structural innovation and theoretical innovation. New network organization, flexible organization, and borderless organization are all innovations of organizational structure, while the transformation from "be organized" to "self-organized" is a fundamental and revolutionary change and an innovation of organizational theory.

References

1. Kunz, H. (2014). *Management: From the perspective of internationalization and leadership*. China Renmin University Press.
2. Spencer, H. (2001). *Sociology research*. Huaxia Publishing House.
3. Fayol, H. (1985). *Industrial management and general management*. China Social Sciences Press.
4. Webber, M. (2004). *The works of Weber, IV: Economic action and social organization*. Guangxi Normal University Press.
5. Ao, L., & Jin, H. (1998). The overall structure of the Three Gorges Project management system (TGPMS). *China's Three Gorges Project, 9*, 22–24.
6. Simon, H. (1988). Administrative behavior. Beijing University of Economics Press.
7. Simon, H. (2013). *Administrative behavior*. Machinery Industry Press.
8. Drucker, P. (2009). *The practice of management*. Machinery Industry Press.
9. Coase, R. (1994). *On the institutional structure of production* (pp. 1–24). Sanlian Bookstore.
10. Bertalanffy, V. (1987). *General system theory: Foundation, development and application*. Tsinghua University Press.
11. Daft, R. L., Murphy, J., & Willmott, H. (2010). Organization theory and design: Cengage learning. Inc., Hampshire.
12. Tian, P. (2008). Social engineering—The basic way to identify the world in modern society. *Chinese Social Sciences, 4*
13. Aristotle. (1965). Politics (p. 199). Business Press.
14. Tonnies, F. (1999). *Community and society*. Business Press.
15. Li, B. (2008). Exploration of engineering community research and engineering sociology—"Engineering Community" research III. *Natural Dialectics Newsletter, 30*(1), 63–68.
16. Wei, S. (2003). The cultural reasons for the differences between individualism and communitarianism, the historical evolution path of social order between east and west. *Journal of Fudan (Social Science Edition), 3*, 1–8.
17. Chen, J., & Xiao, D. (2007). Analysis of enterprise organization innovation based on entropy theory. *Economy and Management, 21*(3), 62–66.

18. Liu, Y. (2013). *Anomalies of engineering ethics in China and engineering ethics education for engineering students*. Hunan University of Technology.
19. Li, B. (2010). The formation, dynamics and disintegration of engineering activity community—"Engineering Community" research IV. *Natural Dialectics Newsletter, 1*, 40–44+16+126–127.
20. Li, B. (2010). *Introduction to sociology of engineering: Research on engineering community*. Zhejiang University Press.
21. Shang, Y. (2003). South to North Water diversion focuses on environmental protection. *Current Affairs Report, 1*, 35–37.
22. Ding, S. (2006). *Project management*. China Building Industry Press.
23. Wang, X., Yuan, G., & Lei, L. (2010). Scientific selection of post-management model for the Three Gorges Project of the Yangtze River--On the integration of dam and reservoir. *Exploration of Economic Issues, 12*, 20–23.
24. Lu, Y. (2002). The practice of project management of Three Gorges Project construction. *China Three Gorges construction, 1*, 3–4+51.
25. Chandler, A. D. (1992). Organizational capabilities and the economic history of the industrial enterprise. *Journal of Economic Perspectives, 6*(3), 79–100.
26. He, X., Wang, Y., & Li, T. (2011). U.S. defense technology major engineering organization management model and enlightenment. In *Proceedings of the 2011 national defense technology industry forum*, 7.
27. Li, T., Kong, D., & Li, C. (2012). Analysis of evolution path of organization and management model for national defense science and technology major projects. *Science and Technology Progress and Countermeasures, 29*(24), 42–46.
28. Jin, H. (2000). Design, development and implementation of the Three Gorges Project management system. *Hydroelectric Power, 6*, 52–54.
29. Liao, F. (2011). Organization communication research of "Two bombs and one satellite" project. Chengdu University of Technology.
30. Chen, J., & Xiao, D. (2007). Analysis of enterprise organizational innovation based on entropy theory. *Journal of Shanxi University of Finance and Economics, 3*, 73–77.
31. Daft, L. (2011). *Organization theory and design* (10th ed.). Tsinghua University Press.
32. Scott, W., & Davis, G. (2011). *Organizations and organizing: rational, natural and open systems perspectives*. Renmin University Press.
33. Zhang, B. (2002). *Theory and simulation of the complex adaptive system for economic management*. Tianjin University.
34. Tan, C. (2007). The subjectivity existence and reality of complex adaptive system. *Academic Research, 4*, 66–71.
35. He, Q., Li, Y., & Le, Y. (2007). Research on information integration and management system of Shanghai world expo 2010. *Journal of Tongji University: Natural Science Edition, 35*(6), 856–860.
36. Fan, J. (2008). *Research on the process, mechanism and evaluation of virtual organization-oriented organization virtualization*. Xi'an University of Technology.
37. Sun, J. (2007). The management concept of flexible organization. *Chinese Administration, 9*, 60–62.
38. Luo, M. (2007). Knowledge innovation and organizational structure. *Contemporary Economic Management, 29*(5), 1–6.
39. Gu, C. (2006). An overview of organizational structure development. *Coastal Enterprises and Science and Technology, 2*, 46–48.
40. Luo, M. (2004). Postmodern prospect of organizational concept. *Management Science, 17*(3), 16–20.
41. Cai, S., Wang, J., Xu, H., et al. (2007). Self-organizing and coordinating mechanism of evolution from regional enterprise clusters to industrial clusters. *Exploring Economic Issues, 3*, 69–73.
42. Luo, M. (2008). On the transformation of paradigm of organizational theory. *Foreign Economy and Management, 30*(8), 18–22.
43. Taylor, F. (2007). *The principles of scientific management*. Machinery Industry Press.
44. Apple Corp. (2012). Supplier social responsibility progress report. Apple Corp.

45. Liu, A. (1995). *Huainanzi-spiritual training*. Hua Xia Publishing House.
46. Li, B. (2006). Several questions about engineers—"Engineering Community" research II. *Natural Dialectics Newsletter, 28*(2), 45–51.
47. Xia, Z. (1999). Cihai. Shanghai Dictionary Publishing House.
48. Ren, H. (2012). *Huge project management*. Science Press.

Chapter 6
Engineering Management Value Theory

Engineering activity is a value-oriented activity process. Its ultimate goal is to get a more valuable world for human society, and its effective implementation is inseparable from engineering management. Engineering management is a series of activities oriented to the value objectives of engineering activities, such as decision-making, planning, organization, command, coordination, and control, aiming to achieve and optimize a particular value target. Therefore, the engineering management process must clearly define its value orientation and set up the engineering values of "people-oriented, unity of heaven and man, collaborative innovation and harmony construction."

Manned spaceflight is the most representative of high-tech engineering. From the practice of human spaceflight engineering in the United States, Europe, and Russia, it is difficult to see its apparent economic return so far. On the development road of human spaceflight in China, we must think soberly about the value of engineering and how to understand the value of engineering correctly. The answer involves the value of engineering, the essence of which belongs to the category of engineering value theory.

Engineering value is the satisfaction of engineering activities and their results to human and social needs. Engineering activities are the process of creating and realizing engineering value. Engineering value has multi-dimensional characteristics, including economic value, social value, scientific and technological value, cultural value, ecological value, talent cultivation value, etc. Usually, the subjects of engineering decision-making are multiple, and the engineering value will vary with the different value orientations of the decision-making subjects. According to the theory of value, value evaluation is the evaluation subject on the possible value, meaning, and results of the evaluation object, which often depends on the evaluation standard system composed of the needs, interests, and preferences of the evaluation subject [1]. Therefore, in major engineering decisions, it is necessary to balance and coordinate the differences in value orientation of multiple engineering subjects according to the needs of the development of the times, social recognition standards, and norms

J. He, *Principles of Engineering Management*,
https://doi.org/10.1007/978-981-99-1168-4_6

of engineering activities, to make scientific and rational decisions, and to realize and enhance the comprehensive value of engineering activities.

In this chapter, we will analyze the evolution of engineering values and the multiple values of engineering, focusing on economic and social values. The cultural and scientific values of engineering will be covered in other chapters of this book, so we will not repeat them in this chapter.

6.1 Value of Engineering

Various views on engineering values have been gradually formed in people's understanding and practical engineering activities. The following will dialectically analyze the evolution of engineering values, the multiple values of engineering, and how to deal with the multiple values in a coordinated way.

6.1.1 Evolution of Engineering Values

Engineering activities started in agricultural civilization society. Along with the continuous development of engineering theory and practice, the core of engineering values has evolved, and its connotation has been enriched and expanded. People's understanding of engineering activities, i.e., values, has also changed [2]. In terms of time, engineering values also have ancient values, traditional values (mainly modern engineering), and modern values.

(1) Ancient engineering values

The Great Wall, Dujiangyan, and Suzhou gardens are the most famous in ancient Chinese engineering and architecture. These projects and buildings are cleverly conceived and precisely constructed with perfect cost and quality management. Based on the systematic summary of the characteristics of ancient projects, the values that dominate such projects are summarized as the idea of "the unity of heaven and man," which is the central idea of ancient Chinese engineering activities. It is a comprehensive expression of the values, ethics, views on nature, and aesthetics of ancient Chinese people.

(2) Modern engineering values

Traditional, modern engineering values were formed before engineering science, lacking systems research and scientific theories support, which inevitably have certain limitations. This limitation is mainly reflected in two aspects: the narrow vision of engineering understanding and the singularity of engineering value objectives [3]. In terms of understanding, traditional engineering values interpret engineering simply as applying specialized technology and consider engineering activities only practical activities of building artificial nature. It often excludes people and social processes composed of people from engineering activities. Abstractly interpreting engineering as a simple relationship between man and nature and society, conquering and being conquered, seizing and supplying results in humankind's unrestrained demand for nature and neglects the impact on the ecological environment in the whole life cycle of engineering construction. It makes engineering activities a direct force to destroy the relationship between man and nature. In terms of core values, traditional engineering management unilaterally pursues the maximization of the project's economic benefits and operational efficiency with four major control objectives: quality, cost, schedule, and safety. Quality control emphasizes the quality of the project; the quality should reach the standard of the construction product. Cost control emphasizes the economic efficiency index and pursues the minimization of the cost required to complete the project. Duration control emphasizes the efficiency index and the minimization of the overall time required to complete the project. Safety control mainly takes construction and structural safety as the core for eliminating all accidents and avoiding accidental injuries. Traditional engineering management considers the above different objectives in an integrated manner and achieves the overall optimum coordination.

(3) Modern engineering values

The concept of sustainable development has been deeply rooted in people's hearts, and people pay more and more attention to the harmonious development of man and nature, man and society; thus, a new understanding of engineering and engineering activities is formed. Different modern engineering values have emerged. Broadly speaking, modern engineering values include the following types: **political engineering values, economic engineering values, ecological engineering values, and social engineering values. The core is based on people-oriented, unity of nature and man, collaborative innovation, and building harmony.**

Political engineering values are centered on the power position, and the main concern of decision makers holding this type of value is the project's political benefits. The initial starting point of political engineering values is often positive and beneficial to the people. The core issue considered at the beginning of engineering construction is how to improve people's livelihood and promote local or regional economic development. But in the practice process, it is often for personal gain, resulting in the so-called "political achievement projects" and "face projects." For example, the international airport in Fuyang, Anhui Province, is a typical engineering value case because it is overly advanced, focusing on political needs and ignoring economic values. It is a waste of manpower and resources, and the former international airport became a sheep farm for local farmers.

Economic engineering values believe that the core of engineering activities is to consider their economics. The project is required to be evaluated fully based on the construction and operation of the project from the perspectives of cost, benefit, and life of the project following the market mechanism. In practice, under the guidance of such values, relevant subjects will blindly pursue personal interests and commercial profits. The quality, safety, and function of the project are often ignored. Today's endless "soya bean dregs" projects are the products of this type of value.

Social engineering values regard engineering as not only a technical activity but also a social activity. In engineering activities, technical and social elements are interwoven, and the change of technical structure will prompt the transformation of social relationship structure to fit with specific technical structure relationships. At the same time, the standards and management norms of engineering activities have to be coordinated with specific cultural and social goals, and specific social goals regulate the mode, process, and characteristics of engineering activities. Under the guidance of social engineering values, engineering is the process of building artificial nature and harmonious society.

Ecological engineering values include ecological claims to engineering management concepts, requiring maximum conservation of resources and protection of the environment in engineering activities, and ultimately achieving symbiosis and harmony between human and nature, human and society. Wang and Li [4] thoroughly summarized the connotation of green engineering values. Ecological engineering values mainly include taking ecological objectives as the guide, taking into account multiple objectives such as economy, society, science and technology, and environment, giving new connotations to quality, cost, duration, and safety, and considering ecological demands such as saving resources and protecting the environment and implementing whole life cycle management.

Taking the Chinese space project as an example. The principle of the development of China's space industry is that "China takes the development of the space industry as a strategic initiative to strengthen the country's economic strength, scientific and technological strength, national defense strength and national cohesion, and as an important part of the overall national development strategy to maintain the long-term and stable development of the space industry." The evaluation of the value of the human spaceflight project cannot be limited only to its economic value. It takes into account its impact on the national economy, science and technology, national defense, national cohesion, and the country's overall development strategy. Briefly, the development of manned space engineering has the following five major values.

First, human spaceflight is a reflection of the country's comprehensive strength. It can be said that manned spaceflight is the technically highly complex and the most difficult project in the world today, relying on numerous high-tech support and strong economic strength as a backing. It is impossible to implement manned space engineering without a strong science and technology support system and strong financial strength. Therefore, manned space engineering can help demonstrate China's comprehensive national power and enhance the Chinese nation's self-confidence, pride, and international influence.

Second, manned spaceflight can promote the progress of science and technology. Manned space engineering involves modern mechanics, earth science, space science, astronomy, aerospace medicine, and other disciplines; it also involves systems engineering, automatic control, communication, remote sensing, new energy, new materials, and many other high-tech fields. In the process of developing rockets, spacecraft, and measurement and control systems, China's manned space program has innovated system engineering theories and methods, overcome a series of core technical difficulties, and achieved a large number of independent innovative scientific and technological achievements. It has promoted the rapid development and overall progress of China's science and technology level.

Third, human spaceflight is conducive to promoting the development of the national economy. Although it is difficult for manned space engineering to promote national economic construction and national economic development directly, the development of manned space engineering has a profound impact on the national economy. On the one hand, scientists can use the unique environment of micro-gravity, high cleanliness, and full vacuum in space to conduct a series of scientific experiments, explore technological innovation and methodological innovation, and provide a reference for ground production. For example, space breeding using space mutagenesis technology can produce high-yielding, high-quality, multi-resistant new varieties of peppers, tomatoes, rice, wheat, and other crops. On the other hand, the development of manned space engineering can directly drive the development of a large number of space-related industries and optimize the industrial structure. For example, manufacturing and service industries related to space communication and navigation have become important derivative industries of manned space engineering. Satellite communications, satellite navigation, space breeding, and new drug research have greatly progressed, affecting and improving ordinary people's lives. Manned spaceflight has become one of the important driving forces of economic and social development.

Fourth, the manned space program has trained a large number of cutting-edge scientific and managerial talents. The development of manned space engineering has cultivated a scientific and technological team. According to statistics, young people under the age of 35 account for more than 70% of the scientific personnel in manned space engineering. The presence of these elite talents provides a solid guarantee for the sustainable development of China's space science and technology industry. During the implementation of manned space engineering, many young and middle-aged science and technology personnel have been trained and grown rapidly, becoming the backbone of the space industry and the leading talents of the national science and technology support system.

Fifth, manned space engineering is conducive to safeguarding national security. On the one hand, the development of manned space engineering has realized the long-cherished dream of observing the Earth from space. People can have a deeper understanding of the whole Earth and its structure, and accurately forecast the occurrence of floods, hurricanes, earthquakes, and other natural disasters to prevent disasters and reduce disaster losses. On the other hand, the comprehensive national power demonstrated by the implementation of manned space engineering can give sufficient deterrence to hostile elements. Some people consider manned spaceflight as the modern Great Wall. The Great Wall represents the advanced productivity of the Middle Kingdom, a symbol of the strong will to keep the enemy out of the country. Its symbolic significance is much greater than the actual utility, but the deterrent power and military value are equally important.

In conclusion, modern engineering values are a new understanding of engineering activities formed under the cross-penetration of contemporary disciplines, reflecting the new trend of contemporary construction engineering culture embracing the whole society, cross-fertilizing and coordinating with economy, culture, and ecology. It is a renunciation and transcendence of traditional values. On the one hand, it expands the connotation and extension of engineering. In terms of connotation, it integrates science, technology, and non-technical elements into a complete engineering activity system and leads engineering activities. In terms of extension, it incorporates ecosystem and social systems into the engineering system. It pays attention to the inner law of nature and the influence of engineering on social structure. On the other hand, modern engineering values are multidimensional. With the development of contemporary engineering, engineering values are influenced by scientific, technological, social, environmental, and ethical factors. They form values corresponding to them, which are interconnected and intertwined and guide engineering practice [3].

6.1.2 *Multidimensionality of Engineering Value*

Modern engineering, especially preeminent engineering activities, often involves economic, scientific and technological, social, natural, cultural, political, and other factors. It invlolves both natural and social aspects. It not only determines the complexity of engineering management objectives but also influences the multidimensional orientation of engineering activities. Zhang and Chen [5] pointed out that the value chosen for any engineering activity results from the game and coordination among engineering's economic, scientific and technological, social, cultural, and political values. Therefore, engineering management needs to comprehensively review and systematically integrate the multidimensional values of engineering practice from a strategic view to achieving maximum engineering value. The multidimensional values of engineering include.

(1) Economic value

The economic value of engineering mainly refers to the fact that products and services are continuously created through engineering practice activities to meet people's needs and thus gain corresponding profits. As an economic organization behavior, engineering management and activities can and must have profitability. Because profit is not only the driving force of economic organization behavior but also the premise and foundation of the existence and development of the economic organization. If engineering management and activities cannot achieve a profit level matching investment cost, not only will the engineering activities be unsustainable, but their social functions will also be affected. It will be more challenging to achieve public welfare related to social functions. For example, as a major strategic project to alleviate the severe water shortage in northern China, the South-North Water Transfer not only greatly relieves the severe water shortage in northern China but also promotes the complementarity and coordination between the south and the north in many aspects, such as economy, society, science and technology, population, resources, environment, culture. It is of strategic importance to expand domestic demand and maintain the rapid growth of the national economy.

(2) Social value

The social value of engineering means that engineering activities should undertake the necessary social obligations, fully consider and respect the public's interests and emotional attachment, and promote the harmonious operation and healthy development of the whole social system. No engineering activity exists in isolation; it needs to be integrated into the intricate social relations and to coordinate and deal with various ties and multiple interests. Modern engineering activities and project management need to fully consider the interests and emotions of the public, eliminate possible conflicts of interest and promote the harmony and health of the whole social system.

(3) Ecological value

The ecological value of engineering refers to the fact that engineering activities should be based on the ecological environment, fully consider and reflect the respect for the ecological environment and natural environment, explore the sustainability of engineering practice, and realize the harmony between engineering and nature, human and nature. With the development of human beings and the progress of society, human beings' ability to understand and utilize the world is increasing. While paying a painful price for the development mode of high energy consumption, high pollution, and high consumption, we need to gradually realize the ecological support and ecological constraint relationship in the development process. To realize the ecological value of engineering, it is necessary to re-examine the overall value of the ecosystem, re-examine the harmony between humans and nature, establish the correct engineering values, and take the concept of "the unity of heaven and man and the construction of harmony" as the guide. A series of famous historical projects, such as Dujiangyan, Great Wall, Beijing–Zhangjiakou Railway show that the best engineering effect can be achieved only when the project is highly coordinated with the laws of nature. A great project must not be at the cost of destroying the ecological environment and ecosystem but to maximize the environmental value of the project from the engineering practice. Considering the environmental value of the project and protecting the fragile plateau ecosystem and ecological environment, the Qinghai-Tibet Railway project has built migration channels for wild animals, leaving the track for Tibetan antelopes returning home, and supports human sustainable development.

(4) Technology value

The scientific and technological value of engineering refers to the continuous scientific development and technological progress through engineering activities: promoting and enhancing the development of engineering science, management science, social science, and other sciences, and achieving significant technological changes and innovation independently [6]. In particular, modern engineering, especially large-scale engineering, requires key support from advanced management science, management methods, and engineering technical support to continuously solve new scientific and technological problems and innovate and develop engineering management science and engineering management technology.

(5) Talent value

The value of engineering talents refers to the training of excellent engineering management talents through engineering management. Engineering management is a highly unified practical process of "developing things" and "training talent." Large-scale engineering activities can create precious material results, which is the process of "developing things" and cultivating various kinds of excellent talents at all levels, which is the process of "training talent." Engineering managers who have experience in large-scale projects tend to be more far-reaching in their thinking. Their ability to overcome difficulties is significantly enhanced, representing the process of "training talent" in engineering activities.

(6) Cultural value

The cultural value of engineering refers to creating iconic engineering achievements through engineering practice, which has three distinctive features: having significant cultural connotations, manifesting the spirit of the times, and withstanding the test of history and time. Any engineering activity and management should inherit human civilization and national spirit, give full play to the inspirational power of advanced culture, and strengthen cultural construction and accumulation. It requires enhancing the level of engineering activities. For example, Great Wall is a rare ancient military defense project in the history of human architecture, a rare treasure and an extraordinary cultural relic, the pride of the Chinese people and humankind, symbolizing the indestructible will and power of the Chinese nation. The Tiananmen Square, built in 1417 in the Ming Dynasty, is not only an architectural masterpiece with five belvederes and nine pillars, showing the dignity of the emperor but also a great representative symbol of the ancient Chinese civilization and the modern civilization process. It has become a symbol of the People's Republic of China and a place of fascination for the Chinese people and the world.

6.1.3 Dialectical Thinking of the Value of Modern Engineering

Major modern projects often involve many fields, such as society, economy, science and technology, nature, and the environment. Given the intricate relationships and contradictions, it is necessary to follow the principles and methods of combining dialectical and systems thinking and handle these dialectical relationships scientifically and artistically according to the requirements of the scientific development concept of comprehensive, coordinated, and sustainable development so as to realize the multidimensional economic, social, natural and scientific values of engineering.

(1) Interactive development of engineering management theory and engineering management practice

Engineering management practice is the starting and ending point of engineering management theory, which is based on the existing artificial nature and carries out a series of engineering activities, achieving engineering project objectives until final decommissioning. The modern engineering management process involves continuously refining engineering management theory and guiding engineering management practice with engineering management theory. The process is to achieve a higher level of combination of engineering management theory and practice in mutual interaction.

Modern engineering practice activities are characterized by high complexity, facing more uncertainties and often bringing higher risks. Therefore, taking scientific engineering management theory as the guide, constantly standardizing and improving the process of engineering management practice, and predicting and controlling the possible results of engineering activities are conducive to reducing the risks of engineering activities. In the past, certain projects, even major ones, lacked multiple value considerations at the early decision-making stage, resulting in these projects failing to achieve the expected results and bringing about serious problems such as natural destruction and ecological deterioration. Accordingly, the appropriateness of theoretical engineering guidance has "a priori" determined the success of engineering activities, directly affecting their future fate. With the rapid progress of engineering activities, engineering management theories need to be constantly innovated to better guide engineering management practices. In the ever-changing modern engineering practice, engineering managers should continuously update engineering management concepts and innovate engineering management theories. Thus, engineering managers can effectively guide engineering management practice, improve the effect of engineering management practice, and achieve the goal of engineering activities [7].

(2) Deep integration of engineering management concepts and engineering management models

The engineering management concept is the basic used by engineering managers to manage engineering activities. It influences the behavior pattern of management at all levels in two different ways, in terms of tangible regulation or invisible penetration, throughout the entire engineering management process. It also affects the specification, mode, method, and effect of engineering management. The engineering management mode aims to realize the engineering management concept, which is mainly revealed through management strategies, techniques, methods, and tools. There is a deep integration and mutual promotion relationship between the engineering management concepts and engineering management mode. Generally speaking, the engineering management concept influences and determines the kind of engineering management mode, method, and technology to achieve specific engineering goals. On the contrary, the comprehensive application of engineering management mode, method, and technology in engineering management practice will further breed new engineering management concepts.

Engineering management activities guide engineering management practice through engineering management theory, achieve the established goals of engineering in a purposeful, planned, and systematic manner, and continuously meet the needs of human and social development. The engineering management concept will influence the whole process of engineering management activities and determine the engineering's occurrence, development, and results. Various effects, such as economic transition, social transformation, and ecological crisis, are superimposed in today's society. The concept of engineering management is shifted from the pursuit of a single economic value to multiple value objectives, which will guide engineering management activities toward the harmonious coexistence of man and nature, harmonious development of man and society, and harmonious life of man and man. In addition, with modern science and technology development, especially information science and technology, useful management technology and management tools for engineering management activities are available. It profoundly affects the change of people's way of thinking and the improvement of their state of mind, which can further enhance the engineering management concept, engineering management mode, and engineering management methods.

(3) Coordination of engineering management system and engineering management details

The engineering management system refers to the combination of various engineering management elements and various engineering management links in a certain order and manner in the process of engineering activities within a certain time and space. Correspondingly, engineering management details are the sum of various constituent elements constituting the engineering management system and each link's division. From the spatial perspective, the relationship between the engineering management system and engineering management details is that of whole and part. The former embraces the latter. From a temporal standpoint, the two are the relationship of process and link, mutually dependent on each other. In general, if the function of a link in the engineering management process is low, it will become the "bottleneck" of the whole engineering management system, weakening the overall position of the engineering management system and producing the "bottleneck." The theory of whole and part and process and link at the philosophical level requires establishing a perfect engineering management system, clear management concept, and strict management mode in engineering management. It also requires standardizing the detailed management to obtain the "perfection" of the engineering management system with the "detailed managing."

Modern large-scale complex engineering and its engineering management system structure has appeared several new features: from a simple structure to complex structure, static structure to dynamic structure, explicit structure to implicit structure, hierarchical structure to network structure, and other directions of continuous evolution and development. In addition, the scope of modern engineering activities has become more extensive, far beyond the traditional purely agricultural or industrial activities, and gradually evolved into large-scale creation of artificial nature. In these large-scale complex engineering activities, the engineering management system often involves multiple heterogeneous elements such as science, technology, economy, society, nature, culture, and ethics. It requires the formation of multiple complex systems with different levels and complementary functions, posing new challenges to modern engineering management. Based on this, modern engineering management must achieve high coordination and unity between the engineering management system and engineering management details. On the one hand, it is necessary to establish a perfect engineering management system from a strategic view. It requires engineering managers to have a high level and far-sightedness, be good at breaking through the traditional shackles, innovative thinking, and updated concepts, be able to keenly discover the deep-seated problems constantly arising from engineering activities, and grasp the main contradiction and the main aspect of the contradiction. On the other hand, detail management is needed to reach the multi-dimensional value goal of engineering. Engineering detail management is not only a practical behavior that can bring objective economic profit and social effect. It is also an attitude that can construct an engineering management organization and form an engineering culture atmosphere. Therefore, engineering detail management can skillfully solve a big problem using minimum strength or resources, which is conducive to realizing the established goals of the entire engineering activities [7].

(4) Mutual promotion of engineering management norms and engineering management innovation

Engineering management specification is the code of conduct followed by engineering managers to achieve engineering values and goals, mainly composed of various regulations, systems, standards, methods, codes, and ordinances. A scientific and reasonable engineering management specification can ensure the normal and sustainable development of modern engineering activities, promote modern engineering management level improvement and achieve the established engineering value target. However, engineering management norms are not and should not remain unchanged. With the development of the times, the deepening of engineering practice, and the improvement of management level, engineering management norms will also be continuously improved and perfected to realize engineering management innovation finally. Engineering management innovation refers to the breakthrough and improvement of the original engineering management concept, organization, system, method, technology, and tools. It can produce engineering management activities that can be accepted by engineering management norms, representing engineering management norms from "broken" to "established," the dynamic adjustment and dynamic changes. For example, in modern engineering practice, it is necessary to integrate economic, scientific and technological, social, and ecological aspects to reflect multiple value objectives, dialectically negating the drawback of the one-sided pursuit of economic benefits in the era of industrial civilization.

Compared with any previous era, the setting of modern engineering value objectives shows plurality more obviously. In the increasingly complex modern engineering practice, respecting and realizing the multiple values of engineering requires careful consideration of the relationship between the priorities of multiple value objectives. It also requires engineering practitioners to practice scientific engineering management norms consciously and then form the "synergy" of engineering construction. It can be said that any engineering management norms are not eternal and unchanging truths, and their rationality and scientificity have certain time and space boundaries. Therefore, engineering managers must continuously improve and perfect engineering management norms according to subject needs, object needs, and technical equipment requirements in modern engineering activities; and promote the dialectical unification and interactive development of engineering management norms and engineering management innovation.

(5) Joint enhancement of engineering management team and engineering management system

Engineering managers or engineering management team is the maker and the practitioner of the engineering management system. The engineering management system can regulate engineering managers' behavior and improve engineering managers' realm. Every improvement of the engineering management can further provide insight into the "quality" of the engineering management system and contribute to the development of the engineering management system. Therefore, engineering managers and engineering management systems have interaction and common improvement. A scientific engineering management system can regulate the whole engineering activities, which can timely deal with and coordinate various relations and contradictions in engineering activities and improve the rhythm and efficiency of engineering activities. In practice, the establishment of engineering management systems needs to be recognized and understood by most engineering practitioners. Such recognition can further promote the self-reinforcement of the engineering management system so that it gradually changes from external compulsion to internal consciousness [7].

With the deepening of engineering management practice and the increasing complexity of engineering practice activities, many excellent engineering management talents are urgently needed. Outstanding engineering management talents requires continuously improving the engineering management education standard in universities and improving the engineering management system. Under the scientific and rational engineering management system, a new high-quality engineering management team can be cultivated and trained. This high-quality engineering management team can stand at a higher level, re-examine and revise the existing engineering management system with a far-sighted vision to ensure its success. This dialectical process precisely reflects the unification of two principles in engineering activities, "all for people" and "all relying on people." and realizes the main purpose of "people-oriented" engineering management.

6.2 Economic Value of the Project

Evaluation of the economic value of engineering projects is an important content of the preliminary work of engineering projects. Objective and scientific evaluation of the economic value of engineering projects plays an important role in improving the scientific level of project investment decisions, reducing and avoiding investment risks, guiding and promoting the rational allocation of resources, and giving full play to the economic benefits of engineering projects [8]. The following will analyze three aspects of the project's economic value, self, external, and national economic value.

6.2.1 The Economic Value of the Project Itself

6.2.1.1 Direct Economic Value and Indirect Economic Value

Value refers to the object's usefulness to meet the needs of the subject, i.e., the efficacy or utility of the properties and functions of the object to meet the needs of the subject [9, 10]. As a means of human exploitation and utilization of natural resources, engineering projects create great material wealth for human society and have important economic value. The economic value of engineering projects can be divided into direct and indirect economic value.

The direct economic value of an engineering project refers to the monetary expression of the economic benefits that the project itself can directly provide to society. Most engineering projects are carried out for profit. For enterprises, engineering projects have investment value only when a certain level of profit is achieved or a certain yield is obtained. Corresponding to the direct economic value of engineering projects is the indirect economic value of engineering projects. Indirect economic value is the monetary expression of social and economic benefits caused by or derived from the implementation and operation of engineering projects. In daily engineering practice, people often pay attention to evaluating the direct economic value of engineering projects but ignore the indirect economic value. The indirect economic value of engineering projects is mainly reflected in four aspects: one is the indirect value on science and technology; the second is the indirect value on the human capital level; the third is the indirect value on the management level; the fourth is the indirect value on enterprise brand.

From the science and technology perspective, the enterprise may invent a new technology or overcome technical difficulties in engineering project construction. The new technology can be applied not only to the project but also to similar enterprise projects. It may produce technology diffusion and spillover effect so that the new technology can be applied to the whole industry. With the increasing number of new technology applications, the benefits are rising, but these benefits are not fully reflected in the direct economic evaluation of a certain time.

From the perspective of human capital accumulation, in completing a certain engineering project, the enterprise can cultivate the talents needed for the project who can bring more value to the enterprise. People with higher education backgrounds are not equal to "talents" in real life. Only when the theoretical knowledge and skills are applied to engineering practice and when the theory and practice are organically combined is it possible to become real engineering and management talents. The construction of engineering projects provides opportunities for those with higher education to practice and enhance their ability to engage in practical work. Through "Learning by Doing," engineers, technical workers, and other general laborers accumulate experience and improve their knowledge and skills in the construction of engineering projects. Therefore, the engineering project itself provides more higher-quality human capital for the further development of the enterprise.

From the perspective of management-level improvement, the construction and operation of every engineering project will bring lessons learned, and experience comes from practice and guides practice. Enterprises can learn lessons from engineering projects. It will promote the scientific, rational, and standardization of internal management and improve enterprises' management level.

From the perspective of corporate branding, enterprises can establish their brand through high-quality engineering projects. An enterprise brand is an intangible asset and an important means for enterprises to participate in market competition. The brand can show the comprehensive image of the enterprise and has a priceless market value. However, if a project fails, it will affect the brand value of the enterprise and put the enterprise in a disadvantageous position in competing with other enterprises for similar projects.

Engineering projects not only need to consider their direct economic value but also consider their indirect economic value. For example, Anhui Huainan Guqiao Mine is a national key project and Anhui Province "861" key inspection project. It is the key project of Huainan Mining Group to implement the development strategy of "building a big mine," "running a big power project," and "doing a capital project," and to build a national coal base and a large coal power integrated energy base. Since the completion of Huainan Guqiao mine, the direct economic value has been prominent. But its indirect economic value should not be ignored, specifically in three aspects: First, through the project, a large number of new equipment, new processes, new technologies, and new materials have been applied. Through the renewal of equipment and technical research, the level of technical equipment of the entire mine has reached the advanced level of the national coal industry, fully reflecting the scientific and technological value of the project. Secondly, during the construction process of the Huainan Guqiao mine, the project implemented the project manager responsibility system, signed a construction responsibility certificate with the project manager, and gave the project department full autonomy in management. The change in the management system greatly improved the management efficiency. The project's implementation has improved the enterprise's management level, cultivated management talents, and provided management experience for the further development of the enterprise. Thirdly, the high-quality construction of the Guqiao mine project has also strongly promoted the corporate brand building of Huainan Mining Group [11].

6.2.1.2 Near-Term Economic Value and Long-Term Economic Value

Divided the length of time, the economic value of engineering projects can also be divided into near-term and long-term economic value. The near-term economic value refers to the income obtained in the current period. Long-term economic value refers to the income that can be obtained several years, decades, or even centuries after completing the project. Engineering projects are developed to meet certain economic and social needs; whether they are under construction or completed, they should meet the initial construction goals. Generally speaking, most engineering projects can achieve their near-term economic value in the short and medium-term. However, a quality project should not only pursue short-term gains and realize near-term economic value but also pay attention to its long-term value. Many large-scale engineering projects in China, such as the Three Gorges Project, South-North Water Transfer Project, highway, and high-speed railroad projects, consider both near-term and long-term economic value. The Dujiangyan water conservancy project during the Qin Dynasty is a typical case.

Case: The near-term and long-term economic value of the Dujiangyan Water Conservancy Project [12].

The Dujiangyan Water Conservancy Project, built during the Warring States period (256 B.C.), is located west of Dujiangyan, Sichuan Province, on the Min River in the western part of the Chengdu Plain. It was built under the auspices of Bing Li, the administrator of Shu County in the Qin Dynasty, and is still in working order today.

(1) Dujiangyan near-term economic value

Dujiangyan is located where the Min River enters the alluvial plain from the valley river channel. Due to the narrow river channel, the Minjiang River and other tributaries will suddenly rise with the outbreak of mountain floods, often flooding into a disaster. When the floodwaters recede, there are thousands of miles of sand and rocks. Due to the obstruction of Yulei Mountain, the situation of "drought in the east and flood in the west" was formed.

During his term of office, Bing Li, summarized his experience in water management through in-depth investigation and research and finally built the Dujiangyan project in Guan County, where the Min River flows out of the mountains into the plain. The economic value of the Dujiangyan project was that it effectively solved the problem of "drought in the east and flooding in the west" and made the Chengdu Plain a grain-producing area with thousands of miles of fertile land. After completing the Dujiangyan Project, the Chengdu Plain gradually became an important economic center of the country.

(2) The long-term economic value of Dujiangyan

The Dujiangyan water conservancy project's long-term economic value mainly includes flood control and irrigation, ecological function, and tourism. The Dujiangyan water conservancy project uses the local natural conditions and geographical environment. It takes advantage of the special topography, water veins, and water potential. It diverts water without dams and self-flowing irrigation, scientifically solving the problems of automatic diversion of river water, automatic sand discharge, and control of inlet flow. It can be said that the Dujiangyan water conservancy project is a great "ecological project" [13]. In addition to the benefits of flood control, irrigation, and water transportation, the Dujiangyan water conservancy project also has a high tourism value. At present, Dujiangyan has become a world cultural heritage, a world natural heritage, a national scenic spot, and a national 5A-class tourist attraction. This long-term economic value was not imagined when the Duhuiyan project was designed.

6.2.1.3 Evaluation of the Project's Economic Value

For the evaluation of the own economic value of the project, the traditional method is mainly the financial evaluation of the project. According to the current national financial system, price system, and relevant regulations of engineering evaluation, the direct cost and direct economic benefit of the project are analyzed and calculated from the financial perspective. Financial statements are prepared, and financial evaluation indexes are calculated [4]. The project's financial feasibility is judged by examining and evaluating the project's basic viability, profitability, solvency, and risk resistance, thus providing scientific evidence for engineering investment decision-making. For the financial evaluation of engineering projects, cash flow analysis, static and dynamic profitability analysis, and financial statement analysis are generally used. Among them, cash flow analysis is to take the project as an independent system and reflect the cash activities of inflow and outflow of the project each year during the construction period and production and operation period, i.e., the amount of cash inflow and cash outflow in each year during the life of the project. The static analysis method analyzes and calculates the total investment expenditure and revenue after the project is put into operation without considering the time factor. The dynamic analysis method is to consider the time value of capital and uses the discounted cash flow analysis method to analyze and calculate. The financial statement analysis is based on the specific financial conditions of the project and the relevant national financial and taxation system and regulations, analyzes the total investment of the project during the construction period and the operating cost and revenue after the project is put into operation, calculating and balancing yearly, and using the statement format to reflect [14].

The direct and near-term economic value of the project is generally analyzed to evaluate the project's economic value. In contrast, the indirect and long-term economic value of the project is not evaluated. The evaluation of the project's economic value includes the traditional financial evaluation of the project and the evaluation of the indirect economic value, and the long-term economic value generated by the project. There are corresponding evaluation methods for assessing the indirect value of engineering projects. First, the technical progress brought by the implementation of engineering projects is an important factor in enhancing enterprises' economic and social benefits. It is the driving force and source to promote the transformation and upgrading of enterprises. The common methods for evaluating the technical progress brought about by the implementation of engineering projects include the index system method, production function method, data envelopment analysis method, and cost discounting method. Secondly, various evaluation methods for cultivating talents brought by engineering projects exist. The quality and quantity of talents of enterprises can be measured through indicators such as the average education level of workers, the ratio of scientific and technological personnel, and the average technical level of workers. Specific talents in engineering projects can also be evaluated through the means and methods of talent assessment. Again, the evaluation of management level can be evaluated by management performance. The evaluation of enterprise brands can apply the national standard of "Commercial Enterprise Brand Evaluation and Enterprise Culture Construction Guide" (GB/T27925-2011). It uses specific indicators to evaluate the value of an enterprise brand from five aspects: brand importance, growth, profitability, stability, and synergy.

An economic evaluation of engineering projects generally encounters two situations. One is single-plan evaluation, i.e., investment projects have only one technical solution or independent project solutions available for evaluation. The other is multi-alternative evaluation, i.e., investment projects have several alternative technical solutions available for evaluation. For single-plan assessment, using financial evaluation can determine the trade-off of the project. In practice, selecting among multiple options is often. Compared with the economic evaluation of a single option, selecting multiple options is much more complex. There are generally three types of economic relationships between alternatives, i.e., mutually exclusive, independent, and other related relationships, as shown in Table 6.1. Different comparison methods can be used to select the optimal choice [15].

The above methods can be directly applied to evaluating and comparing multiple plans. However, suppose we want to make a comprehensive evaluation of the economic value of the alternative, we should also consider the indirect and long-term economic value. Some public welfare projects, such as large water conservancy projects, need to pay more attention to the long-term value. If an alternative is not the best, but its indirect and forward economic value is much higher than the best one, it can also be considered. For example, in China's high-speed railroad project, there was a debate between the two options of a money-saving efficient railroad electrification transformation program and a costly high-speed railway program at the early stage of construction. In the short term, the engineering cost of railroad electrification is lower, while the high-speed railroad is costly. The debt burden of the high-speed railway project is so heavy that the revenue in the short term can hardly cover its cost, and the current economic value is not much. However, let's consider its long-term economic value. The high-speed railroad construction will significantly shorten the space–time distance between regions, cities, and rural areas. It promotes the rapid

Table 6.1 Alternative types and evaluation methods

Plan type	Evaluation method		
Single plan evaluation	Direct evaluation using indicators such as payback period, investment profit margin, and capital profit margin		
Multi-plan evaluation	Mutual exclusion	Same lifespan	Individual analysis method, incremental analysis method, differential internal rate of return method, minimum cost method
		Different life spans	Least common multiple methods, study period method, net annual value method
	Independent relationship	Mutual exclusion alternative combination method, net present value rate ranking method	
	Co-relationship	Mixed solution	Two-way sorting equilibrium method, Weingartner optimization selection model
		Complementary alternative	Choose an evaluation method based on whether the alternative is symmetrical
		Cash flow-related scheme	The alternative is combined according to the relationship between the alternatives, and then compared and selected according to the evaluation method of the mutually exclusive alternatives

flow of various factors between regions, cities, and rural areas and promotes the coordinated development of the regional economy, which has great long-term economic value. It is precisely the reason why the railroad authorities insist on selecting the high-speed railroad. In multiple alternatives selection, the direct value of the alternative, the near-term value and the indirect value, and the long-term economic value should be considered comprehensively.

6.2.2 *The External Economic Value of the Project*

Engineering projects bring the economic value of externality, i.e., the construction and operation of engineering projects have positive (negative) impacts on society, other organizations, and individuals who do not bear direct benefits (costs). The externality of engineering projects contains both positive and negative externalities, and neither positive nor negative externalities can make the resource allocation reach Pareto optimal state. The external economic value of the project is the economic value caused by the construction and operation of the project that is not included in the project benefits and costs. Solving the externality problem of engineering helps improve engineering projects' resource allocation efficiency.

6.2.2.1 Positive Externalities of the Project

Take the positive externality of engineering as an example. In Fig. 6.1, the horizontal coordinate indicates the output level, the vertical coordinate indicates the output price, MR_H is the marginal revenue curve, MR_E is the marginal external revenue, and MR_S is the marginal social revenue curve. Where $MR_S = MR_H + MR_E$. The project subject does not consider the impact of its actions on others when making production decisions but only calculates its own costs and benefits. For the project subject, its optimal output level is Q_H. At this output level, the marginal benefit equals the marginal cost. Still, the output level of Q_H is not the socially optimal output level and does not consider the external benefits caused by the production of the product. For society, the output level that is consistent with the social optimum is Q_S, and at the output level of Q_S, the marginal social benefit equals the marginal cost, and the project subject bears the external benefits of the decision.

The positive externality of the project is the external benefit, which refers to the benefit that can be obtained free of charge by other subjects other than the project investment and operation subjects [16]. Taking the Three Gorges project as an example, besides the direct monetary benefits brought by its power generation, its positive externalities are mainly reflected in the following aspects: first,

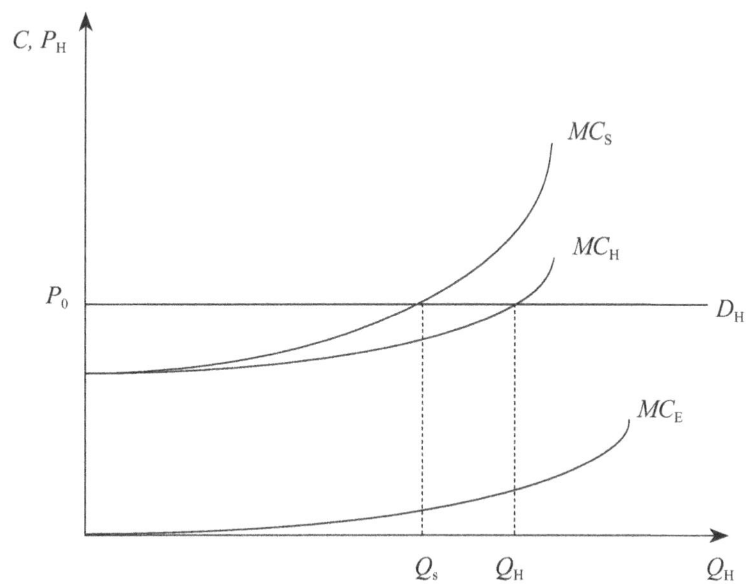

Fig. 6.1 Basic principles of positive externalities of production

flood control benefits. The flood control capacity of the Three Gorges reservoir is 22.15 billion cubic meters, and it can effectively control the upstream floods through the storage of the Three Gorges reservoir. Second, environmental protection benefits. Replacing many thermal power units, the Three Gorges Hydropower Station can save 50 million tons of coal consumption annually and effectively reduce the emissions of pollutants such as sulfur dioxide and carbon dioxide. Third, the reservoir will form 1150 square kilometers of water surface. In addition to the navigation channel, there are still nearly 700 square kilometers of water surface. The flow rate becomes slower, water quality becomes fertile, and the water surface layer turns warmer, suitable for shrimp, shellfish, fish, geese, ducks, and turtle growth as a freshwater aquaculture base. Fourth, improve shipping conditions. The Three Gorges Project is located at the junction of the upper and middle reaches of the Yangtze River, which can effectively improve the navigation conditions between Chongqing and Wuhan and save transportation costs [17]. Organizations, enterprises, and individuals that obtain the above external benefits do not reimburse the related costs to the investment and operation subjects of the Three Gorges Project. Therefore, these benefits belong to the positive externalities of the Three Gorges Project.

The positive externality of engineering projects can benefit people outside the investment. In contrast, the engineering investment and operation subjects do not get the due benefit, and the market mechanism cannot achieve the best resource allocation. There are two basic ideas to solve the positive externality of engineering projects: the "Pigou school" based on government control and the "Coase school" based on market regulation. The former school favors the government and administrative intervention; the latter prefers to leave the problem to the market and advocates ownership autonomy in negotiation.

(1) "Pigou allowance"

The British economist Arthur Cecil Pigou (1877–1959) introduced the famous theory of "Pigou's Allowance" in his "Welfare Economics" (1920). This theory addresses what should be done to maximize total social welfare in the presence of positive externalities. "Pigou's Allowance" theory suggests that government subsidies can be used to reduce private supply costs to maximize social welfare, i.e., the portion of welfare spillover from positive externalities can be used to compensate enterprises' loss. In the case of the Three Gorges Project, the government can subsidize the positive externalities of flood control, environmental protection, breeding, and shipping to reduce the construction and operation costs of the Three Gorges Project. It encourages social capital to participate in the construction and operation of projects with strong externalities and public welfare.

(2) Nationalization

Nationalization is one of the main ways for the government to intervene directly in externalities. Combining the private benefit of engineering project subjects and social returns can effectively solve the problem of insufficient supply in the market of positive externality engineering. The business objectives of state-owned positive externality engineering projects are more inclined to pursue the goal of maximizing social returns so that the quantity and scale of their supply can reach the socially optimal level [18].

(3) Property transaction

In the analysis of property transactions, "Coase's theorem" points out that as long as the property rights are clear and the transaction cost is zero, the final market equilibrium result is efficient no matter who is given the initial property rights. In engineering management, we can solve the problem of engineering externality through market mechanisms with the help of "Coase's theorem." It is often difficult to clarify property rights in real economic activities, and the transaction costs are not small. Different rights definitions and divisions will lead to different resource allocation efficiency, and the setting of the property rights system becomes the basis of optimal resource allocation. For example, in some public resource protection projects, the problem of improper use of shared resources may arise, resulting in the "tragedy of the commons." An important reason for the "tragedy of the commons" is the lack of property rights or unclear property rights. Suppose the property rights of common resources can be assigned to individuals through clear property rights; in that case, the misuse of public resources can be effectively prevented, and the "tragedy of the commons" can be prevented.

6.2.2.2 Negative Externalities of the Project

Take pollution caused by the construction and operation of engineering projects as an example. In Fig. 6.2, the horizontal coordinate indicates the output level, the vertical coordinate indicates the cost and the price of the product, D_H is the product demand curve, assuming that the market is perfectly competitive and D_H has infinite elasticity, MC_H is the marginal cost curve, MC_E is the marginal external cost curve and MC_S is the marginal social cost curve, then we have $MC_S = MC_H + MC_E$. When making decisions, the project subject does not consider the indirect effects caused by its actions on the outside but only calculates its direct costs and direct benefits. At this point, the optimal output level of the project subject is Q_H, because, at this output level, its marginal cost is equal to its marginal benefit. However, since social costs are not considered, the output level of Q_H is not the socially optimal output level. The real socially optimal output level is Q_S, at which the marginal social cost equals the marginal benefit, and the project subject bears the social cost of the decision. Since the output level Q_S is lower than Q_H, the final pollution level caused by Q_S is less than the pollution level caused by Q_H.

The negative externality of the project, i.e., external cost, refers to the social cost that falls outside the project investment and operation main body. It cannot be compensated by the investment and operation main body but is borne by external organizations and individuals without compensation or in unequal value [19]. For example, while the Three Gorges project brings many positive externalities, it also brings negative externalities in the following aspects: first, after the Three Gorges reservoir is impounded, the water level in the reservoir area increases, the water flow

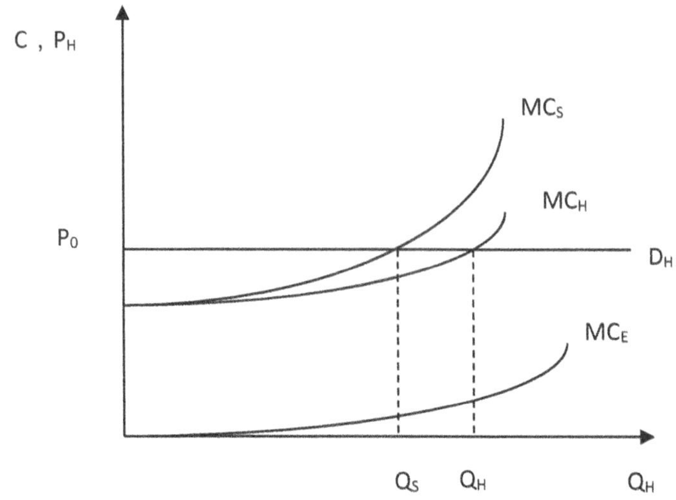

Fig. 6.2 Basic principles of negative externalities in production

slows down, the diffusion capacity of the water body is weakened, and the retention time of pollutants becomes longer. If sewage and garbage are discharged directly into the river without effective treatment, they may cause infectious diseases while polluting the ecological environment. Secondly, the Three Gorges Reservoir Project impacts biodiversity, with the most severe impact on migratory fish. Thirdly, although the Three Gorges project has made great efforts to resettle migrants, the migrants left their hometowns, and their production life and socio-cultural psychology will be more or less adversely affected. Fourth, after the Three Gorges reservoir was impounded, some thousand-year-old towns along the Yangtze River were submerged, such as Fuling Baiheliang, Zhongxian Shibaozhai, Yunyang Zhangfei Temple, Fengdu Ghost Town. The solution to the negative externality problem can start from both government forces and market forces, and there are four main methods.

(1) "Pigou tax"

In Welfare Economics (1920), Pigou was also the first to propose the "Pigou tax." It is a tax on projects that generate negative externalities in order to discourage projects that generate negative externalities. In the case of emissions, for example, taxes are levied according to the degree of environmental harm caused by the emitter. The amount of tax is equal to the difference between the social cost of production and the private cost of the enterprise. The "who pollutes, who treats" policy implemented by China to control corporate pollution is a specific application of the Pigou tax. For example, for the water pollution caused by the Three Gorges Project, the government can levy a certain amount of tax on the Three Gorges Project according to the degree of pollution, which can cover part of the costs the society bears for the project. For the resettlement of immigrants caused by the Three Gorges Project, the government can require the main body of the project to pay a portion of the resettlement costs to improve the living conditions of the immigrants. The "Pigou Tax" is a measure commonly adopted by governments from legal responsibility to consider negative externalities.

(2) Public regulation

Public regulation means that the government can make relevant decisions with its administrative power to control negative externalities in the market. The negative externality problem in engineering projects can be solved by government control. For example, for the negative environmental externality caused by the emission of "three wastes" from chemical enterprises, the government can set the emission standard and punish the enterprises that exceed the emission standard or even order them to close. However, to a certain extent, public regulation lacks efficiency and cannot completely solve the problem of negative externalities.

(3) Negotiations

When there is a division of property rights, low transaction costs, and a small number of participants, people can solve the problem of negative engineering externalities through private negotiations. For example, the sewage discharged by a chemical plant brings negative externalities to the lives of nearby residents. The nearby residents can negotiate with the plant and demand a certain pollution treatment fee or compensation fee to cover the pollution treatment cost borne by the nearby residents for the plant. However, the initial definition of property rights and high transaction costs will influence the negotiation results.

(4) Business consolidation

Merging economic units that impose and receive external costs is an important means of addressing negative externalities, which may arise either from voluntary transactions between the engineering subject and those affected by negative externalities or from administrative intervention by the government. For example, suppose a chemical plant brings negative externalities to nearby agriculture in the production process. In that case, the pollution caused by the chemical plant will be paid for by the farm, which will not be willing to take responsibility for pollution control in the absence of an enterprise merger. However, if the chemical plant is merged with the farm, the combined enterprise will bear the cost of pollution. The decision maker of the enterprise will consider the revenue of chemical products and the cost of chemical product pollution and control the pollution to maximize profit. The level of pollution after the merger of companies will be lower than the level before the merger.

6.2.2.3 Evaluation of the External Economic value of the project

In recent years, to solve the problem of mutual coordination between the project itself and various aspects of social relations, the external economic evaluation of the project has gradually received attention. For example, the World Bank attaches great importance to the external benefits of engineering for developmental investment projects in developing countries. Scientific and rational evaluation of the external economy of engineering is an inevitable trend of human civilization and industrialization, which corrects the traditional drawbacks of over-emphasizing financial evaluation. Therefore, external economic evaluation of large-scale projects has gradually become a focus.

The traditional engineering evaluation only evaluates the project's economic benefits and ignores the assessment of the externalities of the project. The external economic review of engineering must consider both direct and indirect costs and the benefits of engineering. The project's indirect benefits and indirect costs are collectively referred to as external effects, which are benefits and costs outside the project and are not accounted for in the project. When calculating the external economic effect of the project, two conditions are met: the relevance condition and the non-pricing condition. The relevance condition means that the engineering project implementation will impact parties that are not directly related to the project. For example, the nearby residents of the Three Gorges Reservoir Project have no direct relationship with the project itself. Still, the project implementation has an important impact on the production and life of the nearby residents. The non-pricing condition means that the external economic effects of the project are not passed through the exchange, and no compensation is required for its effects [20]. Still taking the Three Gorges Reservoir Project as an example. After the storage of water in the Three Gorges Reservoir, many thousand-year-old towns along the Yangtze River were submerged, and the economic value of this negative externality was not determined through an exchange. No one paid for this negative external economic effect.

It is important to note that not all externalities lead to resource misallocation. Depending on the mechanism of action, externalities can be further subdivided into technological and financial externalities. Technological externalities are formed due to technical linkages and are unrelated to the market price system. Financial externalities are developed mainly thanks to the transmission of market mechanisms and are directly related to the market price mechanism [21]. Since the financial externality is mainly reflected through price instruments and can still function as a market mechanism, this externality does not lead to market failures. Since technological externalities do not reflect the results of price changes and do not work through the market mechanism, they may lead to market failure. If the project's investor is the government, then a series of national economic benefits to be achieved by the project should belong to the "internal revenue" that the government should consider before deciding to invest in the construction of the project. At this point, the externalities of the project have been internalized and will not cause market "failure." However, if the investors of engineering projects are mainly private capital, the above national economic benefits are not included in the private accounts; at this time, attention should be paid to evaluating the externality of engineering projects [22]. Currently, most technological externality evaluations are unable to achieve great precision quantitatively, and a combination of quantitative and qualitative approaches can be performed.

6.2.3 National Economic Value of the Engineering

Any country or region in the modernization of its national economy needs to construct some major engineering projects according to the overall needs of socio-economic development to solve key problems of general nature in socio-economic development. Major engineering construction plays a pivotal role in national economic development, and financial evaluation and national economic evaluation of the project should be conducted before making decisions [23]. In addition to financial assessment, it is necessary to analyze the industrial economic value, regional economic value, and scientific and technological economic value of engineering projects from the perspective of the country and the public. Making a correct national economic evaluation and judging the reasonableness of engineering projects in terms of resource allocation. Take China's high-speed railway project as an example. We should analyze not only the economic benefits of the project itself but also the impact of the high-speed railway on the socio-economic development of the cities it passes through, the impact on other modes of transportation, and the many different impacts on the financial market and stakeholders.

6.2.3.1 Industrial Economic Value of the Project

The construction of some important basic engineering projects is a necessary prerequisite for the development of the industry. For example, the development of the transportation industry or logistics industry must rely on the construction of highways, railroads, aviation, and other engineering projects. The development of water conservancy and the hydropower industry must depend on the construction of dams and the installation of related equipment. In short, the construction activities of basic engineering projects are the basis for the survival and development of the industry.

The construction activities of engineering projects promote the development of the industry and drive and promote the further development of other related industries. For example, a construction project can effectively drive the growth of more than 50 associated industries, such as building materials, metallurgy, chemical industry, machinery, decoration, furniture, and home appliances. China's high-speed railroad—Beijing-Shenzhen high-speed railway, with the longest operational mileage, a total length of 2,439 km, connects the Pearl River Delta, the middle and lower reaches of the Yangtze River city group, and the Bohai Sea economic circle, promoting the development of logistics, commerce, tourism and other related industries along the economic circle.

From the perspective of the relationship within the industry, some major engineering projects have substitution relationships with other projects of similar nature in the industry, and the enterprises are highly competitive. In particular, some engineering projects are particularly huge. Their commissioning will form a major impact on the market within the industry. It will directly lead to a decline in product prices throughout the industry and affect the entire industry's profitability; even due to excess capacity and the formation of vicious competition, it will jeopardize the development of the industry sector as a whole. For example, some industries with excess capacity in China are often caused by the lack of necessary macro management of the relationship between engineering project construction and industrial economic development. From the perspective of inter-industry relationships, there may be complementary, competitive, and substitution relationships between engineering and existing projects. Lacking effective coordination and management of various transportation projects may result in one kind of transportation mode substituting another transportation mode. Such as high-speed rail to the highway and civil aviation. It is rational for a single mode of transportation, but the efficiency of the overall transportation system may not be reasonable. Suppose various transportation construction projects are effectively coordinated and managed; in that case, various modes of transportation will compete but form a complementary relationship, thus improving the overall efficiency of transportation. The relationship between engineering project construction and industrial economic development involves the relationship between upstream and downstream industries. Upstream industries often hold certain resources needed for the project, such as raw materials like minerals or core technologies. Some industries have high barriers to entry. If other countries control the upstream industry, or if the upstream industry's core technology is missing, they will restrict the development of engineering or even the whole industry. The downstream industry of engineering is at the end of the entire industry chain. In some industries, companies often control the sales market through brand barriers. Suppose there is a lack of management of engineering projects from the perspective of upstream and downstream of the industry; as such, upstream and downstream enterprises simultaneously exploit the profits of engineering projects, and they may be locked in the low end of the industrial chain. Therefore, engineering management must focus on the industry's upstream and downstream relationships.

6.2.3.2 Regional Economic Value of the Project

The construction of large-scale engineering projects has important significance to regional economic development, and local governments at all levels in China have taken the construction of major engineering projects as an important element of regional economic development to plan and implement. Especially in less economically developed areas, government departments often aim to increase employment, improve residents' income, increase tax revenue and drive regional economic growth through the construction of large-scale engineering projects.

The construction of major engineering projects often becomes a new growth point for the regional economy. On the one hand, major engineering projects drive regional economic growth through fixed asset investment. On the other hand, major engineering projects drive the development of related industries and promote regional economic growth through industrial forward, backward, and lateral linkage. In some less developed mountainous areas, a major transportation infrastructure project can often open the bottleneck that restricts regional economic development, benefiting industrial development and driving regional economic growth significantly. At the same time, the increase in investment will multiply the regional GDP through the multiplier effect of investment. Therefore, the construction of major engineering projects is the booster of regional economic growth and an important "fulcrum" to pry regional economic development.

The construction and operation of major projects cultivate new tax sources for local governments. Major investment projects not only involve many taxes and fees during the construction period and require paying relevant taxes and fees but also bring a continuous stream of tax revenue after the projects are implemented. At the same time, major engineering projects also drive the development of many supporting industries, which increases enterprise profits and residents' income, thus increasing tax revenue.

The construction and operation of the project can provide more jobs and raise the income level of residents. The increase in the local employment rate means higher income and better living conditions. The projects also boost the income of employees in other industries through developing other industries. At the same time, new investment projects also improve the management level of entrepreneurs, improve the quality and skills of employees, and allow employees to accumulate more experience, which will further increase the income level of employees through improving human capital.

The construction of some major environmental infrastructure and livelihood projects can enhance the environmental carrying capacity of the region. It will not only become the advantage and brand of local investment attraction and drive regional economic development but also improve the living environment and the residents' quality of life. For some heavy industrial engineering projects, attention should be paid to the regional environmental capacity in the implementation and operation. They should focus on environmental friendliness and resource conservation, strictly implementing national environmental standards, minimizing the damage to the local resources, environment, and ecology, and maximizing the comprehensive benefits [24, 25].

However, large-scale projects may cause the insufficiency of local infrastructures such as electricity, water, and raw materials. It may cause the price rise of local products and living goods. Land acquisition and relocation, arable land occupation, labor use, and migration may also create a series of conflicts between local governments and residents, which can affect the smooth implementation of the whole project. Sometimes, there may be a separation of investment and development place and enterprise registration place as large-scale engineering projects can occur across provinces. There can be problems such as tax collection and the contradiction of benefit distribution brought by the separation of investment and development place and enterprise registration place. Therefore, the relationship between engineering construction and regional economic development should be included in the engineering management framework.

6.2.3.3 Economic and Technological Value of the Project

The development and construction of engineering projects have an important role in promoting technological innovation. On the one hand, science and technology have a significant supporting role in constructing engineering projects. On the other hand, the construction of modern engineering projects has put higher requirements for developing science and technology, which has injected power into the development of science and technology. For example, our country implemented manned space engineering and achieved good technology economy benefits. Manned spaceflight engineering incorporates high and new technologies. These high and new technologies cannot be obtained through the technology market and only rely on independent research and development. The construction of manned space engineering projects has strongly promoted related basic scientific research and applied technology research in China.

6.2.3.4 Evaluation of the National Economic Value of the Project

(1) Evaluation of the industrial economic value

The construction of engineering projects has an important impact on industrial development. Industrial development covers many aspects, such as industrial-scale expansion, industrial performance improvement, industrial technology development, and structure optimization, and changes dynamically with time [26]. The industrial economic value evaluation indicators of engineering projects are mainly three: scale, efficiency, and technological progress: ① Industrial scale indicators. From the aspect of industrial scale, five indicators are mainly considered: total industrial output value, total assets, number of employees, industrial per capita capital investment, and the average size of enterprises and these indicators reflect the development process of the industry through the reflection of industrial scale and capital scale. ② Industrial efficiency indicators. The cost profitability index is used to reflect the business results, i.e., the unit cost profitability demonstrates the efficiency and profitability of industrial development [26]. ③ Technological progress indicators. Mainly include the output and added value of the high-tech industry and its proportion to the output and added value of all industries.

(2) Evaluation of the regional economic value

The main indicators for evaluating the regional economic value of engineering projects are ① Contribution rate of economic growth. What is the contribution rate of a certain engineering project to the local economic development through its construction and operation, which is used to measure its impact on the regional economic growth? ② New employment rate. It is used to assess the impact of the construction of major projects on the local employment level and whether the project's construction has brought new jobs. The ratio of the difference between the number of jobs expected to be added and the number of jobs expected to be reduced after the implementation of the project and the number of local employment before the project implementation is used. In the construction of engineering projects, some major projects occupy a large amount of arable land and other natural resources. They not only do not bring new jobs to the local area but also cause an increase in the unemployment rate, resulting in a series of social problems and bringing a negative impact on the construction of the project itself. Therefore, the expected new employment rate of the project is used to assess the impact of the construction of the project on regional employment. ③ Change of residents' income level. The difference between the expected average income level of the residential population after the implementation of the project and the average income level of the existing residents before the project implementation is used. ④ Fiscal revenue. The construction and operation period of the project will pay taxes to the government department where the project is located, and the amount of taxes paid by the project can be calculated through investment estimation and financial evaluation.

(3) Evaluation of the economic value of science and technology

Evaluating engineering projects' scientific and technological economic value can be considered in technological innovation activities at the enterprise level and technological progress at the industry level. The evaluation indexes can be divided into ① Enterprise R&D intensity. The level of enterprise R&D intensity determines the technological innovation ability of the enterprise. It is generally expressed by the proportion of R&D investment to the enterprise's sales revenue or the proportion of new product sales revenue to its total sales revenue. ② Total Factor Productivity (TFP). The growth rate of total factor productivity is usually regarded as an indicator of technological progress, which can measure the role of technological advancement in the construction of engineering projects. It should be noted that the sources of total factor productivity growth include technical progress, organizational innovation, management innovation, and specialization. ③ The number of patents obtained, including invention patents, application patents, and design patents, which are used to reflect whether technology development activities are active.

6.3 Social Value Engineering

Engineering people need to form certain relationships and carry out purposeful and organized social actions in the process of understanding nature, adapting to nature, using nature, and building artificial nature [27]. Modern engineering activities have an important and far-reaching impact on the surrounding nature, ecology, and social environment. While improving people's material and cultural living standards and realizing the economic and ecological values of engineering, engineering activities must consider the social values of engineering and realize the harmony and unity of humans, nature, and society. The core value of engineering management is "people-oriented, unity of heaven and man, collaborative innovation and harmony" [28]. The realization of engineering social value requires the division of labor and joint efforts of engineering community members. The process of realizing engineering social value is also the process of fulfilling the social responsibility of engineering community members. Along with the rapid development of China's economy and society, the interaction between engineering and society is becoming increasingly complex, which requires us to examine the interaction between engineering and society from a broader perspective and evaluate the social value of engineering more systematically and comprehensively.

6.3.1 Engineering Community Social Roles

Engineering activities are carried out in a specific spatial and temporal context and a specific social environment. Any engineering activity involves many social factors, from conception, design, and feasibility analysis to implementation, evaluation of the project operation, and decommissioning. Engineering activities as purposeful and organized collective social actions; the engineering community is the basic organizational form of engineering activities [29]. The social role, a behavior pattern that conforms to one's social status or identity and rights and obligations, usually also represents a social expectation. Society always expects actors to act according to their social status or identity. The actors usually require themselves this way and try to make the behavior conform to the "role expectation." The "heterogeneous" members of the engineering community have different social roles, and such social roles are not static. Many engineering community members are constrained by the role norms and pursue their own interests and needs as well. This dual existence is the root cause of many contradictions and social problems in engineering activities [3].

Engineering activities are embedded in certain social structures and social relations. They are influenced by them. At the same time, engineering activities, as one of the components of the large social system, also affect and drive the changes in social structures. Therefore, studying engineering activities requires analyzing the social roles and interaction patterns of each member of the engineering community in the framework of specific rules. Based on this, various contradictions and problems within engineering society can be revealed, and engineering activities' development and change laws can be better grasped.

6.3.1.1 Engineering Community Constituents

The process of human development is the process of relying on nature, adapting to nature, understanding nature, and constructing artificial nature reasonably and appropriately. Engineering activities bridge the gap between scientific discovery, technological invention, and industrial development [30]. The engineering community occupies an important place in the research related to engineering management theory. However, there is a long-standing misconception in the engineering community that engineering is the activity of engineers, and therefore, the engineering community is a group of engineers. However, it isn't easy to complete any project only with an engineering group. Real engineering activities need different types of members, such as owners, contractors, supervisory units, financial institutions, government departments, and the public, to perform their respective duties, take responsibility, and work together. Thus, the engineering community is a complex system of engineering activity subjects characterized by hierarchy, multi-role, division of labor, and multiple interests [29, 31]. It is a complex system composed of "heterogeneous" participating subjects such as owners, contractors, supervisory units, engineers, financial institutions, government departments, the public, and other interest groups. Every subject plays a different social role.

(1) Owner. In engineering activities, the investors, including the government, enterprises, and individuals, will set up special organizations or assign special personnel responsible for project management in the owner's capacity. The owner is the owner of the project; in the process of project implementation, the owner's identity can be subdivided into fund-raiser, manager, and controller of the whole process (investment control, schedule control, quality control, and safety control).

(2) Contractor. The specific implementer of the project refers to the enterprise with certain production and operations capacity, technical equipment, personnel and capital, with corresponding engineering qualifications and business qualifications, and can provide the owner with the required products and services.

(3) Supervision unit. It is an organization that provides paid professional service activities. To control the investment, construction period, and project quality of engineering construction, based on laws, regulations, and relevant technical standards, this unit monitors the quality, schedule requirements, and safety management on behalf of the construction unit and coordinates the relationship between all relevant units.

(4) Engineer. Engineers have a special status in the engineering community and play multiple roles. First, engineers have the professional technical knowledge and practical experience, and they are the "technical authority" in the engineering community. Secondly, engineers need to assist the project managers and formulate effective project implementation standards and technical management systems to achieve the expected goals of the project. In addition, due to the uncertainty of engineering activities, engineers need to choose the best solution among different alternatives and implement it according to the specific environment and conditions during the project implementation. Therefore, engineers are always in the engineering activities, organically integrating labor, technology, management, and other production factors.

(5) Financial institutions. Modern large-scale engineering projects with large capital needs require financial institutions' support, including financial support, credit support, etc.

(6) Government departments. Engineering construction needs to meet certain social objectives. Government departments mainly perform inspection functions from the social perspective on the project establishment, implementation process,

and engineering quality, performing social management, supervision, and inspection functions.

(7) Social public. It refers to the public groups directly or indirectly affected by the engineering activities, including the original residents, the surrounding community, and end users.

(8) Other stakeholders. It refers to the other participating subjects of engineering activities other than the above seven types of subjects. For example, engineering-related non-governmental organizations (NGOs), non-profit organizations (NPOs), news media, etc.

6.3.1.2 Social Network Relations of the Engineering Community

Engineering activities are engaged in social activities in the form of collective or community activities. The modern engineering community comprises investors, contractors, engineers, managers, and other stakeholders. Each participating subject is an actor with cognitive behavioral ability. According to various roles, they have their own interests and demands, which are realized through engineering activities under the constraints of rules and regulations, resulting in a complex social network relationship among the participating subjects of the engineering community. First, enterprises or individual investors will entrust the owner to exercise the function of the owner in the form of the entrusted agency relationship. The ultimate principal is the public for public works invested by the government. Secondly, the government exercises administrative functions in the project approval and the establishment of the engineering community while supervising the whole process of engineering construction. Again, in the specific construction process of the project, the owner entrusts the contractor to be responsible for the specific implementation work in the form of bidding and at the same time, entrusts the supervision institution to supervise the contractor. Financial institutions provide financial support for investors and contractors. In addition, engineers are employed by specific engineering activity communities. Engineers may have multiple roles as technical experts, consultants, and managers and are responsible for engineering design, technical consultation, technical guidance, and management of engineering activities. Some scholars call engineers in engineering communities "marginal men." Finally, the public affected by the project will participate in the pre-project decision-making through the project demonstration and hearing and supervising the implementation process and the output and social impact of the project. In addition, NGOs, media, and other stakeholders will also supervise the project community and provide feedback on the results to the relevant government administrative departments. The interaction among the participating subjects of the engineering community is shown in Fig. 6.3.

An important feature of major complex projects is that they are susceptible to political, economic, and social influences, and decision-making, planning, and management are processes in which many stakeholders interact [32]. Due to the complexity of the external environment and the limited cognitive ability of human beings, it is difficult to accurately ex-ante forecast the results of major engineering activities [33],

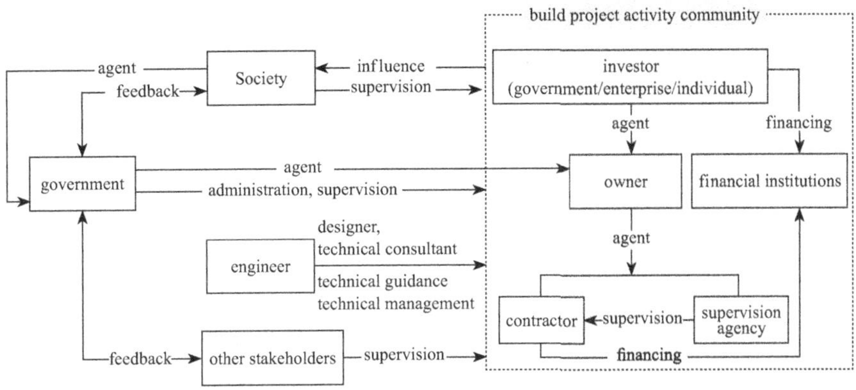

Fig. 6.3 The interactive relationship among the participants in the engineering community

which leads to many uncertainties in the implementation process of major complex projects. These uncertainties easily lead to contradictions and conflicts among stakeholders, which then evolve into social problems (land acquisition, demolition, engineering migration). The complexity of the engineering community is manifested in the existence of complex internal relations and the complex relations between it and the external society.

(1) Internal social network relationship of the engineering community

Among the complex internal relations of the engineering community, the key point is to deal with the relationship between engineers, workers, and investors. First of all, in the engineering community, engineers are designers, organizers, implementers, and managers of engineering activities, realize the transformation of scientific knowledge in the engineering field, transform scientific activities into direct productivity, and play an important role in engineering activities. In addition, as an important communication medium, engineers use their professional knowledge and practical experience to convey engineering information in time, helping other members of the engineering community understand engineering better. They assist investors, decision-makers, and the public makes better engineering decisions. Secondly, there is a close connection and interaction between workers, engineering investors, managers, and engineers. Investors provide employment opportunities to workers while relying on them to complete the design goals and plans of the project, and investors derive their economic benefits from workers' labor. The relationship between engineering managers and workers is mainly a leader-executor relationship in engineering technology operations. The relationship between workers and engineers exists in the technical specification as mentors and mentees. Any engineering activity must have certain capital input, and there can be no real engineering activity without investment and investors. However, investment in large-scale engineering projects is characterized by a large amount of capital, a long time, and many uncertainties, easily influenced by many factors such as politics, economy, environment, and society. Economic and ethical factors must be considered simultaneously when making investment decisions and weighing multiple values, such as economic, ecological, and social values.

(2) External social network relationship of the engineering community

In the external social network relationship of the engineering community, special attention should be paid to properly handling the relationship between the engineering community, engineering users, society, and government.

1. Engineering community and engineering users

The practical needs of engineering users constitute the basis for the existence of engineering projects. The project construction is to meet users' needs, and the design and function of the project reflect the needs of engineering users. As a project often has multiple project users, the interests of different project users may differ, making the interests of varying project users conflict with each other. Due to certain engineering projects' environmental and operation risks, losses may be brought to engineering users. Therefore, information exchange channels should be opened in engineering decision-making, design, implementation, and operation and maintenance. It is important to establish a reasonable mechanism for expressing and participating in the interests of users and guarantee the right to know, the right to speak and the right to choose project users.

2. Engineering community and the public

With the development of network information technology and the advancement of democratization, the engineering community, as the executor of changing nature and promoting social progress, has a very close relationship between the engineering community and the social public. For any organization to survive and develop better, it is essential to establish a good organizational image in society and the public. "It is the ethical responsibility of the engineering community to put the safety, health, and welfare of the public in the highest position [34]. The relationship between the engineering community and the public is manifested as follows: (1) the engineering community is not only the applicator of science but also the benefactor of the public, taking the honor of creating value for the public and accepting the public's inspection of the work results; (2) the public products provided by the engineering community are closely related to the public's life and accept the public's supervision; (3) the public, as taxpayers, enjoys the right to know about the target planning, project decision, construction process and final results of the project, and has the right to be informed of the process of engineering design, production, operation, and product service, as well as the impact of the project on the environment and society or the efforts invested in it.

With the development of modern network information technology, emerging media play an important role in the interactive communication between engineering communities and the public to further improve the public image of engineering communities. Firstly, it is necessary to strengthen the interactive communication between the engineering community and the social public and make the public understand engineering better through direct face-to-face communication (engineering demonstration meeting, hearing meeting), mass communication media, etc. Secondly, improving the public's engineering awareness and enhancing public engineering technology literacy through various training and popular science activities is necessary.

3. Engineering community and government

Firstly, the government is the organizer and manager of modern society's operation. It is extremely important in engineering planning, decision-making, design, implementation and evaluation of some social and major projects. Secondly, the historical status, social role, working environment, and compensation of important members of the engineering community, such as investors, engineers, and workers, are closely related to the government. Once again, with the advancement of the new public administration movement, the scientific and democratization process of decision-making is accelerated. The government adopts the channels of presentation, demonstration meetings, and hearing meetings, as well as NPC, CPPCC, and news media at all levels in the decision-making mechanism of major projects. There is a close connection between the engineering community and government in the major decisions, and the right of participation, information, and consultation of the engineering community is increasingly respected.

The government creates a good social environment (political environment, legal environment, investment environment, etc.) to better serve the engineering community, especially by reforming the administrative approval system, improving the approval efficiency, strengthening the management and supervision of the bidding link, and creating a fair and just environment for the contracting and implementation of the project. Secondly, the government creates conditions for improving the professional quality of engineering community members, such as increasing workers' pre-job training and skill training, reforming the technician title evaluation system, etc. Again, the government provides the necessary infrastructure and conditions for the engineering community; for example, the government improves the living conditions of workers by renovating shantytowns and building affordable and low-rent housing. In the South-North Water Diversion Project, the government set up a migrant relocation and resettlement command to properly solve the problems of housing, transportation, medical care, and education for the relocated migrants.

6.3.1.3 Life-Cycle-Based Social Interaction of Engineering Communities

The community of engineering activities arises from the purposeful human process of building artificial nature. If an engineering activity is compared to a drama, the engineering activity community is the "collective of actors" performing the drama. The drama requires a process of ups and downs, and the actors appear in a certain order [35]. The life cycle of the community of engineering activities can be roughly divided into the conception and preparation stage, construction stage, operation and maintenance stage, and disintegration stage.

First of all, in the pre-preparation stage of engineering, because engineering activity is a purposeful social activity process of human beings, engineering activity needs to choose a suitable engineering activity target in the "possible world" first. The person who proposes the engineering activity target first is the project's initiator. For example, in 1802, Albert Mathieu-Favier, a French engineer, proposed building the Channel Tunnel project. The scientificity and feasibility of the project were evaluated by government departments, investors, consulting organizations, and the public. In November 1973, the British and French governments signed a treaty on constructing the Cross-Harbour Tunnel and proposed a concrete plan. In November 1984, the British and French governments reached a basic agreement, and in May 1985, the two governments finally selected a concrete plan. In the South-North Water Diversion Project, Chairman Mao Zedong proposed the great idea of South-North Water Diversion in October 1952, and relevant departments and units did a lot of planning and demonstration work [36].

Second is the construction preparation stage of the project. Enterprise or project department is the most common type of engineering activity community organization form in modern society. For the life history of a specific engineering activity community, establishing an engineering project department means formally establishing an engineering activity community. For example, at the beginning of the "South-to-North Water Diversion" project, the South-to-North Water Diversion Project Construction Committee was established and led by the Premier of the State Council. A full ministerial level was set up: the Office of the South-to-North Water Diversion Project Construction Committee of the State Council. At the same time, to ensure the smooth implementation of engineering activities, the legal project person must coordinate the relationship between contractors, survey and design, supervision, construction, consulting, and other construction business units through contract management.

Then, the construction stage. The "whole process" of project implementation includes basic stages of "decision making, design, implementation, installation, project operation, project abandonment." In different stages of engineering construction activities, to complete the corresponding tasks, various interactions will occur among members of the engineering activity community, such as engineering contractors, suppliers, design units, supervisors, and the public. Attention is paid to the control of the engineering construction process, timely solutions to various technical problems, and special attention to the quality of the project. In the Anglo-French Channel Tunnel Project, to solve the ventilation problem, reduce the risk of undersea construction, and improve the reliability of operation and maintenance, instead of using a common tunnel for a large-span two-lane railroad, two railroad tunnels were excavated. In addition, another service tunnel was opened between these two main tunnels to connect the main tunnel. In the South-North Water Transfer Project, to comprehensively strengthen the project's quality management, the lifelong system of quality responsibility implementation rules has been formulated, which has greatly improved the quality of construction.

Finally, the completion and acceptance stage. After completing the project, government departments and the owner will organize the complete acceptance of the project; meanwhile, the owner will settle the project payment with contractors, suppliers, design units, etc. In addition, the completion of the project and the end of the construction period also mean the community's dissolution of specific engineering activities. For example, the British-French Cross-Harbor Tunnel was officially opened to traffic on May 6, 1994, after more than 8 years, costing about 10 billion pounds and the hard work of 11,000 engineers and technicians. The first phase of the South-North Water Diversion Project East and Central opened in 2013 and 2014, respectively.

6.3.2 Social Responsibility of Engineering

Responsibility and values are two different concepts; responsibility implies right, due, and obligation, while values include goodness and merit. However, there is an inherent unity between responsibility and value. Responsibility is included in the scope of value. Engineering value is a special value form created by engineering activities, which concentrates on the degree of satisfaction of engineering activities and their achievements to human social needs. In contrast, engineering responsibility is a kind of engineering value selection behavior based on correct values. The wide scope of engineering activities and diversified interest subjects determine the multi-dimensional characteristics of engineering value. Therefore, from the perspective of value theory, responsible behavior in engineering activities is a kind of value choice, and the fulfillment of responsibility is the premise and foundation for generating great external value.

6.3.2.1 Concept of Engineering Social Responsibility

Engineering social responsibility refers to that in the process of engineering activities, based on the responsibility for society, the environment, and the future, and through the choice of engineering value, the engineering community tries to reduce or eliminate the negative impact of engineering activities. The size of engineering social responsibility affects the social value of engineering, and engineering social responsibility is closely related to engineering social value. The core of engineering social responsibility is "people-oriented, the unity of heaven and man," which ultimately realizes the harmonious coexistence of man and nature and the harmonious development of man and society. Unlike technical and corporate social responsibility, as modern engineering activities involve economic, scientific and technological, social, cultural, and natural aspects, many social responsibility subjects in engineering activities require cooperation among all participating subjects. On December 22, 2010, the "Guidelines on Social Responsibility in China's Foreign Contracting Industry" was officially released, mainly covering project quality and safety, rights and interests of customers, owners, and employees, and natural environment protection [37].

6.3.2.2 Components of Engineering Social Responsibility

In engineering activities, each participating subject, including government, owner, contractor, supervision unit, engineer, financial institution, public and other stakeholders, assumes different roles and functions, and their social responsibilities should be different [38]. The social responsibility of engineering should be an organic integration of the group responsibility of the engineering community and the individual responsibility of each community member. The following is a systematic analysis of the components of engineering social responsibility from two aspects: engineering social responsibility of direct stakeholders and engineering responsibility of indirect stakeholders, in which the subjects of engineering social responsibility of direct stakeholders cover owners, governments, engineers, and contractors, and engineering social responsibility of indirect stakeholders include the public, media and non-governmental organizations (NGOs). The structure of engineering social responsibility is shown in Fig. 6.4.

(1) The owner's engineering social responsibility

As the project investment decision maker, the social responsibility that the owner should undertake mainly includes the following aspects: first, eliminate or reduce all kinds of negative impacts during the whole life cycle of the project. Second, to establish an open and transparent communication mechanism and platform so that the public can be informed of the objective and true information related to the project. Third, considering technology's "double-edged sword" characteristics, actively use technology to solve various development problems and minimize technical risks. Fourth, pay attention to international, global, and inter-generation resource and allocation issues [40].

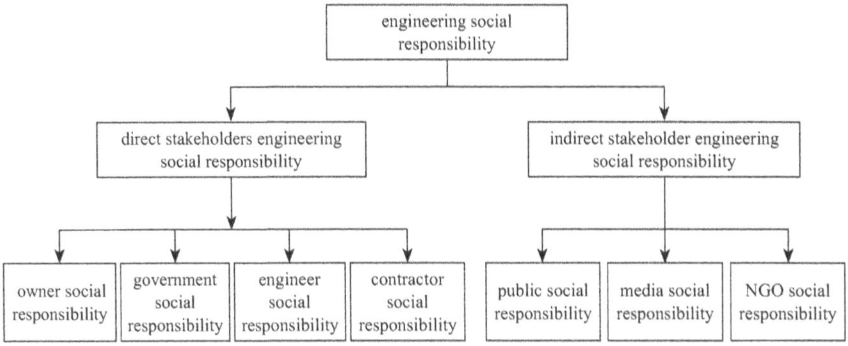

Fig. 6.4 Engineering social responsibility structure chart [39]

(2) Government's engineering social responsibility

As the engineering planner, organizer, and supervisor, the government's engineering social responsibility runs through the whole process of engineering construction. Especially in government-type engineering projects with strong public welfare and social characteristics, the scientific and democratic decision-making of project establishment, legal compliance of approval process and effective supervision of the implementation process constitute important contents of engineering social responsibility [36]. With the deepening of the new public administration movement, the importance of the government's engineering social responsibility is further highlighted as an important element in enhancing the image of the government and its prestige. The government's lack of awareness and implementation of engineering social responsibility has produced some "performance projects" and "face-saving projects" that go against the laws of nature, which easily induce the government's trust crisis and governance crisis [41]. It also reflects the mutual deviation between political and social values in fulfilling the government's social responsibility in engineering [42].

(3) Engineering social responsibility of engineers

Engineers play a key role in the engineering community as designers, decision-makers, and executors of engineering activities [43]. The multiple roles played by engineers mean that they take on more responsibilities, including responsibilities to employers, peers, the public, the environment, and society [44]. The fulfillment of these responsibilities forms multiple restrictions on engineers' behavior. It forces engineers to seek a balance of interests among multiple subjects, often putting them in an ethical dilemma. In addition, in modern society, to ensure scientific and democratic engineering decision-making, engineering stakeholders are required to participate in engineering decision-making. It requires the communication media function of engineers to disseminate engineering-related information to relevant subjects in a timely, objective, true, and complete manner to ensure their right to know and participate, which in turn expands the scope of social responsibility of engineers.

(4) Contractor's engineering social responsibility

The engineering social responsibility of contractors and supervisory units refers to the responsibility to investors (owners) and other stakeholders in engineering construction to achieve sustainable development from economic, social, and environmental aspects. Contractors who actively undertake social responsibility can win the trust of customers and government support, reduce operating costs, improve operational efficiency, enhance corporate visibility, improve the project's reputation, and attract outstanding talents to join. Thus social responsibility is the basis for contractors to achieve sustainable development. To sum up, in the engineering management process, engineering contractors should integrate social responsibility into their own operation and development, take into account economic interests, environmental interests and social interests, and make the engineering construction beneficial to the public, the environment, and the whole society.

Workers are the operators and executors of engineering project implementation. The social responsibility of workers in the process of engineering activities also constitutes an important content of contractors' engineering social responsibility. Workers are required to strictly follow and comply with the relevant regulations in the process of implementation and operation of engineering activities and operate according to the appropriate protocols to avoid the responsibility of environmental disasters caused by misoperation. Secondly, workers are the earliest and most direct perceivers of environmental and social problems that may be caused by engineering activities [3]. Therefore, they should put public safety, health, and welfare first, detect and report various problems early, and nip various risk factors in the bud [45].

(5) Engineering social responsibility of indirect stakeholders

In the process of engineering construction, the important elements of the media's social responsibility for engineering are timely, truthfully, objectively, and completely disseminating relevant information and knowledge, establishing an effective communication platform between all parties in the engineering community, and reducing information asymmetry. The public also has certain responsibilities for engineering supervision and reporting. Non-governmental organizations (NGOs) and the non-profit sector (NPO) staff, in the whole process of engineering planning, design, implementation, and evaluation, should take the position of an objective and impartial third party to truthfully reflect and monitor the problems in the project and should be responsible not only for their actions but also for the public and other stakeholders.

In summary, engineering social responsibility refers to the process of engineering construction based on the attitude of being responsible for society, the environment, and the future, fully considering the real and potential impacts caused by the project to the stakeholders, and minimizing or eliminating all possible hazards to human beings and the environment caused by engineering activities.

6.3.2.3 Multi-stage and Dynamic Nature of Engineering Social Responsibility

The coordination between the engineering system and social system, and the natural system is an inevitable requirement for modern engineering to develop in the direction of systematization, which requires the relevant participating subjects of the engineering community to have a strong sense of social responsibility. Engineering is a complex human practical activity. Engineering construction can be divided into several stages: project decision, planning and design, construction and implementation, operation and maintenance until project abandonment, etc. Engineering social responsibility presents the characteristics of stages and dynamics, and the influence of different stages and subjects on engineering social responsibility has significant differences [39]. The project decision-making stage should consider scientific, democratic, and fair decision-making. The construction and implementation stage should focus on humanization and ecology. The project operation, maintenance, and evaluation need to consider various benefits. Engineering social responsibility plays a restraining and regulating role in the behavior of engineering activity subjects in different stages. On the other hand, because modern large-scale engineering activities involve many participating subjects, engineering social responsibility contains multiple levels and dimensions. Carroll [46] believes that corporate social responsibility mainly includes economic, legal, ethical, and discretionary. Large engineering projects also include technical responsibility, management responsibility, environmental responsibility, ecological responsibility, etc. Therefore, engineering social responsibility is characterized by multi-stage, dynamics, and multi-dimensionality.

(1) Engineering decision stage

The whole life cycle of the project starts from the project decision stage. The first task of this stage is to determine the project objectives, scientific, rational, and feasible objectives related to the project's success or failure. The government needs to play a guiding and supervising function in this stage, strictly control the project approval, follow scientific and democratic decision-making procedures, and fully consult experts and stakeholders to ensure that the project can consider economic, environmental, and social value. In addition, engineers are involved in the project decision-making process as important consultants and thus also bear important responsibilities. For example, the South-North Water Diversion Project emphasizes decision-making consultation. It has set up an expert committee to consult on major technical, quality, economical, and management issues involved in the project and ultimately make scientific decisions.

(2) Engineering planning and design phase

In order to achieve the expected goal of the project in a specific time and space environment and under certain resource constraints, the planning and conception of the project, the concept of the project in the world of thinking are expressed through engineering language and symbols, forming the project design and drawings and other documents. The role of engineers in the planning and design stage is particularly critical. To meet the functional needs of engineering projects, designers must solve many technical problems. At the same time, the design must reflect the unity of science and humanity and achieve the harmonious coexistence of man and nature. For example, engineers have solved many world-class technical problems in the South-North Water Diversion Project. The first phase of the East Line Project has 13-step pumping stations, the most concentrated modern pumping station group in Asia and even in the world. The Central Line Project uses shield structure, which is a precedent at home and abroad (Fig. 6.5).

(3) Engineering construction implementation phase

The construction and implementation of the project is a series of activities to transform the design intention into an engineering entity and form the final product. At this stage, the project's social responsibility is mainly on the engineers, managers, and workers. While paying attention to construction safety, we should pay attention to the impact of construction on ecology and surrounding residents and strictly ensure construction quality. During the South-North Water Diversion Project's implementation phase, to fully guarantee the construction quality, the South-North Water

Fig. 6.5 South-to-north water diversion middle route crossing the yellow river project. *Source* Xinhuanet

Diversion Project Construction Inspection Brigade, Supervision Center, and other units are established to focus on inspecting quality issues. Rapid quality certification is carried out to scientifically assess the impact of quality problems on the structural safety of the project and define the nature of quality problems. Implement a quality accountability system to increase the penalty for quality problems.

(4) Engineering operation and maintenance phase

Engineering activities are group activities for building artificial nature, operating artificial nature, and decommissioning artificial nature. In the operation and maintenance stage of the project, the using unit of the project is the main bearer of social responsibility. The user unit can support improving the ecological and social environment while gaining economic benefits through engineering results. The public users in this stage also have certain supervisory responsibilities for engineering projects and products.

Based on the above whole life cycle analysis, the social responsibility of each member in the engineering community differs in different stages of engineering activities, thus, engineering social responsibility is a multi-stage, multi-subject, and multi-level concept, as shown in Fig. 6.6.

Engineering activity subjects bear certain social responsibilities, and their behavior is subject to the necessary constraints and norms. Engineering social responsibility is not only an inevitable requirement to meet the sustainable development of engineering but also an inherent requirement to build a harmonious society. On the one hand, as a kind of collective and comprehensive practical activities, engineering activities integrate various factors such as economy, science and technology, society, culture, and ethics, which require the division of labor and cooperation of all members of the engineering community. On the other hand, as a social existence, engineering activities cover various aspects of engineering projects, such as decision-making and project establishment, planning and design, operation and maintenance and effect evaluation, etc. All these series of aspects involve social factors. Therefore, any engineering project exists in a certain social structure and social relations. It is subject to certain time and space restrictions and limitations. Therefore, the social value evaluation of engineering projects cannot be neglected when analyzing

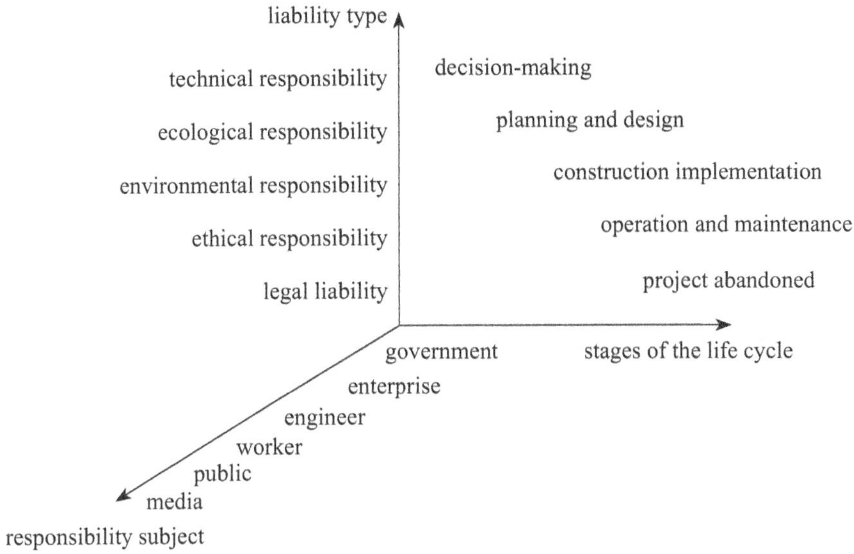

Fig. 6.6 Engineering social responsibility from a multi-dimensional perspective

engineering activities, especially the systematic analysis of social risks and various social impacts that may be generated by engineering projects [47].

6.3.3 *Evaluation of the Social Value of the Project*

As a market economy activity, engineering community members are "rational economic people." They may pursue the maximization of personal interests, pay less attention to the social benefits of engineering activities, and transfer the environmental and social losses caused by the project to others and the future, resulting in negative externalities and directly affecting the social value of the project. Therefore, engineering community members should be fully aware that they are both "economical people" and "social people." Engineering management must pay attention to social value objectives, fully consider public interests and emotional requirements, and assume the necessary social responsibility and moral obligations to nip possible conflicts of interest in the bud and promote the harmonious development of society [48]. The evaluation of engineering social value is indispensable for the realization of engineering social value objectives.

6.3.3.1 Concept of Project Social Value Evaluation

(1) The connotation of engineering social value evaluation

Unlike economic, environmental and ecological impact evaluation, social value evaluation insists on being "people-oriented." It emphasizes social analysis and public participation, incorporates the impact of engineering activities on people and society into the value evaluation framework, and comprehensively applies theories and methods of sociology, anthropology, economics, management, and other related disciplines to identify, monitor, and evaluate various social impacts or social risks through systematic investigation and collection of factors, data, and information related to engineering activities. Carrying out social value evaluation of engineering is conducive to optimizing the implementation plan of engineering construction, improving stakeholders' participation and support of engineering activities, avoiding or reducing social risks of engineering activities, reducing social conflicts, maintaining coordination between economic and social development, better promoting harmonious development of human, nature, and society, and realizing "the unity of heaven and man" [49, 50].

To comprehensively reflect the interaction between engineering activities and social development, engineering social value evaluation not only focuses on the sustainability of the development of engineering itself but also highlights the harmonious operation of the whole society. It systematically analyzes engineering activities' various social impacts or risks (direct and indirect, short-term and long-term, real and potential, positive and negative) [51]. Theoretically, social value evaluation should be carried out for all engineering activities. Considering the high cost involved in engineering social value evaluation, in practice, comprehensive and systematic social value evaluation is generally carried out only for those major engineering activities with obvious social development goals that have complex social factors, long-term social impacts, major social conflicts, significant social risks and severe social problems [47].

(2) Subject and object of engineering social value evaluation

First, the evaluation subject. In the process of engineering social value evaluation, the evaluation subject is required to stand in the society's position as a whole and take the social value standard as the basis for judgment, which is different from individual evaluation and group evaluation. In the evaluation process, the evaluator should pay attention to the organic integration of "individual identity" and "social identity" and give a fair and objective social value evaluation.

Second, evaluation object. The evaluation object of engineering social value is diversified engineering social value. We need to focus on the following aspects to systematically analyze the characteristics of engineering social value evaluation objects: First, the macroscopic and multidimensional evaluation objectives. Engineering social value has multiple social development goals, specifically economic growth, social stability, income increase, unemployment reduction, environmental protection, and cultural heritage. Secondly, the difference in evaluation criteria. Due to the dynamic nature of the environment, the complexity of influencing factors, the multidimensionality of social goals, and the diversity of social benefits, the evaluation standard of engineering social value vary somewhat among different industries and regions. Thirdly, the participation of stakeholders is emphasized. Stakeholders of engineering activities refer to the groups that can influence or be influenced by engineering activities. In the process of engineering social value evaluation, the relationship between engineering activities and stakeholder groups should be analyzed in depth and systematically. The analysis should be carried out according to the influence path of "engineering activities bring influence → stakeholders are influenced → stakeholders react → society has an influence on engineering activities → engineering activities are influenced." The mutual adaptability between engineering activities and local society and the social sustainability and risks of engineering activities can be analyzed by referring to this logical framework.

6.3.3.2 Content of Project Social Value Evaluation

From the connotation of engineering social value, it can be seen that it is a prerequisite and basis to improve the social value of engineering for all engineering community members to effectively fulfill their social responsibilities. Engineering social value evaluation is to maximize the overall social benefits of engineering activities as the starting point, analyze the social impacts of engineering activities (such as promoting economic and social development, social equity, and sustainable development) in an in-depth and systematic manner, objectively assess the mutual suitability of engineering activities and local environment, systematically identify the various adverse impacts and social risks that may arise in the process of engineering activities, and better promote the harmonious development of man, nature and society.

(1) Social impact analysis

First, it is necessary to define the regional scope of the social impact of major engineering activities to analyze the social impact of major engineering activities. Secondly, to identify the individuals or organizations affected by engineering activities, including direct and indirect effects, active and passive effects, real and potential effects, etc. Again, analyze various possible social impact effects, including direct and indirect, primary and secondary, positive and negative, etc. In addition, engineering social impact analysis can be divided into national, regional, and community levels in terms of content, specifically including positive and negative effects on income level, employment level, quality of life, infrastructure, public services, social equity, etc. [52].

(2) Analysis of social adaptability

Social adaptability analysis focuses on examining the mutual adaptability relationship between engineering activities and the local social environment, human conditions, customs, and habits, as well as the degree of participation and support of the local government, people, and various institutions in engineering activities. Specifically, it includes: firstly, using the stakeholder analysis method to objectively evaluate the importance and influence of each stakeholder, determine the main stakeholders of engineering activities, and analyze the interest demands of stakeholders; secondly, analyzing the attitudes and recognition degree of different stakeholders towards the construction and operation of the project, analyzing the attitude of local government towards the project and the strength of collaboration and support, then analyzing the local people's attitude towards the project and the degree of their participation, and also pay attention to the attitude and support of other local social organizations (e.g., environmental protection organizations); third, analyze whether the social environment, human conditions, religious beliefs, ethnic relations, and custom systems in the area where the project is located can meet the construction and operation needs of the project [53].

(3) Social risk analysis

Social risk analysis identifies and evaluates various social risk factors that may affect engineering activities. It analyzes the causes that lead to significant social risks in engineering activities and propose management measures to cope with social risks, to reduce or eliminate possible social conflicts and potential social risks in engineering activities. The analysis of engineering social risks needs to pay special attention to the following aspects: there are relatively serious social equity problems in the area where the project is located; the area where the project is located faces large-scale unemployment due to industrial structure upgrading and adjustment; engineering activities are expected to have significant negative impacts, such as involuntary migration, environmental pollution, ecological damage, etc.; engineering activities will change the behavior, customs, and values of local residents. Successfully implementing engineering activities is highly dependent on community support.

6.3.3.3 Methods for Evaluating the Social Value of Engineering

Engineering social value reflects the responsibility and contribution of engineering to society. Unlike engineering technology program evaluation, economic value evaluation, and ecological value evaluation, engineering social value evaluation not only focuses on the sustainability of the project's own development but also requires a strategic level analysis of social development, comprehensive and systematic analysis of the impact of engineering activities on local economic and social sustainable development, and emphasizes stakeholder participation. Based on the existing theoretical and methodological evaluation systems, engineering social value evaluation emphasizes the comprehensive influence of multiple factors, the selection evaluation benchmarks, and the analysis of stakeholders.

(1) Multi-attribute integrated evaluation method

Due to the fact that there are many influencing factors of engineering social value, including direct and indirect influences, current and future influences, current and potential influences, especially some qualitative evaluation contents which need to use experts' knowledge and experience, there exist fuzziness and uncertainties.

Therefore, engineering social value evaluation is a typical multi-attribute, comprehensive evaluation problem. In simple words, fuzzy hierarchical analysis is an evaluation method that combines hierarchical analysis and fuzzy comprehensive evaluation methods. The method integrates the advantages of both hierarchical analysis and fuzzy comprehensive evaluation methods, determines the evaluation index weight set through hierarchical analysis, combines qualitative analysis with quantitative analysis, and makes the evaluation results more scientific and reasonable.

For the issue of engineering social value evaluation, first of all, a set of systematic and complete comprehensive evaluation index systems of engineering social value is constructed by following the principles of systems, scientificity, relevance, feasibility, and a combination of quantitative and qualitative. Secondly, use the hierarchical analysis method to calculate the weights of each evaluation weights. Expert surveys and statistical analysis can be adopted to evaluate the qualitative and quantitative indicators, and the evaluation value is normalized to obtain the fuzzy evaluation matrix. Finally, according to the hierarchical relationship of evaluation indicators, a fuzzy hierarchical analysis model is established, and the indicators at all levels are assembled to get comprehensive evaluation results.

(2) "Baseline" survey method

In order to analyze the impact of engineering activities and evaluate the social value of engineering scientifically and objectively, it is necessary to provide a basis for comparison. The "baseline" survey method is a comparative analysis method. It adopts various survey methods and means to conduct a comprehensive and systematic survey on the current situation of regional economic and social development affected by engineering activities before the implementation of engineering activities to provide a comparative basis for the subsequent evaluation of engineering social value. In the process of using method, the "baseline" survey method can systematically sort out the historical process of economic and social development of the regions affected by engineering activities by searching the literature, collecting relevant data and statistics, and comprehensively grasping the basic conditions and characteristics of local society, economy and culture. It can include demographic characteristics (residents' income, employment, ethnicity, religious beliefs, customs, and values), relevant infrastructure situation (culture, education, health, housing, transportation), and community social structure (social organization of production, community organization). At the same time, special surveys can be conducted through expert interviews, questionnaires, etc., to understand the expectations, attitudes, recognition, and participation of each target group in the project.

As a comparative analysis method, the baseline survey method includes whether there is a comparison (with or without engineering) and before-and-after comparison (before and after engineering construction). Whether the baseline is accurate or not is very important. Considering the long construction and operation period of large engineering activities, during which changes in policies, environment, institutional mechanisms, and the start of construction of other projects will cause changes in socio-economic conditions, making the conclusions of the initial survey change.

The project may not cause such changes, so the social impact of engineering activities should be considered from a dynamic perspective, with special attention to the "baseline adjustment. In addition to the initial "baseline" survey, information should be collected to timely assess the possible changes of local socio-economic conditions within the project impact time frame, and "baseline adjustment" should be made from a dynamic development perspective [54].

(3) Stakeholder analysis method

The stakeholder analysis method mainly identifies various stakeholder groups that affect or are affected by engineering activities, clarifies the relationship between engineering activities and different interest groups, and analyzes the reactions of different stakeholder groups to engineering activities. In specific engineering management practice, we can construct a stakeholder group analysis table to systematically reflect the mutual relationship between engineering activities and stakeholders from four aspects (see Table 6.2). These four aspects are the classification of stakeholders, their interests in engineering activities, their reactions to engineering (attitudes and requirements), and their influence on engineering (size and manifestation).

(1) Classification of stakeholders. According to the definition of stakeholders to comprehensively sort out the stakeholders of engineering activities, and according to the influence path of engineering on different stakeholders, divided into two major groups in terms of direct influence and indirect influence. In addition, the two groups of stakeholders can be subdivided to account for the differences in characteristics of different stakeholders (such as gender, age, ethnicity, occupation, religion, customs, social status, personal ability, etc.).

(2) Analysis of the different stakeholders. The project's impact on different stakeholder groups varies greatly, with some people benefiting and some people losing, thus forming different stakes, which in turn will affect the interests and attitudes of

Table 6.2 Analysis of stakeholder groups

Stakeholder classification	Interest in engineering	Reaction to engineering	Influence on the engineering project
...			

Source Ma [55]

different stakeholders towards the project. The beneficiary group will hold a favorable attitude towards the project and actively support and participate in the project. In contrast, the damaged group will hold an opposing attitude toward the project. The reaction and attitude of both beneficiary and damaged groups to the project will depend on the compensation package.

(3) Analysis of the influence of different stakeholders. Due to the differences in power and status, resources and ability endowment, different stakeholders have different influences on the project activities. Here, special attention should be paid to protecting the interests of vulnerable groups such as women and ethnic minorities to avoid social conflicts and contradictions caused by social injustice.

6.3.3.4 Case Study of Social Value Evaluation of South-North Water Transfer Project

The cross-sectoral nature of engineering activities and the reality of diversified interest subjects determine that the project contains multiple social values. The South-North Water Transfer Project is by far the largest, longest distance, most populous, and widest beneficiary water transfer project in the world. The project spans four major basins, including the Yangtze River, the Huai River, the Yellow River, and the Hai River, generating huge benefits in water supply, drought relief, shipping, and flood removal. The social value of the South-North Water Transfer Project will be analyzed here from four aspects, including promoting regional socio-economic development, enhancing the environmental and ecological value of the project, setting an example of respecting history and preserving cultural relics, and achieving harmonious relocation [56].

(1) Strongly promoting the economic and social development of the region where the project is located

The South-North Water Diversion Project is divided into three lines, namely East, Central, and West, to solve the water shortage problem in the northern part of China, especially in the Yellow and Huaihai basins. It builds a large water network of "north–south deployment and east–west mutual aid." The overall water supply area of the project controls an area of 1.45 million square kilometers, and the project plans to transfer 44.8 billion cubic meters of water annually, benefiting about 500 million people in the planning area. With improved water resources conditions, China's northern industrial and agricultural output will increase by 50 billion yuan annually. It will create more opportunities and space for regional economic and social development, industrial structure adjustment and upgrading, and the urbanization process. In addition, the South-North Water Diversion Project will also generate various comprehensive benefits such as agricultural water supply, flood control, shipping, and water drainage benefits.

Fig. 6.7 Landscape engineering along the south-to-north water transfer project. *Source* Xinhuanet

(2) Huge environmental and ecological values

After implementing the first phase of the South-to-North Water Transfer Project, the East and Central lines, it effectively alleviates the groundwater overdraft problem in the receiving water area and protects the carrying capacity of local water resources. It also curbs the trend of water ecology deterioration in the northern region and helps improve and restore the ecological environment. The South-North Water Transfer Project has learned the basin's pollution management lessons. The project actively explored pollution management and environmental protection and proposed the general principle of "water conservation before water transfer, pollution treatment before water supply, and environmental protection before water use," which has led to a new way of scientific planning, advance pollution treatment, and ecological protection. It established a benchmark for pollution management in other basins. It has greatly improved the water quality and water sources along the project line and the surrounding environment along the project line, making the South-North Water Diversion Project more extensive in terms of benefit area and fully reflecting the project's social value (Fig. 6.7).

(3) Setting an example of respecting history and culture

The South-to-North Water Diversion Project, its East and Central lines, are located in areas with a high concentration of Chinese culture and history and the cultural relics are highly concentrated, so the protection of cultural relics has attracted great attention from society. The protection of cultural relics in the South-to-North Water Diversion Project follows the principle of "key protection and key excavation–beneficial to both capital construction and cultural relics protection" and strives to protect the important historical and cultural heritage of the country and minimize the damage to cultural relics and sites caused by the project construction. During the demonstration stage of project selection, the national cultural heritage protection sites were deliberately bypassed. They include Anyang Yinxu and Zhenghan Forbidden City in Henan Province, and the provincial cultural heritage protection sites such as Beipinggao, Shanyang City, and Jiangwu City. A long section of the Eastern Line uses the ancient canal to transport water after repairing, dredging, and widening. It fully maintains the shipping, water transport, ecological, and landscape functions of the Grand Canal. It is also an important means to protect the economic and social value of the heritage. During the implementation of the project, the western Zhou city site of Chenzhuang was discovered in the eastern line, and the channel was rerouted for it. Engineering construction "makes way" and "reroutes" for important cultural relics, which fully reflects the responsibility for history and culture and becomes a model of modern engineering respecting the culture and protecting cultural relics, which is an important embodiment of the social value of the project.

(4) Effectively solving large-scale migration and achieving the goal of harmonious relocation

The South-to-North Water Diversion Project has large-scale immigrants. The Danjiangkou Reservoir Dam Raising Project in the middle line involves the resettlement of 345,000 people. To solve the resettlement problem scientifically and reasonably, we follow the principle of "people-oriented and pragmatic," formulate targeted migration policies and related support policies, improve the compensation standard for land acquisition, combine resettlement with local economic and social development, and raise funds through multiple channels by local governments to improve the infrastructure construction of new villages for migrants. We also increase the supply of public medical care and education services so that immigrants can integrate into the local society as soon as possible and achieve the goal of harmonious relocation [36].

6.4 Realization and Enhancement of Engineering Value

The core meaning of engineering activities is value creation. Engineering and value are closely connected. Engineering activities carry value realization, and the development level of engineering activities reflects the value pursuit and value creation ability of engineering subjects. As discussed in the previous section, engineering value refers to the economic value of engineering. It includes the scientific and technological value, natural value, social value, talent value, and cultural value of engineering, showing diversified characteristics [57]. In the new period, in the diversified value system of engineering, the value guidelines of different dimensions may be coordinated or in conflict with each other. There may even be a state of partial coordination and partial conflict. The realization and enhancement of the diversified value of engineering cannot be achieved without a series of engineering management innovation activities and innovative thinking.

The essence of engineering activities lies in innovation, including time, content, and degree. Innovation is the fundamental driving force of engineering development, which is conducive to achieving harmonious and sustainable engineering development. Engineering activities that leave the guidance of innovative views can only be low-level repetitions. This section takes the innovative view of engineering management as the guide and elaborates on a series of innovative activities to promote the realization and enhancement of diversified engineering values from two aspects of engineering management top-level design and engineering management practice.

6.4.1 *Top-Level Design of Innovative Engineering Management*

6.4.1.1 Innovative Engineering Management System

The Sichuan-East Gas Transmission Project is a key project of the 11th Five-Year Plan of the country. A major strategic decision is made considering the objective requirements of comprehensive, coordinated, and sustainable development of humankind and nature. To ensure the project's smooth development, the "Project Management Model under the Integrated Leadership of the Command Department" has been built, which implements integrated decision-making, integrated organization, and integrated control of engineering activities, realizing institutionalization, standardization, and standardization of engineering activities. In the past three years of engineering construction, we have created 51 domestic engineering technology and safety and environmental protection standards for developing and constructing ultra-deep and highly acidic gas fields and created 62 new records and high indexes. It formed a construction technology series for building long-distance and large-diameter pipelines in complex mountainous areas and areas with dense water networks and gradually built a corporate standard system for exploration and development, production and construction, transportation and distribution, and safety and environmental protection of ultra-deep and highly acidic gas fields, and gradually developed a natural gas industry chain integrating exploration, testing, development, production, purification, transportation, sales, and use.

The engineering management system aims to realize the core functions of engineering management, which mainly includes: engineering management theory, engineering management methods, engineering management organization, engineering management tools, and engineering management practice. The engineering management system can be regarded as a bridge connecting engineering management theory and engineering management practice. Its construction and innovation are about: taking advanced engineering management concepts as a guide, changing engineering development mode, innovating engineering management theory and method, guiding engineering management practice activities, and improving engineering practice effect [28].

(1) Updating the concept of engineering management

The process of engineering activities is an innovation process, and engineering management is also innovation management, which mainly includes: innovation in many aspects such as engineering management concept, organization mode, system design, technical change, theory, and method. Engineering management activity is a dialectical process from practice to understanding, from understanding to guiding practice, and then to practice and to understand again; thus, engineering management practice is both the logical starting point and final destination of the engineering management theory system. Since ancient times, people have taken the central position in engineering activities, and they are both the direct promoters and the final destination of engineering activities. "Everything for people, everything depends on people" is the centralized embodiment of this concept [28]. To deeply reveal the essence and connotation of engineering activities, it is necessary to rely on developing an engineering management philosophy, an effective weapon to guide engineering practice [58]. With the deepening of theoretical research and the development of engineering practice, engineering management ideas are gradually developing from pure engineering thinking to philosophical and ethical thinking, forming engineering science, engineering ethics, and engineering philosophy. Engineering thinking mainly uses systems theory and methods and optimization theory and methods to organize activities and achieve engineering goals. Ethical thinking mainly emphasizes fairness and morality, responsibility and credit, and pursues the balance of interests and risk sharing of engineering communities. Philosophical thinking mainly uses dialectical, development, and systems views to coordinate the process of engineering management practice and achieve sustainable development of engineering activities. It is necessary to establish a new engineering management concept with scientific development concept as the guiding ideology, engineering philosophy as the overall leader, and sustainable development as the ultimate goal. This new engineering management concept must reflect "people-oriented," the core of the harmonious development of man and self, man and nature, man and society. It must reflect the systems concept, ecological concept, values, social concept, ethical concept, and cultural concept of engineering [59]. "People-oriented, unity of heaven and man, collaborative innovation and harmony" is certainly both the content of engineering activities and the core of the engineering management concept [28].

The engineering management concept runs through the whole engineering activities, penetrates all stages and links of engineering activities and its management, is the starting point and destination of engineering activities, and the soul of the whole engineering activities. Only advanced engineering management concepts can help us consciously follow economic, natural, and social rules, as well as moral, fairness, and justice, to achieve sustainable engineering development.

Throughout the development history of engineering activities, with the continuous development of engineering management theory and engineering management practice, the engineering management concept is also in the process of innovation and growth. It can be roughly divided into three historical eras: listening to God, conquering nature, and harmony between heaven and man [60]. "Listening to God" is mainly the engineering management concept formed by ancient engineering, and it is a fatalistic management idea. It believes that people and engineering activities can only be arranged by nature and cannot form an effective human dynamic activity of understanding nature and building artificial nature. The concept of "conquering nature" is a modern engineering management concept, which embodies the desire of man to overcome and conquer nature, opposes man with nature, exaggerates the subjective initiative of man, and overemphasizes the role of engineering in conquering nature for man. "Harmony between man and nature" is the modern engineering management concept that fully expresses the harmonious development of man and nature and is the basic view of scientific development. Under the guidance of different engineering management concepts, engineering management practices often show different fates. Under the guidance of the engineering management concept of "listening to God," it is difficult to achieve significant engineering practice results because it cannot give full play to people's subjective initiative. Naturally, it cannot serve people well, neither "all for people" nor "all rely on people." Under the guidance of the "conquering nature" engineering management concept, it leads to the irrational expansion of engineering activities and increases the development of natural resources and the ecological environment, which inevitably brings about the predatory exploitation of natural resources and great damage to the ecological environment. Humans have failed to conquer nature, behaved more vulnerably, and will continue to be punished by nature. The "harmony between man and nature" engineering management concept not only reflects the main active position of man but also emphasizes that the use of nature by man can only be limited. The purpose of human engineering activities is not to conquer nature but to achieve harmony between man and nature and the harmony between engineering and nature so that the multivalued objectives of engineering can be realized in the end.

(2) Transformation of engineering development mode

Modern engineering management is the process of scientific and rational planning, organization, coordination, and control of engineering practice using advanced management concepts, standardized procedures and methods, and modern management techniques. To change the way of engineering development, it is necessary to establish a correct idea of engineering development and transform the concept of engineering management in the first place.

Engineering practice should be examined macroscopically from different angles, from the perspective of philosophy, to transform the idea of engineering development [55]. We need to explore the economic affordability and scientific and technological matching ability of engineering activity bearers from the economic and technical perspectives, examine the impact of engineering activity on the social public and the harmony between engineering and society from the perspective of the social public, examine the impact of engineering activity on natural ecological environment and the harmony between engineering and nature from the perspective of natural ecology, and to examine the fairness and justice of engineering activity from the standpoint of government departments. The concept of harmonious engineering management includes the harmony between engineering and nature, the harmony between engineering and society, and the harmony between engineering communities. In the past, engineering communities had different statuses in engineering. They possessed a different amount of engineering information, easily leading to adverse selection and moral risks due to information asymmetry. In addition, there was an antagonistic relationship between engineering communities, resulting in an imbalance of interests of relevant parties, causing the inefficient allocation of market resources and destroying the principle of social fairness and justice, which made it difficult to realize the diversified value of engineering. The concept of harmonious engineering management requires the engineering community to jointly participate in engineering management to realize the goals and interests of all parties involved, share the results of engineering activities and share the risks of engineering activities.

From the perspective of existing engineering practice, on the whole, China's engineering activities have made great achievements and provided a strong impetus for national economic and social development, raising the level of economic development, improving national livelihood, and enhancing national competitiveness. However, most of my country's engineering activities lack the ability of independent innovation, repeating the high-input, high-energy-consumption, high-pollution engineering development model and road, resulting in a huge waste of resources, causing vicious damage to the environment and weak sustainability of engineering activities. Even some projects, such as face-saving projects, tofu-dreg projects, etc., seriously contradict the spirit of the scientific concept of development, resulting in extremely bad social impact. For this reason, it is necessary to change the development mode of engineering further, promote the transformation of engineering activities from extensive to intensive, and take the new road of industrialization. Firstly, with the support of scientific and technological progress, we need to improve the level of engineering management, promote the level of engineering practice, improve the quality and efficiency of engineering, and realize engineering activities to serve people better. Secondly, introduce "circular economy" and "green economy" into engineering management, achieve the goals of saving resources and protecting the environment, reduce the negative external effects of engineering activities and the damage they bring to the natural environment and ecosystem, and realize the harmony between engineering and nature. Thirdly, improve the participation of the public in engineering, bring into play the positive external effects of engineering, promote social equity, and realize the harmony between engineering and society.

(3) Innovation of engineering management theory

China has a long history of engineering management practice, but it has stayed in the traditional management mode and low-level development stage for a long time. With the change in social system, scientific and technological progress, and socio-economic development, engineering practice's scale, level, and complexity are continuously changing. Engineering is an organized and purposeful group activity, a complex system with multiple objectives, variables, parameters, and disturbances characterized by huge investment, large scale, complex structure, integrated function, long time, and uncertainty [61]. Modern engineering also has the distinctive characteristics of being "based on high technology and driven by innovation," integrating various resources, emerging technologies, and creativity, developing in the direction of knowledge-intensive and technology-intensive, with significant integration and constructive characteristics. With the development of engineering management practice, the existing engineering management theories can no longer meet the needs of modern engineering management practice. Innovating engineering management theories and methods and establishing a corresponding engineering management theoretical system is necessary.

In the new period, especially since the reform and opening up, China has carried out many large engineering practice activities and accumulated rich engineering management experience. These engineering management experiences have been summarized and sublimated by the theoretical and engineering circles, forming a large number of application theories of concepts, models, and methods with Chinese characteristics and producing some influential engineering management theories, such as: "lean, flexible and agile" production system theory, meta-decision theory and application, intelligent interactive, integrated decision support system. However, a complete engineering management theory system has not yet been established [62]. The engineering management theoretical system is based on people's profound understanding of engineering management practical activities through systems and meticulous thinking activities, regulating the constituent elements of engineering management theory and its combination mode. Its core function is to clarify the development of engineering management theory research, define the content and scope of engineering management theory, reveal the internal logic between the elements within engineering management theory, and guide the healthy and efficient development of engineering management practice.

Innovation and development of engineering management theories and the establishment and improvement of engineering management theoretical systems need to rely on the depth of engineering management practice, application of engineering management theoretical methods and technologies, and support of basic theories and methods under the overall guidance of scientific development concept and engineering philosophy. The development of engineering management theory and the establishment of an engineering management theoretical system cannot be separated from the guidance of engineering management philosophy. The engineering management philosophy is to think about the fundamental problems of engineering management activities, explore the basic issues and basic laws of engineering management activities, and research engineering management ontology, engineering management epistemology and engineering management methodology through engineering dialectical view, engineering development view and engineering systems view. The philosophy of engineering management mainly explores the essence of engineering from a philosophical angle. It explores the concept, worldview, and methodology of engineering management, which helps to form the theory of engineering management and guide and promote the development of engineering management practice. In engineering management activities, practice is the logical starting point. Without engineering management practice, engineering management theory will not be born, and it is difficult to establish a perfect engineering management theory system. Engineering management theories, methods, and technologies are formed in engineering management practice, crossing with basic theories and methods, such as quality management, safety management, etc., which summarize and enhance engineering management practice and guide engineering management practice. The basic theories and methods, including engineering, management, economics, mathematics, and information science, provide the necessary basic theories and methods to support the generation, development, and improvement of engineering management theories.

6.4.1.2 Innovation Project Management System

Qinghai-Tibet Railway is a landmark project to implement China's western development strategy and one of the four major projects in the new century. The entire project took nearly 60 years since the decision was initially made in the 1950s, and the whole line was opened to traffic in 2006. The Qinghai-Tibet Railway Project has adopted a strict contract-based management system, using contracts such as design and consulting contracts, survey and design contracts, scientific research contracts, construction contracts, supervision contracts, procurement contracts, labor contracts, and insurance contracts. The main methods include a well-planned contract system, revision and improvement of contract terms and conditions, and comprehensive ensuring of contract objectives. In the contract management process, we clarify departmental responsibilities to improve the efficiency of contract management, conclude contracts per the law to ensure the contract's legality; strictly review the contract to ensure its completeness. The project management implements dynamic management and contract tracking systems. The project construction mainly overcomes the three major problems of "multi-year permafrost, plateau hypoxia, and ecological fragility," becoming the world's highest altitude and longest line plateau railroad.

(1) Clarify the property rights system of the project

To a certain extent, the innovation of an engineering management system is the process of redistribution of responsibilities, rights, and benefits of the project, which is prone to the lack of management subjects, absence of supervision, and policy defects. Therefore, it is necessary to start by clarifying property rights, clarifying the project ownership and use rights through the property rights division, determining the management subject or legal entity subject, improving the efficiency of project management, and achieve the goal of multiple values of engineering [63].

The engineering community mainly includes the owner, contractor, supervision unit, engineer, financial institution, government department, public, and other stakeholders. The owner (construction unit), as the financier and initiator of the project, is the owner of the project rights and the person in charge of the project. The owner assumes the responsibility and risk of the project investment and has the right to decide the overall vision of the project, the functional positioning of the project, the scale of the project investment, the goal of the project management, the operation mode of the project and other engineering communities of the project. According to the principal-agent theory, project management can carry out incentive-compatible system design from bidding, construction, and operation to reduce rent-seeking, rent-setting, and project quality and safety issues [64]. The incentive-compatible institutional procedure involves the interest distribution between project issuers and agents. The latest engineering management theory believes that the goal pursued in engineering construction is not simply the realization of project cost, quality, schedule, and safety goals but further the satisfaction of each engineering community and the unification of responsibility, power, and benefit of engineering community [61].

(2) Optimization of engineering organization structure

The organization of engineering is to optimize the allocation of resources, improve management efficiency, and achieve engineering goals through certain organizational principles; so that each engineering community can form a combination through the division of labor, collaboration, and the division of different responsibilities, rights, and benefits. It includes engineering organization structure and engineering organization behavior. Engineering organization structure should consider establishing lean, rational, and efficient static systems and the stable interrelationship established by the internal composition of the organization and each part. Engineering organization behavior aims to study dynamics about how to bring out the best effect of the organization. An authoritative leader is needed in engineering organizations to lead, focus on, motivate, and guide organizational members, communicate effectively, resolve various conflicts and disputes, and unify the thoughts and actions of the organizational members to achieve an organizational vision and engineering goals. It is necessary to consider the form of engineering organization structure and behavior model from aspects of external and internal factors and to design and optimize the engineering organization structure based on engineering goals, engineering content, engineering organization goals, and engineering organization work content.

As far as the form of engineering organization structure is concerned, it can be roughly divided into functional, project, matrix, and composite types. A functional engineering organization structure is composed of a team in a single functional area and set up corresponding functional departments so that the engineering organization becomes part of the regular organization. A project-based form of engineering organization structure is a form of organization structure in which the engineering organization form is independent of the functional engineering departments, and the engineering organization itself is independently responsible for its main work. Matrix engineering organization structure is a structure between functional and project type. It is divided into weak matrix type, strong matrix type, and balanced matrix type. A compound type is a form of engineering organization structure compounded by functional, project, and matrix types. Each kind of organization structure form has its prerequisites, advantages, and shortcomings, and there is no universal organization structure form that can be adapted to all projects. Table 6.3 gives the comparison results of different engineering organization structure forms. For projects under construction, the best organization form must be chosen according to its characteristics. In addition, according to the changes in objective and subjective conditions faced by the project, the form of the engineering organization structure should be adjusted or even reengineered. When reengineering the engineering organization, it is necessary to consider carefully: firstly, to take into account the interests of the engineering community while not harming the interests of the owner; secondly, the new organizational structure has higher management efficiency and can solve the problems existing in the original organizational structure; thirdly, to grasp the timing of adjusting the organizational structure to maintain the continuity of the project progress.

Table 6.3 Comparison of engineering organization structures

Comparison items	Functional style	Matrix	Project style
Customer demand	Failure to reflect and satisfy in time	Quick response and satisfaction	Quick response and satisfaction
Resource utilization	Reasonable use is not available	Reasonable use is available	Repeated resource allocation and waste
Organization size	Smaller	Medium	Larger
Use of technology	Standard	Complex	Newer
Complexity	Lower	Medium	Higher
Uncertainty	Lower	Higher	Higher
Information transfer	Lack of communication	Complex information communication and information loop	Information communication

(3) Innovative project management mode

Innovative engineering management concepts can promote the development of engineering management theory and innovation of engineering management systems but also promote the improvement of the engineering management system and change of engineering management mode. For example, it is possible to develop from competitor to partnership, from scientific development concepts to sustainable engineering development [61]. Summarizing the development history of engineering management, we can find its general rule. It generally went through the main development stages, such as the budding period, empirical management, administrative management, project management, and comprehensive innovation. Engineering management gradually moved from experience to science and from tradition to modernity.

It can be seen from ancient Chinese engineering construction management practices that engineering management and administrative management are integrated [65]. Ancient China's social and economic system was the most typical agricultural economic society, and administrative management was the most important management mode in social management. Therefore, any project construction and its related management activities are often carried out in the name of the government to achieve the corresponding results. The members of its think tank mainly consisted of scholars of Confucianism and Taoism. The administrative management process fully reflected their main ideological views, which influenced the results of ancient engineering management practice activities. From the founding of New China to the 1980s, due to the implementation of the planned economy system, projects were mainly invested in and constructed by the state, and government departments were both owners and administrative departments. This management mode mainly has shortcomings, such as a single investment subject and a lack of supervision mechanism. Since the reform and opening up, especially since the establishment of the market economy system, the basic engineering management system has been gradually formed in China, mainly including a legal person responsible system, bidding system, investment and financing system, project supervision system, and contract management system. It has changed the planning mode of managing engineering by the government and greatly improved the level and effectiveness of engineering management.

To a certain extent, the engineering management system determines the engineering management mode. The engineering organization management mode can be roughly divided into four types: government-led, government-led + enterprise participation, government-led + enterprise subject, and enterprise-led. The selection of different engineering organization management modes is mainly influenced by: economic system, social environment, technical level, and engineering objectives. The scientific and rational selection of engineering organization management modes is beneficial to realizing engineering goals [66]. Different modes of organizational management differ in mode characteristics, operation mechanism, and degree of marketability. For details, see Table 6.4.

With the development of engineering management theory and practice and the change in the engineering management system, there is a need to continuously explore the emerging engineering management mode that meets the characteristics

Table 6.4 Comparison of engineering organization management mode

Comparision items	Government-led	Government-led + enterprise participation	Government-led + enterprise subject	Enterprise-oriented
Mode characteristics	Administrative command and technical coordination, the government is the owner	Administrative command and technical coordination, with the government as the owner and large enterprises as contractors	Administrative supervision and technical coordination, large enterprises as owners	Administrative supervision and technical coordination, large enterprises as owners
Operation mechanism	Administrative command and technical command are parallel, and the government is responsible for the entire engineering activity	The state takes the lead and the ministries and commissions take charge, and the government contracts the project to large enterprises	Ministries and commissions are responsible for guidance and coordination, and large enterprises are responsible for the entire engineering activities	The government approves engineering projects, and large enterprises are responsible for the entire engineering activities
Degree of marketization	Low	Low	Medium	High
Classic case	Two bombs and a one-star project	Manned space engineering	Three Gorges project, south-to-north water transfer project	Coal mining engineering, petrochemical engineering

of the times. The innovation of the engineering management mode aims to coordinate the interests of all parties of the project, stimulate the efficiency of engineering management, improve the engineering management level, and maximize the value of the project.

6.4.1.3 Innovative Engineering Management Technology

Petrochemical projects are managed integrated, adopting a new management model with Chinese characteristics. That is an integrated project management team + general engineering contractor + construction supervision contractor, strengthening the engineering construction concept of "eliminating bottleneck constraints mainly by technical transformation" and promoting engineering science and technology innovation. By combining engineering technology development with major petrochemical engineering construction, a series of complete sets of technologies have been successfully developed and applied in many contemporary world-class petrochemical projects, such as China National Petroleum Oil and Gas Development Technology and China National Petroleum Gasoline Quality Upgrade Core Technology. Engineering management technology innovation of petrochemical engineering depends on engineering technology innovation, while the purpose of engineering management technology innovation is to promote engineering technology innovation further and realize the goal of engineering diversification value.

(1) Rely on independent innovation to promote the progress of engineering science and technology

The rapid development and independent innovation of engineering science and technology are conducive to the promotion of resource-saving and environmental protection-oriented industrial structures, economic growth patterns and residents' consumption patterns, and are conducive to the construction of ecological civilization, which are decisive for the ultimate realization of scientific and sustainable economic and social development [58]. Engineering science and technology can transform scientific knowledge into real productivity, which directly improves the output efficiency of engineering activities and realizes the engineering value goal, and can promote the scientific and technological progress of the whole society and promote economic and social development.

Engineering science and technology progress have a variety of realization channels. Relying on independent innovation to promote engineering science and technology progress is the historical mission given to China's engineering activities, which has important strategic significance for China's economic development and national security. We should always emphasize major science and technology and strive to build an independent innovation system that integrates "industry, academia, research, and application." To master the unique core technology with independent intellectual property rights, effectively enhancing the independent innovation capability of engineering activities and driving the overall improvement of engineering science and technology. Engineering management has "industrial relevance." It is mainly manifested in product innovation brought by engineering science and technology progress. Product innovation is not only the power source to enhance the competitiveness of enterprises but also the concrete embodiment of the transformation of scientific and technological achievements into productivity. In a period in the future, the role of engineering science and technology will be more direct [67], which is mainly reflected in: firstly, in the field of advanced manufacturing, through major engineering activities, it will continuously promote scientific and technological progress, provide a guarantee for industrial upgrading and enhance comprehensive national power; secondly, in the field of information technology, through major engineering activities, it will conquer the core technology of information engineering, build information highway network and continuously benefit people's livelihood; third, in the field of energy security technology, through major engineering activities, develop new energy sources, develop alternative energy sources, improve energy efficiency, optimize energy structure, build an energy supply system with diversified structure, and ensure energy security.

In petrochemical engineering, creating a strong atmosphere of technological innovation and promoting improving engineering technology content. Achieving the improvement of engineering performance and alleviating and mitigating the increasingly serious environmental problems. In the first half of 2013, China's haze weather, with a large impact range and long duration, seriously affected the air quality. Among the many solutions, there is no doubt that reducing the sulfur content of vehicle exhaust and improving the quality of gasoline and diesel are among the feasible solutions. The PetroChina's gasoline quality upgrading core technology developed independently by the Petrochemical Research Institute marks that the PetroChina entering a comprehensive quality upgrading stage, successfully achieving the leap from the National III stage to the National V stage and effectively breaking the dominant pattern of foreign oil technology in terms of market share. The rapid improvement and development of this engineering technology level will be conducive to promoting the comprehensive upgrade of China's oil processing level and gasoline quality and promoting the green development of China's social economy to realize the harmonious coexistence of engineering and nature.

(2) Establish and improve the engineering management technology system

Engineering management technology is a series of methods and means to reasonably allocate the resources involved in engineering activities and optimize the system integration of engineering technology to achieve engineering goals, which is the unification of engineering technology and management art. Engineering technology is the material carrier of engineering management technology, and engineering management technology is the methodology of engineering technology [68]. Engineering technology is the basis of engineering activities. Leaving the progress of engineering science and technology is difficult to realize not only the multivalued goals of engineering but also the basic engineering activities. However, engineering activities are not purely technical but a comprehensive integration of technology and economic, social, cultural, natural, environmental, and other factors. Making breakthroughs in major projects is inseparable from engineering technological innovation and development and engineering management technology innovation and development. Engineering technology innovation refers to the "hard technology" innovation in engineering practice, mainly including technology creation, technology invention, technology change, technology transfer, etc., which can be realized through independent innovation and technology introduction. Engineering management technology innovation refers to the improvement of engineering management methods, means, and tools under the guidance of engineering management theory to improve the efficiency of engineering management, which mainly relies on the experience of engineering managers and the research of theoretical researchers. In China's engineering management practice, compared with engineering technological innovation, engineering management technology innovation is relatively backward because engineering management technology innovation needs experience and technology of engineering management and the academic research outcomes from researchers.

New engineering management technologies have continuously emerged in China's engineering management practice. Typical engineering management technologies have been formed, such as the "integrated work platform" information management in Sulige Gas Field. The core technology is an "intelligent, digital, and modular" production management system. It realized "automatic data entry, automatic generation of alternatives, automatic alarm for abnormalities, automatic operation control, single-well electronic well patrol, and data sharing." It enables the integrated management of the entire gas field development, construction and production management process based on information technology. The format management of Guangzhou Zhujiang Huangpu Bridge, mainly through the formatted engineering construction business, formatted construction business, and formatted supervision business, clearly defines the business scope and division of responsibilities among the three parties as owner, construction, and supervision in highway construction. It uses more than 1,400 business forms applied to all aspects of highway construction. It plays an important role in standardizing highway construction activities and improving the management level [69]. These typical engineering management techniques embody the "people-oriented and human-integrated" concept, which fully reflects the core concept of "all relying on people and all for people" in engineering activities.

Our existing engineering management concept, engineering management system, and engineering management technology are proposed in the specific historical period and specific historical stage to meet the needs of continuous transformation of economic and social development. They are inevitably characterized by "temporary," "short-sightedness, "one-sided," and other characteristics. Innovative engineering management technology is to make overall planning for engineering management technology from a macroscopic view, eliminate the contradiction and conflict within engineering management technology, and establish a harmonious and unified engineering management technology system.

Since engineering management is the management in an open environment with complexity and variability, Chinese engineering practice urgently needs an engineering management system with Chinese characteristics. The basic objectives of engineering management involve quality, safety, progress rate, cost, sustainability, etc., which are related to the effectiveness of engineering activities. At present, engineering management theories mainly carry out research work on interrelationship, integration and optimization, coordination, and control among engineering management objectives from a methodological level. They lack re-examining and innovative thinking on engineering management objectives from a philosophical perspective to promote the harmony and unity of engineering management objectives and realize the "harmony and win–win" of the engineering community.

6.4.2 Innovative Engineering Management Practices

6.4.2.1 Innovative Engineering Quality Management

In order to improve the domestic product rate of the Qinshan II project, the key equipment pressure vessel of Qinshan Unit 2 was awarded to a domestic manufacturer for manufacturing. In July 2003, when the pressure vessel was ready for installation, it was found that there were quality problems in the welding, which might lead to serious safety hazards. Safety is the life of a nuclear power plant, and in October 2003, the Qinshan II project formally signed a rework contract with Westinghouse. Until March 2004, Unit 2 was connected to the grid for power generation. It was delayed than the original plan and brought serious economic loss to the project: the direct economic loss was 1 million RMB per day of loan interest and 6.5 million RMB per day of late power generation.

(1) Expand the connotation of engineering quality

Engineering quality is the basic morality and basic requirement of engineering activities, which is related to the success or failure of engineering activities and has one vote veto power. Engineering quality not only integrally marks the core quality of engineering enterprises, covering the realm of engineering leaders, the quality of engineering teams, and the level of science and technology, but also reflects the supervision level of government administrative departments to a certain extent. The traditional engineering quality view only involves the narrow scope of engineering entities such as engineering decision, project establishment, project scheduling, engineering bidding, implementation, and acceptance, which is essentially reflected as the engineering economic value target. Under the guidance of this engineering quality concept or value goal, it is inevitable to fall into engineering activities without considering the constraints of natural resources and the ecological environment, regardless of the influence of society and the public, just blindly pursuing the economic value of engineering activities. It will inevitably damage the engineering and realization of other value goals, resulting in the inability to develop harmoniously between engineering and nature, engineering and society.

The quality of engineering and social harmony are inextricably linked. Engineering quality directly affects the realization of a series of engineering multiple value objectives such as social operation, economic development, and people's lives

[28]. With the deepening of engineering theory research and the development of engineering practice, it is necessary to use the new engineering management concept and theory, examine engineering quality from the angle of engineering philosophy, and establish a new engineering quality view or a broad engineering quality view. The modern engineering quality concept should be in line with the engineering management concept of "unity of heaven and man" and reflect the requirements of scientific development, harmonious development, and sustainable development. The concepts of engineering quality and engineering harmony are material and include spiritual and cultural aspects. Modern engineering quality concept is essentially a multi-dimensional quality concept of sustainable development, with two distinctive features of multi-dimensionality and sustainability. Multi-dimensionality means that in the spatial dimension, engineering quality should perform well through the scope of physical engineering quality; it involves the quality management of the all-around and whole engineering process. Sustainability means that in the time dimension, green, ecological, energy-saving, and sustainability are incorporated into the scope of engineering quality management.

Qinshan Phase II project brought certain economic loss due to the project delay, which affected the quality of the project. However, this economic loss was compensated by eliminating engineering safety hazards, avoiding the possible nuclear leakage accident, and minimizing the damage to the natural environment. According to the modern concept of engineering quality, "sacrificing the small" to "exchange for the big" is conducive to achieving harmony between engineering and nature.

(2) Innovative engineering quality supervision system

China's overall engineering quality level is not high, which is not commensurate with its scale development, so it is necessary to realize the improvement of engineering quality level through engineering quality supervision. The implementation of the modern engineering quality concept of multi-dimensional sustainable development cannot rely on advanced engineering management concepts and conscious behavior of engineering subjects alone but must be controlled through the relationship of interests, legal responsibility, and moral constraints in the engineering quality supervision system, and constantly innovate and develop the engineering quality supervision system in the engineering management practice.

Engineering quality control is different from general product quality control and has its own characteristics: firstly, the project has a huge investment, a long construction period, is extensive and influential, and has great risk; secondly, the quality characteristics of engineering projects are numerous, and multi-dimensionality and sustainability are the significant features of modern engineering quality concept; thirdly, there are many factors affecting engineering quality, both economic and political factors, and sometimes also influenced by the international economic and political environment. The sustainable improvement of engineering quality requires the engineering community to establish the modern engineering quality concept and the concept of continuous improvement of engineering quality, advocate the atmosphere of learning and innovation in the organization, clarify the working mechanism of continuous improvement of engineering quality, establish the incentive mechanism to encourage breakthrough and innovation, encourage the members of the organization to apply the conclusion of engineering quality analysis consciously and gradually improve the engineering quality standard, to realize the continuous improvement of engineering quality enhancement.

6.4.2.2 Innovative Engineering Safety Management

At the end of the 1990s, Shenhua Shendong Coal Group Company first carried out transformation experiments in Daliuta Mine using the "Four Transformations" model (production scale, technology and equipment updating, informatization of management means, and professionalization of team). By the end of 2008, it had set nine world records in the iconic indicators of the coal industry. It achieved important results in the management and technological innovation of the mining project. Under the guidance of the "unmanned" engineering safety management concept, the safety factor of coal production has been greatly improved, and the accident rate of coal mining has been reduced. The new modern engineering construction mode of "reliable system, excellent equipment, competent personnel, and efficient management" and the progress of engineering science and technology provide important support for the concept of "unmanned."

(1) Update the concept of engineering safety management

Engineering safety is the most basic requirement of engineering activities. Engineering safety management is an important link in engineering management, which should be attributed to unsafe human behavior, unsafe material state, unsafe environmental influence, and blind areas of safety management. It is important to update the concept of engineering safety management and establish "people-oriented" engineering safety management. "Unattended," that is, it is safe to have no operators or reduce operators in the dangerous situation of coal production. This engineering safety management concept has not only important practical significance but also important theoretical significance. Firstly, this concept reflects the reverse thinking of the innovation of safety management mode in a high-risk industry, highlights the function of engineering science and technology in engineering safety protection, avoids the unsafe factors of people themselves as far as possible, and cuts the chain of engineering safety accidents from the source. Secondly, this concept reflects the role of engineering science and technology progress and engineering management innovation. It establishes a trinity of "environment, quality, and responsibility" engineering safety management system, undergoes a transformation from emphasizing people away from danger to danger away from people, and greatly improves the initiative of engineering safety management. Thirdly, this concept reflects the humanistic concern of the enterprise, aims to improve the production conditions of workers, respects the life safety and physical and mental health of workers, and shows the core concept of "people-oriented" in engineering activities.

(2) Innovative engineering safety management system

To adhere to people-oriented, engineering safety management should be institutionalized, standardized, and systematized, follow the basic policy of "safety first, prevention first, comprehensive management," and continuously innovate and improve the engineering safety management system. In engineering safety management the ultimate goal of engineering safety management can be realized by updating the concept of safety management, establishing a safety production system, improving safety management technology, implementing safety production education, and carrying out safety production inspections. Traditional engineering safety management is more concerned with the safety condition of the project itself, and the focus of safety management is placed on the level of physical results of engineering activities, ignoring the safety protection of the personnel involved in the project. Transforming the concept of engineering safety management, innovating engineering safety management mode, strengthening engineering safety management control, and improving the engineering safety management system is crucial to achieving the effectiveness of engineering safety management. To establish the concept of "people-oriented" engineering safety management is to carry out engineering safety management in line with the principle of "all for people and all relying on people." On the one hand, "all for people" means that in engineering safety management activities, the main body is people. Engineering safety management activities should focus on the safety of people in which economic interests or other interests unconditionally obey human safety. On the other hand, "all rely on people" means to give full play to people's subjective initiative, guide people in the safety activities for self-education, self-improvement, self-management, and form "I want to be safe" active engineering safety management atmosphere.

Looking at engineering safety management issues from a comprehensive perspective is necessary. Engineering safety accidents bring serious losses to the project itself, may endanger the outside of the project, and cause serious safety hazards to society. For this reason, the common participation of engineering community members is required in engineering safety management activities.

Looking at engineering safety management issues from a developmental perspective is necessary. Engineering activities generally have a long construction period, far-reaching impact, and other characteristics, which require not only the implementation of engineering safety management during the construction process but also the implementation of supervision of engineering safety after the construction of the project. For this reason, long-term and dynamic implementation of engineering safety management is required in the engineering safety management system.

Viewing engineering safety management issues with an open eye is necessary. Engineering activities are generally carried out in an open system environment, which brings a lot of uncertainties to engineering activities and significantly increases the risk factor of engineering activities. For this reason, in the engineering safety management system, it is necessary to consider the impact of changes in risk factors on engineering safety and assess the sensitivity of engineering safety risks.

The practice of "unattended" engineering safety management shows that, in addition to the reduction of staff and increase of efficiency supported by engineering technology, a professional team can improve the comprehensive quality and skills of personnel, standardize human operation, and reduce engineering safety accidents caused by the human operation. It concentrates on the concept of "people-oriented" engineering safety. The informationization of management means, through the summary of coal mine management experience and the standardization of management process, has further improved the digitalization and automation level of engineering safety management, reflecting the need for long-term and dynamic engineering safety management.

6.4.2.3 Innovative Engineering Risk Management

For engineering construction and management, engineering risk refers to all uncertainties that may affect the achievement of engineering value objectives and the possible losses they bring. There are many types of engineering risks, including economic and management risks such as capital and contract, environmental risks such as natural disasters, and technical risks such as engineering surveys. Compared with other products, engineering construction is characterized by many subjects, large scale, long duration, sensitive environment, complex technology, etc. The complexity and uncertainty of engineering construction are increasing, and it has more significant risks than general products. In particular, some large-scale projects have a series of significant impacts on socio-economic development and the natural ecological environment, leading to the coexistence and complexity of multiple risks in engineering activities.

Engineering risk management refers to understanding engineering risks through risk identification, risk measurement, risk assessment, and risk control, and the comprehensive application of various risk prediction models, risk control techniques, and risk management methods to effectively prevent and control project risks. In China's engineering risk management practice, there are problems such as backward risk management ideas, obsolete risk management methods, and a lack of risk management tools, so engineering risks often occur out of control. With the development of the national economy and the deepening of the economic system reform, it is necessary to adopt market means to solve the problems arising in engineering risk management, mainly applying economic means to regulate the behavior of each market entity, forming an engineering risk management system mainly by economic means and supplemented by administrative means, and innovative engineering risk management is the call of the times.

(1) Understanding the characteristics of engineering risk management

All projects have risks, and the adoption of advanced, rational, and easy-to-operate methods for engineering risk management is the basic guarantee to achieve the expected value goal of the project. Compared with other risks, engineering risk has objectivity and uncertainty characteristics and variability, relativity, and stages, determined by the characteristics of engineering projects. The variability of engineering risk is mainly manifested in the change of risk nature, risk consequences, and risk factors. The relativity of engineering risk mainly has two aspects: the relativity of the engineering subject and risk size. The stages of engineering risk are characterized by obvious temporal characteristics, mainly in the risk brewing stage, risk occurrence stage, and consequence stage. They have different expressions in different stages, such as planning, design, construction, and operation.

Based on engineering risk characteristics identification, it is necessary to further distinguish the types of engineering risks, divided into three categories: engineering physical risk, management capacity risk, and environmental element risk [70]. Firstly, in terms of engineering physical risk, mainly includes construction, technology, material, and equipment. Second, in terms of engineering management capability, it mainly includes decision risk, compliance risk, and liability risks. Thirdly, in terms of environmental elements, it mainly includes natural, economic, social, political, and institutional risks. In the practice of engineering management, these different types of risks are often intertwined to form a complex network of interrelated and mutually influencing risk relationships.

(2) Improve the engineering risk management system

Currently, China is at the peak of engineering construction, and many large-scale engineering constructions have distinctive features such as large-scale, high-cost, and complex technology. Engineering management's theory and practice have made great progress. Although the liability system of engineering legal person, diversified investment of engineering, capital fund system of engineering and bidding system of engineering have been introduced continuously. However, compared with developed countries and regions, China's engineering construction and management are still in the initial stage. The engineering risk management level is still backward, and the engineering risk management system is not yet sound. Regarding the management system, as China is still in the early stage of the market economy, participating engineering subjects are not clear, and various systems such as credit mechanisms, engineering guarantees, and engineering insurance have not been fully established. It is easy to form risks for construction projects.

With the development of engineering management theory and practice, the government's management of engineering is gradually changed from a management mode based on direct intervention by administrative means to a management mode based on indirect regulation by economic and legal means, and it is urgent to establish a method of engineering risk management with an organic combination of various management means based on economic, legal, and administrative means, and continuously establish and improve the engineering risk management system. Emphasize that the government should play an active role in engineering risk management but should not interfere too much, mainly through economic and legal means of engineering risk management. For example, in engineering risk management, the U.S. government mainly adopts the measures of classification and management, controls the risk management of public works according to the law, protects the safety and health of construction workers according to the law, and underwrites workers' compensation insurance and political insurance for overseas projects [71]. According to Zhou

[72], an engineering guarantee and engineering insurance system should be established as soon as possible, referencing international practice and in line with China's national conditions, mainly including a bidding credit guarantee system, performance credit guarantee system, advance payment credit guarantee system, warranty guarantee system, engineering insurance system. Therefore, improving and innovating engineering risk management systems is necessary. While actively researching engineering risk management theory and technology, it should vigorously cultivate composite talents who can skillfully master and comprehensively apply professional knowledge in engineering technology, risk management, finance and insurance, and legal system to realize effective management of engineering risks.

(3) Establish an engineering risk management system

With the in-depth research on engineering risk management, a set of mature qualitative and quantitative engineering risk management methods has been formed, from risk identification to risk analysis. At present, in engineering risk management, there is an urgent need to establish a perfect engineering risk management system, which mainly includes: first, establishing engineering risk management subjects including owners, contractors, intermediary consulting agencies, industry associations, and governments; second, establishing engineering risk management mechanisms such as risk avoidance, risk transfer, and risk retention; third, establishing engineering risk management methods including risk identification, risk measurement, risk control, and risk prevention.

For the effective management of engineering risks, comprehensive and integrated management concepts and methods can be adopted [70]. In practice, it is mainly based on mature engineering risk management theory, combined with methods and technologies from multiple disciplines, such as knowledge management, organization management, comprehensive evaluation, and artificial intelligence. The objective is to finally build a framework of engineering risk management composed of the inner operation mechanism system of the engineering risk system, engineering risk management methods and technology system, engineering risk management process and organization system, and engineering risk management information systems to realize continuous and dynamic risk management for the whole life cycle of large projects.

In the practice of engineering risk management, it is necessary not only to base on general risk management principles and methods but also to take into account the special features of engineering risks and give corresponding engineering risk management measures. First, it is necessary to set up the engineering risk management objectives to accurately measure engineering risks through engineering risk identification and risk assessment. Secondly, establish an engineering risk warning system and provide engineering risk control measures. Finally, establish the corresponding risk-sharing mechanism to prevent and control risks through risk avoidance, risk transfer, risk dispersion, risk self-retaining, and other operations.

6.4.2.4 Innovative Engineering Sustainability Management

Yupu Wang made a case study of oil recovery engineering in the Daqing oilfield and deeply studied the dialectical relationship between oilfield exploitation and economy, society, and environment [73]. Practice shows that facing diminishing oil resources, it is possible to achieve sustainable engineering development as long as the scientific concept of development can be implemented and engineering philosophy can be applied comprehensively.

(1) Expand the connotation of engineering sustainable development

Sustainable development reflects the basic requirements of the scientific concept of development. It pays more attention to the long-term development of the economy and society, which can not only realize and meet the needs of the present generation but also take into account the development requirements of future generations, mainly including human sustainability, economic sustainability, social sustainability, ecological sustainability, and natural sustainability.

The concept of sustainable development is introduced into engineering practice activities, and the proposition of sustainable engineering development is put forward. As engineering activities involve many aspects such as human, social, economic, science and technology, culture, nature, and environment, it is necessary to expand the concept and connotation of sustainability and define sustainable engineering development. Engineering sustainability includes the sustainability of engineering economic value, engineering natural environment, and engineering social responsibility, and the sustainability of engineering talents' growth, engineering science and technology innovation ability, and engineering cultural heritage. Engineering activities are "all for people" because the results created by engineering activities serve human beings and the cultivation of engineering talents by engineering practice activities. The continuous growth of excellent engineering talents (engineering management talents and engineers) is an important guarantee for developing new engineering activities, which reflects the primary demand of "all relying on people" in engineering activities. Engineering activities, especially large-scale engineering activities, carry a lot of social responsibility, including the social value of engineering and the negative external effects generated by engineering activities. The sustainability of engineering social responsibility is mainly manifested in the continuous realization and enhancement of its social value, continued reduction of negative external effects of engineering activities, winning a good social reputation for engineering activities, and finally realizing the harmonious coexistence of engineering and society. The sustainable economic value of engineering is the basis for the survival and development of engineering. Otherwise, engineering activities will be stagnant or even shrink. Engineering activities can produce materialized results, promote scientific development, and achieve technological progress. The sustainable capability of science and technology innovation is the need for social development and engineering activities' development, which is conducive to improving the output efficiency of engineering practice and realizing the multiple value objectives of engineering activities. The process of any engineering activity is inevitably influenced by the cultural atmosphere of a specific historical period, and the sustainability of engineering cultural inheritance is manifested in two aspects: on the one hand, the cultural construction and inheritance of the engineering activity community; on the other hand, the inheritance and promotion of existing culture as a result of engineering activity, which will affect the efficiency of engineering activity and the realization of engineering value objectives. The sustainability of the natural engineering environment mainly shows that the results of engineering activities should not only realize and meet the needs of the present generation but also not damage the ability of future generations to meet their needs, and finally realize the harmony of engineering and nature.

(2) Rely on the progress of engineering science and technology to achieve engineering sustainability

In the era of the natural economy, major projects are not common, and the problem of sustainable development of engineering is not yet prominent. With the establishment of the market economy system and the accelerated pace of industrialization, large projects and even mega projects have emerged one after another. Due to the continuous transformation of the economy and society and the need for economic and social development in specific historical stages, the concepts of engineering management, engineering management system, and engineering management technology are often "temporary" and "one-sided." Engineering managers tend to see only the project's economic value, but ignore other values of the project, often seeing only the current value of the project; it is difficult to see the sustainable development value of the project, showing "short-sightedness." Under the guidance of such narrow engineering management concepts and ideas, there are many one-sided engineering innovation ideas and engineering evaluation standards, which have led to many unsustainable development problems in engineering management practice.

The principle of universality of linkage tells us that the engineering concept under the guidance of sustainable development needs to focus not only on the harmony between humans and nature and the coordination between humans and society but also needs to put special emphasis on the mutual complement and coordination among the economic, social, scientific and technological, cultural, and ecological functions of engineering [74]. Engineering science and technology play an irreplaceable role in engineering activities and are a direct driving force for the development of productivity and economic growth. In human history, every industrial revolution is the product of great progress in engineering science and technology, and every industrial revolution promotes a great leap in social productivity. The result of engineering science and technology progress is inevitable to bring industrial structure upgrading and economic structure optimization. Due to engineering science and technology's limitations, the advancement of engineering science and technology has also brought negative consequences to nature and the environment while providing a strong impetus for engineering and economic development. We need to fully understand and actively promote the positive effects of engineering science and technology, gradually reduce and overcome its negative effects, and correctly apply the achievements of engineering science and technology progress for the benefit of humankind. Engineering science and technology progress are important to achieve sustainable engineering development. As far as the sustainable economic value of engineering is concerned, the progress of engineering science and technology can realize the economic growth mode from rough to economic and continuously realize the economic value of engineering. Engineering science and technology progress provide a powerful tool to use renewable resources sustainably in engineering natural environment sustainability. As for the sustainability of engineering social responsibility, the progress of engineering science and technology can better exert the positive external effect of engineering and provide the possibility to realize the social value of engineering.

The engineering practice of Daqing Petroleum Group Company shows that as long as the scientific concept of development is taken as the guiding principle, advanced engineering management concept and engineering management methods are taken as the guidance, correctly rely on the progress of engineering science and technology, the harmonious coexistence of engineering and nature, and engineering and society can be realized. The sustainable development of engineering can be achieved.

References

1. Zhang, X. (2006). Engineering value and evaluation. *Philosophy Trends, 12*, 42–47.
2. Zhao, J., & Ding, T. (2011). Environmental value and humanistic value of engineering. *Studies in Dialectics of Nature, 27*(5), 73–78.
3. Mao, R., & Jia, G. (2011). *Sociology introduction to construction engineering*. Tongji University Press.
4. Wang, J., Li, Y., & Xiao, Z. (2010). Green project management and value engineering study. *Journal of Engineering Management, 24*(1), 41–45.
5. Zhang, X., & Chen, L. (2009). Engineering essence and value choice in the perspective of modernization. *Journal of Southwest Petroleum University (Social Science Edition), 2*(1), 59–63.
6. Liu, H. (2006). The value of construction projects and the ways to enhance the value. *Value Engineering, 25*(10), 62–64.
7. Yang, S., Huang, Z., & Ren, X. (2012). Engineering management dialectical thinking. *China Engineering Science, 2*, 14–24.
8. Xu, T., Qu, C., & Yang, S. (2012). *Engineering economics*. Science Press.
9. Gao, J., & Yue, Y. (2010). Project's economic value in terms of tourism resources. *Economic Research, 90*(16), 168–169.
10. Hu, Z. (2004). Judgment on the evaluation of environmental resources value. *Economic Management, 17*, 31–33.
11. Research Group. (2009). Guqiao coal mine of Huainan mining group. In *National large-scale coal mine construction site meeting and forum for promoting large-scale modernization of coal production*. Ordos.
12. Li, K., & Li, P. (2004). Dujiangyan—The bright pearl of China's traditional water culture. *China Water Resources, 18*, 75–78.
13. Chen, M. (2008). Feeling about Dujiangyan. *Urban and Rural Construction, 9*, 77–78.
14. Liu, L. (2008). *A study of construction project economic evaluation based on components*. Harbin Institute of Technology.
15. Xu, J. (2005). *Project financial evaluation method and application in project loan*. Shanghai Maritime University.

16. Ma, X., & Ma, Y. (2005). Economic and policy evaluation of landfill gas power generation project. *Renewable Energy, 3*, 79–81.
17. Zhou, D., & Jiang, L. (2011). From the Jingjiang flood diversion into gorges—The achievements of water conservancy construction in Hubei. *Learning, 15*, 11–13.
18. Zhou, X., & Zhou, Y. (2010). Analysis of the internalization of positive externalities. *Journal of Hotan Teachers College, 29*(4), 14–15.
19. Li, M. (2011). *Engineering economics*. Chemical Industry Press.
20. Feng, W. (2012). *Engineering economics*. Peking University Press.
21. Du, K., & Gong, Q. (2006). Analysis of externality of regional economic activities. *Exploration, 12*, 1–3.
22. Lin, X. (2002). Application of transportation externality theory in project evaluation. *Quantitative Economics and Technological Economics Research, 19*(12), 93–96.
23. Yang, S. (2012). *Engineering economics*. Wuhan University of Technology Press.
24. Xie, X. (2010). *The regional economic impact of major construction projects*. Chongqing University.
25. Lao, C. (2007). Regional economic evaluation methods for major construction investment projects. *Southwest Finance, 10*, 33–35.
26. Du, L. (2007). *The evaluation index system of industrial development planning*. Dalian University of Technology.
27. Li, B. (2008). Engineering community and engineering sociology-research on engineering community. *Dialectics of Nature, 30*(1), 63–68.
28. He, J. (2013). On the core of engineering management theory. *China Engineering Science, 15*(11), 4–11.
29. Li, B. (2010). *Introduction to engineering sociology: Research on engineering community*. Zhejiang University Press.
30. Yin, R. (2007). *Engineering and philosophy*. Beijing Institute of Technology Press.
31. Zhang, X. (2009). Social functions of engineering community. *Studies in Philosophy of Science and Technology, 26*(2), 90–95.
32. Duan, Y., Li, Y., Le, Y., & Qian, L. (2012). The social structures of complex engineering community networking and management. *Construction Economy, 10*, 79–82.
33. Vidal, L. A., & Marle, F. (2008). Understanding project complexity: Implications on project management. *Kybernetes, 37*(8), 1094–1110.
34. Wei, X., Gang, E., Wu, X., & Weng, D. (2003). *Environment and development: Engineering, ethics and the environment*. Tsinghua University Press.
35. Li, B. (2010). The formation, dynamic change and disintegration of the engineering activity community-the fourth study of the "Engineering Community". *Dialectics of Nature, 1*.
36. Feng, J., & Liu, H. (2008). Research on the subject structure of engineering social responsibility. *Science and Technology Management, 28*(12), 269–271.
37. Du, J. (2011). Guide to social responsibility of China's foreign contracted engineering projects. *WTO Economic Herald, 1*, 63–63.
38. Harris, C. E. (2006). *Engineering ethics: Concepts and cases*. Beijing Polytechnic University Press.
39. Liu, H. (2009). Analysis of social responsibility of water resources engineering community. *China Rural Water and Hydropower, 8*, 63–65.
40. Bao, O. (2010). Engineering investors from the perspective of engineering sociology. *Studies in Dialectics of Nature, 26*(6), 48–53.
41. Du, C., & Li, B. (2006). *Engineering research: Engineering in interdisciplinary perspectives*. Beijing Polytechnic University Press.
42. Dong, D., & Zhang, Y. (2012). On the separation of public goods and the current value orientation—The "Projects for People" and "achievement for performance." *Nanjing Normal University (Social Science Edition), 6*, 20–26.
43. Li, B. (2006). Several questions about engineers-second research on "engineering community." *Journal of Dialectics of Nature, 28*(2), 45–51.
44. Robinson, S. (2007). *Engineering, business and professional ethics*. Routledge Press.

45. Xiao, X. (2009). On environmental ethical responsibility of engineering community. *Ethical Studies, 6*, 65–70.
46. Carroll, A. B. (1979). A three-dimensional conceptual model of corporate performance. *Academy of Management Review, 4*(4), 497–505.
47. Tian, P., & Chen, S. (2013). An analysis of the "embeddedness" of engineering and social action—The research status of engineering sociology. *Engineering Research—Engineering from an Interdisciplinary Perspective, 5*(4), 344–353.
48. He, J., & Wang, M. (2008). View of philosophy in engineering management. *Technology Progress and Policy, 25*(10), 1–3.
49. Wang, C., & Li, K. (2004). Investment project social assessment seminars—Chinese investment projects social assessment. *China Engineering Consulting, 1*, 43, 44.
50. Ding, L. (2009). Project management: The social dimension of engineering. *Construction Economy, 5*, 8–10.
51. China International Engineering Consulting Corporation. (2004). *Guidelines for social evaluation of Chinese investment projects*. China Planning Press.
52. Ma, X. (2010). Social impact analysis-past and present of strange names (Part 1). *China Investment, 6*, 100–103.
53. Wang, C., & Li, K. (2004). Investment project social assessment seminars. *China Engineering Consulting, 8*, 50–51.
54. Guo, F., & Zhang, F. (2002). Several important issues and countermeasures in social evaluation of construction projects. *Journal of Changsha Railway University: Social Science Edition, 3*(1), 100–103.
55. Ma, X. (2011). Analytical method of interest groups-analysis tool for social evaluation of investment projects (Part 1). *China Investment, 5*, 96–99.
56. Mai, S., Wang, K., & Kuang, H. (2011). Evaluation of social value of enterprises. *Research Management, 32*(6), 100–107.
57. Yang, S., & Zhong, J. (2012). Dynamic decision theory and methods of complex product development project management. *China Engineering Science, 12*, 25–40.
58. Xu, K. (2008). Development of engineering philosophy and implementation of scientific outlook on development. *Journal of Beijing Normal University (Social Science Edition), 1*, 90–92.
59. Yin, R. (2008). Engineering from the perspective of philosophy. *China Engineering Science, 10*(3), 4–8.
60. Yin, R., Wang, Y., & Li, B. (2007). *Engineering philosophy*. Higher Education Press.
61. Wang, Z., Yang, Z., & Ding, J. (2011). Modern engineering management theory and the framework of knowledge engineering (Part 2). *Journal of Engineering Management, 25*(2), 132–137.
62. He, J., Wang, M., & Wang, Q. (2009). Engineering management theory and framework. *Technology Progress and Policy, 26*(21), 1–4.
63. Gong, M. (2008). Research on rural water conservancy management system reform from sociological perspective. *China Rural Water and Hydropower, 2*, 8–10.
64. Hong, W., & Wang, Z. (2008). Game analysis and countermeasures of rent-seeking behavior of project supervision. *People's Yellow River, 30*(2), 66–69.
65. He, J. (2013). Characteristics and management thinking of ancient Chinese engineering buildings. *China Engineering Science, 15*(10), 4–9.
66. Li, T., Kong, D., & Li, C. (2012). Analysis of the evolution of the management model of major defense science and technology projects. *Science and Technology Progress and Policy, 29*(24), 42–46.
67. Xu, K. (2010). The role of engineering science and technology will be more direct. *People's Daily*, 6–6.
68. Zhang, S., & Wang, M. (2010). On the systematization of engineering management technology. *Science and Technology Progress and Policy, 27*(19), 23–26.
69. Wang, M., Zhang, S., & Cheng, Q. (2010). Studies on major projects standardized management. *Technology Progress and Policy, 27*(6), 23–26.

70. Zhang, D., Tong, J., & Cheng, S. (2014). Research on risk management system of complex engineering. *Engineering Economy, 11*, 44–49.
71. Wu, Z., Liu, C., & Cheng, Z. (2000). The role of the US government in engineering risk management. *International Economic Cooperation, 1*, 38–40.
72. Zhou, Q. (2002). On the establishment of engineering risk management system. *China Engineering Consulting, 5*, 23–26.
73. Wang, D. (2008). New progress in engineering philosophy: A summary of the "seminar on engineering and engineering philosophy." *Philosophy Trends, 3*, 106–108.
74. Wang, Y., & Wang, H. (2005). Engineering science and engineering philosophy. *Studies in Dialectics of Nature, 21*(9), 59–63.
75. Chen, P. (2008). *Investment projects comprehensive benefit evaluation method research and application*. Nanjing Polytechnic University.

Chapter 7
Engineering Management Innovation Theory

The core of engineering activity is "making things" or changing the nature of things; through these activities, we achieve a specific purpose [1]. Engineering is unique, and each project is carried out in a particular environment with individual goals and values. Every project is inseparable from innovation in technology, organization management, or investment methods. It says that innovation runs through the whole life cycle of engineering activities and is the key to determining the project's success. Engineering innovation is the use and integration of natural resources and social resources by human beings. It is the process of selecting, synthesizing, and integrating such elements as capital, human resource, culture, politics, science and technology, law, etc. Evidence shows that this process is not achieved spontaneously but by the innovation subject through a series of organizational and institutional arrangements, relying on scientific management tools. As Dr. Weber, the Apollo project chief director, once said, "We do not have a technology that others do not have; our key technology is scientific organization and management." [2] Thus, it can be seen that scientific organization and management are crucial to achieving the goal of engineering innovation, and engineering management theory must include engineering innovation into its research scope.

The research on engineering innovation in engineering management theory mainly concerns the essence and characteristics of engineering innovation, which is the logical starting point of engineering innovation theory. Any effective management must be based on a comprehensive and scientific understanding of the management object, and a different understanding of the connotation of engineering innovation leads to different engineering management concepts. For example, we consider engineering innovation statically as creating a new "artificial object" to meet people's needs of production and life, or dynamically as the whole process from engineering decision, design of engineering scheme, selection of technical means, the detailed implementation of engineering, operation of engineering, social evaluation of engineering to decommissioning of engineering. If it is the former, along with the completion of the engineering project, the task of engineering management is finished. If it is the latter, it is necessary to make dynamic and systematic process considerations

J. He, *Principles of Engineering Management*,
https://doi.org/10.1007/978-981-99-1168-4_7

for all aspects of engineering innovation. Engineering innovation management will continue until an engineering artifact is decommissioned.

Secondly, the research on engineering innovation in engineering management theory should pay attention to engineering innovation goal constraints and organizational forms. On the one hand, leaving the consideration of engineering innovation goals, it is difficult for engineering management to find the core elements that stimulate engineering innovation to happen, and the management will be aimless. No matter what kind of form any engineering innovation finally presents, its goal is to meet people's material and spiritual needs, so people-oriented should be the goal of engineering innovation. On the other hand, if we ignore the research on engineering innovation organizing, it is challenging to realize the integration of heterogeneous elements in engineering innovation through management and then realize successful engineering innovation. Thus, the essence of engineering innovation is integrating all kinds of technical elements and non-technical elements. Therefore, collaborative innovation is bound to become an important organizational form of engineering innovation.

The realization and enhancement of engineering innovation on engineering value has been discussed in Sect. 6.4, so it will not be repeated in this chapter.

7.1 Connotation and Characteristics of Engineering Innovation

7.1.1 Connotation of Engineering Innovation

Innovation was first introduced as an economic concept by the Austrian-American economist Schumpeter, who defined innovation in his "Theory of Economic Development," published in 1912, as "Innovation is the introduction of new combinations of factors of production in the production system." It consists of five aspects: introduction of new products, new processes, the opening of new markets, control of new sources of supply of raw materials, and establishment of new business organizations [3]. Among them, introducing new products and processes can be collectively called "technological innovation" and is considered the fundamental factor of economic development. Schumpeter's theory of innovation has attracted the attention of Western economists. Theoretical research on technological innovation has been carried out, and many schools of technological innovation theory have been formed.

Compared with the well-known concept of technological innovation, the concept of engineering innovation is not very common. It was put forward by the Chinese engineering community and engineering philosophy community under the conceptual framework of the national innovation system. Academician Ruiyu Yin once pointed out: "in the process of implementing the strategy of building an innovative country, engineering innovation is a key, and engineering innovation is the main battlefield of innovation." [4] Professor Bocong Li also emphasized the importance of engineering innovation in the national innovation system. He pointed out: "the national innovation system is complex with rich content and very complex components. Suppose we compare the overall innovation activity of a country to a national level and national-scale innovation; on the battlefield of this innovation battle, there are the outpost battlefield, the logistics battlefield, and its main battlefield. Engineering innovation is the main battlefield of national innovation activities. When examining and evaluating the success of a country's national innovation system and the process of building an innovative country, the key is to see how well the country has fought and succeeded in the main battlefield of engineering innovation." [5] Because of the core position of engineering innovation in the national innovation system, the research on the engineering innovation concept has attracted academia's attention. In summary, the current definition of the concept of engineering innovation includes the following views.

Interpret engineering innovation from the perspective of epistemology. Yin et al. summarized the knowledge accumulated by human beings in understanding nature and developing productivity as a complex knowledge chain with network characteristics: science-technology-engineering-industrial knowledge chain. Looking at engineering innovation from the perspective of the knowledge chain, every industrial innovation is always accompanied by the generation of new knowledge. Once generated, this new knowledge is transformed into a certain production condition, forming a new production function and serving the purpose of constructing artificial nature. Therefore, from the perspective of knowledge theory, engineering innovation is the process in that engineering knowledge condensed in artificial nature is incorporated into the production function and gets the first commercial application [6, 7].

Engineering innovation can be defined in three latitudes: content, degree, and time. During the construction period, the typical feature of engineering activities is to create a thing that did not exist in the world, from the generation of ideas to the formation of engineering objects, which contains a series of innovations. In terms of content, engineering innovation includes management, technological, process, equipment, material, etc. The basic performance and specific characteristics of innovation vary in different engineering fields; even in the same field, each project will be different because of special initial conditions, boundary conditions, and target requirements, leading to differences in engineering innovation. From the viewpoint of levels, engineering (including engineering management) innovation includes engineering law discovery, technical principle innovation, technical invention innovation, and technical application innovation. Engineering law discovery should belong to the innovation at the top of the tower. At the same time, technology principle, technology method, technology integration, and technology application are the innovation below the top of the tower. From the perspective of time, engineering innovation is developed following the process of engineering activities, go through the innovation of decision-making, implementation, and operation [8].

From the perspective of engineering philosophy, Bocong Li believes that engineering innovation's essence and basic contents are the choice and construction in the innovation space. First, as a purposeful activity of human beings, engineering innovation is an intentional selection and construction activity carried out by innovators for all its elements and processes. In other words, engineering innovation activities are continuous selection and construction activities in the innovation space. Secondly, engineering innovation includes “society embedded” activities and its contributions need to be accepted by society for “re-selection”, and they are embedded, integrated, and construct into the whole society. The former is to study the selection and construction of engineering innovation from the innovator’s perspective, while the latter is to study the selection and construction of engineering innovation from the society’s perspective (including the market’s perspective) [9].

From the above research on the connotation of engineering innovation, we can draw the following understanding. Firstly, engineering innovation integrates natural and social resources, which is the best manifestation of the human subjective initiative. Natural resources mainly include land, water, minerals, and artificial resources such as roads, bridges, factories, and mines. Social resources mainly include politics, economy, culture, technology, and law. Secondly, engineering innovation aims to construct a new artificial nature suitable for human survival and development. Therefore, engineering innovation must always adhere to the value pursuit of “people-oriented”. Finally, engineering innovation is the unity of technological and management innovation. Engineering innovation provides solid support to engineering innovation. Without the innovation of raw material, technological product innovation, and production process innovation, engineering innovation will become a source of no roots and water. Only relying on technological innovation cannot accomplish the goal of engineering innovation because the integration of various natural and social elements is not automatic. Only through organization and management can they be linked, integrated, and coordinated with each other so that the engineering innovation process can proceed smoothly. Therefore, management innovation is the linking condition of engineering innovation and the soft support of engineering innovation.

Through the above analysis, it can be considered that engineering innovation is the main body of innovation. It includes investors, designers, implementers, managers, and other participants to meet human needs as the driving force. The harmony between man and nature performs as the constraint. In this process, the use and development of natural and social resources make purposeful selection and construction in a certain innovation space and create a new artificial natural process through the integration and social operation of heterogeneous elements such as technology and non-technology.

7.1.2 Engineering and Technological Innovation

In order to have a deeper understanding of engineering innovation, it is necessary to explore the difference and connection between technological innovation and engineering innovation.

(1) Engineering innovation and technological innovation are inseparable

On the one hand, engineering innovation is based on technological innovation. Ancient water conservancy engineering, civil engineering, modern aerospace engineering, railway engineering, every engineering innovation takes technological innovation as the basic unit. Technical innovation is the basis of engineering innovation, which directly affects the degree and level of engineering innovation, especially the original technical innovation will bring about great changes in the engineering field. For example, the emergence of 3D printing technology will lead to fundamental changes in the manufacturing industry, and the development of cloud computing technology will bring radical changes in communication engineering. Advanced technological innovation is the decisive factor for advanced engineering innovation. Still, it does not mean that the more advanced technological innovation is more beneficial to engineering innovation. The primary purpose of engineering innovation is to create or change the properties of things. The safety, reliability, and effectiveness often have more competitive advantages than the advanced nature of engineering. Therefore, for engineering innovation, in most cases, it is the choice of suitable and mature technology, not necessarily the most advanced innovative technology, and the choice is crucial for engineering innovation. On the other hand, engineering innovation is the integrated embodiment of technological innovation achievements. Engineering innovation is the practical base of technological innovation activities, and the ultimate goal of technological innovation is to transform potential productivity into real productivity. While production in modern society is networked existence, discrete and individual technological innovation cannot meet the needs of production development. Therefore, technological innovation in various fields and industries needs to be integrated through engineering innovation to meet the higher production and living needs of human beings.

(2) Engineering innovation is not equal to technological innovation

There is a natural connection between engineering innovation and technological innovation. In some people's minds, engineering innovation is technological innovation, and the regularity of technological innovation can be directly transplanted into engineering without the need to discuss it specifically. This is incorrect in the practice of engineering innovation. We cannot confuse technology with engineering, what's more, technological innovation should not be confused with engineering innovation. In the specific engineering innovation practice, the success of technological innovation does not necessarily lead to the success of engineering innovation. For example, in the 1980s of the last century, Motorola invested heavily in the Iridium satellite system. It is a typical case of great technical success but miserable engineering failure. Regarding technological innovation, the Iridium satellite communication technology is advanced, perfect, and impeccable. It realizes the "myth" that humans can communicate with each other anywhere on the earth and is considered a milestone technology in modern communication technology. However, from the perspective of engineering innovation, due to the deviation of the market positioning of the Iridium communication system in the process of operation, the poor matching of technical systems, and the lack of innovative integration of economic and management elements, the Iridium communication project was declared bankrupt after only 15 months of operation.

The case of the Iridium satellite system warns us that in handling the relationship between technological innovation and engineering innovation, we must recognize the close relationship between the two and pay attention to their essential differences. Any engineering innovation must systematically integrate technical and economic elements, political, resource, management, social and institutional elements, ethical, psychological, and other non-technical elements. According to statistical data, technological innovation impacts the entire project. The contribution rate of innovation success is only 10%. Other innovations, such as changes in investment and financing methods, breakthroughs in marketing models, and innovation in organizational management methods, play an important role in engineering innovation. The key to engineering innovation is the synthesis and integration of multiple innovations.

7.1.3 Engineering Innovation Features

Innovation is an intrinsic characteristic of engineering. Engineering innovation is the key to determining the merits of engineering, even the success or failure. The Qinghai-Tibet Railway project has taken 50 years, from the beginning of survey and design in 1956 to the opening of the whole line in July 2006. The ultimate success of this plateau railroad project, which has the highest altitude, longest line, and tough natural conditions in the world, cannot be separated from a series of engineering innovations.

From overcoming technical problems in road construction in frozen soil areas and salt lake road construction to the addition of personnel, machinery, equipment and engineering materials, living supplies, medical labor security, and the organization and coordination of various departments and regions, innovative engineering practices are required. Engineering innovation is a complex system that integrates technical, economic, cultural, political, environmental, and other elements. It is a continuous, open, and dynamic process. Negligence may cause the failure of the entire project. Practice without theoretical guidance is blind and can easily lead to failure. Therefore, a deep understanding of the nature and characteristics of engineering innovation and carrying out engineering innovation activities according to the regularity of engineering innovation is necessary to achieve the goal of engineering innovation.

(1) Engineering innovation is carried out in the process, and the process is its essential feature

Engineering innovation is the continuous, purposeful selection and construction of engineering innovation subject to a certain time and space, technical environment, social environment, and natural environment. Engineering activities are process-oriented, from the goal-setting of engineering innovation to the initiation and completion of the innovation task, through innovation in engineering decision making, engineering design, investment and financing system, engineering construction and organization management, engineering self-measurement control and inspection, engineering use and maintenance, engineering social evaluation, as well as innovation in engineering waste disposal and withdrawal mechanism, etc.

The Three Gorges Project of the Yangtze River is undoubtedly a major innovation in the history of human engineering development. It is a complex system engineering in the giant system of nature and human society. It involves the natural ecology, human environment, politics, economy, project construction, and related disciplines. In order to achieve the expected engineering goals, engineering innovation is carried out in the following three stages: the first stage is innovative thinking. In this stage, through extensive and in-depth investigation and research, scientific experiments, and design demonstrations, we can accurately understand all aspects of nature and the objective world, reveal the essence of things, and complete the decision-making process on these foundations. The final argument of the Three Gorges Project is divided into 14 topics, namely hydrology, geology, earthquake, sediment, flood control, ecology and environment, water and soil, electrical and mechanical equipment, power systems, shipping, immigration, investment estimates, comprehensive economic evaluation, and comprehensive planning and water level selection. After conducting scientific and detailed discussions on 14 topics, 412 senior experts concluded that the Three Gorges Project is technically feasible and economically reasonable. It is better to build than not to build and build early than later. The second stage is the innovation of project implementation. In the implementation phase of engineering innovation, the quality, capital investment, and schedule of the engineering innovation process should be systematically controlled. In this stage, there should be a strict and clear process based on the approved design documents; the following is included: the decision of major technical solutions, the development of high-quality standards, the development of various plans such as the financing plan, the overall progress of the project, the implementation plan of sub-projects, the establishment of an information system for digital management, the rapid feedback of engineering innovation information at all times, and the strict control of the whole process in stages; In the third phase is operation and management stage. In order to achieve the expected goal of engineering innovation, the economic and social benefits of the project should be maximized through a series of organization and management. The Three Gorges Power Plant of the Three Gorges Project has 14 sets of 700,000 kW units in operation, with a total power of 9.8 million kW, generating 1461 kWh of electricity by December 2006. The safe operation of the units has formed strong support for the power supply of China. The navigation capacity of the ship locks has reached 45 million tons per year.

(2) Engineering innovation activities are the integration of various technical elements and non-technical elements, which are systematic and integrated

From the perspective of epistemology, engineering knowledge is different from science, technical knowledge, and humanities and social science knowledge. It is an interdisciplinary knowledge and practice system formed based on engineering practice. Engineering innovation involves complex elements such as science, technology, economy, culture, politics, system, psychology, ethics, ecology, etc. These elements are integrated in engineering innovation activities with certain procedures, rules, and structures, to realize the goal of engineering innovation in selecting and constructing the main body of engineering innovation. Engineering innovation exists systematically, and the function of the engineering innovation system is realized by integrating various heterogeneous elements. From a philosophical point of view, integration is a dynamic process in which the elements of a system are organically combined to become a whole under the constraint of the system's goal. Integration is not a simple superposition of components. Still, to achieve a certain function, a synergy of various heterogeneities around a certain goal, according to certain rules, through certain organization and management. The integrated characteristics of engineering innovation are mainly reflected in two levels: the first level is technical elements. Engineering innovation activities require the selection, organization, integration, optimization of multiple disciplines and technologies on a larger time and space scale. The second level is optimizing technological elements' integration with economic, social, and management aspects under certain boundary conditions in engineering innovation activities [10].

First, engineering innovation is the integration and connection of subject and object elements. The innovation subject carries out any engineering innovation in a certain objective environment and the best match between subject elements and object elements is required throughout the innovation process. Secondly, engineering innovation integrates heterogeneous elements such as scientific and technological elements, natural and environmental elements, social and humanistic elements, etc. Whether engineering innovation can be carried out smoothly depends on whether the above elements coordinate. Finally, from the perspective of engineering innovation subjects, the subjects involved are also heterogeneous, including project investors, engineers, project managers, workers, and other stakeholders. They have different positions and responsibilities in engineering innovation, and their rights and obligations differ. The success of engineering innovation depends on whether stakeholders' interests are reasonably divided and whether they can be integrated into a community of interests under a unified engineering innovation goal.

Engineering innovation is an integration process of heterogeneous elements. When faced with complex real-world engineering problems, any single discipline is insufficient. We need to integrate scientific knowledge, technical knowledge, financial knowledge, marketing knowledge, legal knowledge, aesthetics knowledge, and even anthropological knowledge into the project. We also need to realize the reconciliation of various interest relations and the integration of various social factors in engineering innovation. We need to achieve harmony between people, technology, and the natural environment in engineering innovation, to create "high-quality projects" that satisfy all parties. In such a process of integrating heterogeneous elements, it is necessary to match various elements, reconcile various requirements, and carry out complex trade-offs.

(3) Barriers and traps will be encountered in the process of engineering innovation, and engineering innovation is risky

Since uniqueness is one of the basic characteristics of engineering activities, it determines that engineering activities must be innovative. In other words, it is the inevitable requirement of engineering activities to be innovative. However, innovators and managers must also be aware that innovation may be successful (including great success); but may also fail (including miserable failure). Many of our state-owned enterprises underwent a technological transformation in the 1980s and 1990s. From the technical level, technical transformation is undoubtedly a manifestation of technical progress and the pursuit of innovation. However, in the wave of technological transformation, many people are confused to see the phenomenon of "not to engage in technical transformation is waiting for death, engage in technical transformation is looking for death". This cannot help but make some people feel confused. Kristensen's book, "The Innovator's Dilemma: When New Technologies Cause Great Companies to Fail," published in 1997, has drawn people's attention to the fact that innovators are not necessarily winners who always wear laurels, but they are also stuck in difficulties. Philosophically and practically, the process of engineering innovation is the process of innovators moving from "possibility space" to "real world," and the road of engineering innovation is a road full of uncertainties, which may lead to success or failure, and is a difficult road with many barriers and traps [11]. In general, engineering innovation risks mainly include technical, social, and environmental risks.

1. Technical risks of engineering innovation

Engineering innovation has to synthesize and integrate various engineering technologies. Still, in the process of engineering innovation, the constructive nature of innovation subjects and the complexity show that technology is by nature an uncertain existence accompanied by risks. Whether biotechnology in biological engineering, network technology in electronic communication engineering, or nanotechnology in material engineering, they have created one miracle after another for mankind in engineering innovation while bringing huge risks to the general public. Reflecting on modern society, the German scholar Ulrich Beck pointed out that the incalculability of their consequences accompanies the growth of the capacity for technological choices, the dangers of highly developed nuclear and chemical productivity, the destruction of the foundations and spheres on which people think and act, such as space and time, work and leisure, and the emergence of unknown and unpredictable consequences as the dominant force in society in a risk society [12]. The nonlinear development of modern technology has plunged humanity into an unprecedented risk environment, and the multiple risks induced by technological development have been cascaded through engineering innovations, triggering broader and longer-term uncertainties, leading to multiple technical, economic, and ethical risks. In modern bioengineering, the introduction of nanotechnology has brought this engineering innovation to an unprecedented level and height, but it also brings many potential risks. Humans can now produce nanoparticles as small as 1/7000th of a hair, and it is entirely possible for such tiny particles to enter the body through the alveoli and skin of the human body by simple diffusion and permeation, causing great harm to human health.

The technical risks in engineering innovation are complex and can be divided into the reasons of the technology itself and the reasons of the use side. The reasons for the technology itself are mainly the limitations of the technology itself, immaturity, failure to develop and safety technology to match the technology, etc. The reasons for using technology are mainly the improper operation of the personnel using a certain technology, technical products beyond the use of the period, the use of technology neglect, carelessness or will not use the security technology with it, etc. But in reality there may be more artificial reasons. For example, the tetracycline incident caused damage to the teeth of an entire generation; pesticides destroyed large tracts of fertile land and polluted the environment. But there is no right to require technical experts to provide completely certain knowledge and completely safe technology, because this is impossible [13]. However, if the process of innovation in engineering, the innovation body can be fully aware of the uncertainty of technology. To control and manage the entire innovation process, technology can reduce the risk of disasters caused.

2. Environmental risks of engineering innovation

Any engineering innovation activity is done in the natural environment and the environment of people. The development of engineering innovation will undoubtedly have an impact on the original environment, and thus many unpredictable environmental risks will appear. The development and utilization of nuclear energy is a major innovation of human energy engineering. Nuclear engineering will produce nuclear waste or radioactive waste while providing clean and efficient energy for human beings, which brings unprecedented environmental risks to the environment of human existence; water conservancy projects bring great economic benefits and social interests to society, but improper design may also destroy natural resources and ecological environment. For example, Sanmenxia project case mentioned in Sect. 4.4.2. In 1964, the Soviet Union built the Aswan Dam on the Nile River for Egypt. Before the dam was completed, the Nile River transported sediment to the Mediterranean every year and made it silt up to the coast at roughly the same rate as seawater erosion on the shore. After the dam was completed, the sediment was trapped in the reservoir, the erosion of the Mediterranean coast was not compensated, and the inherent natural balance was destroyed.

3. Social risks of engineering innovation

Generally speaking, the economic benefit is one of the core goals of engineering innovation. Engineering innovation projects will generate a lot of investment and with investment will bring economic benefits. However, it should also be noted that engineering innovation is closely linked to the interests of the public. To some extent, engineering innovation is a social activity, and engineering innovation is ultimately for people. "People-oriented" is the basic concept of engineering innovation. The Three Gorges Project, Qinghai-Tibet Railway Project, South-North Water Diversion Project, high-speed railroad are based on the concept of "people-oriented and for people", and the interests of all stakeholders are fully considered in the decision-making, design, construction, and operation of engineering innovation, thus getting the support and cooperation of most people. On the contrary, some projects start only from political achievements, regional interests, and local interests, which damage the interests of other stakeholders trigger mass incidents, and cause social risks. In recent years, some regions have seen outbreaks of mass incidents caused by the siting of large infrastructure projects. Some of the most influential ones include the resistance of Xiamen citizens to the location of PX (dimethyl benzene) chemical project in the city and the resistance of Guangzhou Panyu residents to the establishment of a domestic waste incinerator near Huijiang Village in Dashi Street, Panyu District. The so-called "neighborhood avoidance projects" are public facilities that bring convenience and welfare to the public, but negatively impact the residents nearby and cause neighborhood avoidance, such as nuclear power plants, chemical plants, coal and waste treatment plants, chemical plants, coal, and waste disposal stations. The Chinese government attaches great importance to preventing social risks of "neighborhood avoidance projects". During the 12th Five-Year Plan period, it successively formulated the "Guidelines for Investment Project Feasibility Study", "Interim Measures for Public Participation in Environmental Impact Assessment," and "Regulations on Expropriation and Compensation of Houses on State-owned Land" which effectively reduce the social risks of "neighborhood avoidance projects" through legislation.

7.2 Engineering Innovation Target and Mode

Human beings carry out practical activities under the guidance of rationality. As an important human practical activity, engineering innovation must consider rationality. Engineering innovation activities under rational guidance must answer what innovation is for. There is no doubt that the goal of engineering innovation is people-oriented to create a more suitable living and development environment for human beings. Modern engineering innovation activities are complex giant systems, with various independent and interrelated innovation elements. The synergy of these elements is the key to determining the success or failure of innovation, and collaborative innovation is the primary mode of engineering innovation.

7.2.1 "People-Oriented"—The Goal of Engineering Innovation

People-oriented is to take people as the goal, pay attention to people's lives and values, and put people in the most respected position.

The "people-oriented" in engineering innovation not only advocates that people are the fundamental purpose of engineering activities but also tells why engineering innovation is pursued and for whom. It also advocates that people are the fundamental driving force of engineering innovation activities and describes how to pursue engineering innovation and rely on whom. "For whom" and "rely on whom" are inseparable. People are the fundamental purpose of engineering innovation activities and the fundamental driving force of engineering innovation activities. Everything is for people, and everything depends on people. The unity of the two constitutes the complete content of people-oriented [14]. All engineering innovations are for people, and meeting people's needs is the basic driving force for all engineering innovations. The famous American psychologist Maslow once divided the needs of people into five successive levels, namely, physiological needs, safety needs, emotional and belonging needs, the need for respect, and self-fulfillment needs. Maslow believed that when people's low-level needs are met, they will turn to achieve higher levels of needs. The other two needs, namely the need for knowledge and ***aesthetics***, are not included in Maslow's hierarchy of needs. He believes that these two needs should be between the need for respect and the need for self-realization. In an in-depth exploration of Maslow's hierarchy of needs theory, the needs mentioned above can be summarized into two needs: people's material and spiritual needs.

Since humankind has had engineering innovation activities, the content of engineering innovation has changed as the forms of engineering innovation become complicated. Still, each engineering innovation's driving force comes from satisfying people's material and spiritual needs. Water conservancy projects were firstly designed to meet people's needs in agricultural production. Housing construction projects were designed to meet people's needs of living. Road and bridge projects were designed to meet people's needs for traveling. Churches and temples were designed to meet people's needs of emotion and belonging. Museums and theaters were designed to meet people's needs of knowledge and aesthetics. In a word, starting from people's needs and aiming at people is the goal and value pursuit for all engineering innovations to occur and be carried out continuously. China's engineering innovation embodies the idea of being people-oriented everywhere.

On the other hand, all engineering innovations are creative practices of human beings, and no engineering innovation can occur without human beings. People in engineering innovation have both individual characteristics and collective characteristics. A community of engineering activities completes any project. They include managers, engineers, and general workers. The engineering activity community's knowledge, ability, and sense of responsibility determine the success or failure of engineering innovation. The engineering activity community ultimately determines the success of engineering innovation. A good engineering innovation first benefits from good engineering decision-makers. In increasingly large-scale and complex modern engineering, the decision-makers awareness of innovation, risk awareness, and decision-making capabilities will have an important impact on engineering innovation. It is difficult to guarantee the effectiveness of engineering innovation if it lacks engineering managers with scientific management knowledge and rich practical experience. The technical design, implementation, evaluation, and engineering operation in engineering innovation must rely on engineers and engineering personnel. China's current engineering quality often fails to reach the ideal state. In addition to the gap between China's technical level and advanced international levels, another important factor is that we lack a community of workers with high skills and good professional ethics. According to relevant information, most of the current projects in China are mainly completed by civilian workers. Therefore, cultivating an engineering innovation community that meets the needs of engineering innovation is an essential subject of engineering innovation in China.

The goal of engineering innovation is "people-oriented." Next, we should continue to ask what kind of "people" is the foundation. In Chinese history, "human" has never been an abstract individual who exists independently, but a "human" between heaven and earth. The idea of harmony between man and nature is the core idea of Chinese culture. Confucianism regards heaven, earth, and man as a holistic system, emphasizing the harmony and unity of heaven and humanity, nature, and man; emphasizing that man should respect, fear, and obey heaven while not denying man's subjectivity but advocating victory over heaven, combining man's subjectivity with respect for the laws of nature, combining obedience to heaven, respect for heaven and love for things with knowledge of heaven, participation in heaven and fight with fate. Therefore, the engineering innovation of "human-oriented" in the Chinese context is the engineering innovation with the quality of "unity of heaven and man."

Whether an engineering innovation is successful and has positive value depends on its innovation result. That is, whether the artificial nature obtained through engineering innovation truly realizes the harmony between man and nature, man and self, and man and society in the process of use.

7.2.2 Collaborative Innovation—A Model of Engineering Innovation

7.2.2.1 The Concept of Collaborative Innovation

Collaborative innovation is not a new concept in Chinese academia. It first appeared in the research on technological innovation. Along with the increasing scale and speed of technological innovation and the increasingly fierce market competition, a single enterprise's technological innovation can no longer meet the needs of its comprehensive competitiveness. Some scholars have begun to explore the issue of combining multiple organizations and departments to carry out technological innovation and then put forward the theoretical proposition of collaborative innovation.

The famous West German physicist Hermann Haken introduced the concept of synergy in his creation of the synergetic study [15]. Synergy refers to the nonlinear and complex interactions between subsystems that enable the whole to achieve effects that cannot be achieved by individuals alone. Since then, synergetics has been further developed in the field of management. The famous strategic management expert H. Igor Ansoff mentions in his book *Corporate Strategy* that the effect of making the overall benefits of the company greater than the sum of the individual components is called synergy and can be expressed as "$2 + 2 = 5$" or "$1 + 1 > 2$". In the book "Activating Invisible Assets," Japanese strategist Hiroyuki Itami defines synergy more strictly, breaking down Ansoff's concept of synergy into two parts: "complementary effects" and "synergistic effects," arguing that synergy is a way to maximize the performance of resources [16].

Thermas Fisher believes that collaborative innovation is systematically optimizing and joining the elements of each innovation agent; collaborative innovation can be analyzed in two dimensions: integration as well as interaction [17]. The integration dimension mainly includes knowledge, resources, actions, and performance, while the interaction dimension mainly refers to reciprocal knowledge sharing, optimal resource allocation, optimal synchronization of activities, and system matching among various innovation subjects. According to the different positions on the two dimensions, collaborative innovation is a process from the communication to coordination to cooperation to synergy [18]. Michael Gibbons believes that collaborative innovation integrates various innovation elements and the unimpeded flow of innovation resources within the system. Collaborative innovation is a value creation process with knowledge value-added as the core and enterprises, universities and research institutes, government, and education departments as the main innovation agents [19]. Chinese scholar Jin Chen argues that collaborative innovation is a large-span integrated innovative organization model developed by enterprises, government, knowledge production institutions (universities, research institutions), intermediaries, and users to achieve major technological innovations. Collaborative innovation is a new paradigm of today's science and technology innovation that promotes enterprises, universities, and research institutions to give full play to their respective capabilities and integrate complementary resources through the guidance of national will and institutional arrangement, to accelerate technology diffusion and industrialization and collaborate in industrial technology innovation and industrialization of scientific and technological achievements [20].

By synthesizing the above scholars' research on collaborative innovation, it is easy to see that collaborative innovation is an inevitable innovation mode in the context of contemporary scientific and technological development. In terms of engineering innovation, facing the explosive growth of knowledge, the increasingly complex innovation environment, and diversified innovation demands, neither enterprises, research institutions, universities, and colleges, nor governments can accomplish the goal of engineering innovation alone. A common innovation goal must lead all innovation participants, openly share, cooperate, support each other, and effectively integrate all innovation resources.

7.2.2.2 Government-Driven and Market-Driven—The Power Structure of Collaborative Engineering Innovation

Collaborative innovation has to integrate innovative organizations with different interests. In essence, from the viewpoint of natural human attributes, there is no inevitable need for individual organizations to cooperate to maximize organizational interests. It is difficult to imagine any organization that can successfully control all the innovation elements to actively seek other partners for collaborative innovation. No person or organization is willing to share the benefits initially theirs to other individuals and organizations. So what is driving the growing number of innovative organizations ready to share their creative work with other innovative organizations for collaborative innovation? Or in other words, where does the impetus for collaborative innovation come from? This requires analyzing in the context of two major areas of human activity. Although there is a wide range of human activities, for example, economic, cultural, political, and religious, they are all in the "private" and "public" domains. The "private sphere" is the field of activity of individual human beings as economic persons to realize their individual interests, and the driving mechanism of action is the market. The public sphere is the field of activity in which human beings, as social beings, carry out activities to meet the needs of the survival and development of the community. The driving mechanism for action is the government.

Judging from the history of Chinese engineering innovation, ancient Chinese engineering is mostly creation activities in the "public domain." Whether it is water conservancy projects, military projects, or road and bridge projects of various dynasties and generations, most of them are to meet the needs of human survival and development. Therefore, many ancient Chinese projects were promoted by rulers on behalf of the country. For example, the famous Great Wall and Dujiangyan. In addition, the palace and tomb projects reflecting the rulers' interests were also carried out in the state's name because, in the era of feudal emperors, the monarch represented the state. With the rise of capitalism, the market economy has become a common mode of economic development. Human activities under the market mechanism have changed from "public domain" to "private domain". Interests have become the driving force for engineering activities, and the market has been used to distribute benefits to become the biggest driving force to promote engineering innovation. As we all know, modern engineering innovation in the West is mainly driven by the market. However, since the capitalist economy has not been fully developed in China's modern history, there has been no real market-driven engineering innovation model. After the founding of New China, engineering innovation promoted by the government became the main engineering innovation model. From the creation of New China until the reform and opening up, China had a planned economy. The essence of a planned economy is to liberate people from the shackles of private ownership and let the government plan people's social practice activities, including engineering innovation activities, on behalf of the interests of all people. China's "two bombs and one star" project is a major engineering innovation activity completed under the government's careful planning and comprehensive leadership. However, as Xiaoping Deng once said, China is still in the primary stage of socialism and must develop a socialist market economy in a long historical period. Therefore, since 1978, China has entered a period of comprehensive development and prosperity of the market economy. In the process of market economy development for more than 30 years, China's engineering innovation activities have also gradually formed a market-driven collaborative innovation model. The engineering innovation activities under the market economy system are based on who invests and who benefits. China's real estate, high-speed railroad, and electronic communication engineering innovations have developed rapidly and played an essential role in improving China's comprehensive strength. However, the market economy also stimulates the pursuit of the private interests of human beings to the maximum extent. In order to realize their own interests, each organization and department involved in engineering innovation forms a community of interests to control the profits of industry innovation, ignoring public interests, resulting in the phenomenon of refusing to actively invest in engineering innovation that does not benefit much in the short term to meet the public interests of society. Therefore, in order to correct the negative effects of the market mechanism, since the 1990s, through the proposal of a national innovation system to the dissemination of collaborative innovation concept, China's engineering collaborative innovation began to change to the dynamic structure of state-driven and market-driven gradually. The collaborative innovation of major projects such as space engineering, Qinghai-Tibet railroad project, railroad speed-up project, Yangtze River Gorge project, etc., are implemented and successfully promoted by these two powers.

7.2.2.3 The Combination of Government, Industry, Academia, Research, and Application—The Organizational Form of Collaborative Engineering Innovation

In September 1992, the Chinese government decided to implement the manned space project and established a three-step development strategy: the first step was to launch a manned spacecraft, build a preliminary experimental manned spacecraft project, and conduct experiments on space applications. The second step was to break through the technology of astronaut space exit activities, space vehicle rendezvous and docking, and launch a space laboratory to solve the problem of short-term manned space applications on a certain scale. The third step was to build a space station to solve the problem of long-term manned space applications on a larger scale. On November 20–21, 1999, Shenzhou-1 completed its mission, thus, China successfully conducted the first manned space engineering flight test. After completing the manned missions of Tiangong-1 and Shenzhou-10 in June 2013, Jing Haipeng and Chen Dong entered space with the Shenzhou-11 spacecraft on October 17, 2016. After flying for 33 days and conducting a series of scientific experiments, Shenzhou XI spacecraft returned successfully on November 18 by the Shenzhou XI spacecraft return capsule. This is another important milestone in the history of China's manned space career and another landmark achievement in building an innovative country. China's manned space program has conducted a total of 11 missions. The engineers and technicians participating in developing and testing overcome various technical difficulties, and adhere to independent innovation. They create a brilliant history of China's manned spaceflight, and achieve the continuous leap of manned space technology. It marks the second step of the manned space engineering mission that has achieved significant results, laying a good foundation for the future development of manned spaceflight and the construction of space stations. Reviewing the successful experience of China's manned space engineering is the adoption of the engineering innovation model of collaborative innovation. Just as the Shenzhou VI innovation team pointed out in the summary of its successful launch, the more highly integrated grand system engineering, the more active the socialist collaboration required. The successful implementation of China's manned space engineering results from the combination of government, industry, academia, research, and the application of collaborative innovation.

The government-industry-academia-research-application cooperation innovation mode indicates different cooperation methods and types of cooperation adopted among government, industry, academia, research, and application cooperation subjects. It is composed of multiple elements and is a complex system with internal structure and function [21], which is an important organizational form of Chinese engineering innovation, especially large-scale engineering innovation.

(1) Guided by the government

We have already discussed that both government and the market drive engineering innovation in China. Still, because China's market economy has not been running for a long time, various innovation resources and production factors are mostly in a state of division. The experience in constructing large-scale public facilities related to people's livelihood is still insufficient, which is difficult to satisfy the demand for public welfare engineering innovation that cannot simply rely on the market mechanism. Therefore, for a certain time period, government's guiding role in engineering innovation should not be neglected, especially for livelihood engineering innovation such as high-speed railroad, south-north water transfer, and waste treatment. The government should not only guide through national development strategy but also need to give financial support through public financial investment. In the collaborative innovation process, the role of government is roughly reflected in macro guidance, policy guidance, interest integration, service guarantee, financial support, etc.

(2) Industry base

Enterprises are the practical subjects of engineering innovation and core elements in the collaborative innovation system. However, the enterprises involved in engineering innovation are not discrete enterprises unrelated to each other. Instead, enterprise alliances are closely connected by industrial chains. Therefore, collaborative innovation in engineering is based on industry, which is perhaps an important difference between collaborative innovation in engineering and technical activities. In the process of technological innovation, each independent enterprise is the main body of collaborative innovation. While enterprise synergy in engineering is based on the synergy of the industrial chain, which connects different types of enterprises to form an enterprise alliance around engineering innovation goals, integrates all kinds of innovation elements systematically by using market mechanism and benefit-sharing mechanism, and realizes the value of collaborative innovation together.

(3) Supported by high technology

Along with the emergence of artificial intelligence, big data, quantum technology, nanotechnology, new energy, and material technology, science and technology in the 21st century are advancing at an unimaginable speed for human beings. The new industrial revolution has come to us; the emerging high technology has become the supporting force of future engineering innovation. New science and technology development requires engineering in the form of collaborative innovation more than ever before for innovation activities. Major scientific discoveries and technological inventions provide more possibilities and ways for a new integration of innovation resources in engineering activities and provide a broader platform for generating and using new knowledge in engineering, and the progress of science and technology promotes the interaction and cooperation among universities, research institutes and enterprises, which become the supporting elements of collaborative innovation.

(4) User-oriented

Collaborative innovation promotes enterprises, universities, and research institutes to give full effect to their respective capabilities and advantages, integrate complementary resources, and realize each party's advantages, accelerating technology promotion, application, and industrialization through government guidance and institutional arrangements. The elements of the collaborative innovation system are not mechanically stacked together. The synergy of the elements needs to be generated from within or outside the system to break the original balance. For engineering innovation, the system equilibrium is broken because the innovation subject discovers or creates "new users" and user needs or potential needs become the driving force for the recombination and action of all elements of the collaborative innovation system. User demands are the root of the synergy of the elements of collaborative innovation. Therefore, user needs are always the directional mark of collaborative innovation and an important initial condition leading to collaborative innovation.

In short, engineering innovation is achieved through collaborative innovation. Engineering innovation is a multi-element integrated system and a dynamic process full of risks. The integration of various elements in the entire process of engineering innovation activities is not automatically achieved. The selection and construction of elements by the subject of engineering activity is also not accidental and random. In engineering innovation activities, the integration of technical elements and between technical and non-technical elements must be effectively organized and managed. Unlike general engineering management, engineering innovation management is the management around the realization of innovation goals or engineering management based on innovation.

7.3 Engineering Innovation Management

7.3.1 Concept of Engineering Management Innovation

Engineering innovation management, in short, is engineering management for achieving innovation goals. Specifically, engineering innovation management is based on engineering innovation. Through a time-limited flexible organization, efficient decision-making, planning, organization, command, coordination, and control of engineering innovation are carried out to achieve the overall goal of engineering innovation. The object of engineering innovation management is the process of engineering activities, and the main body of engineering innovation management is all actors participating in the management of engineering innovation activities. At the macro level, the government mainly supports and regulates engineering innovation activities through policy tools. At the meso level, enterprises, universities, research institutions, and public service institutions of science and technology innovation manage engineering innovation activities by constructing various public innovation platforms, incentive mechanisms, and cooperative organization systems. At the micro-level, there are managers of specific engineering projects who mainly realize the goal of engineering innovation through the management of particular engineering innovation projects. Engineering innovation management should have the following characteristics:

(1) Engineering innovation management is the whole process management

Engineering innovation activities start from the engineering decision, through design, construction, and engineering operation until the decommissioning of the project. The process nature of engineering innovation makes the management of engineering innovation must be a whole process, not just single link management of engineering technological innovation, engineering investment innovation, and engineering management innovation. The process nature of engineering innovation management requires implementing dynamic whole process control of engineering innovation management. The essence of whole-process engineering innovation management is to enhance the core competitiveness of engineering innovation as the goal, take innovative engineering concept as the guide, take engineering technology innovation as the basis, take the integration of various innovations (input innovation, management innovation, decommissioning innovation, evaluation innovation and decision innovation) as the means, and through effective management mechanism, methods and means, strive to achieve that every innovation subject in engineering innovation activities actively pursues innovation in each link and each period.

(2) Engineering innovation management is a complex system management

Engineering innovation is a collective selection and construction activity of the engineering innovation community in a complex spatial and temporal structure. In the whole activity process, various innovations such as science, technology, capital, policy, system, culture, resources, and innovation subjects form a complex net-like structure. The scale of engineering activities can be huge, such as the Three Gorges Project, South-North Water Transfer Project, and Manned Space Shuttle Project. They often involve tens of billions of capital investment and tens of thousands or even hundreds of thousands of human resources investment. They require integrating knowledge resources from multiple disciplines such as science, technology, finance, management, society, culture, policy, and law. Facing such a complex engineering innovation system, any linear management thinking and management mode cannot effectively manage engineering innovation. Therefore, it is necessary to adopt a complex system management method, establish a systematic engineering innovation management platform, integrate human resources, scientific and technological resources, natural conditions, ecological environment constraints, culture, policies, systems, and other elements of engineering innovation through engineering innovation public service institutions using information technology, build engineering innovation information network system, and most effectively realize the synergy and integration of various engineering innovation resources.

(3) Engineering innovation management is risk management

The engineering innovation process is full of barriers and traps, and high risk is an important feature. Engineering innovation risk includes decision risk, technology risk, market risk, organization risk, capital risk, information management risk, policy risk, external environment risk, etc. In this way, engineering innovation management is actually risk management in the engineering innovation environment, which includes innovation risk identification, risk assessment, innovation risk warning, and risk prevention.

7.3.2 Methodology of Engineering Management Innovation

Engineering innovation management is to take engineering innovation as the object and carry out highly efficient decision-making, planning, organizing, commanding, coordinating, and controlling activities for engineering innovation through the time-bound flexible organization in order to achieve the overall goal of engineering innovation. Since engineering innovation is complex, high-risk, and social, it determines that engineering innovation management methods cannot stay only in the specific methods unfolded from a purely economic-technical perspective. Engineering innovation management is not only technical–economic activity but also social management activity, which contains a profound philosophical connotation in engineering innovation management methods and substantially guides and influences the practice and development of engineering innovation management. Therefore, it is necessary to think philosophically about engineering innovation management methods, raise it to the level of philosophical methodology, and extract some regularities from it.

(1) Systems harmony theory of engineering innovation management

Engineering innovation management is a giant complex system, including engineering technology, social, ecological environment, and policy subsystems. Each subsystem is composed of multiple heterogeneous elements. Harmony and unity must be achieved between each subsystem and the elements of each subsystem to realize the overall optimal engineering innovation goal. Harmony is the basic requirement of engineering innovation management. The systems harmony of engineering innovation management is mainly manifested in the following aspects.

1. Harmony between different innovation subjects

The subjects of engineering innovation are diverse. They belong to different subsystems and are all stakeholders of engineering innovation activities. Their interest demands are not the same, and most of them tend to maximize their own interests. To ensure that engineering innovation is generated, each participant should have a consensus and expectation that the innovation results can be reasonably distributed while the innovation brings new value-added. In the current engineering innovation, problems such as mismatch of innovators' rank and mismatch of reward and risk are significant. Engineering subjects should fully use legal guidelines, industry conventions, and mutual trust among stakeholders to breed harmony. In particular, a harmonious value allocation mechanism should be developed to match the possible contribution of each participant and the degree of risk they take to mitigate the negative impact of phenomena such as damage to the interests of vested interests and disruption of the balance of the original organizational structure that innovation may bring. In this way, each participant with high satisfaction may form a harmonious force through effective checks and balances and improve the management effectiveness of engineering innovation.

2. Harmony between different innovation subsystems

Engineering innovation management aims to achieve innovation goals through the optimal combination of subsystems. In this process, the harmony of the engineering technology system with social and natural environment systems is the basic principle of management goal. Successful engineering innovation should consider the advanced technology and the degree of support of economic and natural ecological systems for technological innovation. Engineering innovation management should make trade-offs among the above three, ensure moderately advanced technology, and consider the economic and ecological costs. Therefore, a comprehensive assessment of risks and benefits of technological innovation and economic and ecological costs must be carried out in engineering innovation. In addition, any engineering innovation should promote social equity and justice, not the other way around. Whether it is power engineering or transportation engineering innovation, it is unjust if it is not to promote the welfare of entire society, but only to benefit some people and some regions. The management of engineering innovation achieves the social harmony of engineering innovation through democratic participation and public post-use evaluation mechanisms. Therefore, effective engineering innovation management is marked by the ultimate realization of harmony between human and nature and human and human.

3. The development of various industries in engineering innovation should be harmonious

The interconnectedness of engineering-related industries in terms of supply and demand, technology, and especially innovation objectively requires that the industries' development must be coordinated and interact in a chain. It is hard to imagine at what stage the development of mining, transportation, and security will be without the innovation of information engineering. The development of medical and fermentation engineering will be restricted without innovation in biological engineering. Industrial and material engineering development will lag without the innovation in energy and mining engineering. Therefore, in order to maximize the comprehensive value of engineering innovation, we should first ensure the innovation in basic scientific theory and technology application. We should establish an open and harmonious network, balance the development speed of each industry, make the cake of products or services bigger in the harmonious background, and then share this cake together according to the pre-defined rules. In other words, the innovation of various industries within engineering should be combined to build a win–win innovation network and learn from each other's innovation achievements as much as possible. To make full use of the amplification effect of innovation achievements and promote the innovation of various industries to achieve the goal of engineering innovation through collaborative innovation.

(2) Comprehensive and integrated theory of engineering innovation management

The idea of comprehensive management originates from the systems management idea. The core of complexity management theory lies in the fact that the object to be managed is a complex system, as such, the relevant theory of complex system is to be used for management. Integrated management is a management concept based on systems methodology. Integrated management does not form general management capabilities through the "linear" integration of management resources, but through the "nonlinear" integration, thus new management capabilities emerge. For example, through the combination of expert systems, knowledge systems, and computer systems, complex engineering management problems can be formed, and the key technology can be tackled through technology integration. Therefore, the essence of integrated management behavior is not the pursuit of mastering as many management resources as possible but the construction of a platform with the ability to manage complexity. This platform provides an effective management organization, management system, and mechanism.

Integrated management mainly focuses on managing system complexity. Considering the main causes of complexity, it is natural that management needs to consider each subject's objectives, the unity of subject, object, and environment, coordination of the relationship between benefits and risks, and the consistency of methodological diversity and comprehensive effect. All these are concentrated on one point: integrated management needs to adhere to the principle of integration from the perspective of objectives, operations, technologies, and methods.

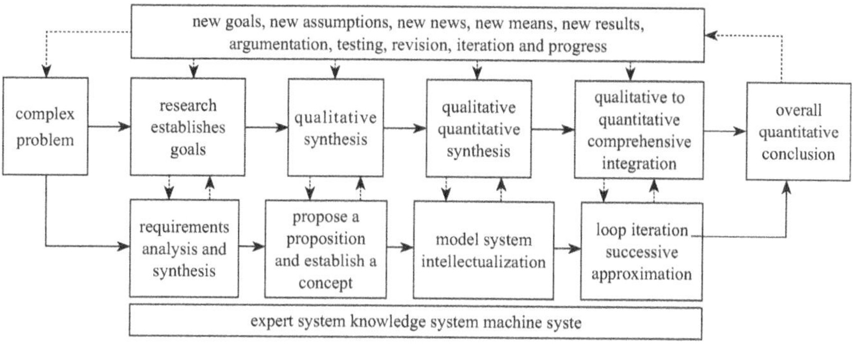

Fig. 7.1 General steps for comprehensive integration from qualitative to quantitative

As mentioned before, engineering innovation is a complex system, and the methodology of comprehensive integration must be adhered to in engineering innovation management. First of all, the subject of implementing management is comprehensive and integrated. It is an organization composed of multiple subjects, and such an organization generally has a core subject. Individual subjects may be replaced at different management stages and within the organization. Through learning and adaptation, a platform is built to develop the management ability to manage complexity. The subject's ability does not lie in its ability to have resources but in the ability to design systems, choose mechanisms, and other comprehensive coordination capabilities.

The comprehensive integration in engineering innovation management is mainly achieved through comprehensive integration from qualitative to quantitative aspects. Qualitative research is the premise of quantitative research, while quantitative research provides the depth of qualitative research. Qualitative and quantitative integration in integrated management is essentially the combination of science and experience, the unification of logical and image thinking, and its general steps are shown in Fig. 7.1 [22].

(3) Dialectical unity of engineering innovation management

With technological innovation as the core, engineering innovation integrates various innovation resources facing increasing complex knowledge and technical creation activities. Reasonable and scientific engineering innovation management can increase engineering innovation practice's scientific and technological effects by the multiplier or even by exponential effect. It can integrate the multiple value objectives of engineering innovation practice at the level of strategic coordination. Engineering innovation management activities have far exceeded the scope of economy and technology and become a complex integrated activity, which needs to examine the problems in engineering innovation management activities with dialectical thinking.

1. The unification of economic and ecological effects in the process of engineering innovation management

Engineering innovation management must meet the natural value effect needed for human survival and sustainable social development. In recent times, engineering is often regarded as an activity to objectify the essential power of human beings. Human beings enjoy the pleasure and thrill of "conquest" in the practice of engineering projects. While taking profit growth as the only goal in engineering construction, people seldom consider the natural ecological background from which human economic activities cannot be separated. Thus the possible ecological effects of engineering activities are underestimated. The ecological effects of engineering activities are not sufficiently estimated, and even the possible ecological costs are intentionally avoided and ignored for local interests. As a kind of economic organizational behavior, one of the nature of engineering innovation management is "profit-seeking" which cannot exceed its economic purpose. However, suppose the economic benefit of engineering innovation is the goal, disregarding the sustainable development and protection of the natural ecological environment and the long-term interests of human future development, engineering innovation activities will gradually lose the support of natural resource base and social environment, and eventually become unsustainable. Therefore, engineering innovation management must change the single economic value pursuit. Engineering innovation management should develop in the direction of harmonious coexistence of humans and nature, harmonious development of humans and society, and harmonious life of humans.

2. The unification of the standardization and management innovation in the process of engineering innovation management

Engineering innovation management norms are the general term for various management regulations, systems, standards, methods, and codes. It is formulated to realize their values and goals in engineering innovation management. It stipulates the content, procedures, and methods of engineering innovation management activities in written form and is the code of conduct for engineering managers. The establishment of scientific engineering innovation management standards plays an important role in ensuring the normal and sustainable progress of engineering innovation activities and improving the level of engineering innovation management in the context of the continuous expansion of modern engineering scale and the increasingly diversified engineering value goals. However, engineering innovation management norms are not static. With the development of the times, technological progress, and management levels, engineering innovation management norms also need to be continuously improved and improved according to time, place, and people. The key point in this process is to correctly handle the relationship between "fixation" and "change", "break" and "establishment". Any management standard should be relatively stable, maintain its continuity within a certain period, and cannot be changed day by day; otherwise, it will cause the employees to be overwhelmed and confused. However, the stability of engineering innovation management norms is preceded by unstable "breaking". The process from "breaking" to "establishing" is essentially the process of continuous innovation of engineering management norms.

3. The unification of the management system and management details in the process of engineering innovation management

Engineering innovation management system is a functional whole composed of various elements and links in a certain space–time structure of engineering management according to a certain combination. Accordingly, engineering innovation management details are various elements and links constituting engineering innovation management system. The relationship between engineering innovation management system and engineering innovation management details is the relationship between whole and part from the spatial point of view. In terms of time it is the relationship between process and link. The two are interdependent. Without parts and links, there will be no whole and process. There is no such thing as parts and links without the whole and process. Engineering innovation management system as a whole and process is an organic combination of all parts and links of engineering innovation management. Its functions should be greater than the sum of all parts and links, and the whole and process have new functions that parts and links do not. When the parts and links form a whole and process in an orderly and optimized structure, the function of the whole and process will be greater than the sum of the functions of the parts and links. When the parts and links form the whole and the process in disorder and non-optimized structure, the original performance of the parts and links will be weakened or mutually offsetting, so that the overall and process function is less than the sum of the parts and links. Moreover, the low functional status of a certain part or link in engineering innovation management will result in the "barrel effect" and become the "bottleneck" of the engineering management system. It will hinder the smooth process of the engineering innovation management system and weaken the overall function of engineering innovation. In sum, the functional state, combination mode of all parts and links of engineering innovation management determine whether the process of the engineering management system is smooth and the function is excellent. Specifically, the most significant feature of the engineering innovation management system is its integrality and dominance. It stipulates the status and relationship of the subject and object of engineering construction and the basic definition of their respective rights and obligations. It also defines its own system operation procedures and ways, and has specific management means and methods compatible with it. The structural composition of the engineering innovation management system focuses on the outline of fundamental principles and the establishment of basic methods, rather than on the settlement of specific technical and operational details. Only when the engineering management system as a whole achieves reasonable structure and smooth process, all the elements and links can be in their right place and generate the systems effect of "overall function is greater than the sum of parts."

References

1. He, J., Wang, M., & Wang, Q. (2013). *The status Quo and the development of chinese engineering management*. Higher Education Press.
2. Xu, Z. (2011). An analysis of military technology progress and its transparency. *Science and Technology Management Research, 9*.
3. Schumpeter, J., & Backhaus, U. (2003). The theory of economic development. In Joseph Alois Schumpeter (pp. 61–116). Springer.
4. Yin, R. Engineering innovation is the main battlefield of technological progress. *Study Times*, 310.
5. Li, B. (2008). Several theoretical issues on engineering and engineering innovation. *Northern Forum, 2*, 101–105.
6. Yin, R. (2006). Understanding of engineering and engineering innovation. *Mineral Exploration, 8*, 21–24.
7. Cai, Q. (2008). Looking at engineering innovation from the perspective of quaternary knowledge chain. *Journal of Northeastern University, 5*, 387–391.
8. He, J., et al. (2008). Research on the interactive relationship between engineering harmony and engineering innovation. *Chinese Engineering Science, 12*, 4–9.
9. Li, B. (2004). A brief discussion on the ternary theory of science, technology, and engineering. In D. Cheng & L. Bocong (Eds.), *Engineering research* (Vol. 1, pp. 42–53). Beijing Institute of Technology Publishing.
10. Yin, R., Wang, Y., & Li, B. (2007). *Engineering philosophy* (pp. 18–23). Higher Education Press.
11. Li, B., et al. (2011). *Engineering innovation: Breaking the barriers and avoid traps* (p. 4). Zhejiang University Press.
12. Beck, U. (2004). *A critical introduction to risk society*. Pluto Press.
13. Zhang, Y. (2010). Research on the risk types of modern science and technology and its game. *Journal of Wuhan University of Science and Technology*, 6.
14. He, J. (2013). On the core of engineering management theory. *Engineering Science in China*, 11.
15. Haken, H. (1977). Synergetics. *Physics Bulletin, 28*(9), 412.
16. Serrano, V., & Fischer, T. (2007). Collaborative innovation in ubiquitous systems. *Journal of Intelligent Manufacturing, 18*(5), 599–615.
17. Duin, H., Jaskov, J., Hesmer, A., & Thoben, K. D. (2008). Towards a framework for collaborative innovation. In *Computer-aided innovation (CAI)* (pp. 193–204). Springer.
18. Gibbons, M. (2003). A new mode of knowledge production. In *Economic geography of higher education* (pp. 243–257). Routledge.
19. Chen, J., & Yang, Y. (2012). The theoretical basis and connotation of collaborative innovation. *Research in Science of Science, 2*, 161–164.
20. Xiong, L. (2011). Collaborative innovation research review-based on the perspective of realization approach. *Science and Technology Management Research, 14*, 15–18.
21. Wang, A. (2012). Research on collaborative innovation model of government, industry, university and research. *Geological Education in China, 4*, 40–43.
22. Sheng, Z., & You, Q. (2007). Comprehensive integrated management: Methodology and paradigm-exploration of Sutong bridge project management theory. *Complex Systems and Complexity Science, 6*, 1–9.

Chapter 8
Theory of Engineering Management Environment

The main research objects of engineering environment theory include engineering, environment and engineering environment, environment theory, and engineering environment theory.

Environmental theory is the sum of concepts, ideas, and theories about the environment and concepts, ideas, and theories about the relationship between human activities and their environment. The theory of engineering environment is a branch of the theory of environment, which is related to and connected with the theory of environment. The environmental theory is holistic and general, and the engineering environmental theory is partial and specific. The theory of engineering environment can be divided into micro, meso, and macro. Micro engineering environmental theory, such as environmental factor theory, mainly discusses the relationship between engineering activities and environmental factors. Meso engineering environmental theory, such as natural environment theory or social and cultural environment theory, mainly discusses the relationship between engineering activities and natural environment or social and cultural environment. Macro engineering environmental theory, such as comprehensive environmental theory, mainly involves holistically and comprehensively the relationship between engineering and the environment. In addition, the theory of the engineering environment can be divided into implicit theory and explicit theory. The former is a non-independent theory, which is included in the ancient comprehensive philosophy and scientific theory system, and presents ideas and thoughts in the form; the latter is an independent theory, which is the result of modern academic circles' special exploration of engineering environment problems, and presents the theory of self-consciousness in the form. The implicit and explicit are not necessarily related to the theory's depth, height, and influence; in fact, many engineering environmental theories with profound influence are implicit theories.

In this chapter, the theory of engineering management environment makes a systematic theoretical summary and theoretical thinking of engineering environmental problems. Theoretical summary mainly combs historical concepts, thoughts, and views, while theoretical thinking mainly focuses on theoretical thinking and

J. He, *Principles of Engineering Management*,
https://doi.org/10.1007/978-981-99-1168-4_8

exploration of practical issues. The core of thinking and exploring is the environment of engineering activities and the relationship between engineering activities and the environment. We should take people's understanding in different periods as the main line, summarize the historical comprehensive theories and special theories, and show the authors' basic understanding of the contemporary engineering environment.

Environment, according to the definition of *Webster's Third New International English Dictionary*, refers to the surrounding things, specifically, refers to the surrounding conditions, influences and forces; they belong to the two categories of nature and social culture, so environment includes two parts of natural environment and social cultural environment [1]. The natural environment includes climate, land, biology and other elements; among them, climate includes air flow (wind), precipitation, dry humidity and other elements; land includes soil, water system, mountain system and other elements; biology includes animals, plants, microorganisms and other factors. The social and cultural environment includes political, economical, custom, cultural heritage, law, and other elements. In order to see clearly, the theoretical summary and thinking in this chapter take the environmental category and elements as the framework, take the natural environment and social and cultural environment as the two main lines, from the evolution of the engineering environment, historical theory, and contemporary theory, to the rethinking of the engineering environment problems, and build the system of the engineering environment theory.

On the one hand, engineering activities are constrained by the environment, and on the other hand, they, in turn, have different degrees of impact on the environment.

8.1 The Evolution of the Engineering Environment View

After entering the twenty-first century, in the global scope, the natural and social-cultural environment problems caused by engineering construction and operation are increasingly prominent, which seriously affect people's healthy life and the harmonious development of society. In order to achieve sustainable development, it is urgent to find or build a sound engineering environment view that is suitable for the times. It neither encourages "inaction" on the environment nor indulges "doing whatever you want" on the environment, but advocates seeking the balance and harmony between the two extremes. This section intends to review the engineering environment view and its basic connotation in different periods to provide a reference for seeking or constructing the engineering environment theory in the new era.

8.1.1 Mainstream Engineering Environment View in Different Civilization Stages

From low to high, human civilization has roughly gone through four stages: hunting culture, agricultural society, industrial civilization, and postindustrial civilization. The engineering environment view presents the following four main types in these four stages.

During the period of hunting civilization, people follow nature. Limited by the development of productive forces, the engineering activities in this period were mainly simple construction, such as building wood as a nest and digging earth as a hole. Its main feature is integrating engineering activities with production activities and survival needs. *Tao Te Ching* said: "for I am abstracted from the world, the world from nature, nature from the way, and the way from what is beneath abstraction [2]. This view ranks people at the end of the chain of nature, heaven, earth, and people. It shows the ancient people's awe of nature and the ancient people's engineering thinking mode.

During the period of agricultural civilization, people conform to nature. With the gradual deepening of the understanding of nature, human beings gradually adapted to nature. They used natural means to carry out more complex engineering activities, building houses, gardens, palaces, temples, and had other engineering achievements. The main characteristics of engineering activities in this period are conforming to nature, meeting the needs of human life, and the needs of faith, governance, and aesthetics. Conforming to the natural engineering environment concept, including many specific ideas, "the harmony of nature and human" is the core concept.

During the industrial civilization period, people conquered nature. Since the industrial revolution in the eighteenth century, the concept of "controlling nature and being the master of nature" has spread in the ideological circles, influencing the global society, guiding people to conquer nature, and creating a new civilization. Human beings began to control, transform and control nature through science and technology and began to carry out engineering activities on a large scale. Many artificial natural objects such as railways, large water conservancy facilities, and nuclear power plants have appeared one after another. Reducing the power of nature and enlarging the capacity of human beings is the ideological basis of conquering the natural engineering environment.

In the postindustrial civilization period, people pursue sustainable development. Postindustrial civilization is a new type of civilization without the drawbacks of industrial civilization. The serious disadvantage of industrial civilization is unsustainability, which is mainly manifested in environmental pollution and unequal consumption of resources between generations. From the process and results of challenging and demanding nature, people gradually find that human power is not omnipotent, and excessive use of the environment will negatively impact human beings. This prompted people to slow down the pace of development, rethink the relationship of "people, environment, and engineering," and start to pursue sustainable development. The concept of sustainable development, initiated in the west, is a modern theory of "the harmony of nature and human," which corresponds to the ancient Chinese idea of "the harmony of nature and human."

8.1.2 Basic Reflection on the Evolution of Engineering Environment View

The change of understanding of nature and the thinking on the results of engineering activities are the basis of the evolution and progress of engineering environmental view.

In the period of ignorance, the human passively adapts to and relies on nature and makes primitive demands on nature, such as taking tree leaves as clothing, caves as a residence, and no engineering activities. In the early stage of civilization, human beings learned to acquire and utilize complex nature and carry out engineering activities on a small scale by looking up, taking things far away, and getting close to them. After civilization's maturity, human beings have a deeper understanding of nature and a better understanding of themselves. They have begun to make extensive use of nature and large-scale engineering activities. Large-scale engineering activities discharge waste and harmful substances into nature. After accumulating over time, they seriously damage the ecological environment, seriously affect human survival and development, and make humans pay a lot. After paying the high cost, we began to find new development ways and strategies and began to find and build unique engineering environment views.

In the relationship of "human, environment, and engineering," the understanding of human power determines the view of the engineering environment to a great extent; therefore, re-thinking the view of the engineering environment can reveal its cognitive basis. In the relationship between humans and nature, underestimating the power of humans, overestimating the power of humans, and even treating humans as the ruler of nature will lead to the engineering environment view harmful to human survival and development.

Understanding nature and its laws also determine the view of the engineering environment to a great extent. Suppose the understanding of nature and its laws is insufficient. In that case, the view of engineering environment based on conquering nature and transforming nature is often not conducive to human survival and development. Fully understanding nature and its laws and taking respecting nature and its laws as the ideological basis, the engineering environment view can better guide the construction of artificial nature and is more conducive to the sustainable development and improvement of humans and society.

Understanding the social role of engineering activities also affects the engineering environment view. The process of engineering activities is often the process of social structure adjustment and social relationship reconstruction. Ignoring this adjustment and reconstruction will lead to an unhealthy engineering environment outlook and then to engineering activities that are not conducive to social health. Only when we fully understand the impact of engineering activities on social structure and social relations can we form a healthy concept of engineering environment, guide people to carry out engineering activities in social development and improvement, and promote social harmony and sustainable development.

8.2 The Historical Theories of Engineering Environment

Taking engineering activities and their relationship with the natural environment, social and cultural environment as the mainline, this chapter investigates, clarifies, and explains the historical theories of the engineering environment in ancient China and the West. It also provides a theoretical and practical reference for improving contemporary engineering management. From the available data, there are more ideas, thoughts, and theories about the engineering environment in ancient China, but less in the ancient West.

The historical theory of engineering natural environment includes climate environment theory, land environment theory, and biological environment theory. As for the historical idea of engineering social and cultural background, including many aspects; limited by available materials, this section only discusses five: political theory, economic theory, custom theory, cultural (heritage) theory, and legal theory.

8.2.1 The Historical Theories of Engineering Natural Environment

Engineering activities can significantly impact the natural environment, and the natural environment can also have a significant constraint on engineering activities. This is the ideological basis of the theory of engineering natural environment. The natural environment consists of three elements: climate, land, and biological. As for the historical idea of engineering natural environment, some related to three elements simultaneously, which is the comprehensive natural philosophy theory; some only related to one major element, which is the branch natural environment theory.

8.2.1.1 Natural Philosophy Theory

1. **Yin Yang theory**

Yin Yang Theory is the core theory of ancient Chinese natural philosophy. According to the definition of the *Great Dictionary of Philosophy*, Yin and Yang "originally refers to the orientation of sunshine," later "refers to two opposite states of Qi," and later used to compare social phenomena [3], represents the most basic opposite relationship of all things. The theory of Yin and Yang is natural philosophy first, followed by social philosophy and humanistic philosophy. This theory, referring to the difference between heaven and earth and sun and moon, divides all things in the world and their elements, attributes, features, functions, states, and principles into two categories: Yang and Yin, including heaven and earth, sun and moon, men and women, hardness and softness, exterior and interior, reality and emptiness, exterior and interior, up and down, front and back, movement and stillness, expiration and inspiration, and rain and shine. The theory of Yin and Yang regards the opposition, coexistence, mutual accommodation, and evolutionary development as the basic law of the growth, movement, and change of all things. As the basic theory of natural philosophy or cosmological philosophy, the theory of Yin and Yang echoes the contemporary theory of contradiction.

Yin Yang theory, in ancient China, is one of the basic theories of engineering natural environment theory. In engineering and the relationship between engineering and the natural environment, the theory of Yin and Yang pays special attention to the balance of Yin and Yang and the combination of Yin and Yang.

The balance of Ying and Yang is the prerequisite for all things in the world. According to *Yi Zhuan*, "one yin and one yang are called Tao" [4]. The ancients believed that "isolated yin means no life, and only yang means not last long" [5]. The balance of Yin and Yang is the basic principle to guide the decision-making, design, construction, and maintenance of engineering and engineering management activities. For example, if the rainfall is Yang and the water storage is Yin [6], in a specific area and within a particular time range, the amount of rain and the water storage capacity must be the same. The balance of Yin and Yang can ensure that the water does not flood and runoff, and the virtuous cycle of cloud generated rain. Under the situation of insufficient water storage capacity, people set up water conservancy projects to balance Yin and Yang. Before the Han Dynasty, people built embankments to enclose lakes and fight for farming lands from the water. In the Han Dynasty, there was a risk of insufficient water storage capacity. Therefore, people began to conceive and implement the Jianhu Lake (gradually silted up after the middle Tang Dynasty) project to increase water volume from sacrificing farmlands and promote agricultural production.

The combination of Yin and Yang is the basic law of the movement of all things in the world. According to the records of *Huainanzi*, "the former Yellow Emperor ruled the world, while Li Mu and Taishanji assisted him. They followed the sun and moon laws, the Qi of Yin and Yang, the four seasons, and the regular calendar [7]." The combination of Yin and Yang is an important concept to guide engineering and engineering management in planning, organization, and resource allocation. In architecture engineering, the ancients took the convex as the Yang and concave as the Yin, the mountain as the Yang, and the water as the Yin. In the site selection and design of the residential courtyard, hills and pools are considered. In site selection and arrangement of large-scale courtyard and forest garden, all considered hills and ponds. This point is still widely valued in the contemporary era. Taking the construction of new campuses of major universities as an example, even if the selected address did not have hills and ponds in the past, hills and ponds would be designed and built, even groups of hills and ponds or called small rivers. Wetland is Yin, and the combination of Yin and Yang brings the architecture layers, rhythm, and vitality.

2. **Doctrine of the mean**

According to the Great Dictionary of Philosophy, the mean is "a kind of cosmology, methodology and moral realm advocated by Confucianism" [8]. The word of the mean comes from *the Analects of Confucius*; *the Doctrine of the Mean* is integrated into the mean. *The Doctrine of the Mean* said: "not biased is called middle, and what is not easy is mediocrity. The middle, the right way of the world, the mean, the theorem of the world." [9] The doctrine of the mean, first of all, belongs to the philosophy of the universe, second to the philosophy of knowledge and philosophy of life. It is the basic theory of ancient Chinese cultural philosophy and the core theory of ancient Chinese management philosophy. Since the Song Dynasty, it has been regarded by Confucianism the thinking method of governance.

The two core ideas of the doctrine of the mean are "neutralization" and "holding the two and using the middle," which are the important thoughts guiding the engineering activities of the ancients. Neutralization that is, neutralization and harmony. "People's emotional activities such as happiness, anger, sorrow, and joy are not expressive, called neutralization. If it is expressed but moderate, achieving a harmonious state [9]". "Making it neutralized, the world will be in the right position, and everything will grow and breed [9]. Neutralization is an ecological idea. It emphasizes the balance of natural ecology and social culture ecology at the macro level, the balance of the internal ecology of all things, and the spiritual ecology of human beings at the micro-level. Engineering activities should consider the improvement of the quality of human life and consider the protection and improvement of the living conditions of all things in the world so that they are in their natural position and can reproduce forever. This requires human beings to pay attention to the diversity and differences of other species in nature in engineering activities, respect their right to live, living environment, and way of living, and let them grow freely and endlessly.

Holding the two and using the middle, namely grasping the opinions of both ends, and then using the moderate ones for the people [10], is an ecological thought, which not only pays attention to the balance of natural ecology but also emphasizes the balance of social and cultural ecology. In the aspect of nature, for natural resources, engineering activities can neither be not using nor over-using without restriction. In terms of social culture, for social resources, especially human resources and money, engineering activities have to use resources; meanwhile, engineering activities cannot use resources without restriction. It is not allowed to build everywhere at any time. We should "pay attention to the economical use of public expenditure, the welfare of the people, and employ the human resources appropriately" [11].

8.2.1.2 Climate Environment Theory

Climate refers to the temperature, humidity, precipitation, tide, air flow, and other conditions and changes in a specific region. In ancient China, there was a special calendar to record climate change. In *the Book of Rites Month Order*, it is mentioned that "in the past, Duke Zhou identified the twenty-four solar terms as a clear natural season, which was divided into 72 periods. So the climate was originated from Hao, and was set by Duke Zhou." [12] Climate change is cyclical, forming four seasons and solar terms, providing an important basis for people to carry out various activities, and providing important guidance for engineering activities. In engineering activities, it is necessary to consider adapting to the climate conditions, such as residential buildings. The climate conditions in the area are the important basis for the design and construction of density, style, roof slope, etc. Secondly, we should consider adjusting and optimizing the regional climate environment, such as town and village construction projects, and designing and building scattered ponds, which are important measures to increase water distribution, improve local air humidity, and regulate local precipitation and temperature. There are many ideas, thoughts, and theories about the engineering climate environment in ancient China, among which the concept of time has the most profound influence.

The Time View is the period and node of hot and cold changes, which requires all kinds of engineering activities to conform to the four seasons. *The Book of Changes on Wenyan* said, Wisdom is as omnipresent as the sun and the moon, and the administration is as orderly as the four seasons [13]. More deeply, the time view includes at least three meanings: according to time, appropriate time, and timely. The three meanings are not only independent but also cross each other.

"According to time" emphasizes that human activities, especially engineering activities, must follow "according to time." That is to say, "according to the time of the day, and human activities are carried out" [12]. On the one hand, "don't start before the time," [12] "when the time doesn't come, you can't force to initiate; you can't take action without researching things thoroughly." [14] To carry out appropriate activities at the proper time. On the other hand, we should not act against time. Ancient Chinese projects, especially large-scale government projects such as urban construction, all paid attention to the use of human resources not against the agricultural time; all engineering activities that violated the agricultural time were regarded as unethical actions of carrying out large-scale constructions and harassing the people and wasting money.

Appropriate time requires human activities, especially engineering activities, to comply with and adapt to time changes. In the book of *Shang Jun Yiyan*, it is mentioned that "if the governance is suitable for the time, it will not conflict" [15]. Here, "governance" is the management of human activities and the management of natural resources and social and cultural resources. Appropriate time view has guiding significance for planning, organization, resource development, and allocation on engineering activities.

"Timely" means that human activities, especially engineering activities, should seize the opportunity within the effective time range. If we miss the chance, we will inevitably have the result of "even if you work hard, it is hard to succeed" [16]. Many ancient projects, especially agricultural projects, were initiated in time. For example, the reservoir construction is mostly carried out in winter for two reasons: first, the winter is the slack season for farming, and the workforce is relatively sufficient;

second, the winter is the dry season, which is convenient for the project development. In the low-temperature area, the frozen situation can save material and workforce and shorten construction.

8.2.1.3 Land Environment Theory

Land is a natural resource on the surface of the earth. According to *Webster's New International English Dictionary*, the land is the earth's surface soil and all its natural resources [17]. Land includes the earth's surface soil and other natural resources on the earth's surface, such as water and mountains. Therefore, the land environment includes soil, water, and mountain systems. There are many historical theories about the engineering land environment. Due to the limited space, this section only discusses the ideas related to these three elements.

1. **Land capability view**

The land capability view is about its bearing capacity and maximum bearing limit. This concept requires that human activities, especially engineering activities and agricultural and animal husbandry production, be carried out within the limits of land carrying capacity. The land can accommodate all things, Zuo Qiuming said: "only the land can cover all things as one, and its things are not lost. It grows everything, raises birds and animals, and then enjoys the reputation and benefits it deserves" [18]. However, the carrying capacity of the land is limited. The number of animals, plants, and houses that can only be built in a year cannot exceed the upper limit. Therefore, according to the land carrying capacity, the ancients carried out engineering activities and agricultural, forestry, animal husbandry, and fishery production activities. For example, due to the limited living load of mountain trees, it is not recommended to deforest trees, and it is prohibited to cut trees in spring when plants grow. This concept first appeared in the ban of the Dayu period, "in spring, axes are prohibited in mountains and forests to facilitate the growth of vegetation" [19]. Later, Guan Zhong, Prime Minister of the State of Qi, also formulated the four prohibitions in spring, summer, autumn, and winter. His "spring ban" explicitly proposed that "cutting down forests and destroying the growth of vegetation is prohibited in spring" [20]. Mencius pointed out that "when the axes cut the forests in proper time, then the wood will be inexhaustible" [21]. Later, Xun Zi pointed out that "when the plants are flourishing, axes are not allowed to avoid damage and stop growing" [22]. For another example, due to the limited carrying capacity of aquatic animals such as fish and turtles in rivers and lakes, the fishing should not be too much. It should be carried out at a proper time. Confucius advocated "fishing with a hook instead of a net" [23], that is, not fishing in large quantities with nets, but only in small amounts with hooks. Mencius also said, "if you don't violate farming season, you will have plenty of grain; if you don't violate fishing season, you will have a lot of fish" [24]. Mencius also said, "if there is a five-acre house full of mulberry trees, people over 50 can wear silk padded clothes; chickens, dogs, and pigs, do not delay the breeding time, people over 70 years old can have meat to eat; a hundred-acre of arable land, as long as without losing the opportunity to cultivate and harvest, a family can have enough to eat… [24].” Exhausting the land capacity will eventually be harmful to human beings’ sustainable survival and development.

2. **Borrowing land view**

The concept of borrowing land refers to the concept of using the power of the land to carry out engineering activities and agricultural and animal husbandry production. In ancient China, the first way to borrow land was to borrow the power of water. “Water nourish all things, it makes all things get its benefits, but it does not conflict with all things” [25]. Water is the best way to nourish and benefit all things [26]. Waterpower has explicit power and implicit power; explicit power can be directly borrowed, and implicit power can be borrowed through special devices. For example, when a waterwheel is used to irrigate farmland, it can only be operated by the power of water.

“Where there is a waterwheel on the riverside, the weir and dam block the flow, around the waterwheel, force the wheel to turn, pull the water into the cylinder, and flow into the farmland. The waterwheel rotates around the clock, and there is no problem in irrigating 100 acres of land. When no water is needed, the waterwheel is tied to stop turning” [27]. With the help of waterpower, the wheel can run round the clock. Otherwise, it must “turn with the power of cattle, or gather several people to step on it” [27]. In addition, water power also promotes human engineering manufacturing. In order to adapt to the surrounding environment of the water area and facilitate daily production and life, the residents near the water area invented and built water tools or water buildings. The ancestors of ancient Yue in the lower reaches of the Yangtze River invented boat making techniques in order to carry out various production activities; [28] In Hemudu, Yuyao, the ancient Yue people created the “stilt architecture” to adapt to the natural environment with plenty of water and humidity [29].

3. **The view of complying land**

The view of complying with land refers to conforming to or using land or geographical conditions to carry out engineering and production activities. Land or geographical conditions include topography, land type, and land property.

First, based on the terrain. Soil, water system, and mountain system all have specific shapes and positions; Based on the terrain, we conform to or utilize the shapes and positions of soil, water, and mountain. When they built projects, especially houses and agricultural facilities, the ancients would conform to or make use of the terrain to achieve the state of human and nature coordination, artificial and natural cooperation. Based on the terrain, the first thing is to use the terrain. For example, most of the ancient garden construction is ingenious, while the ingenious things are based on terrain, [30] not only the garden base is selected based on the terrain, but also the layout, building, and scenery are designed based on the terrain. The second is along the terrain, especially the mountain. Most houses, in the middle of the mountain, at the edge of the mountain, or at the edge of the water, are built along the mountain or at the edge of the water. They are arranged in rows and layers, such as the Diaojiao buildings in the mountain areas of Northern Guangxi, Western Hunan and Western Hubei, which are generally located in the west to the east or in the east to the west to coordinate with the mountain system. Finally, adapting to the terrain, especially the flow of water. Houses and production facilities can be built according to the trend of water flow, and can also control or intercept the water flow; however, control or intercept, in a short time, there will be a flood, which will not only submerge or damage the built things but also affect the life of surrounding residents. The ancients opposed controlling or intercepting the water flow, advocated to control the water according to the trend of water flow, open up ditches or dig and dredge the existing river bed, to make the reasonable water storage and smooth flow [31]. In this way, the river water can achieve "never overflow" [32] and avoid flood disaster.

Second, based on the type of land. Land types have diversity, different functions, and different carrying capacities in animals and plants. "Plants and land (including soil and terrain) are interdependent, and each has its best symbiotic objects" [33]. In ancient China, engineering and production activities were carried out based on respecting land diversity. In terms of operation, use the land according to the land function, carry out building construction, grain, tree planting, and aquaculture; for example, the grain planting is not carried out in the non-arable land, and the building construction does not occupy the arable land. Therefore, do not arbitrarily change the pond into paddy fields or dry land, nor do residential projects such as filling the pond to build houses. In terms of principles and objectives, we should strive for "harmony is precious" [34], and harmony is the harmonious coexistence of people, people and things, and things and things, which is the most valuable relationship [35].

8.2.1.4 Biological Environment Theory

The biological environment, an important part of the natural environment, refers to the organic system composed of animals, plants, microorganisms, and other life forms. Chinese ancients attached great importance to the biological environment of human activities, especially engineering activities, and put forward some theories, including the following four influential ones.

1. **View of integration**

The view of integration emphasizes the coexistence of human beings and all things in the world, each having its own place, each in its own place, and integrating into a whole. This is a good state of "harmony of nature and humans." According to *Qi Wu Lun* written by Zhuangzi, "Heaven and earth coexist with me, and everything and I is one thing" [36]. Ancient large-scale engineering projects, from design to construction and then to operation, all pay attention to the coordination of human life and the survival of all things around to achieve the unity of things and self. Dujiangyan, a water conservancy project, is a typical case. This kind of project has one thing in common: abandon all extreme measures, optimize the ecosystem, make all things live forever so that human beings live forever. Su Shi expressed the elegant taste of living environment with "prefer to eat without meat, not to live without bamboo" in *Yu Qian Seng Lv Jun Xuan*.

2. **Harmony view**

The harmony view advocates coexistence, mutual promotion, and common prosperity of diversity and differentiated organisms. *Guoyu Zhengyu* said, "Harmony is the principle of creating things, and the diversity guarantee continuous and permanent" [37]. This ecological concept, which focuses on biodiversity and diversity, had a significant impact on the engineering activities in ancient China. The construction of large-scale gardens and large-scale buildings pays attention to species protection and even takes measures such as closing hillsides to facilitate afforestation for a long time to protect all kinds of creatures in the forest from growing safely and keeping growing.

3. **Cherishing life view**

The view of cherishing life advocates that human beings love living things and accumulate good fortune. To recognize the right and value of the existence of lives of non-human beings and to cherish and treat them kindly is to treat human beings themselves kindly. When the Confucians expound "benevolence," they often extend the moral category to all the living bodies in nature, from "being kind to people" to "being kind to creatures." Cherishing Life, in ancient China, was found in daily life, in agriculture, forestry, animal husbandry, fishery production, and more in engineering activities. Humans should cherish all living bodies, kill less or no wild animals, avoid killing livestock and animals in the breeding season, and not indiscriminately cut wild plants. It is to accumulate blessings for human beings and future generations.

8.2.2 Historical Theory on the Social and Cultural Environment of Engineering

Engineering activities not only interact with the natural environment but also with the social and cultural environment. The social and cultural environment can provide social and cultural resources for engineering activities. It can also affect engineering activities as structural factors penetrate into the results or products of engineering activities. For example, the Great Wall, the Forbidden City, and other engineering achievements reflect the social and cultural environment before their emergence and become an important part of the social and cultural environment after their emergence. Chinese ancients attached great importance to the social and cultural environment of engineering and put forward many theories related to politics, economy, law, and other dimensions.

8.2.2.1 Political Theory

The term "politics" was first used in the book of Shangshu Biming: "If the way of governance is right, politics can be well governed; if the benefits are spread, the people can live and work in peace and contentment." [10] According to *the Encyclopedia of China*, politics is the mutual relationship between classes, nations, and social groups based on their fundamental interests and related activities. Politics and engineering management are closely related. On the one hand, politics can influence engineering activities through engineering policies; on the other hand, engineering activities and their achievements can significantly impact the political situation. For example, the construction of agricultural, military, and water conservancy projects in the Wei Dynasty provided a solid material basis and a strong military supply guarantee for its economic recovery and development and subsequent military attack on Wu Kingdom, which to a certain extent determined the change of the political pattern in the later three kingdoms [38]. There are abundant engineering political thoughts in the ancient Chinese political thought system, among which there are two far-reaching ideas.

1. **The View of Engineering Decision-Making**

In ancient China, the concept of engineering decision-making was mainly formed around two cores: decision-making category and decision-making power.

First, the theory of "Baigong runs the country" advocates the diversity of engineering types. The general preface of *Zhouli0020· Kaogongji* points out: "there are six kinds of positions in the country, Baigong is one of them, … It is the responsibility of Baigong to fully understand the shape and performance of natural materials, and to apply labor to make them useful and to use them by the people according to the characteristics of the materials [39]." In the Eastern Han Dynasty, the note of Zheng Xuan to *Kaogongji* says: "Sikong managed Baigong. Sikong, in charge of the city, built the city, built the state temple, built the palace transportation facility and equipment, and supervised Baigong [40]." Baigong is the name of the official in charge of construction and manufacturing in the Zhou Dynasty, who manages all kinds of craftsmen, or skilled labor, in the country. Baigong is a kind of engineering talent and a kind of engineering category. It emphasizes the diversity of engineering types and engineering talents to meet the needs of the government, society, and the public.

Second, the official decision theory advocates that the decision-making power of the project, especially the government project, should be handed over to the government officials. Taking the capital construction as an example, the decision-making power is the power of location selection first, *Zhou Li* records that "the responsibility of the Dasitu is to take charge of the maps of the land of all states in the country and record the household registration of the people, to help emperor settle down in all states in the country [41]." The location selection of the capital is the responsibility of the head of the local officials, Dasitu. After the site selection of the capital city is determined, the facilities, project, and land allocation will be arranged by the "Liangren" in "Xiaguan." As for the important facilities in the capital, some corresponding officials are responsible for the establishment and arrangement. For example, the "xiaozongbo", the second in the "chunguan" series, is in charge of the position of the throne of the founding of the country and the sacrificial altar [42]." Engineering decision-making power is in the hands of officials, but the supreme decision-making power of major projects is in the hands of the supreme ruler. In addition to the record of *Dong Guan Kao Gong Ji*, the first sentence at the beginning of each chapter in *Zhouli* is: "the emperor builds the country, determines the direction and order, delegating the officials' positions, and benefiting the people."

2. **The view of engineering etiquette**

Etiquette, in ancient China, is not only a political system but also a moral standard. Etiquette, to "the order of seniors and juniors and superiority and inferiority in the imperial courts of the emperors and subjects, or the level of civilians' clothing, food, shelter, transportation, weddings, and funerals [43], has detailed regulations. The etiquette system can be seen in daily life, social life, and political life, which extends to engineering, forming an engineering etiquette system and constraining various engineering activities. The engineering etiquette system has the biggest restriction on construction activities.

First the etiquette system of capital palace construction. The site selection of ancient capital must conform to the "educational" principle of "stabilizing the country, the people, and entertaining the guests." The scale of the capital and the layout of the facilities in the capital must comply with the requirement of "politics," i.e., obeying the country, all civilians follow the right path, and full considerations of all." Infrastructure or living facilities in the capital, such as market and ancestral temple, should meet the requirements of "governance" and "etiquette." General urban construction also meets the requirements of "national prosperity, raising people, and growing everything." Not only that, the etiquette system has "the distinction of building", which has strict hierarchical regulations on the cities, offices, and houses of local rulers.

Second is the etiquette system of temple construction. There are also strict hierarchical regulations for altar and temple used for sacrifice. According to *the Shi Ji Qin Shihuang Ben Ji*, "the ancestral temple of the emperor was seven temples, offering sacrifices to the ancestors of seven generations, five temples for princes and three temples for officials, which could not be abolished even after the ages [43]."

Engineering etiquette system, like etiquette system, is an important means to maintain social stability. Not only that, objectively speaking, it is also an important means to protect resources, maintain ecology and promote sustainable development of society. Sima Qian affirmed the etiquette system, "Etiquette is the top priority in life. If the ruler follows the rules of etiquette, the country will be ruled and be safe. Otherwise, it will be disorder and danger [44]."

8.2.2.2 Economic Theory

Engineering activities are affected and restricted by economic development and promote economic growth and development. Ancient Chinese thinkers, politicians, economic scientists, and engineering scientists paid close attention to this interaction and interdependence and put forward many economic ideas. Among them, the concept of enriching the country, enriching the people, frugality, and making the best use has had a significant impact on engineering activities.

1. **Enriching the country view**

Mozi absorbed Confucius' thought of enriching the people and extended it among the ancient thinkers. He first theorized the theory of enriching the country and put forward it as the country's theory. "Prosperity of the country" is an important goal of governing the country. He pointed out that "those who are in charge of the state all pursue the prosperity of the country and the population growth." Later, Shang Jun school, Guan Zi, Xunzi, Sima Qian, Ligou, and others enriched the enriching country's theory.

The concept of enriching the country advocates increasing the total amount of national wealth, taking wealth as the foundation of engineering construction, and increasing wealth as the goal of engineering activities. Engineering activities are dependent on the material and financial resources of a country. The regulation and implementation of engineering activities need to consider the economic conditions of a country, while national prosperity and engineering activities promote each other, improve productivity and people's living standards, and enhance national strength. The concept of enriching the country is embodied in two aspects: first, the thought of attaching equal importance to agriculture, industry, and commerce. Agriculture, industry, and commerce are indispensable areas to solve the national economy and people's livelihood. They have their own ins and outs. According to Wang Fu of the Western Han Dynasty, "if those who enrich people take the branch for the root, the people will be poor [45]." It affirms the role of industry and commerce as engineering activities in social production and regards industry, commerce, and agriculture as important sources of social wealth. The common development of agriculture, industry, and commerce aims to solve people's food and clothing problems and provide the material basis for implementing engineering activities. Second, we should take measures that combine politics, economy, and military affairs. "Military order goes well along with internal affairs," "help people's living," "soldiers, peasants, workers, and merchants" do their own jobs and make efforts respectively"; "military affairs and civil affairs follow different management rules [46]," and the people works as peasants during the busy season for farming, and practice military training during the slack season of agriculture. In this way, the country can maintain a strong economic, political, and military strength, improving engineering activities and economic development together. It can be said that the concept of enriching country embodies the harmonious and unified development concept of engineering construction and economical construction, which has a strong practical significance.

2. **Enriching the people view**

The concept of enriching the people advocates increasing the national wealth, which is the basis of national governance. "The way of governing a country is to enrich the people first, and the governance is easy if the people are wealthy [47]. Enriching the people is the inevitable foundation of national governance and the premise and foundation of engineering activities. Only when the people have no worries about food and clothing and the basic living materials are guaranteed can the engineering activities be carried out, and the society will be stable during the engineering activities.

To be more specific, enriching people and engineering activities are interrelated in two aspects. First, take agriculture as the foundation. To develop agricultural production and achieve the goal of enriching the people and strengthening the country. Xunzi advocated "strengthening the foundation," strengthening agriculture, and vigorously developing agricultural production because agricultural production is the source of wealth and goods. The agriculture mentioned by Xunzi is big agriculture, including "fruits, peaches, plums, and other fruits," "vegetables, livestock, and poultry," and even "insects and other living things," which is today called agriculture, forestry, animal husbandry, fishing, and hunting. In ancient times, agriculture represented the level of productivity. Developing agriculture and revitalizing the economy was the cornerstone of engineering activities. Second, we need to recuperate, build strength, and reduce the tax burden. The concept of enriching the people advocates thrifty and benefiting the people, effectively lightens the people's burden to maintain the people's strength, cultivates the financial resources, and ensures that the farmers have enough land. Making farmers own land is the key to the development of agricultural production and the necessary condition to enrich the people.

3. **Frugality view**

In the ancient Chinese context, the view of frugality mainly refers to saving government expenditure. The concept of frugality can be seen in Confucianism and Mohism. The Analects of Confucius advocates "saving and loving [48]." Mozi developed this idea, clearly put forward the economic concept of "frugality," and advocated saving economic and financial resources in palace building, agricultural production, and other specific engineering activities, and no extravagant waste. XunYue developed the concept of "frugality" and proposed to save the people's resources and financial resources. "recognizing the rulers; rulers should meet the people's living needs. If the people exist, the country will exist. If the people don't exist, the country will be destroyed [49]."

There are three basic propositions of frugality view in engineering activities. First, to benefit the people. The principle is that engineering activities should serve and be used by the people. "The king is the ruler, and the orders and activities are issued for the people [50]." The king's administration, issuing orders, building projects, and using people's power and property is beneficial to the people. Second, it is set up according to needs. Mozi said in *Ci Guo*: "when the ancients didn't know how to build houses, they were close to the hills and dug holes to live in. The underground was wet and hurt their bodies. Therefore, the holy monarch built houses for." When the ancients didn't know how to build cars and boats, the heavy burden couldn't be carried and the distance couldn't be reached, so the Holy monarch came to make cars and boats for the convenience of the people [51]. "Engineering, whether it is construction or manufacturing, aims to facilitate life, rather than viewing and enjoying, which is the basic need. Beyond basic needs, higher needs are luxury and waste. Therefore, if the engineering activities are consistent with the population, economy, resources and living basis, the country will be peaceful and the people will be safe; otherwise, if the country is not peaceful, the people will be uneasy. After the unification of the six kingdoms, the emperor of Qin Shihuang built a large number of buildings, palaces, and Great Walls, which made a large number of migrant workers move to construction sites, seriously affecting agricultural production, and made social materials and social consumption not balanced. Therefore, the powerful Qin Empire only survived for 20 years and then collapsed. Third, enough is the degree. Mozi put forward that "it is easy to govern with the frugality of its people, and it is easy to forecast with the frugality of its monarch [52]." In Mozi's view, if the rulers are extremely luxurious, which leads to emulating, not only destroys the people's customs, but also consumes the state's financial resources, and eventually inevitably leads to extinction.

Frugality does not mean no need to use, but cannot be extravagant, which involves the issue of "degree." The concept of frugality has contained the modern concepts of sustainable development, nature protection, and ecological balance.

4. **The view of making the best use**

The view of making the best use advocates giving full play to all the available factors of materials, and making full use of them, and recycling them. For the engineering activities and agricultural production in ancient China, the concept of "use as much as possible" is mainly reflected in three aspects. First is the reuse of domestic waste. Wastes include human and animal excrement, vegetable roots and leaves, surplus food, etc. Human and animal excrement can be used as crop fertilizer; vegetable roots and leaves can be used as crop fertilizer and can also be used to feed livestock; surplus food can be used as livestock feed. *The book of Fan Shenzhi* in the Han dynasty recorded that people in the Shang Dynasty used excrement as fertilizer for crops, which showed that the concept and practice of making the best use of everything in agricultural production had appeared in the Shang Dynasty. Second is the chain development of agriculture, forestry, and animal husbandry. Planting fruit trees or mulberry trees on the fishpond bank will associate planting with breeding to form a virtuous circle production chain. The third is the recycling of waste and old things in housing construction. In the process of reconstruction or building the new in ancient times, almost all construction waste, old bricks and tiles, wood, stone, are reused. It reduces the consumption of new materials and effectively solves the problem of material transportation and waste disposal. For materials that cannot be directly used because of their damage, almost all of them can be reused: wood, which is converted into firewood; bricks and tiles, which are crushed and added with mud, or crushed used for pavement; stones, which are crushed and used as pavement, and even burned into other materials. For example, in the Han Dynasty, people burned white marble into lime to achieve secondary utilization. In rural China, the reuse of waste and old things of buildings has almost become a tradition that continues to this day.

The concept of making the best use, which is consistent with the core concept of modern circular economy, plays an active role in promoting sustainable development, natural protection, and resource conservation.

8.2.2.3 Custom Theory

Customs, referring to habits and customs, are gradually formed in the history of life, abiding by regional residents. They are the earliest social behavior norms in human social life [53]. The theory of custom is about the generation, change, form, significance, and function of custom. One part is the concept gradually formed in the process of history. After investigation, the other part is the thought put forward by scholars of all dynasties. The theory of custom influences and restricts the engineering activities in ancient China in many dimensions or levels. The influential concepts or thoughts are the group living together view, the ancestor view, the combining Yin and Yang view, the order of superiority and inferiority view.

1. **The group living together view**

Group living together means that people live in groups. It is not only a custom but also a concept. In ancient times, people used trees to build nests, find natural caves, avoid the wind and rain and the invasion of animals, or resist the attack of the outside world, forming a group living and lifestyle. From primitive tribes to clan communes to villages, the custom of gathering and living in groups has been followed up to now.

Geographical conditions affect the way of living in groups and affect the construction of residences to a large extent. For example, the northern Siheyuan was originally built to meet the needs of four or five generations of large families living together. During the Republic of China, due to the change of family structure, the large family was replaced by the small family, leading to multiple households living in one house. Although it is multifamily, it is still a group. In addition, the circular Tulou in the South and west of Fujian is designed to meet the needs of large families living in complex environments. The Tulou is huge in scale and complex in structure. Each building has four to five floors, and each floor ranges from a dozen rooms to dozens of rooms. Each household's door points to the center of the circle, which indicates that each household of the big family is united, fighting together, overcoming environmental difficulties, and creating a better life together.

2. **The ancestor view**

The view of heaven and ancestor supremacy is a kind of concept that worships heaven and ancestors. The ancient Chinese considered the world to be related to heaven from the beginning [54]. "Among the pre-Qin philosophers in China, especially in the words of Confucianism and Mohism, "heaven" refers to the existence of the most original source and the domination of the universe beyond all things. Confucius inherits the tradition and uses the metaphor of "heaven" to refer to the origin and supreme master of all things. He also believes that "heaven" is the creator of four seasons and all things and the determiner of human destiny." Confucius said, "because of heaven, there are four seasons, and all things grow [55]." Zhuangzi believed that "heaven" has mystery and naturalness. The sage should take heaven as the master and follow the choice and orientation of heaven. In *Zhuangzi Tian Xia*, Zhuangzi said: "take heaven as the supremacy, take virtue as the foundation, take Tao as the gate, and know the profound and unpredictable changes, which is called the sage [56]."

The heaven and ancestor supremacy view directly guides ancient Chinese engineering activities, especially in housing construction. For example, Hakkas in Meizhou, eastern Guangdong Province, have a strong sense of clan. They live together. The Weilongwu embodies Hakka's unique concept of ancestor worship. The semicircle pond in front of the house and the semicircle house in the back form a circle, and the three halls in the middle and the horizontal houses on both sides form a square. Surrounded by carved beams and painted rafters, the ancestral hall is spacious and solemn, and the ancestral niches are resplendent, showing the Hakka people's respect for their ancestors.

The ancestor worship in engineering is also an expression of the view of heaven and ancestor supremacy. In ancient China, engineering practitioners usually regarded the founders of the industry as their ancestors and gods, believing that they had the supernatural power to bless the industry and its practitioners. For example, the construction industry regards Luban as its founder and God, the paper industry regards Cailun as its founder and God, and the metallurgy industry regards Taishang Laojun as its founder and God. Ancestor worship strengthens the professional consciousness of the industry and strengthens the professional ethics of the industry.

3. **The order of superiority and inferiority view**

The concept of order of superiority and inferiority is a kind of concept to confirm the position or rank order. The order of superiority and inferiority exists in society, family, and all social organizations, including industry organizations. In the ancient Chinese concept, the order of superiority and inferiority still exists in space position and all things in the world. The spatial position is respected by the five directions in the east, south, west, north, and center; the center is superior; the position is also respected by the eight trigrams of qian, dui, li, zhen, xun, kan, gen, and kun. Qian is super. This concept has significantly impacted engineering activities, especially urban construction and housing construction activities. In the Shang Dynasty, there was a concept of "central", and there was a theory of "five sides" and "center" in the inscriptions. At the beginning of Zhou Dynasty, Luoyi city, the center of the world, was used as the capital, the concept of "center superiority" was clearer. *Lu's Spring and Autumn Period, Shen Shi* said: "the ancient king, choose the center of the world to build the country, choose the center of the country to build the palace, choose the center of the palace to build the temple [57]." The concept of "taking the center as the most important" first affected the construction site selection of the city. The center of the land is the center of the region or the whole country. The capital city is built in the center, which is convenient for governance.

The reason why the concept of order of superiority and inferiority affects engineering activities, especially urban construction and housing construction, is that through long-term observation, the ancestors realized that there are differences in land orientation and the length of day and night: "In the south, the scenery is short and hot; in the north, the scenery is long and cold; in the east, it is windy; in the west, it is cloudy [58]." But the center is "the combination of heaven and earth and the combination of the qi in the earth and the qi in the sky. As a result, there is no more heat and cold, no more wind and rain, no more Yin as Yin and Yang is combined. The four seasons of wind, rain, cold and heat are the combination of heaven and earth [58]." The ancients attached great importance to the center and the central axis. Looking at ancient Chinese cities, in the plane configuration of the capital city, it is not only inclined to place the imperial palace on the central or central axis of the city but also to highlight the imperial palace through a series of ways such as color, volume, combination, etc., making it the most prominent part of the central or central area. For example, the imperial palace of Beijing is in the center of Beijing at that time. The overall layout of the palace is symmetrical about the axis. The whole palace is arranged along a north–south axis, symmetrical about the left and right. The emperor and empress live in the palace on the axis. The courtyard building in Beijing takes the courtyard as the center and is surrounded by houses, so it is called "Siheyuan." Siheyuan adheres to the traditional concept of "the center is the most important" and emphasizes the axial symmetry layout. It usually has a central axis running through the whole courtyard, facing south from the north, symmetrical left, and right.

8.2.2.4 Cultural Heritage Theory

Cultural heritage is a part of the creation of our ancestors, both intangible and material. Cultural heritage has many values, such as history, science, technology, and art. The cultural heritage involved in engineering activities is the material cultural heritage in the environment on the surface and the intangible cultural heritage in the deep. In engineering activities, it is important to deal with everything left by their ancestors. The ancients had a lot of knowledge and ideas. Among them, there are two groups with profound influence.

1. **Observing, preserving, and imitating morality view**

The observing, preserving, and imitating morality view refers to the concept of observing, preserving, and imitating the morality of the ancestors. Specifically, when the descendants worship and show respect to the ancestral temple and relics of the ancestors, they can see the good morality of the ancestors and keep this morality in their hearts and consciously imitate it in the later social practice. "A man with morality view will surely get the status he deserves, the salary he deserves, and the life he deserves [59]." "The temple of the seven generations can reflect the morality of human beings. The emperor built the seven generations' temple, and the ancestors were all virtuous people whose temples would not be destroyed. So it can reflect morality." Li Daoyuan's *Shui Jing Zhu* in the Northern Wei Dynasty lists the names of the prefectures and counties where the watercourse passes and records the situation of the site of the old city, with notes and textual research attached. The names of places, places of interest, and historic sites where the watercourse passes are recorded in detail [60]. In order to present the ancestral morality to the future generations and make them advocate and imitate the ancestral morality, on the one hand, the ancients respected and paid attention to the protection of ancestral temples and relics in the engineering activities to protect the cultural heritage, on the other hand, they presented the ancestral morality through design and construction to continue the cultural tradition.

2. **Knowing, understanding, and undertaking destiny view**

In the context of Chinese culture, destiny mainly refers to the mission given by nature, that is, the various missions that human beings must undertake. The concept of knowing, understanding, and undertaking destiny is about knowing, understanding, and consciously assuming the mission according to the changes of the times. At the beginning of *The Doctrine of the Mean*, it said: "destiny is natural, nature is Tao, cultivating Tao is education," "destiny is mission [9]." Dajia said, "the reason why I have virtue is that heaven has given me [9]." For engineering activities, the concept of knowing, understand and undertaking destiny contains two meanings: one is to regard the protection and maintenance of material and intangible cultural heritage including the achievements of previous generations' engineering activities as a natural mission; the other is to creatively promote the development of engineering technology or technology according to the changes of the times, and create new engineering achievements to meet the needs of the times. For example, the construction of palaces and temples focuses on the combination of repair and expansion. The specific method is: to repair the original buildin, and expand the new building on or around the original building base. For example, in the Spring and Autumn period, Lu AI ordered people to build a temple on the basis of Confucius' former residence, and later generations repaired or expanded it. By the Ming Dynasty, it had basically formed the existing scale. Then in the Qing Dynasty, Yongzheng ordered the temple to be overhauled. There were nine courtyards in the temple, with the north and south as the central axis. They were divided into three rows, the left, the middle, and the right, 630 m in length and 140 m in width. There were more than 460 halls, altars, and pavilions, 54 gateways, and 13 royal stele pavilion, gradually formed today's Qufu Confucian Temple building group [61].

8.2.2.5 Legal Theory

Chinese law has a long history, originated in Xia and Shang Dynasties, and flourished in the pre-Qin era. The Chinese ancients paid attention to moral guidance in governing the country, but they did not give up the legal binding force, which is also true for engineering activities. The ancient Chinese attached great importance to the formulation and implementation of laws and regulations on engineering activities. Many theories about engineering laws and regulations involve many dimensions and levels. This section mainly discusses the theory of Hong Fan, the legal theory in building construction, and the concept of the engineering rule of law.

1. **The theory of Hong Fan**

Hong Fan is a part of *the Book of History* and the lawbook's name. According to *the Zhuan* in *the Book of History* written by Kong Anguo, "Hong, means big; Fan, means the law; Hong Fan means the great law in the world [62]." The contents of *Hong Fan* are composed of two parts: the compendium and the articles. There are nine laws in total, which are called nine categories in *Hong Fan.* The nine laws are as follows: "the first is five elements. The second is to do five things seriously. The third is to strive to implement eight kinds of government affairs. The fourth is to use five-time recording methods. The fifth is to use the rules of the king. The sixth is to use three kinds of good morality. The seventh is to establish officials in charge of divination. The eighth is to pay attention to the use of various symptoms. The ninth is to encourage people with five kinds of happiness and warn people with six kinds of misfortunes [63]."

The five elements theory is the basic guiding ideology of *Hong Fan. Hong Fan* five elements are water, fire, wood, metal, and earth; "water is flowing downward, Fire is the flaming upward.

Wood indicates right and wrong, metal is characterized by changing, earth means harvests grain [64]." In the minds of the ancient Chinese, the five elements of water, fire, wood, metal, and earth are the elements of all things and the fundamental resources of production and life. The five elements make all things develop and change. The balance of the five elements can bring benefits, while the imbalance of the five elements can bring disasters. Therefore, the ancient people have both worshiped and awe to them. In *Hong Fan,* the five elements are put first in the nine laws, reflecting the worship and awe and showing respect for nature and its laws.

The nine categories in Hong Fan, as the "great law of world" in ancient China, are the law of national governance and the law of engineering management. As a major law of engineering management, *Hong Fan* plays its role in two ways: one is to directly guide or influence engineering activities; the other is to give birth to subordinate laws and regulations to guide or influence engineering activities. After the Zhou Dynasty, the rule of law system continued to develop and improve, and the engineering rule of law gradually formed.

2. **The theory of building legal form**

The theory of building legal form refers to the construction activities regulated and restricted by standards and regulations. "Form is the law [65]." According to *the Xin Tang Shu. Xing Fa Zhi*, "the form is the law that is usually observed [66]." According to the Song Shi. Xing Fa Zhi, "what makes people abide by is the form [67]." A form is also a form of law. In ancient China, the theory of building a legal form appeared very early, until the Northern Song Dynasty, it was integrated into *the Ying Zao Fa Shi.* "(building a legal form). The *Ying Zao Fa Shi,* compiled by Li Jie, is a specification for architectural design and construction, which is promulgated and implemented by the government. The book takes the construction engineering as the object, technology as the core, combines technology and management, and guides or restricts engineering activities. The *Ying Zao Fa Shi* strictly formulates various architectural planning, design, and construction aspects.

The Ying Zao Fa Shi is formulated and adopted with the modular system for the planning and design of construction projects. The book details the "material system" and takes "material" as the basic unit of trade-off. The size, depth, length, and straightness of the house are all based on this system [68] so that all kinds of buildings can be built in accordance with the technical standards. Although there are strict regulations in various engineering construction systems, the *Ying Zao Fa Shi* does not specifically limit the layout of the building group organization and the plane size of the single building. All of them "add or subtract as appropriate" [69] give the designer a larger creative space so that the construction can be carried out according to local conditions and materials.

For the construction process, the *Ying Zao Fa Shi* summarizes and inherits the construction technology experience of the previous generation, which is of great help to the later engineering construction. Many traditional building technologies of the previous generation can help future generations avoid risks. For example, every column has "Cejiao" and "Shengqi" [69], which makes the structure incline inward, greatly increasing the stability of the overall structure. The Ying Zao Fa Shi stresses harmony and unity in terms of structure and decoration. For large and small wooden works, stone works, brick and tile works, and color paintings, there are detailed stripes and specific drawings. There are not only technical regulations on the shape, size, and construction method of the fixed column, bucket arch, and other component mechanisms, but also artistic processing methods.

In engineering management, the *Ying Zao Fa Shi* also has strict requirements. In order to put an end to corruption and waste, the book uses a lot of space to describe work limits and material examples. There are different calculation methods for labor quota and work type construction; in material examples, there are detailed and specific quotas for the use and consumption of various materials. These regulations are of positive guiding significance to project cost budget, project construction, project supervision, project acceptance, etc.

For engineering activities, the *Ying Zao Fa Shi* is a macro management regulation and a micro technical specification, which plays an important role in the history of China's engineering rule of law.

3. **The engineering rule of law view**

In China, the thought of "the rule of law" appeared in the Warring States period. Legalists believe that: compared with rites; the law is more objective, fair, and more able to reflect the meaning of "public"; in terms of effectiveness, the law is more significant than morality and rites, which can "divide and stop disputes, promote merit and stop violence," promote the unity, and enrich the country and strengthen the army [70]. For engineering activities, the rulers of all dynasties in China attach great importance to formulating and implementing laws and regulations. Engineering laws and regulations are quite complete, and every level and link of engineering activities are subject to laws and regulations. As China is a farming country, both the rulers and the common people are very concerned about the projects closely related to agriculture. Therefore, in agricultural and related projects, laws and regulations are the most detailed and abundant, among which water conservancy and operation and maintenance are the most representative.

The emergence of the water conservancy law is a sign of the development of water conservancy and the management of water conservancy projects. There were already water conservancy laws and regulations in the pre-Qin period—the book of *Li Ji Yue Ling* says: "In the months of spring when the rain will fall, the water will rise. Follow the fields around the country and the city, build levees, lead to ditches and roads without blocking them [71]." In Qin Dynasty, there were water conservancy projects or relevant engineering laws, such as *Gong Lv, Yao Lv, Si Kong Lv,* etc.; in Han Dynasty, *Jiu Zhang Lv* included water conservancy projects or relevant engineering laws, such as *Xing Lv*; in Tang Dynasty, there were water conservancy project laws and regulations, such as *Shui Bu Shi, Tang Lv Shu Yi, Ying Shan Ling, Trial Rules for Water Use in Dunhuang County, Shazhou,* etc.; in Song Dynasty, there were *Nong Tian Shui Li Yue Shu*; in the Jin Dynasty, there was the *He Fang Ling*; in the Ming and Qing Dynasties, there was a special volume of *He Fang* in the *Gong Lv*.

Among them, the Tang Dynasty *Shui Bu Shi*, issued by the central government, is the management regulations of water conservancy and water conservancy projects, including the management of farmland and water conservancy, the setting of rollers and the provisions of water consumption, the management and maintenance of navigation locks and bridge ferries, fisheries management and urban waterway management [72]. For example, in the construction and control of canals, the *Shui Bu Shi* stipulates that "Doumen shall be installed at the places where water is used for irrigation, such as the Jin, Wei, Bai River and other major canals, and stone and wood shall be piled up against the wall to make it firm, and it is not allowed to build weirs on the surface of canals. If there is a height of ground under the water of irrigation canals, they shall not be used as canals to build weirs, and Doumen shall lead them at the place with high upflow [73]." The *Shui Bu Shi* also stipulates that the quality of irrigation management will be an important basis for the assessment and promotion of relevant officials [72]. From these terms, the rulers pay attention to standardizing, restricting, and supervising engineering activities through laws and regulations.

In Tang Dynasty, there was a special building law called *Ying Shan Ling*. The *Ying Zao Fa Shi* compiled by Li Jie in the Northern Song Dynasty is also a law for building, which has strict regulations for architectural design and construction. In order to strengthen the management of the construction industry in the Qing Dynasty, in 1734, the Ministry of Engineering compiled and published a *Gong Cheng Zuo Fa Ze Li* as the basis for controlling the budget, practice, and materials of official engineering projects. The book's content is divided into two parts: the example of building and the quota of material estimation. It is the standard for craftsmen to build houses and the basis for the authorities to determine the acceptance procedures and verify the funding.

The theory of Hong Fan established the ideological main line of the rule of law in engineering. Later, laws and regulations such as *Ying Zao Fa Shi* followed the concept of the rule of law and implemented the idea of the rule of law in engineering to all levels and links of engineering activities. China's concept of the rule of law in engineering has been developed and improved continuously.

8.3 Modern and Contemporary Theories on Engineering Environment

With the development of history, the theory of engineering environment is developing. The environmental problems of engineering are highly concerned around the world. Government agencies, social organizations, academic groups, and individuals invest time and energy in theoretical and practical research to form a modern and contemporary theoretical system. The contemporary theory of engineering environment includes two parts: the theory of engineering natural environment and the theory of engineering social and cultural environment.

8.3.1 Modern and Contemporary Theories on the Natural Environment of Engineering

The modern and contemporary theories of engineering natural environment mainly include the theory of climate environment, that is, the view of only one earth, the view of the greenhouse effect and the view of climate change; the theory of land environment, that is, the view of transformation, the view of protection and the view of development; the theory of biological environment, that is, the view of ecological balance, the view of environmental friendliness, the view of harmonious development between human and nature, and the view of sustainable development.

8.3.1.1 Climate Environment Theory

As mentioned earlier, there is a mutual influence between engineering and the climate environment. Engineering activities that conform to the laws of nature can optimize the climate environment in a specific area; conversely, engineering activities that blindly pursue economic development at the cost of polluting the environment can cause adverse climate effects. With the development of society and the deepening of industrialization, more and more high-consumption and high-pollution engineering activities driven by economic benefits far exceed the self-regulation and purification capabilities of the environment, resulting in atmospheric pollution in varying degrees worldwide. In this context, the protection of the climate environment is the mainstream awareness of contemporary people on the relationship between engineering activities and the climate environment.

1. **The view of only one earth**

The view of only one earth says that human beings must take into account the environmental tolerance and the long-term interests of human beings when using natural resources for engineering and other activities. From June 5 to 16, 1972, the United Nations held a conference on the human environment in Stockholm, Sweden.

They announced the *Declaration of the United Nations Conference on Human Environment* (hereinafter referred to as the *Environmental Declaration*). The *Environmental Declaration* believes that in industrialized countries, environmental issues are generally related to industrialization and technological development; people must consider their consequences for the environment more carefully when carrying out various activities, and at the same time, for this and future generations considering the natural resources on the earth, including air, water, land, plants, and animals, must be protected through careful planning or proper management [74]. In engineering, this idea has important guiding significance for engineering decision-making, design, and implementation. Engineering activities cannot only take into account the immediate economic interests but should also fully consider the long-term impact on the environment and the interests of future generations of human beings after the implementation of the project.

2. **The view of the greenhouse effect**

The view of the greenhouse effect emphasizes "overall control of greenhouse gas emissions," especially preventing further deterioration of the greenhouse effect caused by engineering activities. This idea originated from the *United Nations Framework Convention on Climate Change* (UNFCCC) adopted at the United Nations Conference on Environment and Development (Earth Summit) held in Rio de Janeiro, Brazil, on June 4, 1992. According to the framework, "climate change" refers to changes caused by human activities directly or indirectly changing the composition of the earth's atmosphere in addition to natural elements. The framework's purpose is to stabilize the concentration of greenhouse gases in the atmosphere at a level that protects the climate system from damage. This level should be achieved within a time frame sufficient to enable ecosystems to adapt naturally to climate change, ensure that food production is not threatened, and sustainable economic development [75]. The framework is the world's first international convention to comprehensively control greenhouse gas emissions, such as carbon dioxide, in order to cope with the adverse effects of global warming on the human economy and society; it is the basic framework for international cooperation in response to global climate change issues; it is also the authoritative international framework to mitigate global greenhouse gas emissions. As for the greenhouse effect, we recognize that it is a necessary condition for life on the earth. However, the enhancement of the greenhouse effect caused by human activities and the climate change caused by human activities will significantly impact the global environment. At the same time, the impact of human activities on the greenhouse effect is very complex, which can be enhanced or weakened. Climate change can be effectively controlled through the joint efforts of human beings [76].

3. **The view of climate change**

The view of climate change said that human activities mainly cause climate change. For the common interests of all humankind, immediate action should be taken to reduce the impact of climate change and improve the current situation of climate degradation. This view stems from the open letter *Climate Change and the Integrity of Science* written by 255 academicians of the National Academy of Sciences in

2010, responding to the "climate gate" event in November 2009. According to the letter, a series of objective facts show that climate change caused by some human behaviors threatens the social and ecological environment in which we live. Although natural factors have always impacted the earth's climate change, the impact of human activities on climate change is more significant now. The rate of change of global warming is unprecedented in modern times. As a result, it will also cause other changes, such as accelerating the rate of sea-level rise, accelerating the speed of water cycle, increasing ocean acidity, etc., which further threaten coastal cities, food and water parks, marine and freshwater ecosystems, forests, alpine environment, etc. Finally, the letter calls for people to act immediately for the public interest, reduce the improper activities affecting climate change, and start to solve the problems that cause climate change [77]. The causes of climate warming are natural factors and human factors. Among them, human beings have a greater impact on climate change, which has become the consensus of most climate researchers. Since the industrial revolution, the concentration of greenhouse gases emitted by human beings has risen steadily. In 1850, the concentration of carbon dioxide in the atmosphere was 280 ppm, reaching 385.2 ppm in 2008. It has increased by 2–3 ppm per year and maintained a steady and continuous increase [78]. In the field of engineering, a large number of trees need to be cut down for land use, excessive energy consumption for major projects, a large number of greenhouse gas emissions, etc., which are all improper actions leading to climate change. In addition, climate change and engineering activities will induce permafrost changes, such as permafrost degradation, active layer thickness increase, permafrost temperature rise, etc. However, due to the differences in geology, geography, soil properties, and climate, the response process of permafrost to climate change and engineering activities is also different [79]. In recent decades, the number and scale of major engineering construction in China have been increasing, climate change, especially the increase of temperature, the rise of precipitation intensity, and frequent occurrence of extreme weather, will further affect the safety, stability, reliability, and durability of the project by affecting the facilities, important auxiliary equipment and the environment of the major project. It has a certain impact on major projects' operation efficiency and economic benefits. Climate change also impacts technical standards and engineering measures for major projects [80]. Therefore, how to reduce the adverse impact of the project on the climate is also the main issue to be considered by project decision makers and designers.

8.3.1.2 The Theory of Land Environment

The contemporary theory of land environment mainly involves three environmental elements: soil, water, and mountain. At present, the attitude towards the land environment has gone through two stages, along with two different theories, thoughts, or concepts. At first, people pursued material benefits, regardless of the consequences of the transformation of the land environment, until later, by environmental retaliation gradually realized that the maintenance of the land environment development is also the first factor to be considered in all engineering activities.

1. **Transformation view**

In the view of transformation, the natural resources that can create wealth or have the potential to create wealth should be transformed and utilized to meet human development needs. To a great extent, the view of transformation is influenced by the concept of "humans can conquer nature." In ancient China, the concept of "human can conquer nature" has always coexisted with the concept of "harmony of human and nature," which was particularly popular in the late twentieth century. In the aspect of the land environment, people carry the thought of "human can conquer nature," "there is nothing that you can think of while I can't do," regardless of the consequences, make great efforts to transform the land environment.

The transformation view once prevailed in the field of agriculture and agricultural engineering. In order to obtain more arable land and harvest more food, people have reclaimed the lake area and the deposition area in the lake. For example, Honghu Lake, Poyang Lake, Dongting Lake, Dianchi Lake, and other lakes are surrounded by reclaimed land, and the lake capacity is reduced. In the past, 800 Miles Dongting Lake was the largest freshwater lake in China. It was a famous place for fishing and a paradise for birds and other wild animals and plants. However, due to the competition of land and water, the wetland of Dongting Lake has shrunk a lot [81]. In addition, throughout the country, a large number of mountains and ponds have been changed into paddy fields, causing damage to hillside vegetation and the disappearance of water storage bodies and reducing trees or shrubs, resulting in soil erosion.

The transformation view was also popular in urban housing construction engineering and highway construction engineering. Urban housing construction projects and highway construction projects are important for the country and the people. However, in a long period, in the planning and implementation of the project, hills and ponds were removed, and materials were used, causing a considerable degree of damage to the land environment. For example, provincial or county-level roads are built to develop tourism in some mountainous areas. In the process of construction, in order to reduce transportation costs, many quarries are built along the road to obtain local materials. These quarries damage the natural landscape of the mountain area and cause natural disasters such as soil erosion and landslides. Moreover, the traffic also greatly impacts the wetland of the valley [82]. For example, in order to meet the needs of various construction engineering stone materials, quarries can be seen everywhere around some cities, which makes the soil layer formed in tens of millions of years disappear, makes the mountains and trees gradually scarce, and creates a patch of "bald spot," the roar of machines directly affects the normal life of nearby residents, the sand dust blocking the sun makes the city air seriously polluted, and makes the local and even further areas frequent sand dust storms [83].

2. **The view of protection**

The view of protection advocates building a harmonious artificial nature in engineering activities, taking into account land environmental protection, and ensuring that all engineering activities and their subsequent impacts are within the tolerance range of the land environment. Facing many natural disasters caused by the destruction of the land environment, people gradually realize that the environmental damage caused by human beings, the more efficient systems are needed to control it [84]. Therefore, in order to slow down or even stop the damage to the environment, people will take action to protect the environment. Under the guidance of this concept, the international community has developed environmental protection strategies. In 1972, the United Nations held a United Nations Conference on Human Environment in Stockholm, Sweden, attended by 113 countries, to discuss the action plan to protect the global environment and adopt the *Declaration on Human Environment*. June 5, the conference's opening day, suggested that it be designated as "World Environmental Protection Day." The Chinese government also attaches great importance to environmental protection. In recent years, the Chinese government has proposed to build a "resource-saving and environment-friendly" two oriented society, gradually putting environmental protection on the important agenda, and putting it into practice in engineering practice to timely treat the damaged environment and take protective measures for the projects to be carried out in advance.

In recent years, the view of protection has gradually penetrated into the engineering activities related to the three elements of the land environment. Returning farmland to the lake has been carried out in a large area in terms of the water system. It is an engineering measure to restore the lake area or the silted land in the lake which has been reclaimed into farmland to the lake. In order to promote the progress of the project, the state or the government has taken certain measures to support and carry out lake protection. For example, Dongting Lake wetland protection, with the support of Chinese government departments at all levels and the international community, has been listed in the Dongting Lake area and even put on the agenda of sustainable development of the national economy in the middle lower reaches of the Yangtze River. Each part of Dongting Lake has been listed as an important wetland in the world successively, and large-scale projects of returning farmland to lake, wetland restoration and protection have been carried out [81], following the road of sustainable development, and combining lake management with mountain and river management [85]. At the same time, to achieve the principle of harmonious coexistence between humans and water, from competing for land with water to actively retreating [86], so as to effectively curb the degradation trend of Dongting Lake wetland ecosystem. In terms of the mountain system, in order to protect the mountain system environment, the United Nations calls on governments to promote erosion control measures and encourage people to use low-cost, simple, easy-to-use, resource-saving, and environment-friendly products to replace daily necessities [87]. The Chinese government has also initiated mountain protection projects by formulating relevant regulations or protection strategies in most areas. For example, Shaanxi Province has formulated the regulations on the ecological environment protection of Qinling Mountains in Shaanxi Province, which explains the mountain system development accordingly, and stipulates the prohibition, restriction and moderate development of engineering projects that are not related to the protection of ecological functions and have great impact on the ecological environment [88]. In addition, the reforestation project advocated in the world also focuses on the protection of mountain systems. The purpose of returning farmland to the forest is to stop the cultivation of farmland prone to soil erosion, plant corresponding trees, and restore forest vegetation according to local conditions. The construction of this project includes two aspects: the conversion of cultivated land to the forest on the hillside; the other is afforestation on barren mountains and wasteland. According to the statistics of the State Forestry Administration of China, from 1999 to 2011, the total construction task of the project of returning farmland to the forest was 28.944 million hectares, which is equivalent to rebuilding a state-owned forest area in the northeast and Inner Mongolia [89]. Many abandoned and old mines in China have not implemented geological environment control and land restoration projects, and the environment in the mining area is poor. In order to solve the environmental problems left behind by abandoned mines, the state provides special policies and funds for governance. Still, it is urgent to put forward a set of design theories suitable for the national conditions from the technical level. On the basis of summarizing the characteristics of geological environment control engineering of abandoned mines, the design principles, indexes, and processes of geological environment control engineering of mines are put forward, and the design theory is applied to the control engineering design of Fengshan limestone mine in Guangxi. The design of the control project focuses on the main factors of the geological environment of the mine and the indicators of future land use. The control measures include the removal of dangerous rocks, slope cutting, slope leveling, anchoring, drainage engineering, retaining wall engineering, landscape and sculpture engineering, etc. [90] In addition, the forest area within the scope of the control project has increased significantly, which has effectively curbed and slowed down the speed of soil erosion and land desertification, and the ecological situation has been significantly improved. Soil protection engineering activities are mainly reflected in the construction of roads, which will fully consider the surrounding land environment. For example, when the Qinghai Tibet railway was built, environmental protection factors were fully considered: many bridges and tunnels were built based on the original design to stabilize the permafrost and facilitate the passage of wild animals. At the same time, in order not to damage the soil vegetation on both sides of the railway, a series of measures are also taken during the construction of the railway. For example, the activity scope of construction vehicles is reduced to the minimum; the surface turf at the site occupied by the construction site line is gradually moved out carefully, transplanted and maintained, and then paved back to the original place for vegetation restoration around the pile foundation; it is better to take a detour rather than borrow soil at the place with vegetation; the waste generated at the site is all stacked at the designated location and transported to Golmud or Lhasa waste plant for treatment [91].

3. **The view of development**

The view of development, which advocates that human needs and development are based on the premise of not affecting the survival and development of the environment, aims to remind people to adapt to the characteristics of the land environment while carrying out engineering activities, make timely, appropriate and appropriate use so that it can be better developed. In the land environment, the development concept is a theory of engineering activities that follows the land environmental resources circulation or continuous utilization law. This is also the focus and reflection of the concept of development in contemporary land environmental engineering activities.

At present, people gradually realize that the land environment resources are not the resources of some people, not the resources of a generation, but the resources shared by all people and the resources of future generations. In recent years, people have often adhered to the concept of intergenerational equality, based on the correct handling of the relationship between land environment and development, with the sustainable development of human society as the ultimate goal, with the vision, thinking, and action for development for various engineering activities. For example, in order to appeal to all countries to follow the idea of development and protect the environment, the United Nations launched the *Rio Declaration on Environment and Development* at the 1992 conference in Rio de Janeiro [92]; the Chinese government proposed the strategy of environmental "sustainable development" at the Third Plenary Session of the 16th Central Committee in 2003. Under the guidance of many policies and strategic ideas, the utilization goal of land environmental resources is transferred to the development that not only meets the current development needs of people but also does not weaken the ability of future generations to meet their needs to achieve the development goals of stability and continuity of land environmental resources utilization.

8.3.1.3 Biological Environment Theory

The biological environment theory emphasizes dealing with the relationship between human and biological environment, human activities and biological environment, to achieve the harmony between humans and nature. As far as engineering activities are concerned, it is mainly emphasized that people should consider the surrounding biological environment and get along with it harmoniously when planning and implementing a certain engineering activity. At present, the theory of biological environment mainly includes the view of ecological balance, the view of environmental friendliness, the view of harmonious development between humans and nature, and the view of sustainable development. These theories have far-reaching practical significance for guiding human activities and correctly dealing with the relationship between human activities and biological environment.

1. **The View of ecological balance**

The view of ecological balance is a kind of thought dealing with the relationship between economic and social development and the ecological environment. It advocates using ecological balance as a principle to formulate social development strategy and treat and evaluate all activities and objectives related to humans and the environment. In 1909, William Vogt, an American scholar, first proposed this idea in his book *The Road to Existence*. He saw the destruction of ecological balance caused by human activities and its serious consequences and believed that restoring ecological balance is the "way of survival" of human beings [93]. This idea has been accepted by many environmental theory schools in the world, which is of great significance to prevent ecological degradation and promote the awakening of human ecological consciousness.

Ecological balance refers to the dynamic balance of each part of the ecosystem (biology, environment and human) through energy flow, material circulation, and information transmission under a certain time and relatively stable conditions [94]. Ecological balance involves not only the balance between organisms, but also human activities. The appropriateness of human activities is related to ecology, which requires people to use practical rationality to guide the relationship between humans and nature. Therefore, human beings should be inspired from the nature, not to treat the ecological balance passively, but to play an active role, break the old balance that does not meet human requirements, establish a new balance suitable for human needs, and constantly improve the quality of human life within the limits of the ecosystem. The biological chain exists in all kinds of ecosystems. It is formed by the mutual dependence of animals, plants, and microorganisms, and it can achieve the chain relationship of material and energy acquisition and transmission. It can also be understood as the food chain in nature. It forms the phenomenon of "everything has its vanquisher" in nature and maintains the natural quantity balance among species. The stability of the biological chain refers to that in a certain period of time, animals, plants, and microorganisms in the biological chain maintain the equilibrium in proportion to each other so that the interdependence between them formed by predation can be continued, and the transmission of material and energy can be continued [95]. In 1906, in the Kaba forest of Arizona, the United States, people killed carnivores for the protection of deer groups, which led to the large-scale reproduction of deer groups without food and endangered. In the late 1950s, Xinjiang Yili introduced Italian black bees for artificial domestication. Forty years later, although the introduction of black bees brought certain economic benefits to the local area, the local Yili black bees were therefore extinct, and the local plants that depend on the black bee to pollinate were also affected. These cases reveal the truth that human beings are just a node in a chain of biological chains, and their understanding of nature is not systematic and comprehensive enough. A large number of human activities often cause damage to the biological chain, which will bring great harm to the ecological balance. The reason is that the stability of the biological chain is closely related to the ecological balance [96]. Generally speaking, the more species there are, the higher the degree of species diversity, the more complex the relationship between biological chains is. The more stable the biological chains are, the more complete and effective the ability to regulate ecological balance. The closer the ecosystem will be to the equilibrium state.

As we all know, water conservancy projects can meet people's water supply, flood control, irrigation, power generation, shipping, fishery and tourism, and play a positive role in ecological maintenance. By regulating the water quantity, water conservancy projects can resist the impact of flood disasters on the ecosystem, improve the ecological conditions of the dry and semi-arid areas, and regulate ecological water consumption. But almost everything has duality. Water conservancy projects may also significantly impact the ecosystem at the basin and regional levels. For their own safety and economic interests, human beings may drain the river, renovate the river, build the dam, significantly change the geography and landform, affect local climate, and greatly change the shape of the river itself, especially to reduce the diversity of river morphology to varying degrees. The consequence of reducing the diversity of river morphology is serious. It will decrease the diversity of aquatic communities, and the health and stability of the ecosystem will be affected to varying degrees.

Therefore, in the relationship between human activities and biological environment, human beings cannot blindly obtain biological resources from the biological environment but should protect organisms and maintain enough biological resources to ensure the balance of biological resources to achieve ecological balance.

2. **The view of environmental friendliness**

The view of environmental friendliness advocates the harmonious coexistence of humans and the environment, and the friendliness of humans to the environment is exchanged for the friendliness of the environment to humans. The view of environmental friendliness comes from the *21 Century Agenda* of the United Nations "integrated management and control of hazardous agricultural substances": "integrated management of hazardous agricultural substances... it is environmentally friendly and promotes agricultural sustainability [97]."

From the view of environmental friendliness, it extends the view of an environment-friendly society. The view of an environment-friendly society advocates the construction of a harmonious symbiosis of humans and nature. Its essential feature is the coordinated and sustainable development of human production and consumption activities and the natural ecosystem [98]. The harmonious coexistence and co-prosperity of the human and natural environment, including the biological world, land, and water, requires the whole society to adopt production and lifestyles conducive to environmental protection and establish a benign interaction between human beings and the environment. On the other hand, a good environment will promote production, improve life and achieve harmony between humans and nature. Therefore, environmental friendliness aims to ensure people's physical and mental health and the environment.

To build an environment-friendly society is to take the harmonious development of humans and nature as the goal, the environmental carrying capacity as the basis, and the natural law as the core. At the same time, it advocates environmental culture and ecological civilization, and pursues the coordinated development of economy, society, and environment [99]. The main point is to regulate production and consumption activities within the limits of ecological carrying capacity and environmental capacity, effectively monitor the whole process of production and consumption, and take various measures to reduce pollution emissions, achieve harmless pollution, and ultimately reduce the adverse impact of the social economic system on the ecological environment system [100]. In the face of many challenges such as resource shortage, ecological damage, and climate deterioration, building an environment-friendly society is undoubtedly a clear path for human development.

This is the demand of ecological civilization and the inevitable choice to deal with the environmental crisis.

3. **The View of harmonious development between humans and nature**

The harmonious development of humans and nature is the eternal theme of sustainable development and the long-term survival of human society. Humans and nature are a whole that is interrelated and restricts each other. On the one hand, humans must depend on nature; on the other hand, nature, as the object of human action, also changes because of human activities. In order to create new and more comfortable living conditions, human beings constantly adapt to and make use of nature. However, human activities cannot violate the laws of nature. Therefore, in the practice of building artificial nature, we must improve and develop ourselves through conscious activities, improve our understanding of and control over passivity, and maintain the coordinated development of humans and nature, to achieve the sustainable development strategy of human beings [101]. From the perspective of environmental protection, it is necessary to oppose both ecocentrism and anthropocentrism.

As a kind of biological existence, human beings, like other species, are an important part of the natural family. In a harmonious natural ecology, the relationship between humans and other species should be equal. Human intelligence enables human beings to change or regulate the functions of natural systems, which does not mean that human beings can control nature at will. The increasingly serious ecological crisis and climate deterioration are the lessons of nature to human beings. Only by changing the traditional model of relationship against nature, building a partnership model between humans and nature, getting along with other creatures equally, and developing harmoniously with nature, can human beings keep their survival foundation.

Artificial nature can play a dual role in the relationship between humans and nature. On the one hand, artificial nature is not only the direct source of the contradiction between human and nature, but also the means of regulating the relationship between human and nature; on the other hand, the improvement of artificial nature can maintain a dynamic balance with natural nature, and achieve the sustainable development of the relationship between human and nature. The basic conditions for the coordinated development of humans and nature are: to continuously promote human's understanding of nature and constantly improve and expand artificial nature [102].

4. **The view of sustainable development**

Sustainable development is a process that "not only meets human development goals but also maintains the ability of the natural system to continue to provide natural resources and ecological services on which human economy and society depend [103]." The concept of sustainable development emerged in the 1970s and rose in the early 1980s. In 1992, it became a global concept promoted by the *Rio Declaration on Environment and Development* of the United Nations Conference on environment and development in Rio de Janeiro. The core of the sustainable development theory of the declaration is that "the development and environmental needs of contemporary and future generations must be met equally [104]," which is the intergenerational equality of development opportunities and resource enjoyment. In general, the view of sustainable development, from a global perspective, focuses on a basic principle: human beings should always survive and develop, not sacrifice the interests of future generations for the interests of contemporary people, not sacrifice the interests of the whole society for local interests, not sacrifice the interests of the entire world for regional interests; otherwise, all the damage will eventually endanger the whole human race.

Sustainable development is a development model based on the coordinated development of society, economy, population, and ecological environment, including sustainable social development, sustainable economic development, population sustainable development, and ecological sustainable development. Ecological sustainable development is the basis of sustainable development of society, economy, and population. Therefore, we must have correct guidance of ecological concepts to maintain the ecological system's sustainable development and ensure the sustainable development of society, economy, and population.

The view of sustainable development requires human beings to fully respect and protect nature in all activities, especially in engineering activities, to avoid adverse effects on the natural environment, especially the biological environment. This has been widely accepted and implemented in engineering all over the world. The repair and repainting of the Bigsweir Bridge in Dean forest, England, is a case in point. Before the start of construction, the repairers carried out a large number of environmental surveys to verify the existence of species that must be protected near the construction site and took mitigation measures to reduce interference with bats [105]. The Itaipu Hydropower Station, the world's first dam known as "the seventh wonder of mankind," is also a case. After completion, the dam changed the local natural landscape. Seven Star waterfall, a scenic spot upstream of the river, was submerged at the bottom of the water, and the fish production downstream was reduced. In 1982, an accident occurred after the reservoir impoundment, which caused a sudden change in the ecological environment system of the basin area of 820,000 square kilometers, and the wild animals were almost in danger of extinction. After learning the lessons, Brazil pays more attention to environmental protection and takes measures to make the fish of the Parana River breed along the spawning channel upstream of the dam. A large number of rare animals are also carefully transferred for care to preserve the population. More than 20 million trees are planted along the dam by ETAP company. A 200–300 m wide forest belt connecting the green and tropical rain forest beside the reservoir was built and protected the water body in the reservoir area [106].

After 1992, the Chinese government decided not to repeat the mistakes of developed countries, let pollution occur before treatment, let damage happen before remediation, and avoid environmental pollution and damage in various engineering activities in advance. The Qinghai Tibet railway and the west to east gas transmission project are excellent cases. The builders consider environmental protection, wildlife migration, and other factors in advance, and take effective measures to maintain surrounding organisms' survival and biological diversity.

The view of environmental friendliness, the view of harmonious development between humans and nature, and the view of sustainable development all emphasize the correct handling of the relationship between human activities and the natural environment in order to achieve the harmonious coexistence between man and nature. It is necessary for human survival to respect the laws of nature, promote the harmonious coexistence of humans and nature, protect the diversity of biological species and maintain the ecological balance.

8.3.2 Modern and Contemporary Theories on Engineering Social and Cultural Environment

Modern and contemporary theories on the relationship between engineering and social-cultural environment mainly include political, economical, custom, cultural heritage, and legal theories.

8.3.2.1 Political Theory

Modern and contemporary theories of engineering and social, cultural environment, mainly from the perspective of engineering decision-making, study the theory and practice of engineering management, mainly including administrative decision-making theory and scientific decision-making theory.

1. **Administrative decision-making theory**

The theory of administrative decision-making is the concept, thought, or theory of many problems about the administration of engineering decision-making. Before the rise of scientific decision-making, it mostly provides theoretical support or explanation for engineering decision-making by administrative means; after the rise of scientific decision-making, it takes a critical attitude to engineering decision-making by administrative means.

Project decision-making, or administrative decision-making in engineering activities, refers to the decision-making by the chief executive. Because of the lack of professional knowledge, the lack of investigation, analysis, discussion, and demonstration, the chief executive "claps his head to make decisions." There has been a period of administrative decision-making in the history of engineering decision-making in many countries, and China is no exception. The history of engineering decision-making in contemporary China has a clear evolution process. From the 1950s to the 1970s, engineering decision-making was not standardized from procedure to subject, and individual decision-making was dominant. In the 1980s, the decision-making mechanism began to change, and collective decision-making gradually replaced individual decision-making. After the 1990s, it gradually entered the "consulting decision-making era," Reform and improvement of decision-making mechanisms and promotion of scientific and democratic decision-making were listed as eight elements of political civilization construction and political system reform [107]. At present, engineering decision-making is developing from administrative decision-making to scientific decision-making.

With the acceleration of reform and opening up, the government and all walks of life are increasingly aware of the necessity and urgency of scientific decision-making. The Chinese Academy of Engineering is speeding up the construction of a national engineering think tank, striving to provide a series of research and consulting services for major engineering science and technology development issues and strategic issues in social development.

2. **Scientific decision-making theory**

Scientific decision-making theory advocates scientific engineering decision-making or engineering decision-making by scientific means. Scientific decision-makers believe that scientific engineering decision-making should generally conform to the following decision-making standards: the engineering goal should benefit human beings and society; the engineering goal can be achieved; the engineering has the sustainable development; the degree of engineering endangering public health, safety, and environment is low [108].

The decision-making body makes the scientific decision-making of engineering activities, which includes government personnel, engineering technology experts, engineering operators, and the public. The government plays a leading and coordinating role in decision-making, responsible for allocating decision-making resources and establishing communication links among decision-makers. Engineering technology experts provide technical support for project decision-making. The public supervises the quality of project implementation to a certain extent, which plays a good role in public opinion supervision of the government, enterprises, and other behaviors. The American Society of Civil Engineers (ASCE) stipulates several principles to implement the protection of engineering activities on the environment and human beings: quantification, communication and risk management; application of integrated system method; implementation of sound leadership and management in decision-making process; adjustment of key infrastructure to cope with dynamic conditions and practices [109].

Expert argumentation is the key to scientific engineering decision-making and the necessary guarantee to achieve scientific decision-making. Because engineering decision-making is complex, it is important to select demonstration experts and establish a corresponding expert demonstration committee before major engineering decision-making [110]. "Experts can demonstrate the feasibility and scientificity of project decision-making from a professional perspective. After the accurate demonstration of experts, an engineering project's decision-making can increase its scientific characteristics and avoid unnecessary waste and mistakes. The investment and management of infrastructure construction projects, welfare construction related to public interests, etc., should be demonstrated by scientific and rigorous experts, and the feasibility report of demonstration and analysis should be provided to decide whether to put it into practice [111]."

In the aspect of professional expert argumentation, the practice of the United States is worth learning. The United States is a large engineering and technology country. The National Academy of Sciences and the National Academy of Engineering are the most authoritative independent consulting institutions in engineering and technology in the United States. Their main functions are to provide authoritative professional consulting services for government departments, Congress, and the public on the investigation, evaluation, policy recommendations, and other aspects of engineering and technology. Although established less than 50 years ago, the US Academy of Engineering has grown to the same scale as the "century-old shop" US Academy of Sciences. It plays an indispensable role in providing policy advisory activities related to science and technology, engineering, and education and is a real national engineering think tank.

Major project decisions have an important impact on people's lifestyles and economic activities. The demands of the public will exert public opinion pressure on the government and other decision-makers through TV, newspaper, Internet, and other ways to intervene in project decision-making. Therefore, scientific engineering decision-making is particularly important from the perspective of public opinion intervention and social stability.

8.3.2.2 Economic Theory

Engineering activity is an important form of economic activity, which is restricted by the basic law of economic system operation. An important problem to be considered in engineering management is the optimization of resource utilization, that is, to obtain the maximum positive benefits with the least consumption. Contemporary scholars have done a lot of thinking and research on the relationship between engineering activities and economic planning and operation and have formed many theories. There are three representative theories: the low-carbon economic theory, life cycle theory, and ecological footprint theory.

1. **Theory of low carbon economy**

The theory of a low carbon economy advocates building an economic development system based on low energy consumption, low pollution, and reduction of greenhouse gas emissions. This theory first appeared in the British energy white paper *The Future of Our Energy: Building a Low-carbon Economy* in 2003, which was put forward against the background of the severe challenge of global warming to human survival and development. The development of a low-carbon economy is an inevitable choice to improve energy utilization efficiency, actively bear environmental protection, build ecological civilized cities, and achieve a win–win situation between economic development and resource and environmental protection [112].

Nowadays, a low-carbon economy has become a fashion, energy conservation and emission reduction, ecological environmental protection in engineering are generally valued. Project decision-makers and managers strive to reduce expenditure, time, and human resources and improve resource utilization efficiency. This pursuit is not only reflected in the process of engineering activities but also reflected in the results of engineering activities. For example, low-carbon buildings with the theme of green and environmental protection have become an important construction industry source.

2. **Life cycle theory**

Life cycle theory is a kind of evaluation theory that focuses on every link of product or service system life. According to the definition of International Standardization Organization, life cycle assessment (LCA) is a systematic method to collect and measure the environmental impact, material, and energy input–output directly related to the function of a product or a service system in the whole life cycle [113].

From the perspective of environmental science, the life cycle theory conducts a comprehensive quantitative study on the life cycle resource utilization, energy consumption, and waste discharge of raw materials (collection, processing, transportation) and products (production, sales, use, maintenance, recycling, scrapping). It mainly focuses on researching products' effects on resources and energy, ecological environment, human health, and other aspects and cost estimates [114]. China mainly applies the life cycle assessment method to construction products, from the perspective of ecological environment, covering the whole process of raw material mining, building material processing, component manufacturing, planning and design, construction, operation and use, maintenance demolition, and recycling. It calculates the corresponding impact indicators within construction products' life cycle system and makes comparative evaluation, finding a reasonable balance among building function, resource utilization, energy consumption, and environmental pollution [115].

It is also necessary to achieve a reasonable balance between building functions, resource utilization, energy consumption, and environmental pollution in engineering activities. Even if the function value of the project is very large, the damage of energy consumption to the environment exceeds its function value; it should not be carried out. For example, the Zhuhai Airport, due to the unreasonable estimation of passenger and freight volume and regional economic growth, the construction scale is expanded blindly, and the principle of "one-time planning and phased implementation" is ignored. When it was put into use, Zhuhai Airport owed 1.7 billion yuan of project funds, all operating revenues were frozen by the court, and some equipment and properties were seized. After five years of operation, the actual passenger flow only reached 6% of the design capacity, and the monthly passenger flow was only equivalent to the one-day passenger flow of Guangzhou Baiyun Airport. The cargo and mail throughput was less than 1/60 of the design capacity. The annual loss was 20 million yuan [116].

3. **Ecological footprint theory**

Ecological footprint, also known as ecological occupation, is a method to measure the degree of sustainable development proposed by Willian E. Rees, a planning and resource ecology professor at the University of British Columbia, Canada, in the early 1990s. It is the most representative quantitative indicator of sustainable development based on land area. It specifically refers to how much regional space with biological productivity is needed in a designated population unit (a person, a city, a country, or the whole human being) to produce the required resources and absorb the derived wastes under the existing technical conditions. Ecological footprints are used to assess the impact of human beings on the ecosystem by measuring the amount of natural resources that human beings use to maintain their own survival [112]. Ecological footprint can convert the resources consumed by each person in each measurement unit into a unified regional area. By calculating the difference between the total supply and total demand of regional ecological footprint—ecological deficit or ecological surplus, different regions' contribution to the current global ecological environment situation can be accurately reflected [117].

From the perspective of ecological footprint theory, engineering activities should follow the principle of thrift, and all engineering construction should minimize the occupation or destruction of nature, ecology, and resources to achieve sustainable development. China adopts simple and easy measures, such as promoting the use of energy-saving light bulbs, promoting the use of solar energy, wind power generation, etc., not only saving resources but also help environmental protection.

8.3.2.3 Custom Theory

As mentioned above, the custom is a habit and custom gradually formed in the life history and abided by the regional residents. It is the earliest social behavior standard in human social life. With the changes of the times, some old customs disappear or are abolished because they cannot meet social development needs, and some new customs appear or are advocated because they meet the needs of the times. For engineering activities, customs have considerable influence or restriction. New custom ideas or theories are very important for improving the social and cultural environment of engineering.

1. **Changing customs theory**

Changing customs means changing bad old customs and establishing good new customs. Changing customs is an important means to improve social quality. According to the *Shiji Lisi Liezhuan*, "Using the law from Shang Yang to change customs, to enrich the people and the country [118]." In the context of engineering, the term "changing customs" refers to a wide range, which can refer to the adaptation of engineering immigrants to the regional cultural customs in the place of immigration, or the overall reconstruction of social customs in the context of engineering civilization.

With the acceleration of urbanization and industrialization, large-scale water conservancy, electric power, and transportation projects have made the scale of project migration continue to expand. In a sense, migration from the original place of residence to the place of resettlement is a movement of cultural migration, integration, and assimilation. Generally speaking, there are differences between the customs and cultures of the original residence and the resettlement place. For example, the Three Gorges Reservoir area residents, whose original residences are mostly built on the mountain, are good for exclusive residences. According to the proverb, "it is hard to buy exclusive villages with money." There is also a folk saying, "Go up the mountain and drill in the fog. Go down the mountain and touch the river. Shout to the mountain. And take half a day to meet." The houses of the Tujia people are built according to the height of the mountain. Most of them are stilts. Their construction methods and use characteristics are especially suitable for mountain houses. However, the houses where they moved are often a group of people with the same family names living together, forming a natural village [119]. However, due to the importance attached to the farming season, the customs in the relocation area are totally different.

The differences in customs make it difficult for immigrants to integrate into their lives in the places where they live, which leads to a great sense of cultural and psychological loss. Therefore, in the process of immigration, it is necessary to fully investigate immigrants' local customs and customs to ensure that the immigration project is carried out reasonably so that immigrants have a good psychological transition.

2. **The township regulations theory**

The township rules and regulations are the basic social organizations, especially the village organizations, which agree or formulate the social behavior norms. They are an important part of the village culture. There are two kinds of village rules. One is "village rules" made by village administrative organizations or guided by them. The other is "people rules" made by villagers through collective negotiation. The main contents include three aspects. First, the maintenance and custody of the resources and facilities of the village, such as the custody of the land and water area of the village. Second, the village society will generally form some clear personal responsibility and rights agreements to reduce the internal disputes of the village. In order to avoid the villagers competing for the public means of production such as land, water source, and river course in the production, the village members will also form some corresponding regulations. For example, the village regulations of a village in Shandong Province are recorded as: "the ownership of all the land in the village belongs to the village, and the villagers must have complete approval procedures and agreements to apply for the approval of the homestead [120]." Third, manage the common life order of the village, including village rules and regulations, sanctions against violations, dispute adjustment, etc. For example, the Dong people in Guangxi, Hunan, and Guizhou hold two meetings every year: "March is about seeding" and "August is about harvest." They formulate and announce the ecological resource protection and taboos in the seedling season and harvest season [121].

Village rules and regulations have a direct, decisive role in constructing rural villages. Some township rules and regulations are directly related to the process of construction project planning, decision-making, implementation, and maintenance. Many rural regulations and rules in ethnic minority areas directly compile the old habits into articles. Without surprise, the villagers continue to abide by the rules that have been observed for thousands of years and hundreds of years, and the implementation is very smooth and adaptive. The rules and regulations of the people play an important role in contemporary rural construction.

The rules and regulations of the villagers have a significant impact on the construction projects of water conservancy, electricity, and roads around the villages, which can promote or hinder the implementation, operation, and maintenance of the surrounding projects. For example, the construction of a wind power station will bring visual and acoustic effects to the natural environment of the village, as well as the daily life and production of the villagers. In order to construct, operate and maintain the wind power station smoothly, we must have a deep understanding of the rules and regulations of the village or related villages. We need to respect the original rules and regulations and guide them appropriately to accept new things according to the changes of the times.

8.3.2.4 Cultural Heritage Theory

Nowadays, cultural heritage is paid more attention in the world. In November 1972, the 17th UNESCO General Conference adopted the *Convention on the World Cultural and Natural Heritage Protection.* Cultural heritage and natural heritage are clearly defined in the convention, including cultural relics, buildings, and sites [122]. The guiding ideology of contemporary engineering activities is to protect and harmoniously integrate the existing cultural heritage.

1. **View of protection**

The concept of protection requires that the construction activities should be carried out to protect the existing cultural heritage constructively and should not be demolished or destroyed at will. Since the 1960s, relevant international organizations and institutions have protected cultural heritage by issuing charters, conventions, and recommendations. UNESCO has issued a series of proposals and conventions on the protection of historical and cultural heritage, such as *the Proposal on the Protection of Cultural Property Threatened by Public or Private Projects* in 1968, which proposes to protect all cultural properties [123]; *the Convention on the Protection of the World Cultural and Natural Heritage* in 1972, proposes that each state should respond to the cultural and natural heritage within its territory. We should make every effort to determine, protect, preserve, exhibit, retain, and bear the relevant state responsibilities [124]; In 1976, *the Proposal on the Protection of Historical Areas and Its Contemporary Role* was issued, in which the constructive protection of historical areas was emphasized [125]. The International Council for Monuments and Sites has also issued relevant charters, such as *the International Charter for the Protection and Restoration of Monuments* issued in 1964, which states that all and part of the monuments shall not be relocated except under special circumstances [126]. "In 1982, *the Florence Charter* was issued, which clearly proposed the maintenance, protection, restoration, and reconstruction of historical gardens [127]; In 1987, *the Charter for the Protection of Historical Towns and Areas* was issued, which protected, preserved and restored towns and urban areas [128]. Regional organizations have also published relevant conventions, such as the 1976 *OAS Convention on the Protection of Archaeological and Artistic Heritage* issued by the governments of OAS member states. In

1982, the Fifth National People's Congress of China passed *the People's Republic of China Law on the Protection of Cultural Relics*, making detailed and clear provisions on the protection of historical relics [129]. In 2003, the 8th executive meeting of the State Council adopted and implemented the regulations for implementing *the Law of the People's Republic of China on the Protection of Cultural Relics.* It was pointed out that the administrative department for cultural relics under the State Council and the administrative department for cultural relics under the government of the provinces, autonomous regions and municipalities directly under the central government should formulate scientific and technological research plans for the protection of cultural relics, take effective measures to promote the popularization and application of scientific and technological achievements in the protection of cultural relics, and improve scientific and technological level of cultural relics protection [130].

It can be seen from the above charters, conventions, and suggestions that great attention has been paid to the protection and restoration of cultural heritage at home and abroad. In order to reduce the impact and damage to cultural heritage as far as possible, engineering decision-making, scheme design, expression methods, and material selection should be carried out under the concept of protection priority for engineering activities. However, there are many cases of inadequate protection or even damage. An example is the Three Gorges Reservoir area geological disaster management and Baidi city landscape protection. In the vicinity of the world-famous Baidi City, in order to prevent landslides due to poor geological conditions, the relevant departments set up a concrete frame grid along the slope to protect the slope; however, every summer, the fall of the water body will expose the whole gray-white concrete framework and yellow soil of the slope, and the protection project is in sharp contrast with the green forest on the mountain, the pink walls and the ancient buildings in the shade of trees not harmonious, making the landscape of "Bai Di Ling Kong" dim [131].

2. **Integration view**

The concept of integration refers to that engineering planning and design respect the labor and creation of predecessors, respect historical buildings and artifacts, respect the environment, integrate with the styles and characteristics of existing buildings and artifacts around, and build an integrated, harmonious, unified and non-abrupt artificial natural object. Mr. Yang Tingbao, the master architect, once said, "when building and expanding in a complete building complex, sometimes it is not necessary to show the individual you designed, but to focus on the coordination of the group [132]." Therefore, the engineering design should consider that the workers' natural objects to be built are consistent with the surrounding cultural environment. The expansion of Tsinghua University Library in the early 1990s is regarded as a successful case. In addition to the innovation in details, the style of the new pavilion is basically the same as that of the old one, which is reflected in site selection, architectural style, application of main materials and treatment methods. In terms of site selection, the new pavilion did not occupy the central position; in terms of architectural style, the image features of Tsinghua garden buildings are still used, namely, red brick walls, sloping tile roofs, partial flat roof parapets, and semicircle arches for main or key doors and windows, reflecting the peaceful cultural atmosphere; in terms of material selection, red brick and grey tiles are basically the same as the old pavilion [133]. The new pavilion and the old pavilion, as well as history and modern contrast, not only increased the historical connotation and free and open atmosphere of the times, but also protected the cultural heritage. However, there are also some engineering cases of old and new conflicts. For example, when maintaining the ancient buildings of Tujia Diaojiaolou, a white edge is drawn on the roof with lime powder, and the cornice is also carved with a dragon head and a phoenix tail, which makes the original simple and beautiful Tujia buildings more meaningless decorations, which is considered to be out of place [134].

8.3.2.5 Legal Theory

In today's world, all countries are constantly improving and enriching the engineering legal system, and strive to guide and standardize engineering activities with a complete system of laws and regulations. The mainstream of modern engineering management is legalization, which establishes a sound legal system of engineering management so that engineering management can be governed by law and into the track of the rule of law. For engineering activities, the representative of the contemporary concept of the rule of law is the view of norms and guarantees.

1. **Normative view**

The normative view advocates that in the process of project construction, all kinds of behaviors of the parties, government departments, and market subjects should be regulated by law. It is the code of conduct for project subjects and stakeholders that laws and regulations determine the boundaries between legal and illegal as well as reasonable and unreasonable.

In contemporary China, the legislative and law enforcement departments, through the system of laws and regulations, regulate the engineering activities themselves and regulate the responsibilities and obligations of engineering activities to the natural environment and social and cultural environment. The latter provides power and support for implementing the concept of environmental protection, the development of green projects, and the sustainable development of engineering activities.

For example, in the aspect of construction law, the 7th item of *the Regulations on the Administration of Energy Conservation of Civil Buildings* says: "encourage scientific research and technological development of energy conservation of civil buildings, promote the application of energy-saving buildings, structures, materials, energy-using equipment, and auxiliary facilities, as well as corresponding construction technology, application technology, and management technology, and promote the development and utilization of renewable energy [135]." In the construction stage, the construction unit shall not require the design unit and the construction unit to design and construct in violation of the mandatory standards for energy conservation of civil buildings; The design unit, the construction unit, the project supervision unit and their registered practitioners must strictly implement the mandatory standards for energy conservation of civil buildings; the project supervision unit has the right to require the construction unit to make corrections if it does not implement the mandatory standards for energy conservation of civil buildings and report in time. Another example is that in constructing the Qinghai Tibet railway, all soil shovels must follow the legal norms, and no construction is allowed without environmental protection measures. This kind of strict environmental management measures promotes the development of environmental protection work in the construction of the Qinghai Tibet railway. There are strict limits for constructing railway access roads on the plateau. The length and width should be marked with red lines. No one is allowed to cross the red lines. Violators will be punished severely. In many areas of the Qinghai Tibet Plateau, the terrain is flat. In the beginning, drivers saw a flat land. Sometimes they didn't meet the limits and damaged the environment, resulting in severe punishment. Once, a driver out of the access road about 10 m when transporting materials on the construction site. When the headquarters of the Bureau knew about it, in addition to punishing the driver, it also imposed a fine of 100,000 yuan on the project department where the driver was working. The project department manager also took people to restore the grassland [136].

2. **Guarantee view**

The guarantee view advocates ensuring the effective implementation of engineering activities and the rights and interests of the parties and stakeholders through laws, regulations, and standards. The guarantee is mainly implemented in three aspects: first, to protect the rights of all parties involved in the project to develop and use resources according to law; second, to protect the rights and interests of stakeholders; third, to ensure the quality of the project.

For resource use, in China, the 9th item of the People's Republic of China Constitution clearly states: "the state guarantees the rational use of natural resources and protects precious animals and plants [137]." It gives the project subject the right and freedom to use resources. The 6th item of *the General Provisions of the Water Law of the People's Republic of China* stipulates: "the state encourages units and individuals to develop and utilize water resources according to law and protect their legitimate rights and interests. According to law, units and individuals that develop and utilize water resources shall have the obligation to protect water resources [138]." It guarantees the subject of water conservancy project to use water resources reasonably.

For example, in China's construction industry, engineering quality is a saying that "quality is the life of construction projects." Engineering quality affects the reputation and image of engineering products in people's hearts and directly affects people's safety and social stability. According to Article 14 of Chapter II of *The Construction Law of the People's Republic of China*, "professional and technical personnel engaged in construction activities shall obtain corresponding qualification certificates according to law and engage in construction activities within the scope permitted by the qualification certificates [139]." It ensures the overall professional quality level of the construction team and avoids engineering quality problems caused by the professional ability of the construction personnel. Chapter six "construction project quality management" has 12 articles, which put forward requirements for all aspects of construction project quality and provide legal guarantee.

For the project quality assurance, the standard is also an effective means. The International Organization for Standardization (ISO) has established the International Standard 9000 (ISO9000) family; the first standard was formulated in 1987 and then formed a series of standards after continuous modification and improvement. As a member of the organization, China has developed a project quality assurance system under 9000 families and formed a quality management system.

8.4 Modern and Contemporary Comprehensive Theories on Engineering Environment

With the progress of engineering science and technology, with the progress and mutual penetration of natural science, social science and humanities, a comprehensive theory has emerged in the modern and contemporary engineering environment theory system, that is, the theory formed by the overall investigation of engineering natural environment and social and humanistic environment, the integration of political, economic, cultural and legal theories and methodologies. Among these comprehensive theories, environmental security theory, environmental justice theory, and sustainable consumption theory are representative.

8.4.1 Environmental Security Theory

The theory of environmental security explores the security issues of individuals, society, and the state in their environment. Environmental security is based on environmental prevention and defense; the environment's collective prevention and defense is the core of environmental security [140]. Environmental security, specifically, refers to the safety of the human living environment and working or production environment. It was a hot topic in the late 1960s and 1970s, mainly in discussing environmental problems and related issues [141]. Environmental security, including engineering environment security, has two aspects. The first is the security issues of the environment in which the project is located; the second is the environmental impact of engineering activities. Furthermore, engineering environment security includes two aspects: resource security and human security [142].

Resource security refers to that in engineering activities, natural resources can be obtained continuously, stably, timely, in sufficient quantity and economically, and will not threaten the environment and human health. Specifically involved in the following two aspects, on the one hand, the use of resources is safe. The project itself does not degrade the human living and production environment, such as human use of water resources will not lead to flooding, thus endangering the environment and life. Another example is that vehicles, trains, and other means of transportation themselves will not pollute the environment and affect human health. In addition, it also refers to the long-term existence of resources, such as coal, oil, and other resources will not be exhausted, and water resources will not be exhausted.

Human security refers to the sustainable development of individual human beings and the eternal life of all human beings, with emphasis on the safety of the environment associated with human beings [141]. First of all, the sustainable development of humans includes the guarantee of human physical and mental health and the natural end of individual life. For example, human health will not be threatened by natural disasters caused by environmental damage; human beings will not often worry about disease due to environmental pollution. Individual human life will not be terminated earlier because of the rampant illness. Secondly, the longevity of all humankind also involves two aspects: future generations have resources to enjoy, and human beings can reproduce themselves. For example, natural resources, including coal, oil, and various animals and plants, can be extended to future generations without exhaustion or extinction, and people will not be seriously affected by the use of plastic products.

8.4.2 Environmental Justice Theory

Environmental justice theory emphasizes that human beings, regardless of generation, nationality, race, gender, culture, class, rich or poor, enjoy equal rights to a safe, clean and sustainable environment and freedom from environmental damage. Environmental justice is a kind of norm and restriction on the relationship between man and nature on the level of morality and quasi law [143]. Environmental justice is not like the law to measure the correctness of human behavior; it is more to make people's behavior subject to moral constraints. If a person in charge of solving environmental problems cannot deal with the environmental deterioration of his country well, he will be condemned for his inaction. Although his omission does not violate the mandatory provisions of the law, it is not appropriate to measure it by the standard of justice.

The concept of environmental justice can be traced back to 1972 when the United Nations held a conference on the human environment in Stockholm, Sweden. To a certain extent, the conference reached an agreement on the issue of north and south equality. In the same year, *the Limits to Growth*, written by Dennis Meadows, pointed out that the earth, which carries the excessive growth of population and use of resources, has been unable to maintain a healthy and satisfactory state of sustainable development. In 1982, the World Conference on Environment and Development held by the United Nations held a more extensive and in-depth discussion on the issues of economic growth, environmental limits, and equality and published *Our Common Future* in 1987. The book holds that inequality is a basic topic of global environmental problems. If human beings want to make real progress in solving environmental issues, they must solve the issues of poverty and inequality behind environmental problems. A large number of facts in the United States in the 1990s show that industrial entities that discharge waste tend to choose areas where land and labor are cheaper. These areas are often less developed sites with a small population. In 1990, Bullard's *Dumping in Dixie: Race, Class, and Environmental Quality* was published, which caused many scholars to discuss environmental justice. At that time, the concept of environmental justice was recognized by American scholars. In 1992, all the World Conference on Environment and Development outcomes dealt with environmental justice. At the United Nations Conference on Environment and Development in 2002, it was mentioned that attention to development needs had reduced the concentration to environmental problems to a certain extent. As a result, a series of works with environmental justice as the core content appear, making environmental justice the focus of people to solve environmental problems again [144].

Environmental justice includes intra generational justice and inter-generational justice. These two kinds of environmental justice can be examined from three theoretical perspectives: narrow environmental justice, broad environmental justice, and ecological justice.

The narrow sense of environmental justice holds that in the process of environmental protection, human beings should strive for the welfare of human needs, hopes, and creation. The existing countries, international organizations, and governments are responsible persons responsible for dealing with environmental problems and achieving environmental justice. Environmental problems can be solved by formulating environmental policies and improving the economic level. Under the condition of breaking through the inherent national restrictions, we can carry out multi-party dialogue, reach a consensus on the moral level in environmental justice, pay attention to the role of international institutions, and accelerate the construction of national and international environmental legal systems. At present, in the development of air pollution prevention and control in China, regional joint prevention and control have become national-level governance means. Joint defense and control have become a consensus, especially in the Beijing Tianjin Hebei region. In the first quarter of 2013, five cities in Hebei Province ranked among the top 10 cities with serious air pollution. One of the reasons is that the development of Beijing needs raw materials from the surrounding cities. Therefore, Hebei province builds a large number of steel plants and burns a large amount of coal to meet the needs of Beijing's development. This behavior is equivalent to giving pollution-free products to others and leaving pollution to oneself. However, it is not a long-term plan to restrict the growth of the heavy industry in Hebei Province. In 2013, Beijing and Tianjin signed cooperation framework agreements with Hebei Province, including jointly promoting the planning of the capital economic circle and other seven key cooperation areas [145]. It aims to promote the sustained and healthy economic development of Hebei Province.

The broad sense of environmental justice emphasizes that the solution of environmental problems needs to go beyond the regional limitations, consider the improvement of social welfare and social service capacity, and focus on improving the current system and economic development mode. For example, in modern times, to improve the global cooperation mechanism, the developed countries should take more responsibility for the vulnerable groups and the groups that are negatively affected by the modernization process. In terms of intergenerational justice, we should protect the basic human survival welfare and improve the quality of life, that is, health, education, social welfare, etc. That is to say that the next generation will enjoy the same freedom and opportunity as our present generation. Nowadays, there is a consensus that the less developed areas face a more difficult situation among the problems global climate change brings than the developed regions because: first, compared with the less developed countries, the developed countries have a higher degree of modernization and are the main emitters of greenhouse gases; second, because of cheap labor and low land occupation fees, some developed countries have built highly polluting factories in underdeveloped areas; third, some developed countries transfer hazardous wastes to underdeveloped countries and lure these countries through small interests to dump these wastes on their land, for burying, burning, or moving. Therefore, from the perspective of environmental justice, the developed countries should become the main responsible party for environmental problems. All countries should carry out positive dialogue to promote effective solutions to environmental issues.

Ecological justice pays more attention to the negative environmental impact, not just the negative impact of human activities on the environment. Specifically, on the one hand, ecological justice advocates that human activities should not destroy the environment; on the other hand, activities of one party should not pollute the environment of others. A typical example is Mongolia, China, Japan, and South Korea jointly controlling the Mongolia sandstorm project. Sandstorms are caused by the destruction of vegetation and the deterioration of ecological environment. Although sandstorms are generated in a certain country or a certain region, the countries with sandstorms affect or may endanger the ecological environment of neighboring countries and regions. Therefore, even though sandstorms are not caused by their own countries or regions, we need to share the responsibility of environmental governance to the good life and production environment. The theory of ecological justice holds that protecting the environment is everyone's right and obligation. It emphasizes that in dealing with environmental problems, it should pay attention to the relationship between people and between countries. At the same time, more active environmental dialogue should be advocated. In terms of intergenerational equity, we attach great importance to the sustainable development of the environment.

Therefore, according to the concept of environmental justice and ecological justice, in the design and implementation of engineering activities, the right and freedom of others to enjoy a good environment cannot be threatened. Here, both the natural and social environments are involved because everyone has the freedom to enjoy a good living environment and the right to enjoy a good social environment.

8.4.3 Sustainable Consumption Theory

The concept of sustainable consumption appeared for the first time in the report *Policy Factors of Sustainable Consumption* published by UNEP in 1994. Sustainable consumption refers to the consumption mode of meeting the basic needs of human beings, providing relevant products and services, and improving the quality of life under the premise of using the least amount of natural resources and toxic materials and not endangering future generations. The concept of sustainable consumption has become the mainstream consumption pattern in the twenty-first century [146]. In today's world, many countries have established governance models based on sustainable consumption to achieve the sustainable development of nature, economy, society, and people from the perspective of consumption.

Sustainable consumption is a kind of modern life concept which embodies a kind of engineering concept and guides all aspects of engineering activities. Any project should not destroy resources at will, consume resources excessively. The implementation of engineering activities should not exceed the limit of natural environment carrying capacity, and consumption should also be conducive to ecological balance and environmental protection. We should consider the sustainable consumption concept of engineering activities from two aspects of the natural and social environments.

(1) The relationship between engineering and the natural environment reflects the correct understanding of the relationship between engineering and nature

The consumption pattern of engineering activities is not the possession of natural resources by human beings, but the mutual benefit relationship between human beings and nature should be maintained. That is to say, in the process of engineering activities, the use of natural resources and toxic materials should be minimized so that the products or projects that serve human beings can produce the least pollutants in their life cycle and the least pollution to the environment. Sustainable consumption requires the recycling of resources, the realization of optimal benefits, and the minimization of environmental pollution and waste discharge. The consumption of natural resources by engineering activities should not exceed the ecological environment carrying capacity limit. It should always be beneficial to the protection of the natural environment and the maintenance of ecological balance.

(2) From the perspective of engineering and the social environment, highlighting the fairness of consumption

Fairness is an important foundation for harmonious coexistence among engineering activities, social resources, and the environment. Survival and development are the basic rights of everyone and the mission of the society. Consumption takes individual survival and social development as the ultimate goal, which is the goal that engineering design and implementation should follow. Therefore, the consumption of contemporary engineering activities cannot damage the consumption capacity of future engineering activities. We should change the implementation mode of engineering regulation to achieve harmony between engineering and the social environment. The "green bus" is a kind of low-carbon transportation tool in modern society, including natural gas, fuel cells, hybrid power, hydrogen energy, and solar power vehicles. Its exhaust emission is relatively low, and it is an environment-friendly transportation tool. It is close to the low-carbon life advocated nowadays. Its use reflect the sustainable consumption concept.

Sustainable consumption is the eternal theme of human development and one of the basic propositions of sustainable development economic theory research. A healthy consumption pattern promotes the stable development of the economy. Without sustainable consumption, there can be no sustainable development. To have sustainable development, we must first have sustainable consumption [147]. The concept of sustainable consumption in engineering activities is the conceptual basis for building a harmonious society and promoting China's economy's sustainable production and development.

8.5 Historical Review of the Evolution of Engineering Environment Theory

In the history of the development of human civilization, different times have different attitudes and ideas about the engineering environment, so there are different ideas or theories. The previous section has sorted out historical and modern natural environmental theories, social-cultural environmental theories, and comprehensive environmental theories. This section will make a comprehensive review and summary of the historical and modern engineering environment theories and sort out the historical mainline of the evolution of the engineering environment theory from the two dimensions of the natural environment and social and cultural environment.

8.5.1 Historical Review of the Engineering Environment Theory

The engineering environment theory, whether the natural environment theory or the social and cultural environment theory, has been changing and improving because of the continuous improvement of human thinking ability and behavior ability.

8.5.1.1 Historical Review of Engineering Natural Environment Theory

The theory of engineering natural environment is mainly concerned with climate, land, and biology.

(1) Engineering environment theory related to climate dimension. There are many concepts and theories about engineering climate environment in ancient China, among which the concept of time and Yin and Yang have a great influence on engineering practice. The concept of time emphasizes that all kinds of engineering activities should not go against the four seasons and the time sequence; that is, engineering activities should not go against the time and do not move against the time. The concept of Yin and Yang is a concept that emphasizes the balance of Yin

and Yang. Yin and Yang are the two most basic contradictory forces. Yang is the upward, strong, and dynamic development state of things, while Yin refers to things' downward, weak and static development state. For example, the precipitation in the climate environment is Yang, and the water storage is Yin. The concept of Yin and Yang emphasizes that in engineering activities, human beings should, on the one hand, maintain and achieve the balance between precipitation and water storage; on the other hand, attention should be paid to increasing the distribution of water bodies and the balance between cloud and rain can be achieved by using the scattered ponds, to regulate the surrounding climate and ecological environment. With the development of society and the deepening of industrialization, more and more engineering activities with high consumption and high pollution are driven by economic interests. The problem of air pollution has appeared in different degrees worldwide. In this context, atmospheric environment protection has become the mainstream cognition of the relationship between engineering activities and climate. The theory of atmospheric environment protection mainly includes three basic concepts: one earth view, greenhouse effect view, and climate change view. These concepts emphasize that people's engineering activities should consider the immediate economic interests and fully consider the long-term impact on the environment after the implementation of the project and try to protect the interests of future generations.

(2) Engineering environment theory related to land dimension. The ancient land engineering environmental theory mainly includes the view of land capability, borrowing land, and complying with the land. According to the view of land capability, land, as a kind of natural resource, can accumulate strength by itself, and people's engineering activities should follow the land carrying capacity. The concept of borrowing land is a kind of concept theory of activities with the help of land. In ancient engineering activities, people mostly borrowed waterpower. According to the view of complying with land, people carry out various activities based on topography and landform. It requires people to consider terrain, tunnel, Yin and Yang, five elements, and other factors in engineering practice. In modern times, people's attitude towards the land environment has gone through two stages, and two different theories have been produced. At first, people put the material interests above the environmental interests and vigorously transformed the land environment: the concept of transformation, such as the 10,000 mu project and the road construction project. Until later subjected to environmental retaliation, people gradually realized that the land environmental resources were not the resources of a certain part of the people, not the resources of a certain generation, but the common resources of all the contemporary people, and even more the resources of future generations. All engineering activities should consider the impact on the land environment. Modern people mainly hold the view of protection and development. The concept of protection requires people to consider economic and environmental interests in engineering activities. The concept of development requires people to look at all kinds of engineering activities from the perspective of development. When carrying out various engineering activities, people should consider the recycling or continuous utilization of land resources.

(3) Engineering environment theory related to biological dimension. Dealing with the relationship between human activities and the biological environment is an important link to improving humans and nature. In project management, it is also very important to deal with the relationship between engineering practice activities and various elements of the biological environment. In ancient times, the engineering environmental theories on the biological environment mainly include the view of integration, the view of harmony, the view of cherishing life, and the view of Mean. The concept of integration emphasizes that humans and nature coexist, and each has its own position. Large-scale engineering practice should conform to the natural law and not destroy natural resources. For example, the Dujiangyan water conservancy project presided over by Li Bing is a typical case of fully respecting natural laws and rationally utilizing natural resources to serve human beings. The concept of harmony advocates "harmony with living things." The concept of cherishing life requires that people love things to accumulate happiness. It emphasizes the diversity and difference of all things. It holds that as long as different things coordinate with each other, they can develop continuously. The view of Mean advocates the balance between the two sides. These theories emphasize that the exploitation of ecological resources in engineering activities should be appropriate and pay attention to the recoverability of natural resources. At present, the theories about the biological environment mainly include the view of ecological balance, the view of environment friendliness, the view of harmonious development between humans and nature, and the view of sustainable development. These theories have far-reaching practical significance for guiding human activities and correctly handling the relationship between human activities and the biological environment. The view of ecological balance is a theory dealing with the relationship between economic and social development and the ecological environment. It advocates that social development strategy should be formulated based on the ecological balance viewpoint, and all activities and objectives related to humans and the environment should be viewed and evaluated. It emphasizes that in dealing with the relationship between human activities and the biological environment, biological resources should not be obtained blindly from the environment but should actively protect bio-diversity. According to environmental friendliness, we should adopt the mode of production and life conducive to environmental protection and establish a benign interaction between humans and the environment. The view of harmonious development between humans and nature holds that the coordinated development of humans and nature is the eternal theme of sustainable development and the long-term survival of human society. According to the view of sustainable development, ecological sustainable development is a very important aspect under the mode of sustainable development, which is the basis of sustainable development of resources, economy, and society. Generally speaking, the view of environmental friendliness, the view of harmonious development between humans and nature, and the view of sustainable development emphasize the correct handling of the relationship between human engineering activities and the natural environment to achieve harmonious coexistence of humans and nature.

The theory of engineering natural environment contains the concept of "human can conquer nature" or similar concept, or the concept of "harmony of human and nature" or similar concept. The evolution of natural environment theory is a game process of two dominant concepts.

8.5.1.2 Historical Review of Engineering Social and Cultural Environment Theory

As mentioned above, the social and cultural environment theory involves politics, economy, customs, cultural heritage, and law. Compared with the natural environment theory, social and cultural environment theory change is more intense.

(1) The theory of engineering environment in the political dimension. The relationship between politics and engineering management is complementary. The engineering environment theory in political aspect mainly focuses on two viewpoints: engineering decision-making view and engineering etiquette view. As far as the concept of engineering decision-making is concerned, the ancient theory mainly integrates the idea of "Baigong runs the country" and the theory of "official decision-making." Specifically speaking, in ancient project engineering, all the workers performed their own duties and jointly completed the project's construction. Craftsmen are the designers and constructors of the project and play a major role. Based on soliciting opinions from all parties, government officials at all levels make the final decision on the project's construction according to the craftsman's project design. Unlike ancient times, modern engineering decision-making mainly includes administrative and scientific decision-making. At present, China's engineering decision-making is developing from administrative decision-making to scientific decision-making. As far as the theory of engineering etiquette is concerned, the ancient engineering etiquette system mainly includes the etiquette system of capital palace construction and temple construction. The capital palace construction etiquette system has made strict grade regulations on the cities and "housing standards" of different classes of rulers. According to the ancient etiquette system of temple construction, only the emperor was qualified to build altars to worship heaven in the suburbs, while the princes and literati were only allowed to build social altars to offer sacrifices to the gods. These rituals make the ancient city layout and architectural form branded with the unique style of Chinese traditional culture. With the disintegration of the hierarchical system in modern society, although the influence of the theory of engineering etiquette in engineering activities is not as great as before, people still need to seriously consider this factor in the construction process of some landmark engineering projects.

(2) The engineering environment theory in the economic dimension. The economy is one of the effective factors affecting the project's operation, and engineering activity is an important form of economic activity. As for the economic environment of engineering activities, an important issue for project managers is to obtain the maximum profit with the least consumption. In the engineering activities in ancient China, the economic thoughts of enriching the country, enriching the people, frugality view, and the view of making the best use were mainly pursued in the engineering activities in ancient China. It can be said that these ancient engineering economic

thoughts, to a certain extent, contain the modern ideas of sustainable development and maintaining ecological balance. Contemporary scholars have considered and studied the planning and operation of engineering activities in the economic environment, and formed many representative theories, mainly including low-carbon economy theory, life cycle theory, and ecological footprint theory.

(3) The engineering environment theory in a custom dimension. In engineering practice, human beings are often easily influenced by customs. The traditional theories of the engineering environment in ancient China mainly include the view of group living together, the view of heaven and ancestor supremacy, the view of combining Yin and Yang, and the view of the order of superiority and inferiority. These custom theories have an important impact on the design and layout of ancient engineering buildings. For example, Beijing's characteristic Siheyuan and the world-famous Fujian Tulou are the buildings influenced by the idea of living in groups. The construction industry takes Luban as the founder and God. The paper industry takes Cai Lun as the founder and God. The metallurgical industry regards the Taishang Laojun as the ancestor god, which is the concrete manifestation of the idea of "heaven and ancestor supremacy." Traditional Chinese buildings have formed a basic plane pattern with a courtyard as the center unit, which combines Yin and Yang. The central axis symmetry layout of Beijing's Siheyuan, which emphasizes the order of superiority and inferiority, inherits the concept that "the center is the most important." The theory of engineering environment in modern customs mainly includes the theory of changing customs and the theory of township regulations. With the construction of various large-scale water conservancy, power, and transportation projects, a large number of project resettlement has become inevitable. In view of the differences in customs and cultures, it is necessary to fully investigate immigrants' local customs and cultural customs in the process of project resettlement to ensure that migrants have a good psychological transition. Township rules and regulations are the customs established in various villages in China's rural areas, which directly involve project planning, decision-making, implementation, maintenance, and other aspects. We should fully respect these rural rules and regulations in the planning project of contemporary rural construction.

(4) The theory of engineering environment in the aspect of cultural heritage. As an influencing factor of the engineering environment, cultural heritage has a certain impact on the project's decision-making, design, implementation, and operation. In ancient China, when dealing with the relationship between engineering activities and cultural heritage, there were mainly two kinds of ideas: the view of observing, preserving, and imitating morality, and the view of knowing, understanding, and undertaking destiny. The ancients believed that respecting and protecting ancestral temples and relics could inspire later generations to respect and follow suit. Most of the ancient engineering activities carried out the idea of observing, preserving, and imitating morality to protect the cultural heritage. In addition, the ancients stressed that they should carry out engineering activities in a targeted way on the basis of knowing their natural mission. The ancient palaces and temples are not entirely rebuilt, but some of them are expanded and extended on the basis of the original buildings, that is, the proper implementation of engineering activities based on knowing

and understanding the destiny. The thought of knowing, understanding, and undertaking destiny respects and protects the existing cultural heritage and inherits and continues to write history and culture. The theory of engineering environment in modern and contemporary cultural heritage mainly includes the view of protection and the view of integration. The view of protection refers to the constructive protection of the existing cultural heritage in engineering activities; the existing cultural heritage cannot be demolished or destroyed at will. The view of integration refers to the respect for buildings with historical value in engineering design and maintaining harmony and unity with the style of surrounding buildings. Although people strongly advocate the protection of cultural heritage in engineering activities, many engineering activities still destroy cultural heritage in reality. For example, Fengjie County, which is located at the entrance of the Three Gorges of the Yangtze River, has a world-famous Baidi city. In order to prevent landslides, the protective works built along the slope are not in harmony with the ancient buildings, which causes the landscape of Baidi city to be destroyed.

(5) The theory of engineering environment in the aspect of law. There are abundant theories about engineering law in ancient China, including the theory of Hong Fan, the theory of building a legal form, and the view of the engineering rule of law. The theory of Hong Fan established the ideological mainline of engineering law and produced some specific legal ideas, such as the concept of five elements: respecting the laws of nature and emphasizing the harmony of nature and human; the concept of five blessings—the unity of reward and punishment, which is based on the people; the idea of suspicion—democratic resolution of doubts. These legal thoughts have played a certain role in engineering legislation and law enforcement. The theory of building a legal form strictly stipulates various design standards, specifications, relevant materials, construction quota, and indicators, etc., and penetrates the legal thought into the engineering activities. It also contains a series of economic design techniques, which is of great guiding value to the design of wooden structures in the Yuan, Ming, and Qing Dynasties and even the current architectural design activities [148]. The view of the engineering rule of law also embodies the thought of legal theory, which contains a lot of good legal content on the basis of the rule of law to improve the engineering rule of law. The mainstream of contemporary engineering management is the legalization of engineering. Laws and regulations play a guiding role in engineering and play a normative and guarantee role. In the process of engineering construction, on the one hand, the law regulates the behaviors of the parties concerned, the government and the market subjects, to ensure the quality of the project; on the other hand, it guarantees the rights and freedom of the parties to develop and use resources according to law. In short, the formulation and implementation of relevant laws and regulations is a strong backing for implementing engineering environmental protection concepts, developing green projects, and achieving sustainable development of engineering activities.

The theory of engineering social and cultural environment may imply the view of respecting heaven and ancestor or similar concepts or suggest the view of destroying the old and establishing the new or similar concepts. To a certain extent, the evolution of the theory of social, cultural environment is a game process of the two main ideas.

8.5.2 Two Main Lines in the Engineering Natural Environment Theory: The View that Human Can Conquer Nature and the View of Harmony Between Human and Nature

In the theoretical system of engineering natural environment, there are always two kinds of concepts or ideas: human can conquer nature or similar concepts, harmony between humans and nature, or similar concept. These two concepts or ideas have been playing games and guiding engineering activities in the game. Humans can conquer nature, and harmony between humans is both the propositions of ancient Chinese philosophy on the relationship between nature and humans. Humans can conquer nature means that humans can use, control, and transform nature by their own power. The harmony of humans and nature means the unity, indivisibility, and harmonious development of humans and nature. The concept of humans can conquer nature or similar concepts at home and abroad, the harmony between human and nature or similar concepts at home and abroad, all point to the relationship between humans and nature. The former refers to the separation of human from nature, and the latter refers to the integration of human and nature. Since ancient times, there has been a game between two concepts in engineering activities at home and abroad.

8.5.2.1 The View of Harmony Between Humans and Nature

The harmony between humans and nature has always been the leading theory in the theory of engineering natural environment in ancient China. The ancient Chinese revered the idea of "nature" in awe. The concept of harmony between humans and nature can be traced back to the time when Pangu created the world. However, its thought originated from the worship of nature, the concept of destiny in ancient society. The concept of harmony between nature and humans originated from "respect" in primitive society, offering sacrifices to heaven or nature. Later, it changed from "fear" to "respect." The myth of Pangu is a symbol of human exploration of natural law. In ancient engineering activities, the idea of the harmony between humans and nature is embodied in that people either conform to the will of heaven or the laws of nature, or achieve the harmony and coordination between the project and the surrounding nature, to achieve the purpose and concept of mutual promotion and coordination between the natural engineering and artificial work. For example, the ancients decided the engineering activities according to the seasonal climate change, and the construction of reservoirs often chose the winter with less rain. On the one hand, considering that there was little agricultural work in winter, there was sufficient workforce to comply with the "human harmony"; and the other was that the dry season in winter was convenient for dam construction activities, which was in line with the "time." For example, Dujiangyan, Zhengguoqu, and other water conservancy projects built in ancient times also give full play to their hydraulic advantages, which have the functions of water diversion, flood discharge, sediment discharge, water diversion, etc. They make full use of the terrain and rational use of hydraulic conditions, which are conducive to people's production and life, benefit mankind, and fully reflect the engineering environmental concept of "harmony between human and nature." In the theory of historical engineering natural environment, the time view and Yin and Yang view in the theory of meteorological environment, the land capability view, the borrowing land view, and the view of complying land in the theory of land environment, and the view of integration, the harmony view, the cherishing life view, and the view of the mean in the theory of biological environment are the concrete embodiment of the thought of the harmony of nature and human in ancient times.

In ancient China, the concept that humans could conquer nature also appeared from time to time, but it has always been a tributary. Most people are in awe of nature. In engineering activities, even in the use and a little transformation to nature, they will not give up the concept of harmony between humans and nature. They will not break the pattern of harmony between humans and nature.

8.5.2.2 The View of Humans Can Conquer Nature

With the progress of science and technology, the ability of human beings to understand, adapt to, and utilize nature has been greatly improved, and the desire to conquer nature has also been greatly expanded. In this context, the concept that humans can conquer nature once appeared and occupied the mainstream position in the theory of engineering natural environment.

After the industrial revolution greatly liberated the productive forces, many people believe that human beings have unlimited ability to conquer and transform nature by relying on science and rationality. Humans can conquer nature, or similar slogans have appeared in global social development and economic activities, including in the field of engineering. In the middle and later twentieth centuries, driven by the concept that humans can conquer nature or similar concepts, humans have created many engineering miracles that cannot appear in the natural state.

In China, the concept that humans can conquer nature once became the dominant idea of engineering activities in the late 1950s. It reached its climax during the "great leap forward" and "Cultural Revolution." Under the guidance of humans can conquer nature and change nature, trees are cut down, grasslands are destroyed, and lakes are reclaimed as farmland. The so-called "convert wasteland at the top of the mountain, do rice seeding at the middle of the lake"; refining iron and steel, mining and smelting by local methods, and running factories by local methods; eliminating four pests (mosquitoes, flies, mice, some birds) and exploiting wild animal and plant resources in a predatory way. In the 1990s, some scholars believed that the view that humans can conquer nature was not a one-sided emphasis on the struggle between humans and nature. The deterioration of ecological environment was not the result of the thought that humans could conquer nature.

8.5.2.3 The Result and Enlightenment of the Game Between the Two Main Lines

Driven by the concept that humans can conquer nature or similar ideas, people have developed and transformed nature at their will. However, soon after the development and transformation, environmental problems have dealt a heavy blow to human beings mercilessly.

In western countries, environmental problems were gradually exposed in the twentieth century, especially at the end of the twentieth century. The black storm in the United States in 1934, the London smog event in the 1960s, the photochemical event in Los Angeles, the Maas River Valley event in Belgium, the worldwide population explosion, the sharp decline of forests, the hole in the ozone layer, etc., are all environmental problems, such as persistent drought or rainstorm, frequent sandstorms, polluted air and ocean, etc. Since the late twentieth century, there have been several major dam failures in the world, such as mare Basse arch dam in France, Vajont Dam in Italy, and Teton earth rock dam in the United States, causing serious loss of life and property. Dam safety has become a major research topic. The Vajont reservoir (the highest arch dam in the world at that time), which was completed in 1960, caused a landslide on the bank of the reservoir in 1963, which was the most serious landslide event in the world. These problems are almost all related to engineering activities. The drastic and uncontrolled engineering activities and the lack of ecological civilization concept in the engineering field have brought huge ecological and social risks. The continuous increase of greenhouse gas emissions, global warming, the emergence of the ozone layer hole, ecological destruction, species extinction, etc., all poses a great threat to the living environment of human beings [149].

In China, environmental problems are also very severe. Pollution, desertification, and over-exploitation lead to a sharp reduction of resources, soil erosion, and environmental problems. Qinghai Lake is a vivid case. The large-scale agricultural development along the Qinghai Lake started from the period of the "great leap forward"; inspired by the slogan "human can conquer nature," the large-scale reclamation activities have turned the best grassland of 750,000 mu around the lake into cultivated land. In the 1980s, a new round of grassland reclamation was carried out around Qinghai Lake. In the 1990s, in order to plant rape on a large scale in the area around the lake, six state-owned farms were set up in a few years to reclaim 300,000 mu of wasteland; local farmers also flocked to reclaim 50,000 mu of wasteland. After large-scale reclamation, the ecological environment of Qinghai Lake and its basin has deteriorated rapidly: water level continues to drop; desertification continues to expand; grassland degradation is serious; fishery resources are on the verge of extinction; rare and endangered wildlife is on the verge of extinction; the land is becoming saltier.

In the face of various environmental problems, people began to reflect. The concept of harmony between humans and nature is higher than that of humans can conquer nature in many aspects such as value, virtue, and will. The concept of harmony between humans and nature or a similar idea is beneficial to nature and humans. It can ensure the continuation of the harmony between humans and nature or develop in a more superior direction, which has a far-reaching guiding significance for engineering activities.

8.5.3 The Two Main Lines of the Theory of Engineering Social and Cultural Environment: The View of Respecting Heaven and Ancestor and the View of Destroying the Old and Establishing the New

In the theoretical system of engineering social and cultural environment, there are always two kinds of concepts: respecting heaven and ancestor or similar concepts, destroying the old and establishing the new or similar concepts. The view of respecting heaven and ancestor emphasizes historical tradition and ancestral regulation, while the view of destroying the old and establishing the new advocates breaking the old rules and customs and establishing new ones. These two concepts or ideas have been playing games and guiding engineering activities in the game.

8.5.3.1 The View of Respecting Heaven and Ancestor

The view of respecting heaven and ancestors advocates awe of heaven, imitating ancestors, and observing ancestry. The view of heaven and ancestor supremacy, the view of the order of superiority and inferiority, the view of observing, preserving, and imitating morality view, the theory of township regulations are all more or less implied in respecting heaven ancestors. Ancestral system has historical and cultural significance and maintains social order and stability to a certain extent, which is of guiding importance for engineering activities. Ancient buildings all follow the ancestral system. For example, according to *the Tang Liu Dian*, "there should be no double-arched caisson for houses below the prince, no more than five rooms and nine frames for halls above grade three, no more than five rooms and five frames for gate houses at both ends of hall buildings; no more than three rooms and five frames for halls above grade five, and no more than three rooms and five frames for gate houses at both ends of halls and buildings, which still serve as the gate of aconitum… [150]." In ancient times, the living space followed etiquette, reflecting the hierarchy and forming the order. For example, the emperor's hall was nine feet, the princes were seven feet, the officials were five feet, and the scholars were three feet [151].

In contemporary China, some people still follow the ancestral system in production and engineering activities. For example, the Hakkas' clan concept, living style, and Weilong house construction in Meizhou, eastern Guangdong Province, have been discussed in the previous chapters, and they are not repeated here. Another example is the worship of gods or ancestors in engineering, which has been discussed before and will not be repeated here.

8.5.3.2 The View of Destroying the Old and Establishing the New

In the engineering social and cultural environment theory, the potential and obvious concept of destroying the old and establishing the new has had a great influence directly or indirectly. Since the 1980s, many cultural relics, blocks and buildings with landmark value have disappeared in the ongoing development projects. Changsha city is a typical case. In recent years, more than ten old buildings with historical and cultural value have disappeared, including Fulu Palace in Pozi street, Yuanyangjing in Xingfu bridge, Zhongshan Memorial Hall in Jiaoyu street, Zuo Zongtang residence in Caie North Road, Zuo Xueqian residence in Fuyuan lane and Xining Hall in Wanglu garden. Many old streets disappeared in a few years, including Zoumalou, Ganziyuan, Zhuhoujie, Beizhengjie and Yingpanjie, leaving only place names for future generations [152]. In foreign countries, destroying the old and establishing the new cases can be seen from time to time, such as constructing the Romanian Parliament building. This building has several floors underground and 12 floors on the ground, which is undoubtedly amazing. However, at the historical centers in Bucharest, including 30,000 houses and eight churches, have to make way for it [153].

8.5.3.3 The Result and Enlightenment of the Game Between the Two Concepts

The two concepts of respecting heaven and ancestor, destroying the old, and establishing the new have always existed in engineering activities. The two concepts sometimes play games and sometimes are compatible. In modern society, the two concepts are both necessary for engineering activities.

For example, in urban construction, on the one hand, it is necessary to build a new city with larger scale and better functions. On the other hand, it is necessary to protect the old town with historical and cultural significance. How to treat "new" and "old" in urban development? Is "innovation" necessarily accompanied by large-scale "dilapidated"? How to deal with the relationship between "breaking" and "establishing" in urban construction? Only by clarifying these problems can we avoid the risk of "pseudo urbanization", constantly improve the quality of urbanization, and make the urban development step into the track of sustainable development [154]. In fact, the sustainable development of modern cities requires the combination of respecting heaven and ancestor and destroying the old and establishing the new. The evaluation of urban construction depends on the level of new area construction and the level of old city protection. The deeper the historical and cultural heritage of the city, the more significant the game between the new city construction and the old city protection. The game's result is not that one side wins, but reaches a balance to achieve the harmonious unity of urban modernization construction and historical and cultural protection. In Europe, Rome, Paris, London, Berlin, Athens, and other cultural and historical cities all follow the road of retaining the old city and developing the new city. In Asia, Tokyo, Seoul, and other historical and cultural cities all adopt the strategy of building sub-city centers or satellite cities and protecting the old city. In China, Luoyang, Xi'an, Pingyao, Lijiang, and other cities respect history, respect nature, cherish the precious resources left by their ancestors, and build their own distinctive style and features by relying on the profound historical and cultural accumulation and the natural scenery of the original ecology, and become well-known "landmarks" far and near.

It is of guiding significance for all engineering activities to consider "respecting the heaven and ancestor" and "destroying the old and establishing the new."

8.6 Rethinking of Engineering Environmental Problems

With the continuous development of society, engineering practice activities become more and more frequent, and the negative impact of engineering activities on the environment is also gradually highlighted. Therefore, it is necessary to consider the engineering environmental problems further.

8.6.1 Problems in the Process of Dealing with the Relationship Between Engineering and Environment

In recent years, in the process of dealing with the relationship between engineering and the environment, people often put the economic interests of the project above the environmental considerations. Some of them enjoy the pleasure of "conquest" in engineering activities and seldom consider the natural ecological background that human economic activities cannot be separated from, so that some engineering projects may have a serious ecological crisis and natural risks [155]. In order to maximize the economic benefits of engineering, some people often ignore environmental interests, which leads to various problems. Among them, the main problems are the serious destruction of the natural ecosystem and the inadequate protection of cultural heritage.

(1) **The natural ecosystem is seriously damaged**

The serious destruction of the natural ecosystem means that people violate the natural law in engineering activities and over-exploit and utilize natural resources, which causes serious damage to the original ecosystem. The damage of engineering activities to the natural ecosystem can be summarized in the following aspects.

First, the destruction of the earth's surface. During the construction of the project, activities such as clearing the surface soil, excavation of earth and stone, changing the river course, mining stock yard and other activities are easy to cause surface vegetation damage and terrain change, accelerate surface erosion, increase surface runoff, increase water and soil loss, change natural water flow pattern, and aggravate water quality deterioration; cause soil structure damage, soil fertility decline, foundation subsidence, etc., and cause serious damage to the natural ecological environment.

Second, permanent damage to animals and plants. With the development of the economy and the needs of human beings, many animals and plants are in danger, and engineering activities play an important role in it. "Due to the increasing demand and pressure on wildlife resources and ecological environment, the destruction of wildlife habitat and the predatory development and utilization environment, many wild animals and plants are seriously endangered. According to the preliminary statistics of a large amount of data accumulated in the scientific survey of natural resources in China, there are more than 300 species of terrestrial vertebrates, and about 410 types and 13 categories of wild plants in China are in an endangered state [156].

The third is the destruction of water ecology. The construction of water conservancy projects is easy to destroy the ecological balance of the regional water environment, affect the normal reproduction and survival of aquatic organisms, and reduce the diversity of organisms. The discharge of production and domestic sewage produced in the project implementation pollutes the surface water and even involves the underground water source. At the same time, the chemical toxins contained in the sewage seriously endanger the living environment of aquatic organisms even generate direct toxicity. It is harmful to aquatic organisms, resulting in large-scale death of aquatic organisms.

Fourth is the destruction of atmospheric ecology. Toxic gases such as carbon monoxide, nitrogen dioxide, sulfur dioxide, and inhalable particles emitted from engineering construction pollute the atmosphere and worsen the air quality. Nowadays, the smoggy weather in major cities is related to unreasonable human engineering activities to a certain extent.

(2) **Poor protection of cultural heritage**

Cultural heritage is the crystallization of human history and culture, with specific cultural value, scientific value, and aesthetic value. Understanding the importance of cultural heritage and protecting the existing cultural heritage is the guiding ideology of contemporary engineering activities. However, in engineering practice, people's awareness of the protection of cultural heritage is weak, intentionally, or unintentionally ignoring the preservation of cultural heritage, making some precious historical and cultural heritage victims of urbanization. Due to the weak awareness of cultural heritage protection, the engineering practice problems are mainly in the following two aspects.

First, the destruction and imitation of ancient buildings. Nowadays, there is a strange phenomenon in engineering activities in China, that is, demolishing "real cultural relics" while building "fake antiques." In recent years, some ancient buildings have been demolished wantonly, and then engineering practitioners imitate the ancient buildings to build new buildings, such buildings, commonly known as "Shanzhai architecture." "Shanzhai architecture" is just architecture with neither historical value nor the unique ancient flavor and historical massiness of ancient architecture and lacks specific historical and cultural significance. For example, the ancient city of Guizhou Zhenyuan is known as "the lock key of Yunnan and the gateway of East Guizhou." There are more than 50 ancient buildings, such as buildings, pavilions, halls, temples, ancestral halls. Ancient houses, roadways, docks, and post roads can be seen everywhere. In recent years, in the development and utilization of tourism resources, human beings have seriously damaged these ancient buildings. Most of the authentic Ming and Qing wooden houses on the old streets were converted into brick and stone structures. The newly built "old town" style did not retain the Diaojiaolou in southeastern Guizhou but was transformed into a horse-headed wall in Huizhou buildings. Another example is the Hongjiang Ancient mall in Huaihua City, Hunan Province. It is the National Key Cultural Relics Protection Unit known as the "living fossil" in the embryonic period of Chinese capitalism. It has nearly 100,000 square meters and more than 380 ancient buildings of the Ming and Qing Dynasties. Today, many beautiful buildings outside the core area have been razed to the ground, and the new buildings and hotels on the rubble are incompatible with the style of the ancient city. Similar to this kind of engineering practice activities of "demolishing the true and building the false" are numerous, which are regrettable and worthy of people's reflection.

Second, the destruction of underground cultural relics. Underground cultural relics include ancient tombs, sites, and remains. In recent years, a large number of underground cultural relics have been destroyed, and a large number of cultural relics have been unearthed. In the process of urban construction, there is often the situation of building first and then investigation. Before the construction, archaeological investigation and exploration of cultural relics are not carried out. As a result, many underground cultural relics are damaged during construction. Wherever the real estate is developed, the cultural relics will be unearthed; wherever the highway is built, the archaeological team will go to "put out the fire"—this is almost the "law" discovered by Chinese archaeology in the past decade or so [157]. There are countless cases of destruction of underground cultural relics in engineering practice, and the protection of cultural relics is worrying. For example, in July 2005, the Central Literature Research Office of Maojiawan in Xicheng, Beijing, renovated the heating ditch. On the way, more than 1.2 million pieces of porcelain from the end of the Yuan Dynasty to the middle of the Ming Dynasty were excavated. As the project is neither in the buried area nor more than 10,000 square meters, it is impossible to carry out underground cultural relics protection before construction, resulting in a large number of porcelain chips lost. For example, in 2008, the Yuanmingyuan station of Beijing Metro Line 4 did not carry out cultural relic protection before construction, and the construction party demolished the paving stone strips, so that a large amount of information about the construction of imperial road in the Qing Dynasty was lost, leaving only the concrete pavement. Another example is that in January 2013, a number of tombs of the Six Dynasties were found in a construction site in Nanjing. However, the construction unit did not inform the cultural relics department. After receiving *the Notice of Ordering Rectification* issued by the municipal cultural comprehensive law enforcement, the construction unit continued to violate the regulations and force construction, eventually destroying five tombs of the Six Dynasties. In addition, in

June 2013, the construction of Guangzhou Metro Line 6 phase II (Luogang depot) caused damage to the archaeological worksite of the Dagongshan site. The tombs of the pre-Qin period are of great importance to studying the tombs of the pre-Qin period.

8.6.2 Countermeasures and Theoretical Basis for Dealing with Engineering Environmental Problems

The following countermeasures can be taken to deal with engineering environmental problems.

8.6.2.1 Adhere to the Concept of Sustainable Development and Promote the Common Development of Engineering and the Environment

The aim of sustainable development is to meet the needs of contemporary people and not to harm the development of future generations. Academician Ding Lieyun proposed that in order to carry out project construction and pursue its economic objectives, the project's impact on social factors such as employment and environment must also be considered [158]. Historical experience has proved that engineering activities under the guidance of the concept of "harmony between human and nature" can promote harmonious coexistence and mutual benefit between humans, nature and society. The natural environment can provide resources and other material bases for engineering activities. For example, the ancient view of time, the view of borrowing land and the view of complying land and so on, can make the best use of things and ensure that the bearing capacity of resources is not damaged to achieve the coordination between natural engineering and artificial. Therefore, in engineering practice, we must correctly handle the relationship between engineering construction and natural environment and social environment, prioritize environmental protection, avoid excessive interference in the natural environment, and wantonly plunder natural resources. Wang Yingluo, an academician of the Chinese Academy of Engineering, once said that we should understand the project as an ecological social phenomenon in the ecological circulation system. The innovation and construction of the project must conform to the law of the ecological cycle [159].

8.6.2.2 Maintain Resource Security and Human Security Under the Guidance of Environmental Security Theory

Environmental security has become a global appeal. Engineering activities should maintain not only resource security, but also human security. To maintain resource security is to ensure the normal reproduction and survival of organisms and ensure biodiversity. To maintain human safety, on the one hand, it is necessary to maintain biological safety as the premise to protect humans from the harm of degraded resources and environment and reduce the negative impact of engineering activities on human production and life. We must put the construction of ecological civilization in a prominent position, respect nature, comply with nature, protect nature, build a national ecological security barrier, and achieve the unity of economic, social and ecological benefits [160].

The concept that humans can conquer nature has great potential safety risks, which is not safe for the project itself and the related people and objects. For example, to reclaim land from lakes and compete with water for land, the first is to speed up the process of lake swamping, the lake area is continuously shrinking, the surface runoff regulation and storage is difficult, resulting in frequent drought and flood disasters, threatening human life; the second is the decline of aquatic animal and plant resources, the deterioration of the ecological environment in the lake area, resulting in the continuous decline of fish species and the decrease in the number of fish, threatening biological diversity.

It is safe to pay attention to the "harmony between human and nature" and carry out engineering activities according to natural conditions. For example, the Dujiangyan water conservancy project, known as the "originator of world water conservancy culture," makes full use of the local geographical conditions of high in the northwest and low in the southeast. According to the special topography, water vein and water potential at the river entrance, the Dujiangyan water conservancy project has no dam for water diversion and self-flow irrigation, which makes the dike, water diversion, flood discharge, sediment discharge and flow control interdependent and forms a common system to ensure the full play of the comprehensive benefits of flood control, irrigation, water transportation and social water use. In contrast to the current flood fighting activities by throwing sand into rivers, the flood brings sand and soil into the river, and the sand for slope protection is used to pad up the river bottom, making the river surface narrower and the riverbed higher. It seems that we have managed the flood every year, but the governance method against the natural law has laid a hidden danger for the next year, so the annual flood is a record high. At the same time, it also creates a large number of suspended rivers. For example, the bottom of the Xiangjiang River is far higher than the surface of the surrounding cities. Once the water volume exceeds the drainage capacity, waterlogging will occur, resulting in "sea watching" and "boating" events in low-lying areas of the city.

8.6.2.3 Establish Environmental Impact Assessment System Based on Life Cycle

Life cycle assessment (LCA), as an international methodology of product and service system evaluation, is of constructive significance to engineering environmental assessment. For example, in the construction of water conservancy projects, the implementation of an environmental impact assessment system requires economic evaluation of engineering projects and environmental impact assessment, and scientific analysis of environmental problems that may arise from development and construction activities, and puts forward prevention and control measures. The environmental impact assessment can provide the basis for the site selection of water conservancy projects and prevent the environmental damage caused by unreasonable layouts. Specifically, the first is to investigate the local weather environment, hydrology, water quality, soil, aquatic organisms, population, etc. The second is to predict the environmental impact according to the survey results, predict the possible impact of the proposed water conservancy project on the local environment, and predict the degree of impact. Finally, the comprehensive evaluation (including the natural environment, human production and living environment) of the proposed water conservancy project construction is carried out, that is, through certain principles and methods, the overall evaluation of the changes and the degree of changes caused by the various elements and processes of the proposed project on the natural environment and social environment, so as to provide the basis for the comparison and selection of plans. In engineering practice, we should make rational use of natural resources for the benefit of mankind and build a harmonious artificial nature. At the same time, we should evaluate the potential hidden dangers behind the engineering activities to maximize the contribution of engineering activities. Engineering experts have proposed to initiate a "green storm" in the construction industry, strengthen the awareness of ecological civilization, and devote to green environmental protection from the design, construction, use, and maintenance of buildings [161]. Jin Zhixin, on the other hand, believes that environmental audit should be strengthened under the condition of environmental protection. In order to accelerate the process of promoting the strategic goal of sustainable economic and social development, the environmental audit should be incorporated into the operation process and become an important supervision means for environmental protection. It is a natural development process of promoting sustainable development by protecting the environment [162].

8.6.3 Establish the Concept of Environmental Care, Cultivate Environmental Virtues, and Achieve Environmental Justice

Human survival and development cannot be separated from the development and utilization of resources. However, the development and utilization of resources lead to a series of environmental problems. In order to make human behavior and the environment harmonious and reduce the occurrence of various environmental problems and ecological crises, we must have environmental care. The establishment of a sound engineering environment theory is conducive to guiding engineering activities so that human beings can enjoy a good environment and enjoy the benefits given by nature while enjoying the achievements of engineering activities. Therefore, in engineering activities, we must deal with the relationship between engineering and the environment, establish environmental care, cultivate environmental virtues, and achieve environmental justice.

8.6.3.1 Follow the Laws of Nature

Nature is a whole. As an integral part of the earth's ecosystem, human beings are naturally connected with nature. On the one hand, human survival and development are inseparable from the development and utilization of natural resources; on the other hand, nature is unconquerable. The stronger the degree of human intervention in nature, the greater the resilience of nature to human beings. Nowadays, the degree of human intervention in nature has been far beyond the scope of natural tolerance. Nature is carrying out all-around retaliation and punishment to human beings in specific ways: soil erosion, land desertification, air pollution, acid rain. Human survival is facing a real crisis. In the face of this harsh fact, human beings must thoroughly reflect on their previous ideas and behaviors, rethink and position the relationship between human and nature, fully respect nature, and consider both utilization and protection of nature.

Specifically, in engineering activities, human beings should respect nature, follow natural laws, make rational use of natural resources, protect biological diversity, and maintain ecological balance. In fact, people have paid more and more attention to the impact of human activities, especially engineering activities' impact on the natural environment. In order to maintain the health of the natural environment, countries all over the world are making a series of efforts, such as formulating environmental protection laws, regulations, and standards, etc., to strengthen environmental awareness and promote environmental protection actions. In China, the constitution of the People's Republic of China clearly stipulates: "the state guarantees the rational use of natural resources and protects precious animals and plants. It is forbidden for any organization or individual to occupy or destroy natural resources by any means [163]." *The Environmental Protection Law of the People's Republic of China (for Trial Implementation)* was adopted and promulgated by the Standing Committee of the National People's Congress in 1979 [164]. And it was revised and adopted at the eighth meeting of the Standing Committee of the Twelfth National People's Congress in 2014. The UK promulgated *The Environmental Protection Act 1990* [165] and *the Water Resources Act 1991* [166]. Many countries restrict engineering activities by formulating a series of environmental protection laws and regulations to promote the rational utilization and protection of natural resources and the establishment of harmonious artificial nature based on respecting natural laws.

8.6.3.2 Cultivate Environmental Virtues

With the rapid development of modern civilization, environmental problems such as environmental pollution, energy crisis, and ecological imbalance are increasingly perplexing human beings. In this context, it is a need of the times to cultivate environmental virtues and strengthen moral care for the environment. Environmental virtue concerns the interests of human beings and the interests of non-human beings.

Geoffrey B. Frasz defines environmental virtue as environmental virtue refers to the mean between two vices. To some extent, the quality of this virtue can lead people to a better living environment [167]. Thomas Hill proposed the environmental virtue of "moderate humility." He said: "people often think that those who will destroy the natural environment must not correctly understand their position in the natural order, so they must be ignorant or have no quality of humility [168]." Frasz believes that in order to be an excellent environmental citizen, in addition to humility, we also need kindness, "the cultivation of environmental kindness is an integral part of excellent environmental protection citizens [169]." At present, environmental virtue has become the core concept of environmental ethics; many Western ethicists and philosophers have carried out in-depth research on it. For example, in 2005, Philip Cafaro, associate professor of philosophy at Colorado State University and Ronald Sandler, associate professor of philosophy at Northeastern University of the United States, published the first coedited anthology, *Environmental Virtue Ethics*. To a certain extent, these related research results provide theoretical guidance for the cultivation of environmental virtue.

From the point of view of modern western environmental virtue, it is not enough to rely on laws or environmental ethics to restrain people's destruction of the environment and make humans truly care for nature from the heart. Human beings must also cultivate environmental virtues. We must know who we are and what kind of people we should be. Only when human beings truly understand their position in nature, change their indifferent attitude towards nature, no longer regard nature as a tool, and become a man of noble moral character, can the natural environment be truly protected, achieve the common prosperity of man and nature, and finally establish a harmonious relationship between human and nature. Therefore, we must not neglect the cultivation of environmental virtue in our contemporary engineering practice. Academician Shen Guofang believes that environmental problems should be solved through the change of environmental consciousness, attitude, behavior, and values [170].

8.6.3.3 Uphold Environmental Justice

As defined by the U.S. Environmental Protection Agency, environmental justice is to ensure fair treatment for people of different races, cultures, and incomes through environmental laws, plans, and policies. In the International Symposium on environmental justice held at the University of Melbourne in 1997, environmental justice was defined as reducing the unequal environmental impacts caused by unequal relations among countries, nations, and generations. There are 17 items of *Principles of Environmental Justice* adopted by the discussion of the Colored People's Environmental Summit held in Washington in 1991. For example: first, respect the earth and the ecosystem. Environmental justice strongly advocates that we should respect the interdependent relationship between the earth, the ecosystem, and all species on which we depend for existence. No ecosystem destruction is allowed; second, human beings should respect and be equal to each other. Environmental justice requires that all public policies should be based on mutual respect and equality of all human beings, and no discrimination or difference is allowed [171].

In the traditional concept of justice, both procedural justice and substantive justice emphasize fairness and justice. However, they pay more attention to the fairness and justice of the individual rather than the whole. As a new concept of justice, environmental justice breaks through the traditional concept of justice to a certain extent. It pays more attention to the overall environmental injustice caused by environmental problems, especially among countries and nations. It can be said that environmental justice means that people from different countries, nationalities, and classes enjoy reasonable rights, undertake reasonable obligations and receive fair and just treatment in all behaviors and practices related to the environment.

If we want to achieve environmental justice, we must achieve two important changes: on the one hand, we should extend our concern for the special "our own descendants" to the universal "next generation of mankind"; on the other hand, we should change the overemphasis on money and wealth to more attention on ecological wealth. Only in this way can we realize the equality between contemporary people and future generations.

Environmental justice emphasizes that everyone and every generation have equal rights to a safe, clean, and sustainable environment and freedom from environmental damage. It requires us to respect everyone's common environmental interests in any social practice. Of course, we should also remember to uphold environmental justice in engineering practice. We should pay attention to protecting the common living environment and seek welfare for human needs in the process of environmental protection.

References

1. Gove, F. (1976). *Webster's third new international dictionary of the English language unabridged* (p. 760). Merriam Webster.
2. Chen, G. (1985). *Annotation and evaluation of Lao Zi* (p. 163). Zhonghua Book Company.
3. Feng, Q. (1992). *Dictionary of philosophy* (p. 709). Shanghai Dictionary Publishing House.
4. Zhou, Z. (1991). *Zhou Yi translation notes* (p. 235). Zhonghua Book Company.
5. Cheng, D. (1992). *Early childhood story* (p. 75). Shanghai Ancient Books Publishing House.
6. Zhang, S. (2007). *Three strategies for water control-China Yixue water resources management project and project management strategy. Editorial committee of china engineering management outlook: Looking around and looking forward to China's engineering management-a collection of papers from the first Engineering management forum* (p. 345). China Building Industry Press.
7. He, N. (1998). *Interpretation of Huainanzi-Lan Ming Xun* (p. 476). Zhonghua Book Company.
8. Feng, Q. (1992). *Dictionary of philosophy* (p. 194). Shanghai Dictionary Publishing House.
9. Zhu, X. (1983). *Four books* (p. 17). Zhonghua Book Company.
10. Anonymous. (2012). *Shang Shu* (translated by Wang, S., & Wang, C., p. 349). Zhonghua Book Company.

11. Zhu, X. (1983). *Four books* (p. 49). Zhonghua Book Company, 49.
12. Sun, X. (1989). *Book of rites* (p. 399). Zhonghua Book Company.
13. Zhu, X. (2009). *The original meaning of Zhou Yi* (p. 41). Zhonghua Book Company.
14. Philosophy Volume Editorial Board. (1992). *Summary of Chinese academic masterpieces philosophy volume//editorial committee of abstract of Chinese academic masterpieces. Summary of Chinese academic classics* (pp. 35–36). Fudan University Press.
15. Shang, Y. (2010). Shang Jun Shu. Shi, L. *Translation notes* (p. 91). Zhonghua Book Company.
16. Sun, X. (1989). *Collection of rites* (p. 965). Zhonghua Book Company.
17. Gove F. B. *Merriam-Webster*. Dictionary: http://www.merriam-webster.com/dictionary/land
18. Xu, Y. (2002). *Collection of Mandarin* (p. 578). Zhonghua Book Company.
19. Huang, H. (1996). *Supplementary notes of Yizhou book* (p. 207). Northwest University Press.
20. Li, X. (2004). *Guan Zi notes* (p. 995). Zhonghua Book Company.
21. Zhu, X. (2010). *Four books* (p. 203). Zhonghua Book Company.
22. Wang, X. (1988). *Xunzi* (p. 165). Zhonghua Book Company.
23. Zhu, X. (2010). *Four books* (p. 99). Zhonghua Book Company.
24. Zhu, X. (2010). *Four books* (p. 204). Zhonghua Book Company.
25. Zhu, Q. (2000). *Laozi* (p. 31). Zhonghua Book Company.
26. Jiang, G. Lecture notes on Laozi. http://blog.china.com.cn/jiangguojun/art/4782162.html
27. Song, Y. *Tiangong Kaiwu* (p. 8). Research Museum of Kyushu University.
28. Zhejiang Provincial Institute of Cultural Relics and Archaeology. (2004). *Bridge across the lake in the Puyang river basin archaeological report* (p. 40). Cultural Relics Publishing House.
29. Zhejiang Provincial Cultural Relics Management Committee, Zhejiang Provincial Museum. (1978). The first excavation report of the Hemudu site. *Journal of Archaeology, 1*, 42–48.
30. Chen, Z. (1988). *Annotations to the gardening* (2nd ed., p. 47). China Building Industry Press.
31. Zhang, S. (2007). *Three Strategies for water control-China Yixue water resources management project and project management strategy//editorial committee of "China engineering management lookout and outlook": Looking around and looking forward to China's engineering management-a collection of papers from the first engineering management forum* (p. 348). China Building Industry Press.
32. Zhu, Q. (2000). *Laozi* (p. 18). Zhonghua Book Company.
33. Li, X. (2004). *Guan Zi* (p. 1096). Zhonghua Book Company.
34. Zhu, X. (2010). *Four books* (p. 51). Zhonghua Book Company.
35. Zhang, S. (2007). *Three strategies for water control-China Yixue water resources management project and project management strategy//editorial committee of "China engineering management lookout and outlook": Looking around and looking forward to China's engineering management-a collection of papers from the first engineering management forum* (p. 347). China Building Industry Press.
36. Chen, G. (1983). *Zhuangzi* (p. 80). Zhonghua Book Company.
37. Annotation by Chen, T. (2013). *Guoyu Zhengyu* (p. 128). Zhonghua Book Company.
38. Wu, J. (2010). Cao Wei's water conservancy project and the change of political pattern in the later period of the Three Kingdoms. *Anhui Agricultural Science, 21*, 114.
39. Wen, R. (2008). *Notes of Kao Gong Ji* (p. 1). Shanghai Ancient Books Publishing House.
40. Zheng, X. (2010). *Notes of Zhou Li* (p. 287). Shanghai Ancient Books Publishing House.
41. Yang, T. (2007). *Zhou Li translation notes* (p. 145). Shanghai Ancient Books Publishing House.
42. Yang, T. (2007). *Zhou Li translation notes* (p. 284). Shanghai Ancient Books Publishing House.
43. Sima, Q. (1974). *Historical records* (p. 1053). Zhonghua Book Company.
44. Sima, Q. (1974). *Historical records* (pp. 789–790). Zhonghua Book Company.
45. Wang, F. (1997). *Qian Fu Lun* (p. 263). Zhonghua Book Company.
46. Guan, Z. (2009). *Guan Zi* (translated by Li, S., p. 128). Zhonghua Book Company.
47. Guan, Z. (2009). *Guan Zi* (translated by Li, S., p. 256). Zhonghua Book Company.
48. Confucius. (2007). *The Analects* (translated by Zhang, Y., p. 12). Zhonghua Book Company.

49. Xun, Y. (1997). *Shen Jian//editorial committee of the summary of Chinese academic masterpieces. Summary of Chinese academic masterpieces·political and legal volume* (p. 148). Fudan University Press.
50. Mo, D. (2007). *Mo Zi* (translated by Li, X., p. 78). Zhonghua Book Company.
51. Mo, D. (2007). *Mo Zi* (translated by Li, X., p. 39). Zhonghua Book Company.
52. Mo, D. (2007). *Mo Zi* (translated by Li, X., p. 80). Zhonghua Book Company.
53. Shi, Z., & Hu, X. (1998). *Chinese cultural customs dictionary (cultural customs)* (p. 1). China International Broadcasting Press.
54. Ge, Z. (2005). *History of Chinese thought* (Vol. 1, p. 19). Fudan University Press.
55. Zhu, X. (2011). *Four books* (p. 168). Zhonghua Book Company.
56. Chen, G. (1983). *Zhuangzi* (p. 855). Zhonghua Book Company.
57. Xu, W. (2009). *Annotations of master Lu's spring and autumn annals* (p. 460). Zhonghua Book Company.
58. Sun, Y. (2000). *Zhou Li Zhengyi* (Vol. 3, p. 722). Zhonghua Book Company.
59. Zhu, X. (1983). *Four books* (p. 25). Zhonghua Book Company.
60. Anonymous. *The protection of cultural relics in ancient China*. http://www.chinabaike.com/article/1/78/433/2007/20070520113145.html
61. Fan, G. (2001). *History of Chinese architecture* (p. 127). China Construction Industry Press.
62. Li, X. (1999). *Shang Shu Zhengyi* (p. 352). Peking University Press.
63. Li, X. (1999). *Shang Shu Zhengyi* (p. 355). Peking University Press.
64. Li, X. (1999). *Shang Shu Zhengyi* (p. 357). Peking University Press.
65. Xu, S. (2012). *Shuowen Jiezi* (translated by Xu, X., p. 271). Zhonghua Book Company.
66. Ouyang, X., & Song, Q. (1975). *Xin Tang Shu* (p. 1199). Zhonghua Book Company.
67. Tuo, T. (2012). *Song Shi* (p. 1201). Zhonghua Book Company.
68. Jiang, K. (1996). *Chinese academic summary-art volume* (p. 962). Fudan University Press.
69. Jiang, K. (1996). *Chinese academic summary-art volume* (p. 963). Fudan University Press.
70. Peking University Law Encyclopedia Editorial Board. (2000). Peking university law encyclopedia history of Chinese legal thought (pp. 206–207). Peking University Press.
71. Sun, X. (1989). *Collection of rites* (p. 432). Zhonghua Book Company.
72. Editor-in-Chief of China Encyclopedia. (1992). *Encyclopedia of China: Water conservancy volume* (p. 437). China Encyclopedia Press.
73. Water Ministry-Style Residual Volume. (1986). *Interpretation of the authentic documents of Dunhuang's social and economic literature*. Bibliography and Literature Publishing House.
74. UNEP. *Declaration of the United Nations conference on the human environment*. http://www.unep.org/Documents.Multilingual/Default.asp?documentID=97&ArticleID=1503
75. United Nations. United Nations framework convention on climate change. http://unfccc.int/resource/docs/convkp/conveng.pdf
76. Cheng, D., & Zhong, X. (2008). The relationship between the greenhouse effect, climate change, and human activities. *Journal of Xianning College, 6*, 86–88.
77. Gleick, P. H., Adams, R. M., Amasino, R. M., Anders, E., Anderson, D. J., Anderson, W. W., ... & Mabogunje, A. L. (2010). Climate change and the integrity of science. Science, 328(5979), 689–690.
78. Zheng, G. (2008). *In-depth study and practice of the scientific concept of development, strengthening public meteorological service capabilities*. http://news.xinhuanet.com/politics/2008-11/04/content_10303182.htm
79. Wu, Q., et al. (2005). Response analysis of permafrost along the Qinghai-Tibet highway to climate change and engineering impact. *Glacier Permafrost, 1*, 50–54.
80. Chen, X., et al. (2015). Impact of climate change on several major projects in China. *Progress in Climate Change Research, 5*, 337–342.
81. Ding, W., & Tan, J. (2007). *Chinese and foreign experts issued the "Dongting Lake Declaration" calling for wetland protection*. http://news.xinhuanet.com/newscenter/2007-12/04/content_7199178.htm
82. Anonymous. (2007). *Shangri-La's mountain road construction destroys the ecology*. http://www.sznews.com/epaper/szwb/content/2007-04/02/content_1004510.htm

83. She, H. (2005). *Stone mining in Shanxi made sand and dust storms endangering the Jing Jin ecology*. http://www.stone365.com/news/info-2212.html
84. Kütting, G., & Herman, K. (Eds.). (2018). *Global environmental politics: Concepts, theories and case studies*. Routledge.
85. Wang, X. (2006). *Investigation and thinking of Dongting lake governance: People give way to water, and water gives way. People's Daily 2006–10–13*. http://58.68.146.102/rmrb/20061013/15
86. Gong, J. (2006). *Time comment: May Dongting lake have no worries. People's Daily 2006–10–13*. http://58.68.146.102/rmrb/20061013/15
87. The United Nations. (1992). *The Rio declaration on environment and development*. The United Nations.
88. Standing Committee of Shaanxi Provincial People's Congress. (2016). *Shaanxi provincial people's congress standing committee announcement (No. 79). [2016–01–14]*. http://www.gov.cn/flfg/2007-12/10/content_829788.htm
89. Zhang, K. (2012). *More than a dozen provinces and autonomous regions in China have requested to restart the project of returning farmland to the forest [2012–11–07]*. http://finance.sina.com.cn/nongye/nyhgjj/20121107/095313601018.shtml
90. Tang, Z., et al. (2013). Discussion on design ideas of mine geological environment control engineering-taking limestone mine in Fengshan County, Guangxi as an example. *Hydrogeology and Engineering Geology, 2*, 123–128.
91. Rong, Z. (2006). *Qinghai-Tibet railway environmental protection, the embodiment of the humanistic spirit [2006–06–15]*. http://news.xinhuanet.com/politics/2006-06/15/content_4700489.htm
92. The United Nations. (1992). *The Rio declaration on environment and development*.
93. Vogt, W. (1948). *Road to survival*. Kessinger Publishing.
94. Lin, W. (2007). *Ecology*. Science Press.
95. Fang, J. (2010). Overview of the world's biological resources. *Journal of Plant Genetic Resources, 2*, 121–126.
96. Wang, X., et al. (2010). Analysis of the international research situation of biodiversity. *Journal of Ecology, 4*, 1066–1073.
97. United Nations. (1992). *United Nations conference on environment & development, Rio de Janerio, Brazil: Agenda 21*. www.un.org/sustdev/agenda21.htm
98. Zhang, G., & Liu, J. (2009). Philosophical thinking on the connotation of environment-friendly society. *Guide to Ideological Education, 7*, 61.
99. Wen, Z. (2014). *Concept and impact of an environmentally friendly society [2014–09–29]*. http://www.cenews.com.cn/historynews/06_07/200712/t20071229_36712.html
100. Mo, F. (2005). *What is an environment-friendly society (connotation) [2005–11–04]*. http://www.southcn.com/nflr/llzhuanti/hjyhxsh/hjzl/200511150510.htm
101. Li, Y., & Xu, L. (2001). The coordinated development of humans and nature is the only way for mankind to achieve sustainable development. *Theoretical Monthly, 7*, 51–52.
102. Zhu, W. (2013). On the practical dimension of the coordinated development of man and nature. *Inheriting Marxism, 2*, 32–33.
103. Wikipedia. *Sustainable development*. https://en.wikipedia.org/wiki/Sustainable_development
104. United Nations. *General assembly report on the United Nations conference on environment and development, Annex 1, Rio declaration on environment and development*. http://www.un.org/documents/ga/conf151/aconf15126-1annex1.htm
105. Institution of Civil Engineers. *Structures and buildings. Featured case studies. Case study: Bigsweir bridge repainting. [2014–01–13]*. http://www.ice.org.uk/topics/structuresandbuildings/Case-Studies---Information/Bigsweir-Bridge-repainting
106. Wang, X. *"The Seventh Wonder of Mankind"-Itaipu hydropower station*. http://www.people.com.cn/GB/paper2515/9692/893080.html
107. Jiang, Z. (2002). Building a well-off society in an all-around way and creating a new situation in the cause of socialism with Chinese characteristics-speech at the 16th National Congress of the Communist Party of China. *Beijing Party Branch Life, 12*, 4–19.

108. Dong, Y. (2009). *Study on the democratization mechanism of engineering decision-making*. Northeastern University.
109. The American Society of Civil Engineers. *Guiding principles for the nation's critical infrastructure*. http://www.asce.org/Infrastructure/Guiding-Principles-for-the-Nation-s-Critical-Infrastructure/
110. Lu, G. (2009). Decision errors in major projects and construction of decision mechanisms for major projects. *China Science and Technology Forum, 4*, 34.
111. Zhao, K. (2009). Decision analysis of local public product supply-taking Guilin city's "Two Rivers and Four Lakes" project as an example. *Theoretical Monthly, 7*, 23–29.
112. Wen, Y. (2011). Research on the theoretical connotation of developing a low-carbon economy. *Productivity Research, 8*, 10–14.
113. International for Standard organization. (1997). *ISO14040 environmental management: Life cycle assessment principles and frame work*. International Organization for Standardization.
114. Haiek, V. G. (1984). *Management of engineering projects* (3rd ed., pp. 61–62). McGraw-Hill Company.
115. Zhang, L., & Liu, C. (2005). Environmental impact assessment of the life cycle of building products. *Journal of Beijing Jiaotong University: Social Science Edition, 12*, 39–44.
116. Chen, W. (2005). *Research on decision-making mechanism of major engineering projects*. Wuhan University of Technology.
117. Xin, Y., et al. (2014). Applying ecological footprint theory in the sustainable development of ethnic areas in China. *China Population Resources and Environment, 11*, 144.
118. Sima, Q. (1982). *Historical records biography of Lisi* (p. 256). Zhonghua Book Company.
119. Compilation Committee of Badong County Local Records in Hubei Province. (1993). *Badong County records* (p. 480). Hubei Science and Technology Press.
120. Shu, S. (2002). *Yiyuan Folklore* (pp. 26–27). People's Daily Press.
121. Jiang, F. (2003). *Ecological Folklore* (pp. 220–222). Heilongjiang People's Publishing House.
122. Intergovernmental Committee for the Protection of World Cultural and Natural Heritage. (1972). *Convention for the protection of the world cultural and natural heritage*. United Nations Educational, Scientific and Cultural Organization.
123. United Nations Educational, Scientific and Cultural Organization. (1968). *Recommendations on the protection of cultural property threatened by public or private projects*.
124. United Nations Educational, Scientific and Cultural Organization. (1972). *Convention for the protection of world cultural and natural heritage*.
125. United Nations Educational, Scientific and Cultural Organization. (1976). *Suggestions on the protection and role of historical areas*.
126. The 2nd International Conference of Architects and Technicians of Historical Site. (1964). *Charter of international historical site protection and restoration*.
127. International Council of Historic Sites, International Commission on Historical Gardens. (1981). *Florence charter*.
128. International Council of Historic Sites. (1987). *Charter for the protection of historical towns and regions*.
129. Standing Committee of the National People's Congress. (1982). *Law of the People's Republic of China on the protection of cultural relics*.
130. State Council of the PRC. (2003). *Regulations for implementing the law of the People's Republic of China on the protection of cultural relics*.
131. Xie, H., & Shen, P. (2007). Exploration of landscape protection and construction methods in geological disaster management of the Three Gorges of the Yangtze River. *Journal of Chongqing Jianzhu University, 2*, 7.
132. Gui, L. (2011). *People create architecture, and architecture also shapes people-Interview with Academician Guan Zhaoye [2011–03–25]*. http://www.tsinghua.org.cn/alumni/infoSingleArticle.do?articleId=10069006
133. Wei, Q., et al. (2012). Guan Zhaoye and Tsinghua library's song devotion. *Architecture, 14*, 77.

134. Zhao, K., et al. (2007). The construction characteristics of Tujia Diaojiaolou—Taking the ancient construction surveying of Pengjiazhai in western Hubei as an example. *Huazhong Architecture, 6*, 150.
135. Ministry of Housing and Urban-Rural Development of the People's Republic of China. Regulations on Energy Management of Civil Buildings. Ministry of Housing and Urban-Rural Development of the People's Republic of China. http://www.mohurd.gov.cn/zcfg/jsbgz/200611/t20061101_159082.html
136. Anonymous. (2005). *Environmental protection story during the construction of Qinghai-Tibet railway*. People's Daily, 2005–12–12.
137. National People's Congress of the People's Republic of China. *Constitution of the People's Republic of China*. Central Government Portal. http://www.gov.cn/zhengce/2014-03/21/content_2643049.htm
138. National People's Congress of the People's Republic of China. *Water Law of the People's Republic of China*. Central Government Portal. http://www.gov.cn/gongbao/content/2002/content_61737.htm
139. National People's Congress of the People's Republic of China. *Construction law of the People's Republic of China*. Legislative Affairs Office of the State Council. http://www.chinalaw.gov.cn/article/fgkd/xfg/fl/201104/20110400338908.shtml
140. Soroos, M. S. (1994). Global change, environmental security, and the prisoner's dilemma. *Journal of Peace Research, 31*(3), 317–332.
141. Kütting, G., & Herman, K. (Eds.). (2018). *Global environmental politics: Concepts, theories and case studies* (p. 57). Routledge.
142. Kütting, G., & Herman, K. (Eds.). (2018). *Global environmental politics: Concepts, theories and case studies* (p. 59). Routledge.
143. Kütting, G., & Herman, K. (Eds.). (2018). *Global environmental politics: Concepts, theories and case studies* (p. 87). Routledge.
144. Kütting, G., & Herman, K. (Eds.). (2018). *Global environmental politics: Concepts, theories and case studies* (pp. 88–90). Routledge.
145. Weekly Newspaper. (2013). *The viewpoint of the week: Pollution that cannot be thrown away [2013–06–15]*. http://news.cntv.cn/2013/06/15/VIDE1371310079869668.shtml
146. Kütting, G., & Herman, K. (Eds.). (2018). *Global environmental politics: Concepts, theories and case studies* (pp. 72–75). Routledge.
147. Kütting, G., & Herman, K. (Eds.). (2018). *Global environmental politics: Concepts, theories and case studies* (p. 77). Routledge.
148. Yang, H. (2014). Research the economizing design ideas in Ying Zao Fa Shi. *Chinese Culture Forum, 12*, 143–147.
149. Xu, K. (2008). Development of engineering philosophy and implementation of the scientific development concept. *Journal of Beijing Normal University: Social Science Edition, 1*, 90–92.
150. Li, L., et al. (1992). *Tang Liu Dian. Chen, Z., proofreading*. Zhonghua Book Company.
151. Yang, T. (2007). *Annotations of rites*. Shanghai Ancient Books Press.
152. Zhen, R. (2011). *The old buildings in Changsha are disappearing, who is erasing the "Memories of Changsha" [2011–08–11]*. http://hunan.voc.com.cn/article/201108/201108110921304836.html
153. Anonymous. (2012). *Ten ugliest buildings in the world [2012–01–13]*. http://business.sohu.com/20120113/n332087125.shtml
154. Li, Q., et al. (2013). Focus on the "breaking" and "building" of urban construction. *Mass, 6*, 81–83.
155. Yang, S., et al. (2012). Dialectical thinking in project management. *China Engineering Science, 2*, 14–24.
156. State Forestry Administration. (2010). *National wildlife conservation and nature reserve construction project master plan [2010–10–12]*. http://www.forestry.gov.cn/main/218/content-452802.html
157. Zhang, R. (2010). *The destruction of many cultural relics during the project's construction highlights the hidden concerns of cultural preservation [2010–02–09]*. http://news.xinhuanet.com/society/2010-02/09/content_12959643.htm

158. Ding, L. (2009). Engineering management: Focus on the social dimension of engineering. *Construction Economy, 5*, 8–10.
159. Wang, Y., & Liu, H. (2005). Engineering science and engineering philosophy. *Dialectics of Nature, 9*, 59–63.
160. Zhang, Q., & Xi, J. (2016). *Respect nature, conform to and protect nature, firmly building a national ecological security barrier [2016–08–24]*. http://news.xinhuanet.com/politics/2016-08/24/c_1119448608.htm
161. Lu, G. (2015). Set off a "green storm" in the construction industry. *Construction Enterprise Management, 4*, 46–47.
162. Jin, Z., et al. (2007). Coal mine environmental audit research. China coal economic research selection committee. 2007 coal economic research selected works (p. 423). Coal Industry Press.
163. National People's Congress of the People's Republic of China. (2014). *Constitution of the People's Republic of China*. Central Government Portal. http://www.gov.cn/zhengce/2014-03/21/content_2643049.htm
164. Central Government Portal. (2014). *The environmental protection law of the People's Republic of China shall come into force on January 1, 2015 [2014–04–25]*. http://www.gov.cn/xinwen/2014-04/25/content_2666328.htm
165. The Stationery Office. (1990). *Environmental Protection Act 1990*. http://www.legislation.gov.uk/ukpga/1990/43/contents
166. The National Rivers Authority. (1991). *Water Resources Act 1991*. http://www.legislation.gov.uk/ukpga/1991/57/section/165
167. Frasz, G. B. (1993). Environmental virtue ethics: A new direction for environmental ethics. *Environmental Ethics, 15*(3), 259–274.
168. Hill, T. E. (1983). Ideals of human excellence and preserving natural environments. *Environmental Ethics, 5*(3), 211–224.
169. Frasz, G. (2005). Benevolence as an environmental virtue. *Environmental Virtue Ethics*, 125.
170. Shen, G., & Han, S. (2014). China environmental and development international cooperation committee's 2013 report: Environment and society. *Environmental Sustainable Development, 2014*(4), 15–26.
171. Peng, G. (1999). On environmental justice. *Nature Conservation Quarterly, 28*, 6.

Chapter 9
Engineering Management Humanistic Theory

Engineering comes along with the origin of humankind. At the same time, it develops along with the evolution of humankind. Whether to meet people's material needs or spiritual sustenance, any project is dominated and built by people. Engineering activities should not only create artificial nature, nor be limited to the satisfaction of functions, but also become a creative practice for human beings to take care of their survival and design their future, and at the same time, shape and enhance human nature. Therefore, the entire process of engineering activities is technology integration and cultural activities. The artificial nature as a product of engineering necessarily embodies a strong humanistic connotation, carrying culture and recording history.

This chapter takes engineering as the "basic point" and humanities as the "pillars" to trace back to the source of engineering management humanistic theory. Firstly, we take the habitat project with a long history, large quantity, and wide range and most familiar to people as the starting point to comb through the engineering history and related humanistic context. Then we discuss the formation, inheritance, and protection of engineering culture, analyze the relationship between engineering and art, and expound the aesthetic characteristics, aesthetic expression of engineering art, and compare Chinese and Western engineering arts. Only by truly integrating humanistic care into the project's construction can we establish the modern harmonious engineering concept of "Unity of Man and Nature".

9.1 Engineering Humanities

9.1.1 Humanities and Engineering Humanities

Humanity is short for human culture. The dictionary Cihai defines it as humanity refers to various cultural phenomena in human society. The term "humanities" was first seen in "The Book of Changes": "Looking at humanities to observe changes

J. He, *Principles of Engineering Management*,
https://doi.org/10.1007/978-981-99-1168-4_9

in time; looking at humanities to transform into the world." In China, humanistic thoughts have a long history and are reflected in literature, art, science, philosophy, and political culture. The benevolent spirit of "do whatever you want, don't impose on others", the political proposition of "benevolence" and "people-oriented" and the philosophical concept of "unity of man and nature" have not been consistent in the development of Chinese history, but they have always belonged to China's mainstream culture.

Humanities have also experienced a long period of change and development in the West. Humanism is rooted in ancient Greek and Roman culture. It was devastated in the Middle Ages, promoted after the Renaissance, and then became the ideological weapon of the emerging bourgeoisie against feudal autocracy and religious theology. Humanism, humanitarian spirit, freedom, equality, and fraternity, these humanistic ideas that originated from the Renaissance and were strengthened in the subsequent religious reforms and enlightenment, have been deeply rooted in Western society. Montesquieu's "innate human rights", Rousseau's "sovereignty lies with the people", Kant's "man is the end", etc., are all expressions of Western humanist ideas. Since modern times, the prosperity of Western philosophy, science, and art have been the fruitful result of the humanistic spirit. Humanities in the West also refers to human culture.

Humanities consist of two inseparable aspects: people and culture. on the one hand, people are the main subject of culture and the creators of culture, and culture is the objectification of essential human power; on the other hand, culture becomes the way of human existence and has a human nature, which is a distinctive mark of why human is different form animals and why human is human. People are not only the creator of culture but also the owner of culture. Without people, there is no culture, and people without culture cannot be called people. The core of humanities is people. Therefore, respecting people, loving people, doing everything for people, recognizing and protecting human rights, putting people first, and so on have become the consensus of human society. Human culture can also be called humanity. The so-called humanity is also the cultural attribute of humans, a symbol of human thought and spirit, and a symbol of human civilization.

Engineering humanities, or the humanity of engineering, refer to the human attributes possessed by projects with "people" as the main body and various cultural phenomena contained in the projects. In a broad sense, engineering humanities refer to all material and spiritual elements related to engineering, with a very wide range. In a narrow sense, engineering humanities refer to the social ethics, literature, and art that are integrated with the engineering function. The scope is relatively narrow, focusing on the spiritual elements of humanity. Whether the engineering humanities in a broad or narrow sense, all of them should reflect the core of "people-oriented".

9.1.2 Engineering Humanities Requirements

The humanistic requirements of engineering should realize the harmony between engineering and people and the harmony between engineering and nature, engineering and culture. Specifically, the humanity of the engineering includes the following three meanings.

(1) People-oriented must be adhered to achieve harmony between engineering and people. It is the purpose of engineering. "People-oriented" is the core idea of engineering humanities and the chapter's starting and ending point. What is the engineering for? It can be traced back to the source and starting point of engineering. The emergence of engineering is to meet the needs of human beings in all aspects of clothing, food, housing, and transportation and create better living and working conditions for human beings. Therefore, engineering's essential function and fundamental purpose are to serve people. From ancient to modern times, China has had many projects for the benefit of the people, all reflecting the value orientation of "people-oriented". For example, the construction of the Qinghai-Tibet Railway and the Yellow River Xiaolangdi Water Control Project is not intended to create the image of "the world's first". The former is to put an end to the lack of trains in Tibet, the only provincial administrative region without railways in China, promote Tibet's economic and social development, consolidate national unity and ensure border security. The latter is to eliminate the flood suffering suffered by the people on both sides of the Yellow River for thousands of years. All in all, projects are for people.

(2) The natural environment must be respected, and harmony between the project and the environment must be achieved. This is the premise of the project construction. Regardless of the role and function of the project, it is in the natural environment and must be bound by the laws of nature. While people use engineering means to change nature and meet their own needs, they must be aware of the project's impact on nature. In the long history of human beings, the fundamental reason why some projects can be immortal for thousands of years is that they respect the natural rules and pay attention to the harmonious coexistence between projects and the environment. The Dujiangyan water conservancy project built in 256 BC is a classic example. On the contrary, many projects today violate the laws of ecology, destroy nature and pollute the environment, so it is an urgent and realistic demand to promote the humanistic spirit of respecting the environment. Engineering is a bridge between people and city, people and nature. People should use nature to respect and understand the environment, create unified engineering with the surrounding environment, and then achieve harmony with nature.

(3) History and culture must be passed down to realize the harmony between engineering and culture. It is the foundation of engineering construction. The beauty of nature lies in the variety of colors, and the world's prosperity lies in the richness of culture. While pursuing the beauty of appearance, engineering products should also pay attention to the inherent beauty of the historical and cultural spirit that they can convey. Engineering construction should adapt to local conditions and the different geographical and cultural customs and life of local people to integrate into the local

history, culture, and the practice of urban development in terms of the continuity of history and culture. This kind of project has long-term vitality. The exquisite ancient Chinese projects such as the Forbidden City, the Summer Palace, the Temple of Heaven, and the famous Western medieval buildings such as the Bishop's Church in Milan and Notre Dame Cathedral are all enduring masterpieces.

9.2 Engineering and Habitat

With the origin, development, and evolution of human beings, the scope, scale, and significance of engineering have also expanded and extended. Engineering projects in different eras and different regions embody and carry different human civilizations. Humans at different stages of development have created engineering projects that correspond with their development stages. Engineering projects in different regions also show strong regional cultural characteristics. Engineering and people are inseparable and interactive. For humans, the essence of engineering is to serve human beings and evolve to adapt to the lifestyles of human beings in different periods and regions.

9.2.1 Essential Characteristics of Engineering

In the first century AD, the ancient Roman architect Marcus Vitruvius Pollio proposed in his classic "Ten Books on Architecture" that architecture should comply with three basic principles: "applicability, firmness, and beauty" [1].

From Vitruvius' time to the present era, these three factors are still the essential elements of good architecture, despite the different changes of the times. It is true for architecture and all engineering as the most basic and essential characteristics.

(1) "Applicability": usage characteristics of the engineering project

The close relationship between engineering and human activities such as "clothing, food, housing and transportation" shows that engineering is born to be used by people. Therefore, "applicability" is the fundamental requirement of engineering and one of the basic elements that engineering should have. On the one hand, the "applicability" of the engineering project refers to the various uses of the engineering project. Various types of projects are used to meet different human needs. Taking the housing construction project that is most closely related to human activities as an example, the main uses of housing construction can be distinguished: houses, shopping malls, schools, hospitals, etc., which are directly related to people's daily lives; various factories, warehouses, power stations, and various docks, stations, airports, etc. which are related to production and transportation; libraries, cinemas, concert halls, exhibition halls, radio and TV towers, etc. which are related to spiritual culture; office buildings, conference halls, hotels, cafes, etc., which are related to social activities. On the other hand, the "applicability" of the engineering project also expresses the need to meet users' comfort and convenience requirements. For example, people will consider whether the internal space size of the building is convenient to use, etc., and pay attention to whether keeping warm, heat insulation, moisture resistance, lighting meet the basic requirements, whether it has the effect of sound insulation or sound transmission, as well as the requirements for convenient transportation and convenient connection inside the building. No matter what, the function of the engineering project is to meet people's needs and serve people.

(2) "Applicability": usage characteristics of the engineering project

Solid, that is, firm, durable and safe. Reflected in engineering, it means that engineering products should have certain load-bearing and compression resistance, have a certain service life, not easy to collapse, and have a certain safety guarantee. "Solid" is the most basic element of the project. To ensure the realization of the use function of the engineering project, the most important premise is the "solid". Imagine that if a building, a bridge, or a tunnel is not strong enough, causing collapses, all their functions will be gone. In recent years, some projects in China, especially in the field of construction, have pursued unilaterally "new, large, strange, and special" architectural effects, resulting in many projects that simply pursue luxury, novelty, and ignore the requirements for solid, durable, and safe construction, threaten people's personal and property safety. In addition, some projects excessively pursue visual impact, ignoring the rationality and firmness of the structural system and performance of fire prevention, shock resistance, and evacuation, which is uneconomical and has hidden safety risks. Therefore, firmness, durability, and safety are the most basic requirements of the project.

(3) "Beauty": the aesthetic characteristics of engineering

For engineering, "beauty" means that the project is aesthetically pleasing and can please people. People have used science and technology to construct projects and have examined the beauty of engineering works for a long time. The beauty of engineering is not only reflected in the external shape of the project but also in the construction process of the entire project.

First of all, the "beauty" of the project is expressed in the external form of the construction, which mainly refers to the artistic embodiment of the engineering project. This is an intuitive, relatively long-lasting, and fixed existence. At the same time, the "beauty" of the project also includes that the construction should be in harmony with the surrounding environment. In the building planning phase for site selection, we should fully consider the surrounding natural environment, respect nature, and follow trends. We should study nature, integrate with nature, and conform to nature in architectural design.

Secondly, the "beauty" of the project is also reflected in the process of engineering activities, which people easily overlook. Beauty is elegant, systematic, orderly, and harmonious, and the core is harmony. In the course of engineering activities, the concept of existence is transformed into actual existence through engineering practice. It is not only necessary to make careful arrangements for the function, construction, operation and maintenance of engineering results but also requires the beauty and harmony in its "form" and the convenience and comfort in operation to give people the enjoyment of beauty [2].

9.2.2 *Evolution of Human Settlements*

Since the development of human society, it has experienced a long history of evolution. From the perspective of social form, it has gone through five stages: primitive society, nomadic and agricultural society, and today's industrialized and information-based society. The projects in each stage are built according to the needs of human survival and development on the basis of the specific level of productivity development at that time. Therefore, the engineering in each social form presents the types, characteristics, and levels of engineering compatible with human production and lifestyle. Since the human settlements projects are most closely related to human beings, the following is an example of human settlements project to illustrate this issue.

(1) Human settlement projects in primitive society: the origin of human settlements

In primitive society, the production level was low, and people's daily life was based on local materials and depended on the gifts of nature. The original residence of human beings, that is, the engineering at that time, had two engineering forms: nest and cave, which were distributed in different regions [3]. The primitive nest dwelling is mainly a form of architectural engineering widely used by residents in the swamp areas of the Yangtze River Basin; because the Yangtze River Basin climate is warm and humid, it is suitable for building ventilated and light nest dwellings. The primitive cave-dwelling is an early residential form widely used by residents in the loess zone of the Yellow River Basin, because the climate in the Yellow River Basin is dry and cold, and there is a fine loess layer suitable for digging holes, it provides good natural conditions for cave-dwelling [4].

(2) Human settlement projects in nomadic society: mobile human settlements

"Nomadic", as the name implies, is a state of continuous migration without a fixed residence. It is a productive lifestyle gradually developed by human to adapt to the ecological environment of arid and semi-arid regions in harmony with nature. There are two prototypes of buildings on the vast grasslands of Mongolia in China. One serves human spiritual life, the fixed religious activity place Aobao Altar; the other serves human material life, the house with nomadic movement -the dome (prototype of yurt). In the long historical process, the shape of these two buildings and the cultural connotation and mission symbol carried by them have been constantly enriched and improved. Until the thirteenth century, the two developed from the shape of a simple conical stone pile and a shack to Aobao and yurt with similar external shapes, diverse cultural symbols, and colorful sentiments.

(3) Human settlement projects in agricultural society: village settlements

The emergence of agriculture has laid the foundation for the evolution and progress of human engineering. With the development of agriculture, human production and life have changed from gathering and hunting to farming, and the original gatherers and hunters have changed into growers and breeders. In this way, people put forward higher requirements for the way of living. Instead of running around looking for food, people can settle down in one place. Housing construction projects began to appear and gradually developed into villages and further developed into architectural towns. After people settled down, they had more energy and time, so the original handicraft industry that adapted to the needs of agricultural production and daily life was gradually developed. People explored pottery and iron-making techniques and methods and had the initial pottery and metallurgical engineering. The development of agriculture is inseparable from flood control and irrigation, and water conservancy projects came into being. The widespread use of iron also makes it possible to build large-scale water conservancy projects.

(4) Human settlement projects in industrialized society: urban settlements

With the development and progress of human society, on the basis of capital accumulation and the development of science and technology, industrial production activities characterized by large-scale machine production have emerged at the historical moment, bringing earth-shaking changes to human society. The large-scale production of machines broke through the constraints of human and animal resources and led to the great development of transportation. A large number of rural people gathered in cities at an unprecedented speed and scale, resulting in the so-called urbanization movement. Technological progress and innovation make large-scale projects appear one after another, such as canals, tunnels, bridges, railways, etc. These emerging projects constitute the basic elements of urban construction and development of industrialized society; that is, it focuses on the people living in the city. The human settlement project in the industrialized city has gone far beyond the original simple demand of "only seeking a place to live", and has shifted to make people in the city feel convenient, fast, and comfortable.

(5) Human settlement projects in the information society—city settlements

In the middle of the twentieth century, with the invention and use of computers and the widespread application of high technologies, especially information technology, human production activities, life, and social activities began to enter the era of informatization, intelligence, and automation, bringing unprecedented convenience to humankind. People call this era "information society". All the projects without exception show the characteristics of contemporary society "informatization", full of "humanization" color, dedicated to the maximum satisfaction of human needs, to create the most convenient and comfortable production and living style and environment for human beings. "Informatization" makes communication more and more convenient and fast so that "decentralized production" and "home office" are possible. People can work, study, make friends, shop, etc., through the Internet, without leaving the house. In this way, people do not have to live in large cities, but can live in a small area in a relatively scattered way. Therefore, more and more small cities and towns have gradually developed.

Throughout the development process of the entire human society, the relationship between engineering and people is complementary and mutually promoting. On the one hand, the development of human society and the increasing material and cultural needs of mankind have promoted the continuous improvement and innovation of engineering; on the other hand, the development of engineering promotes the continuous progress of human society, and makes human needs increasingly diversified and reaching high-end. No matter how human society evolves and develops and how the depth and breadth of engineering expand, as engineering meets the basic needs of humankind, it must always accompany human development, in the past, present, and future.

9.2.3 Natural Factors of Human Settlements

Engineering must consider natural factors, and engineering should be integrated with nature, that is, the integration of human existence with heaven and earth, or the integration of human and nature. Therefore, different natural conditions inevitably create different human settlement projects. The following will analyze the impact of natural factors on human settlement projects from the four aspects: natural environment, climate, terrain, and materials.

(1) Differences in the natural environment

Man is an organic part of the natural environment. Emphasizing the harmonious coexistence between man and nature is an important thought of "the unity of man and nature" in traditional Chinese philosophy. Generally speaking, "Feng Shui" is to describe the living environment and ecological environment on which people and all things depend, including the understanding of ecological conditions such as sunshine, air and water and the deduction of operation law, the investigation and utilization of mountains, rivers and geographical features, the adaptation to the four seasons of spring, summer, autumn and winter, and the observation and application of astronomical phenomena of sun, moon and stars, etc. The so-called Feng Shui means a good ecological environment. For example, for people living in the northern hemisphere, the ideal residence must face south, with negative Yin embracing Yang, back to the mountain and facing water, and surrounded by hills on the left and right, that is, the so-called "left Azure Dragon, right White Tiger, front Vermilion Bird and rear Black Tortoise".

(2) Differences in climatic conditions

The climatic conditions are arguably the most basic and universal factors affecting the engineering style. Generally speaking, the difference between natural and social factors will lead to different engineering styles in different regions. However, projects in different regions in the same climate zone show great similarity.

For example, in the tropical rain forest, in order to meet the needs of ventilation, rain protection, and sun shading, the roof of local buildings is designed to be more publicized, while the wall of the building almost retreats below the roof, showing convergence and concealment, which forms an architectural style with local characteristics. In the cold northern forest and alpine environment, most buildings adopt thick log structures to resist the severe cold of flying snow. The building walls are designed to meet the needs of thermal insulation, while the roof is built smooth and robust to facilitate snow hoarding, which forms the architectural style of these areas.

(3) Differences in topography

Whether single building or settlement group building, the topographic and geomorphic characteristics directly impact the form of architectural engineering. It can be said that topography is another basic factor that affects and determines the characteristics of architectural style.

Settlement geographers divide settlements in different topographical environments into three types: linear, circular, and cluster. Linear settlements generally occur in environments with clear boundaries and directions, such as rivers, coasts, canyons, etc. Circular settlements are usually presented in relatively open and boundless scenery. Cluster settlements usually appear in natural environments with a concentration tendency, such as basins and hills, mountain cities in Tuscany, which are typical representatives of cluster settlements [5].

(4) Differences in raw materials

Since the first human shelter was built in primitive society, architecture was closely connected with local materials and natural resources. Because of the strong local monopoly, the characteristics of local materials and resources have created conditions as well as set restrictions for regional architecture, which has affected the formation of regional architectural style on a material basis.

The Yellow River Basin is the birthplace of Chinese culture, with dense forests in ancient times. Soil and wood have become the main materials used in ancient Chinese architecture, so Chinese architecture is known as "civil architecture". The typical representative building in Rome, the dome building, is made of natural concrete prepared from local volcanic ash resources.

9.2.4 Cultural Factors of Human Settlements

Different natural environments can naturally create engineering projects with different styles. However, will the engineering characteristics also be similar in similar natural environments? The answer is No. The engineering project is not just a structure, it is created by human beings, with human nature, condensing people's consciousness, value orientation, so it is a cultural embodiment. Cultural differences will inevitably lead to differences in engineering characteristics.

In this section, we integrate the social organizational structure, religious beliefs, traditional customs, and other elements contained in culture into two aspects: regional cultural differences and national cultural differences. Then we explore their impact on the engineering project.

(1) Regional cultural difference

There are different ways of engineering expression from the cultural differences between China and the West due to different understanding of regional characteristics and cultural spirit. Take courtyard houses as an example. The social order of traditional Chinese society is usually based on patriarchal blood, and the concept of family permeates and affects all aspects of social life. A family is a small society. Chinese courtyard is the direct product of this concept, which can make people feel distinct in respect and inferiority, the order in terms of age, and differences between men and women. In comparison, in the West, the blood family concepts are relatively weak, advocating personality equality and freedom of individuality among family members, which is different from the quadrangle style in China.

China has a vast territory, and different regions have also formed distinctive local culture, which profoundly impacts the local engineering characteristics. Take the construction of contemporary railway stations as an example. Excellent architectural design of railway station comprehensively reflects the cultural factors such as regional characteristics, humanistic characteristics, era style, and traffic characteristics with architectural language, which is specifically reflected in the conception, styling, and space consideration of the station architecture. China has a vast territory with huge regional, terrain, and climate changes. Local culture has distinctive characteristics, which is bound to impact engineering culture significantly. With the rapid development of China's high-speed railway, railway stations all over the country have become local landmark buildings, and their architectural styles are trying to inherit and develop the regional culture of the city. For example, Beijing South Railway Station absorbs the architectural elements of Temple of Heaven to build its shape. The dome design at the top of Hohhot Station gives people the feeling of yurt. Lhasa railway station and Potala Palace share the same architectural language. The detailed design of the Suzhou Station echoes with the characteristics of Suzhou. The design of the Changsha Station reflects the undulating curve of the mountains, and the platform canopy looks like water and waves, reflecting the unique regional style of "Shanshui Island City" of Changsha. Wuhan is known as the "hometown of white clouds and yellow cranes". The architectural modeling of Wuhan Station adopts "Millennium crane return". The protruding roof in the middle of the hall symbolizes the "rise of the middle region", and the wavy roof means "river city Wuhan". The surrounding nine eaves are arranged concentrically, showing the image of double eaves of traditional Chinese architecture, symbolizing the geographical location of Wuhan as the "major juncture of nine provinces" and the status of a transportation network hub connecting the whole China and radiating the surrounding areas (Fig. 9.1).

Fig. 9.1 Exterior view of Wuhan high-speed rail station

(2) National cultural differences

Both China and Japan belong to the East. In terms of culture, they belong to the Confucian cultural circle. Japanese culture is deeply influenced by Chinese culture. The civil architectural projects clearly show the cultural identity of China and Japan, and also reflect the differences between the two countries. As representatives of the traditional houses of the two countries, Chinese courtyard and Japanese washitsu are the product of the synthesis of their respective philosophical cultures, national customs, religious regulations, and living habits. Although the two countries' geographical environment, living habits, and socio-economic development have many connections, the architectural styles of China and Japan have different characteristics because of the cultural differences between the two nationalities. As residential houses, Chinese courtyard and Japanese washitsu reflect different aesthetic interests in terms of structural forms and interior decoration. It can be seen that the architecture of different ethnic groups in the same cultural circle also presents characteristics that fit with each ethnic culture.

China is a multi-ethnic country. The people of all ethnic groups present the characteristics of "large mixed living and small gathering". In the process of exchange and integration, each ethnic group retains its own culture with national characteristics, which is obvious in its respective buildings, especially residential ones. Religious belief and culture are an important part of a national culture, which will naturally appear significantly in their engineering projects.

9.3 Engineering and Culture

Generally speaking, culture is the sum of the material and spiritual achievements created by human beings and is the result of human accumulation in long-term historical activities. In turn, human material and spiritual activities are influenced and constrained by their own cultural inertia. Engineering culture has the commonality of general culture and also highlights the individuality of its own culture. Engineering culture is different from Chinese and Western, and there are also regional differences. Engineering culture witnesses the development of history, deposits human emotions, engraves the memory of culture and follows the future trend. In China's current rapid urbanization process, it is significant and urgent to inherit and protect the engineering cultural heritage created by our ancestors.

9.3.1 The Concept of Engineering Culture

All material products created by human beings are culturally significant. Similarly, engineering products also have cultural meaning and condense the wisdom of human beings. Wu Liangyong, a famous Chinese architect, believes that "the problem of architecture must be considered from the perspective of culture." In fact, people have long linked the construction and culture together. Hugo once said that "architecture is the history of stone", and Goethe thought that "architecture is frozen music". From such a perspective, engineering should belong to the category of culture, because engineering has special themes and characteristic behaviors and changes people's way of thinking, behavior, and lifestyle in many aspects [2]. Engineering and culture are both the crystallization of human wisdom. They are not independent and nonintersecting "parallel lines", but have very close connections. Culture is converged in the core of engineering activities and can be "appeared" through engineering activities and results.

What is engineering culture? According to academician Yin Ruiyu and others, "engineering culture" consists of two concepts: "engineering" and "culture" [6]. Zhang Bo advocates in "*Engineering Culture*" that "engineering culture is a manifestation of culture and is the fusion of 'engineering' and 'culture'. "Engineering" and "culture" have both commonalities and differences. Culture in the broad sense contains engineering. Culture not only carries engineering as a background but also permeates the whole engineering activity like air. Engineering activity has its own unique and important role in broad culture as a relatively independent social activity. Engineering culture permeates the whole engineering activities and directly affects the progress and development of human civilization.

How to give a short definition of engineering culture? Suppose you give a brief definition of engineering culture. In that case, engineering culture refers to the total of customs, habits, systems, norms formed by the engineering community in engineering activities and accepted by everyone, as well as the sum of the material culture reflected by the engineering entity. Engineering culture also contains the history, art, function, quality, and other elements embedded in the engineering itself. The core of engineering culture is the engineering values contained in these cultural elements. Various systems and material carriers and spiritual phenomena in engineering activities are the expression and refraction of engineering values.

The engineering community is the main body of the engineering culture. It refers to the social groups related to different subjects, including the project owner, the designer, the construction party, the supervision party, the user, and other related parties. Engineering culture is formed in engineering activities and developed with the progress of engineering. The advanced engineering culture influences the engineering people, and they will naturally be energetic and positively affect engineering activities. This directly influences and determines the nature of the engineering, such as whether the project is energy-saving and protecting environment, whether it is humanized.

No matter what kind of engineering or engineering activities, they are actually the practical activities of human beings using nature. Engineering culture should reflect the connotation of "people-oriented", "the Tao of nature," and "the harmony of heaven and man". First of all, engineering culture is produced and formed in the engineering activities which are both "for man" and "man-made". Therefore, engineering culture should be human-oriented and aim to achieve harmony between engineering and man.

Secondly, as a practical activity for humans to know nature and leverage the power of nature, engineering should follow the laws of nature. Nature is the heaven, and the law of nature is the will of God. We can have good weather and prosperity only by the veneration of heaven. Engineering creators should pursue the harmony between engineering and nature, man and nature. They should also cultivate a unique engineering culture according to the value orientation of "the Tao of nature" and "the harmony of heaven and man". Therefore, the high-level engineering culture should be human-oriented, conform to the laws of nature, respect nature, and strive for harmony between heaven and human nature.

9.3.2 Characteristics of Engineering Culture

Since engineering is the most timely, regional, and social product or activity, it will show different cultural characteristics in different times, regions, and industrial environments. Therefore, engineering culture has the characteristics of periodicity, locality, and industry. First, it is epochal. Culture has the imprint of the times, and engineering culture also arises and develops in a certain time and space. Consequently, the ancient and modern engineering is inevitably marked with the times, showing the cultural characteristics of different historical stages. Secondly, locality. The engineering project is developed in a specific area that has its unique natural landscape and regional culture. Hence, engineering activities are bound to be influenced by the local geographical environment and customs, namely the regional culture. Thirdly, industry. Engineering belongs to specific industries and sectors and is inevitably influenced by the industrial environment, such as the market, law, technical level, and personnel quality [8]. These three characteristics of engineering culture significantly impact the formation of engineering culture. Suppose an engineering project has certain historical and cultural value, or has iconic features in a specific region, or its technology is representative in the industry, or its quality and function are leading in the industry. In that case, it will produce a strong brand effect, become a brand project, and form a corresponding engineering brand culture.

Cultural architecture is the real vitality of architecture. The cultural nature of architecture can emphasize the historical value and guide the fashion direction. Meanwhile it can express a deep understanding of the region and nationality.

9.3.3 *Formation of Engineering Culture*

Any engineering culture is not determined by a single factor but formed by the construction process, operation process, and collective aspects of people or things related to the project.

(1) Culture formed by engineers in the construction process

The engineering construction process is transforming and utilizing natural resources consciously, purposefully, and oriented by the construction subject. The role of the subject cannot be ignored. Once a project is decided to be implemented, for the common goal, under the unified command, in the process of engineering activities, all relevant personnel will abide by the common behavioral norms and operating procedures, thus forming a common cultural orientation. Therefore, engineering culture is a kind of crystallization of group wisdom and an embodiment of group culture. Due to the specificity and diversity of engineering practice, different "engineering styles" are expressed in these practices, forming different engineering culture styles. Different engineering construction enterprises thus have different engineering cultures, each has its distinctive characteristics even in the same construction unit, or in different engineering projects, or different stages of the same project. The main subjects of engineering activities include owner, designer, constructor, and supervisor, who form corresponding derived cultures in engineering practice [2].

Owner. The owner's wishes, concepts, and needs directly determine the project's value orientation and influence the engineering culture's formation.

Designer. The cultural literacy, professional skill level, values of designers and the way they handle the relationship between engineering and environment, society, economy, etc., all directly influence the formation of engineering culture. On the one hand, engineering designers should have basic engineering culture knowledge. Designers should understand general technology, engineering science knowledge, and local knowledge about engineering projects, such as national customs and customs, so they can accurately grasp the characteristics of the times and local features. On the other hand, engineering designers should also have a certain cultural background. This mainly includes the designer's social background, religious beliefs, values, cultural literacy, aesthetic taste, etc. It can be said that the designer's own culture directly dominates the ontology culture of a project, and the cultural taste of a project is directly related to the cultural level of the designer.

Construction party. In the construction process, the cultural and moral quality of the construction party is also an important factor affecting the engineering culture. For example, whether the engineers have formulated scientific engineering construction standards and engineering management systems; whether the workers have complied with the practice norms, labor discipline, and production procedures; whether the logistic staff has provided safety facilities and livelihood security; whether the whole team has cohesion, etc. All of the above become part of the engineering culture. The quality of a project after completion directly depends on whether the construction party has formed a high-quality engineering culture during the construction. In the construction process, the construction party should reflect the principle of "people-oriented". On the one hand, the main body of the construction party is the workers, and the project is associated with workers' hard work; the attitude of the construction party towards the workers will affect their working mood, which directly affects the progress and quality of the project. Therefore, the management of workers should be "humanized", so that they can enjoy good treatment and feel equal care. On the other hand, the construction party should consider the impact of the construction process on the surrounding environment. For example, during construction, install protective barriers and nets around the site to ensure the safety of pedestrians and moving vehicles; try to avoid construction at night to reduce noise, etc.

Supervising party. In construction, the supervisory party is involved in the whole construction process and mainly plays the role of control and coordination. Therefore, the supervisory party's role in forming engineering culture should not be ignored. The supervising party has to supervise and control the project in terms of investment, schedule, quality, and safety and coordinate the relationship between all parties in the construction process. Whether the project is economical, efficient, practical and safe, and whether the whole project activities are harmonious, stable and orderly will be impacted by it. These are the concrete embodiment of engineering culture.

(2) Culture formed by users during using

As mentioned above, the ultimate purpose of engineering construction is to serve human beings and meet their needs. The characteristics of engineering users as the target group are bound to have an important connection with the formation of engineering culture. Such connections can be large or small. Large can be a country or society's need for the development of agriculture, military, and science and technology to generate continuous improvement and perfection of water engineering, military engineering, and aerospace engineering. Small can be a resident for the individual needs of housing. All of them will have an impact on all aspects of the engineering project, thereby forming a particular engineering culture.

As the result of engineering, artificial nature, while being put into use, is a cultural product in the minds of users. The builder gives this artificial nature a corresponding cultural connotation from design to construction. However, since it is a cultural product, different people will appreciate it benevolently and wisely. In other words, the user will give the artificial nature a corresponding cultural meaning. Although this endowment will be influenced to a considerable extent by the builder's ideas, the users are a complex group. The larger the engineering project, the larger the group of users will be, and the more diverse their culture, customs, and habits will be. In the process of using it, it is natural to form a user culture that is not exactly the same as that of the builder. The TV tower in Guangzhou is different from other towers because it has a twisted body and a beautiful appearance. It is loved by people who visit it. The builders gave it a resounding name: "Maiden Look Back". However, Guangzhou residents have given it a strong local name, called "Slim Waist", which seems more intimate, lovely, and catchy.

In the engineering application process, the cultural characteristics present two aspects: on the one hand, different building projects offer different styles. Users can directly determine the type and attributes of a building project by its exterior styling characteristics. Government buildings, schools, hospitals, churches, and temples have different architectural styles that are instantly recognizable and don't need to pay attention to the text signs to identify. On the other hand, the applied function of the engineering projects varies. Different projects are designed to meet different needs of people, such as housing for living, hospitals for medical treatment, schools for studying, office buildings for working, and theaters, museums, and libraries for meeting people's spiritual and cultural needs. In the process of using the engineering project, it is necessary to consider the project's practicality and comfort. For example, the barrier-free access in public places and the residential district with complete facilities for living, employment, schooling, medical care, fitness, and entertainment can provide a great convenience for users, thus forming the cultural characteristics of "people-oriented" in the process of use.

(3) Culture formed by related parties

In addition to the engineering construction side and users, the social public and environment related to engineering may also influence engineering culture. On the one hand, the public influences the formation of engineering culture. Social public, especially social media's judgment or report on engineering, such as whether the project is beautiful and reasonable, whether it has economic value, social value or ecological value, will indirectly influence or promote the society's understanding of engineering, thereby becoming the part of engineering culture. To make the public have positive influence on engineering culture, two prerequisites are needed. First, the public should understand engineering. This requires the public to have certain scientific literacy and ensure the public's right to know about the project. Second, the public should participate in the project. In many engineering activities, the public is both the "audience" and the "actor", and the public participation is conducive to weighing interests and exchanging various values, and injecting new cultural elements into the project.

On the other hand, the project's surrounding environment will also influence the project culture. Any project is in a certain natural and social environment, and the project's advancement must conform to the ecological laws of nature and the operation principles of society and maintain harmony with the surrounding environment and style. The project should comply with natural ecological laws, avoid damage to the natural environment, and reduce the impact on the social environment, such as reducing noise and dust pollution. The project should also strive to make the engineering practice and the surrounding environment develop in harmony with each other. The new tendency of engineering culture, such as ecology, environment, and sustainable development concept, is produced under the influence of the environment.

Among the above points, the strong "rebound" of nature caused by human damage to nature is the most important source of engineering ecological culture. There are examples in both China and other countries. For example, real estate development will follow the terrain. Considerations must be given to local terrain. Measures such as "dig the mountain" or "fill in the ditch" would be considered as inappropriate. Only to accommodate local conditions can the building be built in a staggered manner. This not only preserves the original natural rhythm and protects the natural environment but also saves resources greatly and creates a better living environment. This is a manifestation of respecting the laws of natural ecology.

In short, from the process of culture formation, engineering is the synthesis of many contradictions, mainly manifested in two aspects. On the one hand, engineering is the accumulation and extension of culture in the past generations. It is a kind of solidified culture that contemporary people can witness. Moreover, it is long-term preservation of the cultural state, such as the buildings that survived in the past generations. On the other hand, engineering is a culture of continuous innovation and development, which requires the owner, the designer, the builder, the supervisor, and the user to follow the laws of nature and society and continuously innovate engineering culture in engineering practice. Specific practices include fully respecting the opinions of the public and fully considering the environmental factors so that the engineering culture can be constantly enriched, diversified, and developed, making the engineering culture the "compass" leading to the healthy development of engineering and the power source of engineering progress.

9.3.4 Inheritance and Protection of Engineering Culture

With the implementation of large-scale urban reconstruction in China, many old buildings in the old streets will be demolished and replaced by new steel and concrete ones. Original urban history and culture will be gradually covered up and worn out. Years could not speak for themselves, but stones can. The years have passed by silently, but the stones stay where they are. Architectures can witness the existence and development of history. Engineering contains a wealth of historical information, as the history book made by stones. Such as Beijing's Little Chrysanthemum Hutong, Zhouzhuang's front street and back river, Wuzhen's Water Grid House, Fenghuang's Hanging House, which are all records of ancestral life and integrated with profound culture. Those records will be lost and can not be recovered if not protected. It can be said that the engineering project's cultural heritage is the common belief and symbol of the city, which sustains the core emotion and values of the city. If protection is insufficient, we lose not only the building, the mechanism of the historical and cultural district, the historical city appearance, but also the belief in traditional culture and the faith in regional culture.

Therefore, we must simultaneously pay attention to the protection and inheritance of historical culture when developing engineering projects. The protection and inheritance of engineering culture include two connotations. One is the discarding of the traditional engineering culture. The other is the protection and inheritance of folk customs and folk styles of the area where the project is located. Only by continuously strengthening the preservation of exotic style, living habits, and folk customs can traditional residential houses be energetic and more vital, which is more conducive to the physical protection of construction projects.

So, how to protect and inherit the engineering culture?

First, protection should be the highest priority, while development takes the subordinate.

In the process of protection-development of the old city and architectural heritage, we must always insist on protection first and utilization second. The heritage can only be used under the premise of protection, rather than "protection" under the premise of utilization. The development must be subordinated to protection to achieve the best combination of social and economic benefits.

In order to achieve the purpose of protection first, sociologists, artists, antiquities experts, historians, architects, and other scholars can be invited to conduct a systematic study and value assessment of architectural monuments in historical and cultural cities. From the perspective of culture, we should take overall consideration, coordinate and communicate, increase the efforts on protection, limit low-level development, avoid duplication of development, outlaw illegal development, and eliminate destructive development.

Second, focus on heritage and overall protection.

The protection should adhere to two concepts. First, focus on the overall protection of the city. Protection does not only mean preserving a bridge, a monument, a house, a store, but the general preservation of an old street, a river, a township. The second is the concept of equal emphasis on the protection of traditional engineering and environmental protection, which includes the physical protection of engineering with historical-cultural value and rich traditional characteristics, and the protection of the cultural heritage of traditional engineering itself and the protection of the ecological and social environment of its location. In addition to the protection of single buildings or group buildings, the protection of urban culture and the combination of new buildings with urban culture is also important to protect historical works or buildings. This requires that in the planning and construction of the city, not only material factors such as function, technology, safety, and economy should be considered, but also the appeal of cultural values such as formal aesthetics, regional characteristics, and national identity symbols. This allows new architectural elements to reflect the heritage of the city's culture well.

The renovation and construction of Xi'an South Gate area have well preserved and inherited the history and culture. The South Gate area of Xi'an is located at the center of the famous historical axis of Xi'an, the Dragon Line of Chang'an. This axis connects the cultural remains of different historical periods, such as Han, Tang, Ming, Qing, and modern times. It is also the core area and important urban node of Xi'an's culture, tourism, commerce and transportation, the most important gateway to the ancient city of Xi'an.

In order to better inherit the history and culture and improve the city's quality, Xi'an has completed a comprehensive renovation of the South Gate area. The basic features of the renovation concept, methods and design results can be summarized by the "Nine Harmonies", i.e. "nine-joint", that is, stitching, joining, fusion, integration, superposition, co-operation, form-fitting, intention-fitting, and god-fitting. By stitching together ancient buildings, it can mend historical space and continue historical veins, truly allowing visitors and citizens to enter history, feel humanities and experience culture. At the same time, it allows visitors to seamlessly connect with modern civilization, achieving a perfect blend of Xi'an's ancient history and modern style. The transformation of Xi'an South Gate area is not only reflected in the protection and restoration of cultural relics and the display of cultural heritage, but also the comprehensive improvement of municipal transportation, ecological environment and living facilities in the South Gate area based on the integration of various social resources and the synergistic enhancement. The renovation of the South Gate area requires not only the "same appearance ", i.e. to ensure that all visible buildings are consistent with the style of the city walls when standing on them, so that it can reproduce the historical appearance of the ancient capital of Xi'an to the greatest extent possible. In addition, the renovation of the South Gate area of Xi'an also pursues "meaningful conformity" and "spiritual resemblance", i.e., it does not just stop at "patchwork" to imitate the appearance, but always focuses on reflecting and enriching the historical and cultural connotation of Xi'an, improving the city's taste and promoting the humanistic spirit. Based on this, it makes the form, intention, and spirit integrated and interchangeable, so as to achieve the humanistic effect that the form agrees with each other, the intention is similar to the spirit, and the form is attached to the intention (Fig. 9.2).

The preservation and inheritance of the South Gate area of Xi'an have successfully solved the contradictions of balancing the ancient and the modern, the new and the old, and the repair and the mending. The work has given a new and youthful appearance to Xi'an's ancient appearance. Guided by the concept of all-around holistic protection, the renovated South Gate area rely on the ancient city wall, highlighting the profound cultural connotation of Xi'an and manifesting the harmony between

Fig. 9.2 Panoramic view after restoration

the city and culture. It integrates the ancient buildings into modern life and lets them become the urban living room of the ancient capital Xi'an.

9.3.5 *Corporate Culture of Engineering Construction*

Enterprise culture is an organic and important part of the social culture system and is a kind of behavior and habit formed by the enterprise in the practice process. It is reflected in three main levels: spirit, institution, and material culture. Engineering construction enterprise culture is formed in the engineering construction activities of enterprises engaged in the engineering, which reflects the industry's characteristics, promotes the industry's development, and highlights the characteristics of the industry's culture. Engineering construction enterprise culture has an important guiding role for the engineering construction of the enterprise. It can directly affect the humanistic value of the engineering projects built by the enterprise, which in turn affects the development space of the engineering construction enterprise.

9.4 Engineering and Art

Engineering and art are inseparable. Engineering art also belongs to the engineering culture. Due to the close relationship between art and engineering, the significance of art to engineering is particularly important. Therefore, it is necessary to arrange a separate section to describe the meaning, characteristics, and performance of engineering art and the differences between Chinese and western engineering art. Generalized art also includes management art. Engineering management needs artistic thinking and methods. Only in this way can there be artistic engineering products.

9.4.1 *The Meaning of Engineering Art*

Art refers to the use of innovative thinking and creative ways and methods to reflect more typical images or social ideology than real images. Art is generally divided into: language art (literature), performing art (music, dance, and acrobatics), comprehensive art (drama, film, and quyi), performance art (personal performance), modeling art (sculpture, painting, calligraphy, seal cutting, image, architecture, environment, gardening, design, and creation), and thinking art (perception, inference, integration, imagination, creativity, discovery, invention), management art (organization, demonstration, speech, mobilization, persuasion), etc.

Engineering embodies art. In a narrow sense, engineering art mainly refers to the art of architectural engineering, that is, the art of buildings and structures after construction. From this perspective, engineering art is equivalent to architectural art.

In a broad sense, engineering art also includes the art embodied in the process of engineering construction; that is, under the guidance of aesthetic principles, engineers flexibly use theory, knowledge, experience, and wisdom to express engineering by shaping and creating the beauty of the project, to achieve the intelligence of the project goal [2]. It should be pointed out that the beauty of engineering can not only be understood as the beauty of housing architecture or garden architecture (of course, these two aspects are typical), but also be reflected in water conservancy engineering, transportation engineering, petrochemical engineering, metallurgical engineering and electric power engineering, etc.

9.4.2 The Relationship Between Engineering and Art

First of all, the project itself is aesthetic and artistic. It is a creative engineering model and beautiful expression art. At the same time, engineering is to benefit humankind and be "used" by people. Therefore, it can be said that engineering is a "practical art". Secondly, engineering is the carrier of art. It contains and embodies many kinds of art, gathers and integrates a variety of art categories such as painting, sculpture, calligraphy, music, light and shadow, and makes these arts interpenetrate and interplay. It is the aggregation and integration of a variety of arts [9]. The art carrier mentioned here refers not only to single buildings but also group buildings, such as a block or even a city. For example, Paris, Venice, Vienna, and Rotterdam are filled with a rich artistic atmosphere and deep artistic feelings.

The beautiful skyline is the five lines of music of the city and the embodiment of the rhythmic beauty of the city. In the ancient city of Dali, China, there are no tall buildings and heavy traffic, but people can see such beautiful scenes. Big houses are covered with cyan tiles stacked one after another; small buildings with different heights on both sides of the street are scattered; each shop has its own shapes and colors, forming different styles. Looking from afar, the dazzling silverware, bright cakes, gorgeous jewelry, exquisite jade, beautiful costumes and colorful patterns all make people feel as if they are immersed in the temple of art. The artistic charm of a city can not be reflected only by some scenic spots and buildings. As the carrier of art, the city is presented in the overall effect of a large area through the materials, colors, shapes, and collocations all over the streets and buildings.

9.4.3 Aesthetic Characteristics of Engineering Art

Engineering art presents its own unique aesthetic characteristics because of its unique characteristics. The aesthetic characteristics of the project can be reflected in all aspects of engineering activities, not only in the pursuit of aesthetic appeal and internal decoration in engineering design and construction but also in the contribution of artificial nature to society and the harmonious relationship with man, society, and nature.

Only when the harmonization of inner function and outer form is achieved such engineering has aesthetic value. Among civil engineering works, the one with the richest objectivity, the widest appreciation range, and the closest relationship with life is architectural engineering. Therefore, the following discussion focuses on the aesthetic characteristics of architectural engineering and summarizes them into four "unity".

First, the unity of material and spirit. Engineering exists in material form, but it also expresses spiritual emotions. It is not only an entity that gives visual impact with its own shape, color, and other appearance, but it also has spiritual connotation. It can also act on people through metaphor, symbol, implication, and other spiritual factors [9]. Second, the unity of practicality and aesthetics. Undoubtedly, practicality is the basic property that engineering must have. People love beauty, and on the basis of satisfying the basic function, people will certainly put forward a higher level of beauty demand for engineering, so aesthetics is also a necessary element of engineering. Third, the unity of technology and culture. Engineering is the unity of technology, culture, and art. The engineering project is built according to science and technology. At the same time, the main body of the project is human. People create it according to their own customs, aesthetic feelings and other subjective wills, and it implies a profound cultural meaning. The engineering projects presented by different regions and different nationalities have different characteristics because of their cultural differences. Fourth, the unity of shape and environment. The aesthetics of the project not only lies in the external shape of the project but also have a relationship with the environment around the engineering products.

9.4.4 Aesthetic Reflection of Engineering Art

Engineering project has poetic beauty. Since ancient times, architecture has often been called "solidified music" and "epic of stone", enduring and everlasting. Different engineering works of art can bring people aesthetic feelings in appearance and meet people's actual needs. They are also the carriers of culture and thought, reflecting the historical and cultural characteristics of different periods, as well as their originality and innovation spirit.

(1) Beauty of appearance

"Beauty of appearance" is the most obvious one, which mainly refers to the beauty on the external level of engineering projects, the beauty created by the engineering law, including the artificial nature itself and the environment. The shape, spatial structure, texture, decoration, and color of artificial nature can be directly perceived through the vision, bringing visual beauty and even impacting people. The environment in which the artificial nature is located includes the surrounding artificial environment and the natural environment. The harmony between the artificial nature and the environment will enhance, dilute or even reverse the visual beauty of the artificial nature itself. The National Grand Theatre itself does have a characteristic shape, since it is located in the solemn Great Hall of the People and the Museum of History, which is often criticized as nondescript.

(2) Beauty of content

Since the project's primary purpose is to "use", its functionality, practicability, comfort and economy can not be ignored. Suppose we break away from the "beauty of content" such as function, practicality, comfort and economy, only devoted to the beautiful appearance, but do not make practical efforts to create architecture practicality and comfort. In that case, the project will have serious quality problems in structural layout, daylight, and ventilation, such as seepage, leakage, and crack. As a result, we would fall into the misunderstanding of "aestheticism".

(3) Beauty of thought

The ideological beauty of engineering is mainly reflected in two aspects. First, the beauty of culture. Through the ages, any engineering projects embody the experience and wisdom accumulated by the creator in production and life, carry rich historical information, contain profound cultural connotation, and reflect the times' unique spirit. All these have formed the cultural beauty of the project. Second, the beauty of innovation. Beautiful things are often not conformist. They are often unique or have some new ideas. For example, every seemingly ordinary Pavilion, rockery, stone, small bridges and flowing water in Chinese gardens can present an extraordinary landscape through the creative design and combination of the creators.

9.5 Engineering Humanistic Spirit

9.5.1 Dujiangyan and Sanmenxia

The well-known Dujiangyan and Sanmenxia are both water conservancy projects with the same good expectations and starting points, but the results are the opposite. Local hydrological and natural conditions guided the former, the latter ignored the "nature" of the Yellow River and blindly worshipped man's own ability to "cut the Yellow River in two"; the former used local materials while the latter used reinforced concrete, one of the products of industrialized society; the former benefited for a thousand years and is still functioning today, while the latter was plagued with problems soon after its completion and is still in a dilemma. The sharp contrast between the two in the results of water control not only reflects the tortuous road of human cognition and utilization of nature but also makes people deeply realize the humanistic enlightenment contained in engineering construction: nature is eternal; local is global; people-oriented is the origin.

(1) Natural is eternal

Man and nature are closely connected and interact. Man has the subjective initiative of understanding and utilizing nature, and constantly adjusts the relationship with nature in production and life to achieve the harmonious unity between man and nature. At the same time, nature will react to human behavior. Dujiangyan is the result of harmonious coexistence between man and nature. It is an example that humans use nature, conform to nature and benefit from nature. In order to control nature, humans must understand and master the laws of nature. In particular, it is necessary to grasp the critical point of fundamental change in nature's behavior or state, make proper use of it, and can not go beyond it at will. There are critical points in the position of the fish mouth of Dujiangyan and the elevation of Feisha Weir Dam. We can achieve great results by taking advantage of the critical point's potential.

In addition to historical reasons, the failure of the Sanmenxia Project was mainly due to the hasty construction of the project without following the natural laws of the Yellow River basin and its hydrological changes, which led to too much attention to disaster prevention and control in the downstream and neglected the serious consequences of "siltation above the Tongguan and the continuous development of silt upstream" as envisioned by Huang Wanli. This approach seems to solve the problem temporarily but diverts and creates new problems.

Therefore, people realize that in the face of eternal nature, one must be in awe; in the front of the rigidity of the laws of nature, one must strictly follow. Only under the premise of knowing and mastering the objective laws, acting in accordance with the laws of nature, and giving full play to people's subjective initiative can people create a high-quality project that will benefit thousands of generations.

(2) Local is global

The reason why Dujiangyan has such a strong vitality also lies in its local nature. The main works of Dujiangyan, such as Dujiangyan fish mouth, Baopingkou, and Feisha Weir, fully reflect the combination of previous rich water control experience with specific projects and show the nativity on which it is based. Therefore, these works have formed the engineering model with a profound culture, long history, and long-lasting characteristics that we see today.

In contrast, the Sanmenxia water conservancy project was led and assisted by Soviet experts. They did not understand the historical origin of the Yellow River Basin and mechanically applied their own water control experience to the comprehensive management of the Yellow River Basin. As a result, the project was not acclimatized. Kololev, an expert from the Soviet Union, argued that "it is impossible to find a reservoir that does not require to migrate population and can ensure flood regulation, and there is no need to study it. In order to regulate flood, sufficient reservoir capacity is required, so it is inevitable to submerge and migrate". These arguments have deprived us of the opportunity to "slow down and demonstrate again". But facts have proved that the Dujiangyan project in ancient China and the later Xiaolangdi project have made "inevitable" become "avoidable".

What is national is global, and what is local is global. The former Soviet Union's water management experience was successful in its own country because it respected its own history and hydrology. The success of the Dujiangyan water conservancy project was also based on respect for the local historical and hydrological characteristics of the Sichuan basin. The success of the Yellow River basin should also be based on its own unique historical, cultural, and natural conditions. The vitality of engineering architecture also lies in highlighting local characteristics and cultural atmosphere so that it will have strong vitality and humanistic value.

Of course, emphasizing localization is not blindly refusing to learn from and absorb advanced foreign ideas, systems, culture, and technology, but being "inclusive" without losing its originality so as to keep pace with the times, not lag behind the times and not make future generations make a fool of oneself as "old fashioned".

(3) People-oriented is the origin

The Dujiangyan project can benefit thousands of generations because it always implements the idea of "people-oriented". Whether respecting the laws of nature, history, or local culture, the ultimate goal is to serve people and meet their needs, not only to meet the needs of contemporary people but also to benefit future generations. The Sanmenxia water conservancy project began to take "people-oriented" as the goal, but the result deviated from the original purpose and did not really achieve "people-oriented" in essence. The real "people-oriented" means to pay attention to the combination of modern science and technology and humanities, environmental protection, coordinated development, and sustainable development, and protect the home on which human beings depend while enhancing economic strength [10].

"People-oriented" is the essence of the scientific concept of development, and is the starting point and destination of all work. It requires that economic development be sustainable, maintain the ecological balance, and be conducive to people's physical and mental health and comprehensive development. It requires getting out of the misconception of "seeing things but not people" and the misconception of using the purely economic value as the standard for measuring everything. People are both the creator and the user of the project. Therefore, the construction process and the project result must start from the people's perspective, respect people, understand people, care for people, and take people as the important subject that can promote the construction and development of the project. We have to be in line with the premise of human society and natural environment development law, to improve and enhance the living standards of the people as the basis, and constantly stimulate human creativity, to create more "humane" projects that can meet the growing material, cultural and spiritual needs of people. Only in this way will society develop a source of power; otherwise, it will go to exhaustion and decay.

9.5.2 Humanity Lost in Urbanization Construction

The above analysis mainly focuses on individual projects. If we analyze group engineering in cities, it naturally brings to mind the current urbanization construction in China. The urbanization process, which changes the country's economy, society and culture, is a long-term and complex system project. It first faces the construction of large-scale infrastructure projects to provide the foundation and guarantee for urbanization, which is a necessary condition for its realization. Therefore, the relationship between engineering and people must be dealt with at the beginning of urbanization to realize the basic demand of people as the core.

Since the founding of New China, especially since the reform and opening up, urbanization has entered a brand new development period. It has been moving forward in a difficult and tortuous exploration. During this period, China's urbanization has made remarkable achievements and developed amazingly. By 2015, the urbanization rate had increased from 17.9% in 1978 to 56.1%, an increase of 38.2 percentage points. However, compared with the 80% urbanization rate of developed countries, there is still a long way ahead.

As the urbanization process continues to accelerate, large-scale construction has brought about tremendous changes in the appearance of cities everywhere. Xi'an's various newly built or renovated urban areas are majestic, a scene of the flourishing Tang Dynasty, reminiscent of the achievements of huge urban construction throughout China. Guangzhou's municipal construction is magnificent, with above- and below-ground partitions of buildings in the financial district, each having its own function. The tall buildings are situated in a sequence, each in its own way. In particular, there are many wetlands in the whole city, and the surrounding scenery shows the southern style, making people linger for a while. Xi'an and Guangzhou, to name a few, have given the whole of China a new look, and experts at home and abroad have marveled at China's municipal construction. Along with such a large scale and fast urbanization, there are also some lost phenomena such as the similarity of urban appearance, similar building shapes, and outdated underground facilities due to inexperience [11]. Although these lost phenomena like a few dead branches and decaying leaves in a garden of a thousand colors, they must be well understood and summarized in order further to improve the quality and level of China's urbanization.

(1) Similarity of urban appearance

Many cities are homogenized and look similar to each other, making people feel aesthetic fatigue.

With the accelerating urbanization process, cities are getting bigger and bigger; roads are getting wider and wider; buildings are getting taller and taller; the appearance of each city has changed dramatically. But if we compare cities across the country, we will find that the "look" of some cities is becoming more and more similar. From the towering office buildings to the busy commercial pedestrian street, from the residential buildings to the wide central square, large and small cities are the same, the north and the south are the same, and the original appearance of many distinctive cities is gradually fading. The reasons for this are many. In addition to objective factors such as technical materials, subjectively speaking, due to the rapid development of China's urbanization, the humanistic qualities of city authorities and designers have not kept pace with the times and reached the right height. From overall planning to the specific design, they generally refer to and imitate the urban construction in developed areas, resulting in multiple cities with the same appearance.

(2) Similarity of building shapes

The phenomenon of "a thousand buildings with one look" is very common in cities. Many buildings are very similar in appearance, such as the same glass curtain wall, the same shape structure, the same towering, the same "high, large, wide" characteristics. This makes it easy for people to get lost in the city because they can't find a direction. In recent years, the skyscrapers under construction and planned to be built around the country are springing up, and they keep refreshing the "newest" and "highest" records. The 2012 Skyscraper City Report shows that in the next decade, China will rank first in the world with 1318 skyscrapers over 152 m (using U.S. standards), four times the total number of skyscrapers owned by the United States today. And the total investment in skyscrapers under construction and planned in China will exceed 1.7 trillion yuan. Various places are competing to build skyscrapers to enhance the image of their cities. However, the construction of skyscrapers will inevitably consume a lot of resources, ultra-high large buildings will also make the city's "heat island effect" aggravated and increase safety risks.

(3) Backward underground facilities

The once-in-a-century flood in Jinan, Shandong Province in 2007, almost paralyzed the drainage system. The "July 21" natural disaster in Beijing in 2012 killed 78 people. The heavy rain in Changsha in 2013 caused a girl to fall into the city sewer. Other events include the flooding of the whole city of Wuhan in 2015 and 2016, and the extreme urban flooding in Xingtai, Hebei Province in 2016 which killed many people. All of these events show that the urban construction phenomenon of "attaching importance to above-ground construction and neglecting underground construction" is a misconception that needs to be corrected. Hugo said that sewers are the "conscience" of cities. Urban policymakers and builders should build cities well with a high sense of responsibility to the people, especially underground projects, including urban sewers, which is "people-oriented".

(4) "Empty city" phenomenon

Blind urban expansion, enclosure, and lack of proper and predictable planning will lead to the fact that although there is land, the builders do not know what to build, how to develop, how to use, and how to fill the materialized industries, resulting in the emergence of some "empty cities". Empty cities have magnificent public buildings, wide roads, and large landscape areas. But there is a large number of vacant residential buildings. In addition to a few cars parked in the magnificent administrative center, there are few cars in the city. In addition to the bustling on the renderings, it seems deserted in reality. Some experts have estimated that it will take a long time to digest all vacant houses.

(5) County "White House"

With the increase of national finance year by year, the office conditions of government departments have been improved accordingly. However, some local governments have built office buildings beyond the standard. Some county government office buildings are built in comparison with the White House of the United States, and the scale of the municipal square is no less than that of Tiananmen Square. These buildings often cover an area of hundreds of acres, occupying the golden area of the city. Some have even become famous local landmark buildings and the "most beautiful scenery" of the place. For example, a municipal government building with a construction area of nearly 400,000 square meters is the second-largest single building in the world and the first in Asia. There is a "magnificent" courtyard in the center, which is the "quadrangle building", second only to the 600,000 square meters of the Pentagon. Building buildings and offices in such a lavish manner, the office standard far exceeds the actual work needs. Obviously, it is putting the cart before the horse and reversing the master and servant, contrary to public finance's concept and spirit.

(6) "Romantic" of foreign designers

China is currently the country with the largest number of theaters in the world. Almost every opera house is a landmark of the city. Most of them have one thing in common: they are designed by foreign designers. China seems to have become a "testing ground" for Western romantic designers. According to statistics, 80% of the world's top 200 engineering companies and design consulting companies have set up offices in China. Foreign designers make a large part of the landmark buildings in China's major cities. Wang Shouzhi, a famous design historian and professor of the School of Design of Los Angeles Art Center, also believes that, "What we call landmarks now without objection are a group of buildings related to a national culture built after the precipitation of the times. For example, the Eiffel Tower is made by French architects, and our Chinese designers also designed Tiananmen Gate. It is built by the state through a group of elite design experts within the nation. Only such buildings can be called landmarks. But now, many buildings around us are designed by foreign star architects. They have nothing to do with the precipitation of traditional national culture, and the growth period of these buildings is too hasty" [12].

9.5.3 The Return Path of Humanism

Urbanization is a natural, historical process and a necessary path for China to achieve modernization. The new urbanization is proposed that China's urbanization should not repeat the old path of the urbanization of Western countries and ignore the lessons they learned. A new path of urbanization suitable for China's primary stage of socialism will be developed. After the Third Plenary Session of the 18th Central Committee of the CPC, the central urbanization work conference indicated that we should follow the laws, optimize the situation and finally make the urbanization naturally and sufficiently completed. Urbanization should follow the principle of "active, steady, and solid" and be carried out according to the requirements of "clear direction, steady steps, and applicable measures". This requires three "reverence" and three "priorities": Reverence for the landscape and priority on the environment, reverence for history and culture, and reverence for humanity and livelihood, respectively.

(1) Reverence for the landscape and priority on the environment

Existence is reasonable. The Earth has evolved for billions of years, so uneven landscapes, including mountains, gullies, and rivers, finally came into being. These features all reveal the mystery of nature and reflect the laws of nature. Mountains, rivers, forests, grasslands, and wetlands provide human beings with the natural conditions for survival, reproduction, and development. They deserve to be cherished and cared for. However, some people have no respect for nature and believe that "the nature could be conquered". They arbitrarily flatten mountains, fill up water, and ignore the "gifts" given by nature, which has been causing great damage to nature.

Engels once noted, "Man must not reveal too much in his victories over nature. For every victory, nature has retaliated." Indeed, every attempt by man to conquer and destroy the natural world is the beginning of nature's revenge on the man.

Driven by various factors, in many construction projects, advanced construction techniques and equipment can easily level and modify the original natural landscape to various degrees, pushing slopes, chopping, filling rivers, and digging mountains at will. After the completion of the construction project, in order to improve the quality and attractiveness of the engineering environment, architects spent a lot of workforce, material, and financial resources on artificial sculptings, such as digging pools, piling up stones, ground paving, artificial turf, etc., which not only wasted resources and destroyed the original ecological landscape but may even bring unpredictable ecological disasters.

China's construction projects consume huge amounts of energy, accounting for 46.7% of total social energy consumption. Every 1% reduction in energy consumption of construction products is equivalent to the power generation of 10 Yangtze River Three Gorges. Therefore, green construction and green construction must be carried out to accelerate the development of green construction and vigorously promote energy conservation as well as emission reduction. The construction process of every urbanization project should reflect "green construction". It is necessary to avoid damage to the natural environment when planning and designing. During construction, it should be insisted on "green construction" and pay attention to energy saving, land saving, water saving, material saving, etc. In the meantime, it is also necessary to ensure that the final building is an energy-saving, low-carbon, environmentally friendly "green building". Therefore, in the construction of urbanization, we should abandon the wrong concept of "man prevails over nature", follow the principle of "the unity of man and nature", adhere to the priority of the environment, and respect the laws of nature. It is vital to place the urbanization in the whole economic, social and ecological system, collectively considering the region's population, resources, economy, society, ecological environment, and other important factors, determineing the development scale and speed and its layout according to the regional environmental carrying capacity, and maintain the balance and coordination between urbanization and economic, social and ecological system.

(2) Reverence for history and priority on culture

Respect for ancestors and traditions is the consensus of all human beings. The United Nations has adopted the Convention Concerning the Protection of the World Cultural and Natural Heritage, the Convention for the Safeguarding of Intangible Cultural

Heritage, and the Convention on the Protection and Promotion of the Diversity of Cultural Expressions, requiring governments around the world to strengthen the protection of the cultural and natural heritage of humanity.

Culture is the spirit and soul of a nation. Promoting the revival of Chinese culture is the historical responsibility for the development of new towns. However, in the current urbanization process in China, some people easily leave behind what has been passed on by their ancestors, not knowing that what they are discarding is the most essential and valuable spiritual and cultural wealth. Culture is also the spirit and soul of a city. The current situation that "many cities with the same appearance" and "a thousand buildings with one shape" have made it difficult for people to identify the local cultural characteristics of the city. Therefore, in the process of urbanization construction, we should not "forget our origin", but "remember where the water is from when we are drinking, inherit and innovate". For the same reason, we should avoid "total westernization" and attempt to "use foreign goods for Chinese purposes". We must respect our ancestors, respect history, preserve regional characteristics, highlight national characteristics, and "use the past for the present". We must adhere to protective development, leave the memory of history, and not self-destruct culture.

(3) Reverence for humanity and priority on livelihood

Ultimately, urbanization is to serve the people living in the city. We should respect humanity and put people's survival and life first. Cities, should make people's lives better and have a higher quality of life. Whenever stopping in a corner of the city, either walking or listening, the urban environment should provide a sense of refreshing comfort. Therefore, in the process of urbanization planning, designing, and constructing, we should always adhere to the "people-oriented" guideline and reflect "humanization" in all aspects. It is necessary to put the comfort and satisfaction of the people living there in the first place.

During urbanization, on the one hand, the needs of the entire human life cycle, such as birth, childcare, schooling, employment, medical care, and old-age support, should be taken into account. On the other hand, it is important to consider people's multifaceted needs, not only basic survival needs such as food, clothing, housing and transport, but also their spiritual and cultural needs. Specifically, the design and construction of each engineering project in the urbanization process should also reflect the concept of being "people-oriented", meeting the needs of people in all aspects and making them feel comfortable and convenient. For example, in planning new residential areas or renovating old residential areas, it is necessary to optimize the design in terms of the residential structure, quality, heat insulation, lighting, and energy-saving. In addition, factors including the beautification of the outdoor natural environment, the quietness and safety of the surrounding social environment should be considered. Urbanization needs to provide space for recreation, entertainment, and communication among residents and fully integrate residents' needs for transportation, shopping, schooling, and medical care. When planning, we can arrange the residential and office areas in the same area, and try the "front factory and backyard" pattern. This can change the status quo of "separation of workplace and residence", which shortens the time people spend on their way to and from work.

This can improve efficiency and reduce the unreasonable travel demand from the source, effectively decreasing traffic congestion. As another example, the establishment of special access for people with disabilities in hospitals, and the operation of "zero-transfer" subway, light rail, taxis, buses, and other means of transport, are the concrete embodiment of "people-oriented" principle. Urbanization should be tailored to local conditions, with corresponding planning and design according to the different production and living styles of people in different areas, avoiding "mass production".

To summarize, in the process of urbanization planning and promotion, we should not be "material-oriented", but "people-oriented", considering the convenience and service for the people, and the improvement of people's quality of life. We should not consider officials' personal preferences and interests but from the perspective of how to expand the city's happiness space. Only in this way can we meet the requirements of "people-oriented" and giving priority to people's livelihood, and truly achieve the fundamental purpose of urbanization for the benefit of human beings.

References

1. Vitruvius. (2201). *Ten books of architecture*. Beijing: Intellectual Property Press.
2. Xu, Y., & Lu, J. (2008). The values of engineering culture. *Engineering—Engineering Science Across the Fields, 4*, 78–88.
3. Ding, D., & Jiang, Y. (2003). *Introduction to civil engineering*. China Building Industry Press.
4. Chen, W. (1990). *Cave culture*. Wenhui Publishing House.
5. Li, W. (2008). *Discussion on the residential model of island-tourism villages and towns in southern Liaoning*. Dalian University of Technology.
6. Yin, R., Wang, Y., & Li, B. (2007). *Engineering philosophy*. Higher Education Press.
7. Zhang, B. (2010). *Engineering culture*. Machinery Industry Press.
8. Wang, H., Sha, L., & Li, Q. (2011). Examining the connotation of engineering culture from the perspective of system science-the construction of engineering culture. *Science Association Forum (The Second Half of the Month), 3*, 187–188.
9. Zhang, X. (2002). Questioning architecture as art from "architectural art"—Discussion on the essential characteristics of architecture, art and aesthetic characteristics of architecture. *Huazhong Architecture, 20*, 42–44.
10. Zeng, K., & Kong, X. (2005). Chinese ideology and culture and "people-oriented." *Climbing, 5*, 146–149.
11. Lu, G. (2016). *Humanism of construction engineering*. China Building Industry Press.
12. Li, X. (2014). Landmarks can not cart before the horse. Guangdong Construction News.
13. Li, S. (2014). Innovative development model promotes urbanization. *Urban and Rural Construction, 1*, 6–8.

Chapter 10
Ethics of Engineering Management

10.1 Introduction

In the age of pursuing development speed and economic efficiency, personal achievement and self-transcendence, the reflection and examination of the essence of human existence and the true happiness of life determine the evolution of human moral and ethical concepts. According to Aristotle, "happiness is a real activity of the soul in accordance with virtue", and "the activity of virtuous realization is necessarily happy" [1]. Aristotle defines happiness in terms of goodness and explains the naturalness and universality of happiness by the purposefulness of goodness, thus providing a value-based theory for developing his theory of happiness. On the other hand, the French philosopher Descartes pointed out that virtue is a prerequisite for being human, and that "without virtue, the greater the talent, the greater the disaster." An ethical person means having good qualities such as goodness, responsibility, honesty, justice, equality, and freedom. By observing and adhering to ethical qualities, people can pursue higher happiness and gain more freedom.

In all ages, morality and ethics have been the patron saints of human social order. The country is governed by moral rule, and the people are honored by high morality. In the strategy of governance, the rule of law is a rigid constraint to maintain the dignity of the state by means of compulsory indoctrination by discipline, while moral governance is a soft constraint to regulate social speech and behavior by means of self-discipline and self-restraint. Morality and ethics are for everyone, manifesting wisdom and refining spirit. As Hegel said, only when the moral power of the spirit plays its potential and raises its banner can patriotic passion and a sense of justice be achieved in reality. As an important component of the overall function of society, people should never be fallen to "material aristocrats, spiritual beggars." They should pay more attention to improving moral consciousness and behavior.

As a "social experiment," engineering must be supported by ethics. However, whether engineering needs ethics is still controversial, which has become a major obstacle to the good performance of engineering management. Integrating engineering and ethics will help achieve the good works of "doing good engineering

J. He, *Principles of Engineering Management*,
https://doi.org/10.1007/978-981-99-1168-4_10

well" and contribute to the sustainable prosperity of engineering ethics. It will also help the engineering community implement the humanistic value of "putting public health, safety, and welfare in the first place."

This chapter begins with the essence of engineering ethics, takes the professional ethics of individual engineers as the starting point, and identifies the three major doubts, including their existence, how to identify and the consequences of ethical problems in engineering. It also provides relevant methodological guidance for engineers to solve ethical problems. Beyond the above, this chapter clarifies seven transformations that must be accomplished for the research of engineering ethics to move from "narrow engineering ethics" under the vision of engineers' personal responsibility ethics to "broad engineering ethics" under the perspective of the engineering community, to achieve the parallelism of ethics and excellence finally.

10.2 The Nature and Current Development of Engineering Ethics

10.2.1 The Nature of Engineering Ethics

The words "Lun" and "Li" have been used since ancient times, and they have already appeared in masterpieces such as *Shang Shu*, *Shi Jing*, and *Yi Jing* respectively [2]. Xu Shen commented in *Shuo Wen Jie Zi*: Lun, originated from man and its pronunciation follows the meaning of generation as well as principles. Li, literally refers to the jade and its pronunciation is related to the carve "[3]". The word "Lun" originally refers to the distinction between generations of people, while "Li" refers to the textures of jade. By extension, "Lun" can refer to categories and order, while "Li" implies the meaning of "principle." The word "Lun Li" was first used in the Books of Rites within the chapter on Leji (music). It implies that all kinds of music were born in the heart. The music itself is engaged with ethics [4]. Ancient Chinese societies often used ethics as the reasoning and rules that should be followed in the relationship of human social life. Ethics were also regarded as the order, the rules, and reasonable behavior of human social life [5].

Although Chinese ethical thought has an early origin and a rich connotation, the term "ethics" was influenced by Western culture. This term began to be widely used after the nineteenth century. In the West, the word "ethics" is derived from the Greek word "ethos" [2]. In Homer the Odyssey, "ethos" means residence, abode [6]. The ancient Greek philosopher Aristotle was the first to use the word in the sense of temperament and character, giving it the meaning of "ethical" and "virtuous" and inventing the word "ethics." Thus "ethics" can mean custom, style, character, and method of thinking, defined as the science that studies the origin, essence, function, and development law of morality.

The interpretation of the meaning of "ethics" in Chinese can also be broadly divided into two forms: one is "ethics," which indicates a discipline that is a theoretical study of human moral phenomena. The other is "ethics," which refers to the code of conduct for people to get along with each other, including people's thoughts and wording about morality [7]. Based on meaning one, engineering must have its inherent ethical dimension as a social practice activity. As Don Wilson said, "engineering ethics are the moral principles accepted by the profession of engineering concerning the practice of engineering." Based on meaning two, engineers should have their unique professional ethics as a profession. This is evidenced by Albert Flores' statement that "engineering ethics is the power and responsibility of those engaged in the engineering profession" [2]. From these representative definitions, it is easy to decipher the practical characteristics of engineering ethics as a social experiment and the normative characteristics as an ethical code.

Mike W. Martin and Roland Schinzinger state that engineering ethics is "a set of moral principles about duties, rights, and ideals that should be agreed upon and justified by engineering practitioners. The discovery of these principles and their application to engineering practice is the central goal of the discipline of engineering ethics" [8]. The definition involves six key elements.

(1) The term "engineering practitioners" implies that the study of engineering ethics is not limited to engineers. The implication is that, on the one hand, engineers are the advisors of engineering decisions, the providers and interpreters of engineering solutions, and the designers, executors, and supervisors of engineering activities. They have the irreplaceable role of "engine." However, engineering is a "collective activity," and the main body of engineering activities is the engineering community composed of engineers, investors, managers, workers, and other stakeholders. Therefore, the research object of engineering ethics is oriented toward the engineering community and its related ethical problems [9]. The engineering community is divided into "engineering activity community" and "professional engineering community": the former refers to the coalition of participants who implement projects, emphasizing the organization of members to carry out engineering activities together. The latter refers to the industry association, which belongs to the professional community category and focuses on maintaining practitioners' legal rights and interests [10]. On the other hand, the starting point of the study of engineering ethics is the professional ethics of engineers—"ethics about engineers." Professional ethics of engineers specifically refers to the study of engineers' responsibilities to employers, the public, the environment, and society in their professional activities based on a narrow perspective. The pluralistic composition of engineering subjects makes it necessary to consider the study of engineering ethics should be considered from a broad perspective, i.e., "ethics of engineering," focusing on the ethics of decision-making, management ethics, and environmental, economic, political, and social-ethical issues of engineering activities of "engineering communities" [9].

(2) "Duties, Rights and Ideals" indicates that engineering ethics helps to achieve "doing good engineering well," which consists of two main levels. The first is to guide engineering in the direction of goodness. Engineering activities emphasize the application of scientific principles to optimally transform natural resources into "artificial nature" for the benefit of human beings. It is not a process of applying scientific knowledge to solving technical problems but a process of exploration and

trial and error, in which ethical issues such as right and wrong permeate. Engineering itself does not directly have goodness in the moral sense [11]. Engineering ethics involves value judgments such as "obligations, rights and ideals," which can provide moral rules for engineers to make ethical decisions when faced with the oscillation between good and evil. Second, it promotes the realization of good engineering. When facing ethical issues, engineers are often embarrassed by the difficulty of correctly identifying ethical issues and are often caught in the dilemma of making ethical decisions. Internalized ethical sensitivity is not only the key to identifying and clarifying ethical issues in engineering in time but also has a subtle influence on engineers' prudent ethical decision-making. Engineering ethics effectively cultivate the "ethical sensitivity and ethical choice" of engineers [12].

(3) The term "ethical principles" refers to engineering ethics containing certain norms and principles. The Oxford Encyclopedia of English-Chinese defines ethics as "moral norms," which are common norms observed in human relationships [13]. Accordingly, engineering ethics must include codes of conduct involving moral responsibility and obligation. Its primary meaning is to "establish the cognitive and practical principles expected of engineers, and the norms of behavior that engineers should follow when interacting with each other or with other members of the group and society." Engineers can apply the ethical principles established by ethics to individual cases through judgment (Urteilskraft), or carefully weigh specific situations based on "practical wisdom" (Phronesis) [14]. The code of engineering ethics states that engineers should use their professional knowledge and practical experience to fulfill their responsibilities to promote the well-being of society and to play an important role in "serving and protecting the public, educating, inspiring, and supporting responsible professionals, and enhancing the image of the profession" [15].

(4) "Discovery" embodies human subjectivity in engineering activities, which can be interpreted from three processes: (1) Subjectivity dominant i.e., disenchantment. Subjectivity is essential "human's self-knowledge, self-fulfillment, and self-transcendence of life movement and its manifested attributes, such as autonomy, selectivity, and creativity." Human engineering activity purposefully realizes the demystification of the world by subjectivity, that is, "the process that the world is moving from sacralization to secularization, from mysticism to rationalism" [16]. (2) Instrumental rationality comes to the fore—disorientation. The exercise of human subjectivity in the process of demystifying the world significantly changed the world's outlook and the people's concepts. Max Horkhaimer pointed out that "nature has been reduced to a means of total exploitation without any rationally set goals and without any limits", which led to the receding of value rationality and the prominence of instrumental rationality, making human beings fall into the worship and fetishization of technology. The supreme creed of "technological supremacy" has unrestrainedly conquered and controlled nature, and the environment of human existence has been greatly threatened, and Erich Fromm sadly cried "The problem of the nineteenth century is that God is dead, and the problem of the twentieth century is that man is dead." (3) Return of value rationality—reenchantment. The extremely degraded living environment and the horrible engineering and technological disasters have made humans look at the dream of "moving from the kingdom of necessity to

the kingdom of freedom" more rationally. The re-enchantment based on the return of value rationality urges people to assume environmental and ethical responsibilities to seek the ultimate ideal of sustainable development. People go through the process of "knowledge-practice-re-knowledge" before they can discover the "principles" that are truly conducive to sustainable development.

(5) "Application to engineering practice" emphasizes that engineering ethics is not only theoretical ethics but also practical ethics [18, 19]. On the one hand, as Denis Goulet states, "True ethics is a practice that thinks critically about the content and meaning of the values of one's social behavior." [20] Engineering ethics integrates moral theory research and practical problem-solving. While contributing to the construction and development of ethical theories, it also improves the moral sensitivity of individuals to ethical choices and value judgments and promotes "responsible engineering practice." On the other hand, engineering ethics does not simply and mechanically apply general ethical theories to practical problems but is built on the interpenetration and integration of engineering practicality and ethical normativity and focusing on the real situation of engineering problems. With the wise moral judgment and strong ethical willpower of engineers, the practical purpose of "engineering is a social experiment" is achieved.

(6) The term "discipline" means that engineering ethics should also be regarded as a science that conducts theoretical research on the ethical issues involved in engineering. "Ethics" includes not only the moral concept of individuals, the value concept of "habitat," and the code of conduct of interpersonal communication but also the discipline meaning, i.e., "ethics," which is a theoretical study of human moral phenomena. In contrast, engineering ethics can be understood as an activity and discipline that guides the moral values of engineering practice, solves moral problems in engineering, and justifies moral judgments related to engineering [13]. Its research focuses on five aspects [21]. (1) To establish the core and fundamental issues of the engineering ethics system; (2) To analyze the value conflicts, moral conflicts, and integration problems faced at various stages of engineering activities, and to draw the attention of leaders, managers, and engineers to ethical topics in engineering activities; (3) To explore typical ethical issues in the field of engineering technology; (4) To ethically examine the process of engineering activities, focusing on the contents involving ethical examination and restraint in the operation of general engineering links; (5) To propose moral qualities and ethical norms that engineering practitioners should possess.

Martin and Sinzinger's research breaks through the restriction of engineering ethics in a narrow sense, that is, the professional ethics of engineers, and tends to explore engineering ethics from the broad perspective of pluralistic engineering subjects–engineering communities. However, the connotation of engineering ethics in a broad perspective is not limited to this. As mentioned above, engineering ethics refers to the discipline in which the engineering community gives ethical considerations and moral decisions on ethical engineering issues in the whole life cycle and conducts systematic theoretical research on them based on the purpose of sustainable development and the responsibility to protect public health, safety, welfare and environment [9]. The emergence of engineering ethics has given positive meaning to engineers' efforts [22], enhanced their ability to deal effectively with complex ethical issues in engineering, and increased their moral autonomy. Its role lies not only in theoretical reflection on engineering afterward but also in making ethical considerations permeating the whole process of engineering construction, thus creating a more friendly artificial nature for the benefit of human society.

10.2.2 A Review of International Research on Engineering Ethics

Engineering ethics originated in the 1960s and is represented by developed countries in Europe and North America. The evolutionary history of engineering ethics research frontier is divided into three stages. The first stage is the formation period of disciplinary foundation (from the early 1950s to the late 1980s). During this period, academics tried to construct engineering ethics from different perspectives. However, it has not yet formed a center-focused and more mature conventional science and is still in the pre-scientific stage of slow development and no unified paradigm. The second stage is the period of disciplinary field expansion (from the early 1990s to the end of the 1990s), during which engineering ethics research has blossomed, formed a certain research scale, and established fixed research teams. The third phase is the period of both disciplinary expansion and deep development (the late 1990s to the present), during which ethical challenges, genetic engineering, and ethical decision-making became new issues of great concern. In addition to the issues that still need to be explored in-depth in the previous stage, ethical practices and implications have become a new round of research hotspots in engineering ethics [23].

(1) Engineering ethics in the US

After decades of development, engineering ethics in the United States has become relatively mature and standardized, forming a relatively perfect ethical charter and stable academic establishment, which has promoted the benign development of engineering in the United States. Starting from the disciplinary paradigm of professional ethics and combined with case studies, American engineering ethics conduct in-depth research on the moral issues and choices faced by engineers in their work practice. The contents of the research include "individual responsibility of engineers," "responsibility in engineering practice," and "responsibility of engineering associations" [24]. The basic goal is to develop a "preventive ethics," i.e., to develop the habit of engineering practitioners to consider ethical issues in practice in advance and to enhance their sensitivity, reflective ability, and response skills to such issues, i.e., moral self-awareness [25]. Eddie Conlon and Henk Zandvoort call this model of "professional ethics" the "individualistic approach." The model revolves around the professional activities of individual engineers. The model revolves around the professional activities of the individual engineer, assuming that the engineer is facing an ethical problem and that they have a decision to make. The study of engineering ethics centered on a code of ethics has responded to the need of professional organizations of engineers in the United States to strengthen ethical codes and engineering educational institutions to enhance engineering ethics education. Its popularity and diffusion were rapid and at once became the mainstream of engineering ethics research in the United States.

American engineering ethics is currently experiencing a shift from micro to macro [26]. This is represented by going beyond the narrow engineering ethics that engineers are responsible to their employers, focusing on the social responsibility of the engineer community and the engineering profession for the present and the future in terms of ecological protection and sustainable development [27]. John Ladd pointed out that engineering ethics should consider the relationship between individual engineers and their clients, colleagues and employers, and the relationship between engineering professional activities and society. Under the concept of "broad engineering ethics", the subject of responsibility is the engineering system, and the object is nature and human beings. This concept requires considering the interaction between scientific and technological progress, engineering changes, and human society [28]. As technological progress is changing rapidly, the impact of engineering on humans and nature is out of control, so it is necessary to construct macro engineering ethics from the group perspective of the engineering community in a panoramic manner. In engineering ethics education in the United States, the "engineering community" has gradually become the main educational object [29]. The cooperative community of engineers and ethicists has also received more attention.

On the whole, engineering ethics in the United States presents new features and trends, which are mainly represented in three aspects. The first is the specialization and establishment of research objects. The second is the diversification and practicability of research methods. The third is the internationalization and synthesis of research trends [22].

(2) German engineering ethics

German engineering ethics is mainly conducted in the framework of technology ethics and applied ethics. German engineering ethics focuses on principles and strategic options for solving ethical problems in engineering and technology, focusing on ethical responsibility and technical evaluation [30]. Max Weber distinguishes between "ethics of responsibility" and "ethics of belief," arguing that the goals of action in "ethics of belief" are irrational in terms of possible consequences and that "ethics of responsibility" is not rational in terms of possible consequences. The ethics of responsibility, on the other hand, requires actions that consider their consequences. The ethics of responsibility takes precedence in the field of action [31]. Hans Jonas developed ethics based on "responsibility." He extended the scope of responsibility to all human beings (especially to future generations) and nature as a whole [32], making the question of responsibility a central issue in the study of engineering ethics. He also reminded humans to think in terms of long-term, negative outcomes and to use the power of science and technology judiciously. On this basis, Hans Lenk's "responsibility" is different from Kant's "instinctive morality and conscience." His view emphasizes "concern for consequences" and even the responsibility for "unpredictable consequences" [33]. In addition to the traditional causal responsibility, he argues that people should also take responsibility for caring for protection and prevention [34]. The "ethics of negotiation" developed by Jurgen Habermas and his collaborator Karl-Otto Apel provides a possible explanation for the real implementation of abstract metaphysical, ethical principles through the practice of inter-subjective interaction. It creates a realistic basis for practical professional ethics [35]. It is a theoretical attempt to solve the problem of what and how human beings should be responsible [35].

In addition to the blossoming academic research, the German Society of Engineers (DSE) is committed to raising engineers' awareness of responsible ethics and regulating their professional behaviors. In 2002, DIE promulgated the "Basic Principles of Engineering Ethics" [36], which aims to help engineers raise their awareness of engineering ethics and provide basic ethical guidelines and standards for their behavior. It provides guidelines for judgment and standards in liability conflicts and assists in resolving disputes over liability issues related to the engineering field to protect engineers. It also requires engineers to be responsible for their professional behavior and its consequences and accountable to professional codes, social groups, employers, and technology users. It requires engineers to respect laws and regulations established by the state that are not contrary to universal ethical principles, clarify their responsibility for technical quality, safety, and reliability, and invent and develop meaningful technologies and solutions to technical problems [33]. The promulgation of this principle means that the "responsibility of engineers" is becoming increasingly urgent and cannot be ignored.

(3) Japanese engineering ethics

Based on the Eastern cultural tradition, Japanese engineering ethics has gradually formed and matured based on the absorption of American achievements. Its ideological development is roughly divided into three stages [37]. The first stage develops the ethical thinking of Machinists (businessmen and industrialists), i.e., the principles about practicing and dealing with people spontaneously formed by business and industrialists. The second stage focuses on studying professional ethics related to business-to-society and within the business. The third stage introduced the American system of engineering ethics, which was successfully absorbed and developed.

Japanese engineering ethics not only has a clear ideological development context but also has unique research characteristics. This is mainly reflected in three aspects. First, Japan does not have a very distinct professional tradition and did not have a typical capitalist model, but with the help of "people-centered" management models such as the "annual merit sequence system" and "lifelong employment system," Japan has weakened the ethical conflict between enterprise employers, engineers and employees, and the public, creating amazing industrial prosperity. Second, Japan's local culture is quite different from the cultural traditions of English-speaking countries. For example, Japanese culture is guided by group values and emphasizes that groups are higher than individuals. Therefore, the main body of Japanese engineering ethics practice is not engineers but employers, large enterprises or companies. Engineers do not emphasize recognition of occupation but recognition of the company that is, obedience and loyalty to the company. However, Japan has successfully introduced the American engineering ethics system and made considerable achievements in imitating the Western professionalism model. Third, Japan focuses on strengthening the concept of practical ethics and encourages engineers to take the practical wisdom that "useful engineering ethics must be able to solve problems at the scene of engineering ethics problems" as the compass of professional behavior [25]. In short, Japan's unique cultural background leads to its "harmony but difference" from the American model of engineering ethics system and finds another way to depict a new picture of cross-cultural engineering ethics practice.

(4) Engineering ethics of the former Soviet Union

The vigorous development of industrialization hatched the rich ethical thoughts of the former Soviet Union. Technological activity is defined as the action of society or all humankind, and the moral evaluation of technology is established as the possibility of realizing the ideal society. Then its engineering ethics was studied in the name of "engineer ethics" [38]. Scholars believe that engineering ethics aims to explore the relationship between engineers and the collective, the relationship between engineers, and the relationship between engineers and workers in the production collective. The objects of engineering labor are production processes and technical processes aimed at using natural substances. This is represented by the book "Engineering Ethics" co-authored by B. Г. Нестеров and others. It is the first book devoted to the ethical view of the professional practice of engineers, focusing on the study of professional ethics of engineers and providing guidelines for their behavior.

The Chernobyl nuclear power plant disaster in 1986 completely awakened the myth of "the spontaneous humanitarian nature of technology under socialist conditions" thought by scholars of the former Soviet Union. They began to reflect on various problems brought about by technological creation (especially ecological problems and engineers' responsibility) and gradually accepted the mainstream thought of western engineering ethics [39]. In the social and cultural background of experiencing a variety of value conflicts, the engineering ethics research has realized the transformation from applying socialist moral principles to the reference and absorption of western engineering ethics theories and methods. At present, Russian scholars define "engineering ethics" as the science of studying engineers' individual behavior and formulating ethical rules to adjust their professional activities. It is a branch of applied ethics. They regard engineering ethics as an independent discipline or the sum of ethical rules for adjusting engineers' professional practice [40].

International engineering ethics research presents the following characteristics.

(1) Regarding maturity, American engineering ethics is relatively complete, occupies the high point of the western engineering ethics ideological system, and leads the future exploration trend. Other countries' research is based on their own reality, absorbs the advantages of American theory, and presents colorful and distinctive prosperity with the help of their own culture.

(2) In terms of research content, it has developed from paying attention to practical problems in engineering to the institutionalization of engineering ethics, gradually improved the theory, strengthened the research on engineering history and

engineering association and its articles, and formed a research direction shift from responsibility ethics to group ethics, from individual ethics to communication ethics, and from rule-based ethics to virtue ethics.

(3) In terms of core concerns, improving engineering personnel's ethical literacy has been paid more and more attention, and countries have gradually strengthened engineering ethics education.

(4) In terms of future trends, the vision of engineering ethics is expanding. More attention is paid to the social ethics problems caused by engineering innovation under globalization, the technical ethics problems brought to humankind by the rapid development of scientific and technological innovation, and the ecological environment problems caused by engineering construction under the situation of climate warming.

10.2.3 A Review of Chinese Research on Engineering Ethics

(1) Chinese traditional engineering ethics

Although engineering activities have a long history in China, the concept of engineering ethics is a very modern fashionable item. For thousands of years before the fifteenth century, ancient China has always occupied a leading position in technology. It is an important technology exporter, with the characteristics of the original technology system and original engineering evolution. In particular, China's four great inventions of papermaking, gunpowder, printing, and the compass have changed the face of the world. In the modern era, with the rise of the western industrial revolution, Europe has become the birthplace of modern civilization, China has gradually become a "catcher," and science and technology were declining. However, the temporary backwardness of the development trend can not erase the value of Chinese traditional science and technology ethics. On the contrary, because of the independent evolution and innovation of engineering development path and development mode in ancient China, Chinese traditional science and technology ethics are unique and outstanding. Among them, the essence still contributes greatly to guiding the research of modern engineering ethics. In this regard, Ilya Prigogine once said: "Chinese civilization has a profound understanding of the relationship between human beings, society, and nature... The Chinese way of thinking has always been a source of enlightenment for philosophers and scientists who wanted to expand the scope and significance of Western science." Ancient China made outstanding contributions in many fields such as astronomy, calendar, geoscience, mathematics, agronomy, medicine, and humanities. Joseph Needham said, "without such contribution, it would be impossible to have the whole development process of our western civilization." Moreover, many inventions and creations embody the rational brilliance of the harmonious development between man and nature and the combination of scientific spirit and moral ideal, such as the world-famous Dujiangyan project.

The content of Chinese traditional ethical thought is broad and profound, has a long history, and shows the unique type of Chinese civilization. It still exists in different forms and degrees today and has become a unique cultural resource and ideological foundation for social development and innovation. With the rapid development of modern high technology and the strengthening of international integration and exchange, the civil engineering industry needs the guidance of traditional ethics. The traditional ethics of science and technology affects the practice of science and technology along the following five paths: knowledge ethics, technology ethics, construction ethics, medical ethics, and ecological ethics [41]. Its basic content has three points: first, the unity of knowledge and morality; second, the combination of natural knowledge and ethics; third, scientific research exists at all levels of society [42]. Its basic spiritual characteristics include the unity of heaven and man, using Tao to control technology and people-oriented and practical application. Among them, the unity of heaven and man is the philosophical basis, Tao and technology are the theoretical core, people-oriented is the return of value, and practical application is the prominent feature [43]. Traditional architecture is the best example of integrating these basic spiritual characteristics. In Chinese traditional cultural concepts, architecture has never been just a simple residence and utensils but also the embodiment of order, power, etiquette, and morality. For example, because Confucianism, which deeply influenced the types and styles of ancient Chinese architecture, advocated a ritual system, that is, taking ritual as the foundation of governing the country and the criterion of individual behavior, buildings such as halls, ancestral temples, altars, tombs and so on came into being. Traditional architecture condenses strong political ethics everywhere, from layout orientation, size, and structural components to decoration design. It highlights the orderly, respectful, and humble national structure, having a strict hierarchy system, and is endowed with the social education function of "to regulate the family and the ruling the state." *Huangdizhaijing* summarizes this: "the house is the hub of Yin and Yang and the track model of human relations" [45]. It can be said that traditional Chinese architecture strongly reflects Chinese traditional ethical culture in almost every aspect.

The active elements of Chinese traditional science and technology ethics have great theoretical reference value for developing modern engineering ethics, ecological ethics, engineering science, and other disciplines. The insights provided by Chinese traditional science and technology ethics are also a good prescription for solving the contradiction between man and nature and a rational choice to overcome the differentiation of human science and technology [46]. With the deepening the negative impact of scientific and technological development on society, some domestic and foreign scholars began to look for the wisdom of scientific and technological ethics from Chinese traditional culture. Their research mainly focused on the following fields: the relationship between heaven and man, the relationship between truth and goodness, scientific and technological values, scientific and technological humanities, and the theoretical system of traditional scientific and technological ethics. In ancient China, although there was only the word technology without the concept of science and scientist, the science and technology developed were commendable. The scientific thinking and scientific and technological morality of many world-renowned scientists and inventors are worth exploring [47]. There was little discussion available on the content of scientific and technological ethics at the macro level in ancient Chinese books. Therefore, the current research is still in the sporadic, micro, and local research stage. The works systematically and comprehensively studying Chinese traditional scientific and technological ethics are still very scarce.

(2) Chinese modern engineering ethics

Chinese modern engineering ethics originates from the philosophical reflection on technology and the introduction of western techno-critical theory. Its research began in the 1980s and belongs to the "latecomer" in the field of engineering ethics. Presently, domestic scholars' research on engineering ethics mainly draws lessons from American engineering ethics research results and makes a more in-depth analysis.

In recent years, the concern about engineering ethics in the Taiwan Province of China has been increasing, especially after the "9.21 earthquake" outbreak. Taiwan took the lead in studying American engineering ethics and gradually formed an engineering ethics education system with Taiwan characteristics, which stimulated the engineers to pay attention to the awareness of engineering ethics. In 1990, Tsinghua University and Zhongyuan University in Taiwan took the lead in teaching the course "engineering ethics." At the relevant meetings in 1991, the participants reached a consensus. They called for colleges and universities to offer "science and technology ethics" and "engineering ethics" courses. After that, Yuanzhi University, Jiaotong University, Fengjia University, Taiwan University, Zhonghua University, and other schools joined the movement [48]. Scholars have conducted extensive research on the aspects of the curriculum, teaching methods, and teaching effect evaluation and focused on the formulation of engineering ethics. The Engineering Ethics Manual was issued in 2007, which defines the responsibilities to be undertaken by engineering practitioners and how to behave in case of conflict, and expounds in detail on the relevant norms of the code of ethics through the case study, guiding the behavior of engineering practitioners [49].

The research on engineering ethics in mainland China shows obvious phased characteristics [50]. In the first phase, which is the enlightenment stage (1989–1999), scholars began to reflect on the doubts caused by technological development. *Technological Ethics—Theory and Practice* (Huang Linchu) [51] and *Technology Will Eventually Get Out Of Control—Thinking About the Victory of "Dark Blue"* (Cao Nanyan) [52] published in 1989 are the starting point, these works attempt to study the problems of engineering technology from the perspective of ethics. During this period, ethical reflection was mainly carried out on the issues brought about by the development of military, ecology, medical treatment, and engineering technology. For example, the Sixth National Symposium on Medical Ethics (1991) was held. The second stage was the stage of exploration (1999–2005). The concept of engineering ethics was gradually introduced. Take *Engineering Ethics* (Xiao Ping) [12] in 1999 and *Research on Engineering Ethics and its Main Problems* (Li Shixin) [2] in 2003 as an important sign, and relevant research began to deepen gradually. Great achievements have been made in the preliminary discussion on the connotation of engineering ethics, engineering ethics norms, and basic principles of engineering ethics. The two tendencies of "absence" of ethics in engineering activities and "forgetting" of moral thinking in engineering have been corrected. The third stage is the steady development stage (since 2005). The Chinese translation versions of *Engineering Ethics—Concepts and Cases* (Charles E. Harris, 2006) [53] and *Engineering Ethics* (Martin and Sinzinger, 2010) [15] were published. The discussion on major topics such as discipline orientation, research theme, and systemic development of engineering ethics has been carried out in an all-around way. Scholars have introduced and translated many classic monographs of foreign scholars, deeply analyzed engineering ethics cases, and actively conducted academic exchanges with foreign scholars. However, the current research on engineering ethics in China mainly faces the following three problems: first, there are more theoretical problems and fewer case studies and studies on specific issues. Second, the dialogue and cooperation between ethics professionals and engineers (managers) must be further strengthened. Third, when introducing foreign theoretical achievements, there is a lack of research on domestic issues with insufficient attention to the effectiveness of practice.

Due to the globalization of engineering projects, engineering ethics formed in the background of the United States has been critically examined by more and more scholars from the United States and other countries. The theoretical progress in engineering ethics in the West has enabled Chinese scholars to avoid a narrow understanding of engineering ethics, to widely and deeply explore various moral problems in engineering activities from an open perspective, and study the applied ethics to handle ethical issues in engineering practice. Based on the preparation in the theoretical foundation, practical needs and the ethical-oriented research of engineering philosophy, the research on engineering ethics in mainland China is mainly carried out in the large system of the philosophy of science and technology, focusing on high-tech fields such as biological genes and computer networks, as well as industries related to the national economy and the people's livelihood such as civil engineering and energy. It focuses on evaluating engineering objectives, the equity of interests involved in large-scale engineering projects, and ecological ethics. Its research perspective can be summarized into two dimensions. The first dimension is reflected in the mutual growth of engineering and ethics. From ethics to engineering—use the perspective and methods of ethics to study engineering ethics and to guide and promote the good development of engineering practice. Then from engineering to ethics—explore the impact of engineering development on ethics and establish new ethical thinking. The second dimension focuses on the practical and professional characteristics of engineering [53]. Engineering ethics is not only the research on moral values, moral problems, and moral decision-making involved in engineering practice but also the unique professional ethics and personal ethics that engineers should strive to possess. Although engineers have become the focus of engineering ethics, the research results obtained from the perspectives of ethical responsibility, ethical education, and ethical knowledge show that engineers are still in a passive ethical role, and their own subjective initiative and self-consciousness have not been effectively brought into play. The future direction of engineering ethics is professionalism, which emphasizes the professional excellence and ethical integrity of engineers, that is, excellent professional technology and the perfect personality [54]. Establishing ethical thinking is one of the important ways for engineers to change from passive to active ethical roles [55].

To sum up, there are two ways to study engineering ethics. One is to regard engineering as the application of technology from the perspective of science and technology. Second, from the perspective of profession and professional activities, engineering ethics is attributed to the professional ethics of engineers. No matter which of the above views is used, the research of engineering ethics is easy to get into narrow engineering ethics. In the narrow sense, engineering ethics takes engineers as the starting point. It determines the initial appearance of engineering ethics research by paying attention to the ethical quality of engineers and the ethical responsibilities that engineers should take. For example, Philip J. Chmielewski believes that "engineering ethics refers to the ethics of engineering researchers and designers, and the most important group of engineering researchers and designers are engineers. They are the main participants in engineering undertakings and the main body of engineering activities". As the only object of engineering ethics, engineers must ensure "promoting a responsible engineering practice." This shows that exploring "how engineers become virtuous people" and solving "how to train engineers to voluntarily choose to be responsible" are unavoidable problems in narrow engineering ethics. However, engineering ethics in the narrow sense pays too much attention to the impact of ethical concepts on engineers' professional behavior, which can neither promote engineers to better understand the new engineering ethical problems in the mobile Internet era with rapid technological transformation nor help engineers completely get rid of the "sword of Damocles" in a dilemma under the traditional situation. The more serious impact of engineering ethics in a narrow sense is that this kind of research limited to the "single subject" of engineers makes the efforts of engineering ethics breakthrough, becoming a futile struggle. Li Bocong pointed out that the research on engineering ethics should move from "narrow sense" to "broad sense," and the research theme should change from "professional ethics of engineers" to "engineering decision-making ethics," "engineering policy ethics," and "practical ethics of engineering process" [56]. The development of engineering ethics urgently needs to change the ethical subject from individual engineers to the engineering community to realize the macro turn of engineering ethics.

10.3 Prominent Ethical Issues in Engineering

Engineering is closely related to human life, which involves the complex relationship between man and nature, man and society, and man and man, and ethical issues are contained in it [57]. With the gradual change of the essential characteristics of engineering, the necessity of people's demand for engineering ethics is becoming increasingly prominent. First of all, technological progress has made humankind master great power, and technology has changed from a simple tool to a sharp tool for building artificial nature and creating an "engineering kingdom." Therefore, it is urgent to carry out an ethical reflection on technology. Secondly, engineers are evolving from a profession that provides professional and technical services to employers and customers to a profession that serves the whole community in a way that is both responsible for society and the environment [58]. The professional ethics of engineers came into being. In addition, engineering has never had such a huge impact on human life and penetrated all aspects of human life. The engineering project contains some results beyond the expectations of the project initiator. Ethical norms must regulate it. Ethical problems in engineering have become an important research field recognized by philosophers and engineers. Before making any professional, ethical judgment, we must first be able to find potential ethical problems. If we fail to be aware of the ethical issues hidden in the working environment, the professional, ethical behavior that can be used as a countermeasure will not occur.

10.3.1 Does the Engineering Project Have Ethical Contents

Although an ethical appeal is an internal provision of engineering activities, there are still disputes about whether the project has ethical problems.

(1) Scientific application view of engineering

The "scientific application view of engineering" holds that engineering is the application of science, which is only divided into advanced and backward technology, and there is no difference between good and bad morality [59]. The basis for this view is reflected in the following two points. First, the task of engineers is to build a harmonious artificial nature, the task of scientists is to understand and explain the world, the premise of building artificial nature is to understand the world, and science is the foundation and premise of engineering. Second, engineering is based on science and its extended disciplines, and the basis for the achievements of engineers is also science. This view is generally accepted. Among them, Thomas Tredgold's earliest and most authoritative definition of engineering reflects this idea: "engineering is the art of using the great power of nature for the convenience of mankind and the practical application of the most important principles in natural philosophy." [60] David R. Reyes-Guerra also believes that engineering is the general name of the profession that selectively applies the mathematical and natural science knowledge obtained through learning and practice to open up ways to use natural materials and natural forces reasonably for the benefit of humankind [61]. Under the guidance of the epistemological root of the "scientific application view of engineering," engineers regard their work as the application of natural science. They believe their work is not related to value judgment or ethics. Therefore, engineers do not need to bear ethical responsibility for the impact of their behavior. As F. Dessauer said: "the technological creation of engineers is to connect the human purpose with a transcendental 'fourth kingdom' that designs solutions to technological problems through 'internal calculation'; engineers have almost no moral responsibility for their own technological activities" [62]. The value that understands engineering construction as the value of "pure intellectual activity" leads to the failure of the subject of technological innovation to emphasize its ethical responsibility for technology.

The view that engineering is the application of science only sees science as an important source of engineering knowledge but ignores the close relationship between engineering and society. In this regard, Mike W. Martin and Roland Schinzinger stressed that engineering is not a solution process of applying scientific knowledge along a straight path to solve an isolated technical problem. On the contrary, it is a process of exploration, trial, and error. Any project has its own characteristics, which can not be fully covered by existing scientific knowledge or other engineering knowledge and previous engineering experience. It requires engineers to give play to intuition and inspiration in the sense of similar artistic creation. Therefore, "failure is inherent in all useful engineering design," and there are inherent risk factors endangering people's life, health, and property safety in the project. "What risks can users and the public accept?", "how to determine the acceptable risk level?" and "who will determine this standard?". These problems are not purely technical problems that can be solved by simply applying scientific theory but difficult problems with ethical nature. There are abundant social-ethical problems in engineering.

(2) Technology autonomy theory

"Technology autonomy theory" holds that technology forms its own system, and its development is not affected by external factors. This view has extensive influence and many supporters in the West. Jacques Ellul, its representative, believes that "technology is characterized by its rejection of warm moral judgment. Technology will never accept the distinction between moral and immoral use. On the contrary, it aims to create a completely independent technical morality." "in modern technology, people no longer have the freedom of judgment and choice," "technology is no longer predictable, and people no longer set goals for technology," and "mankind will eventually fully surrender to technology." Norman Levitt also believes that "the development of technology is independent and will not be affected and restricted by social, political, cultural, ethical and moral factors. As a result, a huge technical structure has been formed and cultivated, which has not been specially designed or planned by anyone" [63]. The theory of technological autonomy denies the existence of technological ethics, and holds that there is no correlation between technology and morality, and ethics and morality can not restrict the development of technology. On the contrary, as a force to promote social change, technology dominates and determines people's moral concepts and social state. Engineers who take technological invention and technological creation as their professional responsibility are more likely to understand the essence of technology and tend to the theory of technological autonomy [64]. They attribute the negative impact caused by technology and engineering activities to the unique, essential attribute of technology and take it as an excuse for not undertaking ethical responsibility and a basis for defending the shirking of responsibility.

Engineering technology development is a process of interaction between technological systems and external society (including ethics and morality). On the one hand, technology development has internal inheritance and continuity, and there are "accumulation" and "self-strengthening" of technology. Compared with human use, control and domination, technology shows some resistance and autonomy. On the other hand, technology is an activity undertaken and carried out by people to adapt to nature and create material wealth. Without people's material desire and spiritual demand as the driving force, there will be no generation and development of technology. Technology cannot exist without people, especially engineers. Therefore, it must be recognized that people ultimately have the initiative in the decision and choice of technology, and people can not shirk their responsibility for the development of technology.

(3) Engineering tool theory

The "engineering tool theory" regards engineering as a means or tool system to achieve a goal and believes that each technology is used to solve special problems or serve the specific purpose of humankind. In the view of instrumentalists, the value of technology lies in achieving the goals set by society. Its representative Karl Theodor Jaspers believes that "technology is only a means, and it has no good or evil in itself. Everything depends on what people create from it, why it serves people, and under what conditions people place it" [65]. Under the guidance of "engineering instrumentalism," engineering technology and artificial nature are only "instrumental" and "functional." As the main body of inventing and creating tools, engineers are not obliged to bear ethical responsibility for the adverse consequences caused by the project. Their responsibility is to complete the project task without considering the impact of engineering technology from the ethical dimension.

"Engineering tool theory" separates the relationship between engineering technology and its social consequences and ignores the complexity of modern engineering practice and its results. Engineers do not need to bear ethical responsibility when the negative consequences caused by technology beyond foresight. This view exposes the limitations and absurdity of "engineering tool theory." What engineering ethics should study is precisely the "preventive responsibility" problem caused by technology's negative effect.

Engineering is value-oriented. "The purpose of engineering activities is to form a 'more valuable' world... Human engineering activities are a process of 'creating' and 'promoting' such value, which is a process with value as the yardstick and index of progress" [66]. Influenced by traditional technical concepts such as "scientific application view of engineering", "technology autonomy theory" and the extended "engineering tool theory", engineers often find it difficult to clarify their ethical responsibilities, regard their professional activities as the application of natural science, think that technology itself and its development are completely independent, and understand engineering as a tool and means to achieve their goals. It leads to the indifference to their sense of ethical responsibility. However, as pointed out by Émile Durkheim in *Professional Ethics and Civic Morality* [67], ethics is an important issue in the process of professional development of specialized occupations, which can not be ignored. Engineering ethics issues have been formally put forward with the establishment of professional societies of engineers, and the identification and dialectical analysis of engineering ethics issues have attracted extensive attention.

10.3.2 How to Distinguish Engineering Ethics

Ethical factors are the inherent implication of the engineering project. There are many complex, open and ambiguous ethical and moral problems in the project. It is not ethicists who forcibly plug the ethical dimension into the engineering project or regard the project as a moral load with ethical color-changing glasses. But, "generally speaking, engineering activities have no 'pure' ethical problems. Ethical problems are often 'closely combined' with other problems. When studying and analyzing ethical engineering problems, we must combine ethical analysis with analysis of other dimensions. Otherwise, the analysis of ethical problems in engineering will inevitably fall into the illusion of 'romanticism' or the style of 'pavilion in the sky' empty talk". Martin even believes that "being able to skillfully identify ethical problems in engineering is the first important purpose of learning engineering ethics and the only way to cultivate and improve moral consciousness".

The ethical problem in engineering practice is one of the contemporary world's main social problems. It is because the technical products or design schemes and other services provided by the engineering project for the society have the characteristics of intermediation, duality, and transition. Intermediation means that the utility value pursued by technology is at the lower level of the value ladder. It is not the final good or purpose but the means to realize the absolute good. Duality means that the project can not only serve a good purpose, but also be controlled by an evil purpose. Simply judging the good and evil of its use-value is difficult. Transitional means that "engineering products are for the final consumption and use by users like commodities" and "engineers regard engineering work as a step to promote to managers or leaders rather than the final destination." The above characteristics make the ethical problems in engineering complex and indirect, which are easy to be covered. Through research, Martin and others found that in the life cycle of a product, from product design, production, manufacturing, finished product, and use to product scrapping, the whole process contains moral and ethical problems. Some scholars also classify engineering ethics issues from the perspective of related content, that is, ecological ethics issues, technical ethics issues, social ethics issues, and responsibility ethics issues related to engineering. There are differences between the "micro perspective" and "macro perspective" on what problems engineering ethics should solve and what are the outstanding engineering ethics problems.

From the micro perspective of "engineer's ethics," engineer's ethics is the basis and focus of research. It is based on helping engineers get rid of moral dilemmas. The ethical issues in engineering focus on three aspects. The first is the relationship between engineers and ethics. Specifically, many engineers should pay more attention to ethical issues, be aware of the complexity of ethical issues in engineering, and avoid falling into an ethical dilemma. The second is the responsibility of engineers, which shows that engineers not only have an obligation to inform the public of the possible adverse consequences of technology applications endangering public safety but also realize the autonomy of moral responsibility given by performing their duties. The third is the role conflict of engineers, which is manifested in how engineers who play complex and multiple roles can resolve the conflict and realize the balance of interests. The ethical issues mentioned above are closely related to professional ethics. As T. A. Long said, "the ethical problems that perplex reflective engineers today are such problems, such as truth-telling, confidentiality and the conflict between responsibilities to employers/customers and responsibilities to the public. These ethical problems have been common to doctors for hundreds of years." Therefore, "if we think that the purpose of engineering ethics is to clarify a set of ethical principles, which are necessary individually, combined they are enough to make engineering ethics different from business ethics, there is little hope to achieve such purpose because engineering ethics in this sense exists only when there are ethical aspects unique to engineering ethics…." [68]

However, the multiple roles of engineers make the problems related to engineering ethics more complex. On the one hand, doctors and lawyers take individuals as the main unit to engage in professional activities, which has a large space for independent professional judgment. Most engineers are employed by companies and have an indissoluble relationship with for-profit enterprises. As employees, engineers need to take actions aimed at the employer's interests, that is, to bear the "loyalty responsibility" to the employer. On the other hand, as the main body of engineering activities, engineers play the role of professionals. They also need to rely on professional knowledge and follow objective laws to create an environment conducive to human survival, assuming "independent professional responsibility" and "real social responsibility." The role of the engineer shows the dual utility of technology and economy. This working environment makes the technical and commercial value closely intertwined, and the engineering and management decisions intertwined. In addition, engineers are obviously in a weak position in the workplace, which often leads to the value conflict between being loyal to the employer and fulfilling social responsibility.

Engineering activities involve huge risks, including unknown and unpredictable consequences. Ulrich Beck pointed out that "although the risk is only a possibility, a confirmation of danger will mean irreparable self-destruction" [69]. The value attribute of artificial nature and the risk of the engineering project show that the project does not have inevitable goodness, and the engineering activities have incidental effects beyond the expected purpose. It can distinguish between two cases. In one case, the pros and cons of the project are borne by the same group, and the ethical issues involved are mainly how to balance the benefits and risks of technology. For example, constructing a nuclear power plant effectively solves the energy supply shortage, but it also lurks in great danger. Another situation is the social non-neutrality of engineering consequences; the technical consequences (including benefits and risks) are not equal to all people in the whole society.

No matter which choice engineers face, they can not avoid the objective fact that "the ethical problems hidden in engineering activities are more complex and changeable." In order to accurately analyze the problems of engineering ethics, we must carry out research from the macro perspective of "engineering ethics." "Engineering ethics" refers to dealing with ethical issues such as the role of engineers in the industry, the ethics of engineers' organizations, the ethics of professional engineering societies (societies), and professional, ethical responsibilities. It covers the professional ethics of engineers, the project's responsibility, the project's ethical norms, the coordination of interest relations in the project, the relationship between various elements of engineering activities, and so on. Other researchers further divided them into five categories according to their attributes [69]. One is the obvious violation of moral "goodness." The second is the ethical problem of non-black and non-white in the gray field. Third, there is no obvious violation of law and morality but "old rules and bad habits." Fourth, problems that do not apply to the law but are not easy to detect or inconvenient to implement according to law. Fifth is the conflict of interests—the dilemma of choice.

Scholars from Taiwan in China have done more specific research on this. Wang Huangsan sorted out the common engineering ethics problems in engineering specialty into two types [70]. One is the problems often encountered in the workplace but not unique to engineers, including 12 problems such as truthful statements, privacy, rebate, and private use of public property. The other is the problems often encountered or highly related to the engineering discipline, including 12 issues, such as competence, business confidentiality, document signing, identity conflict, etc. Feng Daowei summed up 27 professional ethics problems that civil engineers are most likely to face, dividing them into three categories: professional, work, and management. Jiang Zhengxian summed up a total of 39 common ethical and moral problems of construction projects. After being repeatedly checked by experts, a total of 16 main ethical issues were obtained, such as project safety problems, project quality problems, project pollution problems, underworld intervention problems, conflict problems of professional technicians' division of labor, pressure problems of interest groups, etc. Lin Tiexiong started with the analysis of the "9.21 earthquake", summing up, there are seven key issues of engineering ethics common in engineering circles in Taiwan [71]: (1) licensing phenomenon, (2) careless construction and inaccurate supervision, (3) Mafia involvement, (4) people's representative's talk about contracting projects, (5) Bai Dao (righteous outlaws) tie up the bid, (6) bid grabbing and (7) the state of mind of civil servants. Lin Tiexiong also pointed out that to ensure the exertion of engineering professional ability, we must have a healthy engineering ecological environment, that is, to build an engineering community composed of engineers and practitioners with professional engineering ability and professional engineering ethics.

Foreign scholars also pay attention to the identification and dialectical analysis of ethical issues. Their research includes the general ethical issues contained in engineering, that is, to investigate the ethical issues involved in engineering as a whole. It also covers the ethical issues in specific engineering fields. For example, Rosa B. Pinkus and others believe that the trade-off between cost, risk, and construction period of engineering projects produces practical ethical problems. In a chapter of *Engineering Ethics*, Martin and Roland Schinzinger indicate that "engineering is a kind of experiment with society as the object, and the risk is the internal attribute of engineering. As such, engineering is rich in profound ethical problems" [15]. Looking forward to the twenty-first century, Wolff believes that the problem of engineering ethics will become more and more important. With the rapid technological innovation and the increasingly complex engineering system, there are bound to be ethical problems that engineers have never encountered in the past. It is not only limited to the appropriateness of the engineer's personal behavior but also related to the appropriateness of the overall behavior of the engineering profession. Under globalization, engineering ethics presents new characteristics, and its research focuses on two aspects. First, engineers are exposed to the international background, that is, the ethical problems they will encounter when carrying out engineering work between countries involving different cultural traditions and levels of economic and technological development. Second, the project's impact crosses national boundaries, such as environmental pollution, military technology, etc. For example, Charles E. Harris emphasized that "perhaps the most important problem encountered by engineers in transnational work is bribery. Bribery should be distinguished from extorting, giving money to people to make connections, and gifts". It can be seen that engineering ethics is not only a key topic in the eyes of researchers at home and abroad but also one of the difficult topics.

10.3.3 Engineering Ethics Makes Engineers Face a Dilemma

The indifference to ethical consciousness is one of the root causes that engineers can not deal with the major problems related to social ethics, which leads to serious consequences. M. Augustine, an American scholar, observed that most engineers who are in trouble on ethical issues are not of bad character, but because they do not realize that they are facing ethical issues. As a result, "they have made bad decisions, tarnished their reputation and implicated themselves for the rest of their lives" [72]. Once engineers lack basic sensitivity to the existence of the above engineering ethics problems, they are easy to face a dilemma. As Martin said, "as in other places, ethical dilemmas will also appear in the engineering project because there are a variety of moral values, which may lead to conflicts."

(1) The conflict between the loyalty responsibility of "company employee" and the professional responsibility of "professional." The status of "company employee" enables engineers to "accept" and "recognize" the conditions and ethical principles of loyalty to the employed company and its employers when receiving the salary. Therefore, "loyalty" has become an important "moral principle" for engineers. On the other hand, as an independent profession, engineers master professional and technical knowledge and are experts in the engineering field. They know the direction and speed of technological development best and what benefits society. They must bear the professional responsibility of "seeking welfare for the public." There is an inevitable conflict between the employer's interest and the public interest. When the requirements put forward by the employer are contrary to the objective law or the

engineer's professional ethics, the engineer will be bullied by the other party if he chooses to blindly be loyal to the employer or assume professional responsibility from a professional perspective, violates the employer's instructions. The engineer will inevitably fall into a dilemma.

(2) The conflict between the independent decision-making of professional authority and the restriction of rights of subordinate status. As the main body of engineering activities, engineers are generally regarded as the authority in engineering decision-making due to their professional knowledge and practical experience [57], and they have the ability of independent decision-making in engineering activities. However, the technical force is concentrated in the upper part of the engineering community's pyramid, which determines that the "basic actor" engaged in specific engineering activities is the enterprise rather than the engineer [73]. As Mario Augusto Bunge put forward, "the goal of applied science and technology research and development is selected by business executives and politicians rather than scientists and engineering technology experts" [74]. Engineers do not have enough "voice" in the engineering organization and have no final decision on their research and development technical achievements. In this regard, Clive Staples Lewis sharply criticized: "people often say that technological development is the growth of human power to conquer nature. In fact, this power is not given to people in the general sense, but only uses nature as an intermediary to increase the power of some people relative to others." In this conflict between professional initiative and subordinate passivity, engineers' decision-making and behavior are difficult to reflect engineers' true wishes. However, public accountability focuses on engineers, making them fall into the power and responsibility imbalance dilemma.

(3) The value conflict of engineering specialty—tug-of-war of value tension between science and business. It is generally believed that professional social services are altruistic, and the pursuit of self-interest is regarded as contradictory to professional identity and social reputation [75]. There is no doubt that the professional behavior of engineers is altruistic in the professional background, but the inseparable connection with for-profit enterprises makes the services provided by engineers have self-interest at the same time. "cost" is usually regarded as the internal standard of "technical result suitability." Engineers need to consider the project's return on investment at any time. From a professional point of view, although economic factors provide a possible professional market for engineers' professional services, engineers' professional image is branded with the mark of "eager for quick success and instant benefit" and "greedy for self-interest," falling into value conflict. The tension between science and business is considered to be one of the most important driving forces to shape the professional role of engineers, but this double-sided role undoubtedly covers an indissoluble shadow of value conflict for the engineering specialty, which makes it subject to the dual accusations of professional ethics and professional moral.

(4) The inherent risk and multi-subjectivity of engineering make engineers fall into the dilemma of assuming expanded ethical responsibility. On the one hand, the engineering project is full of innovation and uncertainty, and its results are never completely positive or negative, but both positive and negative effects. Engineers

often can't predict the final results when they "construct according to the drawing." Moreover, in the process of using professional and technical knowledge to reduce risks, in case of engineering accidents caused by force majeure factors, people mostly attribute the fault to the engineer and force the engineer to bear "the responsibility beyond the ability of the engineer." On the other hand, the personnel involved in engineering activities are not limited to engineers. Enterprise executives, managers, and government regulatory departments are vital in engineering activities. Especially when the degree of specialization is getting finer, engineers can only play a minor role in the overall engineering activities. It is what Hans Jonas described: "what each of us does, compared with the overall behavior of the whole society, can be said to be zero, and no one can play an essential role in the change and development of things… 'I' will be replaced by 'we,' the whole and the high-level behavioral subject. Decision-making and behavior will 'become a matter of collective politics.'" It is the due meaning of each member's profession to bear the corresponding ethical responsibility, but the subject of responsibility cannot be limited to the group of engineers.

10.4 Methods to Solve Engineering Ethics Issues

10.4.1 The Research on Engineering Ethics Provides Methodological Guidance for Engineers to Solve Ethical Problems

There are two main research methods of engineering ethics: "professional ethics paradigm" and "engineering practice-oriented paradigm," which are generally accepted in academic circles. "Professional ethics paradigm" is a theoretical analysis based on the characteristics of engineering practice and the relevant concepts, norms, and principles in ethics, to formulate engineering ethics principles and norms to restrict the behavior of engineers. For example, in the book *Ethics in Engineering*, Martin proposed to use utilitarianism, right ethics and obligation ethics, virtue ethics, and other basic ethical theories to analyze and explore the common concepts of risk and safety, responsibility and right, honesty and deception in engineering, and point out their ethical connotation and value orientation [15]. The "engineering practice-oriented paradigm" is a case study method of typical real events, which is described and analyzed in combination with the relevant theories of engineering ethics to show the ethical problems inherent in engineering. Through real case studies, we can analyze the ethical problems with uncertainty and risk involved in engineering activities and help engineers improve their ability to identify, express and solve complex ethical problems. The book *Engineering Ethics: Concepts and Cases* by Charles E. Harris [52] is a model of this research paradigm.

Engineering ethics research includes both theoretical analysis and descriptive case studies. Theoretical-analytical studies generally give ethical norms directly and lack theoretical proof for the norms themselves, and thus have been criticized by some scholars. For example, L. Winner criticized the practice of case education around engineering ethics codes in the U.S. engineering ethics community as dwelling on the details and ignoring the big picture, like seeing the trees but not the forest [76]. On the other hand, the "descriptive case study" method is less general and applicable because of the specificity and concrete nature of typical cases, which causes methodological limitations. Despite their shortcomings, these two approaches are consistent with the Weberian "ideal model," which can focus on the opposing viewpoints within a certain range. They are not mutually exclusive but tend to merge, i.e., we can describe engineering ethics cases while conducting theoretical analysis, and theoretical analysis studies also use cases to verify the reliability and enhance persuasiveness. Which model is better depends mainly on its being more conducive to finding a combination within a certain range.

Many scholars have different opinions on how to study engineering ethics. In his book *Think Like an Engineer*, Michael Davis argues that both philosophical and natural science methods can be used in the study of engineering ethics, and there is no uniformity of methods. The research methods of engineering ethics should be a collection of methodologies, with at least five methods available: philosophical, inquisitive, technical, social, and professional [77]. Although he is a fan of professional approaches to research, he also believes that the five approaches are complementary and all contribute to engineering ethics and even engineering philosophy. Scholars also put forward that the research methods of engineering ethics can be divided into three categories: one is the "facing the fact" phenomenological method and the "linguistic analysis" method. The study should "face the fact itself and the life world," "face systems reality and social reality," and "face society and the subject characteristics," as three basic principles of theoretical principles and methodology. For example, Dr. Joe Amato from the State University of New York described the historical development of the engineering profession in the United States since 1944, studied the theory of engineering design from an ontological perspective, and explained technology. However, this narrative context description only exists in the historical investigation, which is too vague concerning realistic problems. The second is empirical research and theoretical research; that is, the two's positive interaction and mutual complementarity are the cornerstones of engineering ethics' "down-to-earth" advance and development. The third is the interdisciplinary research method, which deals with the relationship between engineering ethics and engineering philosophy, engineering economics, engineering sociology, engineering history, and other disciplines. Rosa B. Pinkus et al. did a case study on the decision, design, and manufacture of the main engine of the US Space Shuttle through an interdisciplinary analysis of its uncertainty and risk assessment. This study emphasized how engineers identify, express, and solve complex ethical problems and pointed out three basic principles: competence, responsibility, and Cicero's second creed, "protecting public safety," as an analytical framework to express and solve ethical problems generated in practice.

Other research methods are also availabe. For example, Robert McKee found and pointed out the value of various empirical methods for engineering ethics topics by investigating the engineering ethics problems submitted by engineering students and engineers at Stanford University in the past five years. Just as theoretical analysis can clarify the debate of specific case studies, the exploration and investigation of the views of engineering students and practitioners can also broaden the focus.

Different definitions of ethics affect the discussion and establishment of research methods of engineering ethics. Traditional western ethics believe ethics is a discipline that studies good and pays more attention to public behavior. Henry Sidgwick believes that ethics should be "science or research on justification or necessity. Of course, this science or research is based on individual voluntary behavior" [78]. Ethics can guide people's behavior in the direction of good and realize the transformation from "discipline others" to "self-discipline." Ethical theory is the crystallization of ethical research. It is a series of inferences that can be used to guide practice with clarity, systematicness, and some degree of abstraction and universality. For example, utilitarianism advocates the pursuit of "maximum happiness" and considers the maximum happiness of the largest majority of people. Virtue ethics support starting from personal self-cultivation, paying attention to a wide range of ethical problems, and infiltrating virtue into the process of solving ethical problems. Caring ethics, known for "flexibility" and "inclusiveness," is also the key to solving engineering ethics problems. The responsibility ethics characterized by "foresight, self-discipline, care, and integrity" provides a yardstick of value judgment for the behavior of engineers in the technical world. Situational ethics has become a new method of moral decision-making in the rapidly changing reality. The complex consequences of the engineering project make it of ethical significance. Combined with the following case, we interpret five main ethical theories in order to provide ideas for engineers to use different ethical theories to solve ethical problems from a differentiated perspective.

Cole Douglas is a salesman for a hardware company. He is preparing a talk for the bidding for the glass curtain wall connector ordering business of a commercial building in New Orleans. The commercial building is a high-rise building with point glass curtain walls on the outside, so the number of connectors required is huge, and the business profit is considerable. The defect rate of connectors produced by Cole's hardware company is 3.5%, which is enough to ensure the firmness and stability of curtain walls in general buildings. However, the special geographical location and natural climate of New Orleans put forward higher requirements for the quality of connectors. New Orleans is a port city in southern Louisiana, close to the Gulf of Mexico. Hurricanes in the Atlantic can easily land. The connectors produced by Cole's company can resist the attack of weaker hurricanes, but they are "tired" in the face of the devastation of strong hurricanes. According to the statistics of meteorological experts, strong hurricanes on the east coast of the United States make landfall on average once every ten years, but occasional bad weather may also cause strong hurricanes to come suddenly. For example, the attack of Hurricane Katrina in 2005 caught people "unprepared." Cole knows the problem, but he wants to win the business. Because if Cole wins the sale, he will receive another $30,000 based on his normal salary. However, if Cole tells the contractor the defect rate of connectors, his hardware company may lose the business to those competitors with higher reliability of connectors [79]. Therefore, the ethical choice facing Cole Douglas is whether to tell the engineering contractor: if a strong hurricane strikes, the connectors produced by his company may fail, and the glass curtain wall may fall into the busy pedestrian street under the commercial building. It may cause serious casualties.

(1) Utilitarianism—"happiness is the end"

Teleology means that an act can be considered morally correct or acceptable if it can achieve the expected results (e.g., pleasure, access to knowledge, career improvement, realization of self-interest or utility) [79]. Utilitarianism, proposed by Jeremy Bentham and John Stuart Mill is one of the forms of teleology. Utilitarians believe that "the behavior or moral rules followed by everyone should bring the greatest benefits (or happiness) to each relevant person" [80]. "Utilitarianism takes happiness as the standard to measure the right and wrong of behavior. The happiness mentioned here is not the happiness of the actor himself but the happiness of all stakeholders. Between his happiness and the happiness of others, utilitarianism requires him to be as strict and fair as an irrelevant and kind bystander." [81]. Utilitarianism takes into account all stakeholders of any moral behavior. However, in order to implement the spirit of utilitarianism and avoid the possible conflict between the "maximum majority" and "maximum efficacy" of the efficacy principle, contemporary utilitarianism usually removes the condition of "maximum majority" and only seeks to achieve maximum efficacy. In the long run, everyone may benefit directly or indirectly [82].

If Cole is a utilitarian, before making a decision, he will analyze the utility of different choices and adopt the scheme that can obtain the maximum utility. In this case, the possible effects include driving the local economy through commercial buildings, creating hundreds of jobs with the help of projects, increasing the income of hardware companies, etc. The cost is that the defect rate of connectors may increase the falling risk of glass curtain walls in strong hurricane weather (this risk is unknown). As a utilitarian, Cole may think that fixing the glass curtain wall with the company's connectors is more effective than notifying the curtain wall contractor that the connectors may fail in strong hurricane weather. Therefore, he will remain silent on the company's connector defect rate.

A related case is that in the 1970s, Ford Motor Company's Pinto was one of the best-selling subminiature cars in the United States. Unfortunately, when another car hits it from behind, its fuel tank is easy to explodes. More than 500 people died when the Pinto they bought caught fire, and many more suffered severe burns. One of the burn victims sued Ford Motor Company for the defective design. In fact, Ford engineers have long been aware of the dangers posed by this fuel tank. However, after a gain and loss analysis, the company's managers believe that the benefits of repairing this fuel tank (including saved lives and prevented injuries) are not worth $11 per vehicle—it is the cost of installing a setting that can make the fuel tank safer [83]. But to measure the full impact of a traffic death on utility, people must consider the loss of the victim's future happiness, not just the loss of Ford's income and the victim's funeral expenses.

(2) Virtue ethics—"what kind of person should I be?"

The discussion of ethics has always been described as the dispute between deontology and utilitarianism. Deontology emphasizes that the right or wrong of behavior depends on the behavior itself. Utilitarianism insists that the right or wrong of behavior depends on the consequences of behavior. Some scholars pointed out that paying too much attention to the dispute between the two just reflects that both of them ignore the importance of human virtue in ethical choice. Virtue ethics points out that the scope of ethics is not the sum of grand, dilemma, and marginal ethical issues, paying attention to ethical issues unilaterally. We must consider the continuity of practitioners' lives, emphasizing "being" rather than "doing" only. The virtue ethics originates from Aristotle's Nicomachean Ethics [84], which holds that "the purpose of life is happiness, and man's basic activity is rationality—a kind of good activity." Virtue ethics emphasizes the development of a "good man" or "virtuous man," not abstract rules, not the result of behavior or rules, unless these things come from good people or virtuous people or can promote people to become good people or virtuous people. The theory holds that the real core of moral behavior lies in the behavior subject to developing good habits or virtues. The development of virtue ethics has its own advantages, which are as follows: (1) virtue ethics create good people. It tries to teach the concept of virtue by encouraging people to do good. (2) Virtue ethics unifies rationality and emotion. It holds that virtue is not only the intention to act in a certain way but also the intention to think in a certain way. Its purpose is to use rationality to urge people to do moral things. (3) Virtue ethics emphasizes moderation and believes that "moderation of all things" is what people should strive to pursue. The revival of virtue ethics can eliminate the disadvantages of moral logicalization and virtualization so that the moral outlook of actors will not fall into the quagmire of egocentrism.

If Cole follows virtue ethics, he may consider virtue elements and tell potential customers about the defect rate of connectors and his own concern about curtain wall falling and casualties. He will not wantonly boast or deliberately hide risks when introducing products. Therefore, he may suggest that customers choose other products or companies to reduce the possibility of the glass curtain wall falling.

A relevant case is that Fritz Todt was a famous civil engineer in Germany. After the Nazis came to power in 1933, he served as the chief of the German highway. He completed the construction of a 3000 km highway within five years, presided over the construction of the "Atlantic barrier" and submarine base on the northern coast of France, built a huge Siegfried defense line opposite the Maginot defense line, and built a row of Oriental walls to confront the Oriental Stalin defense line. In March 1940, Todt was appointed the Minister of Armaments and Quartermaster of Nazi Germany. The "Todt Group" led by Todt was the engineering construction organization that Hitler relied on most. He was also considered to be the commander of the construction affairs of the Third Reich with the highest authorization. As an engineer with high professional quality, Todd presides over many bridges and service stations, which are known as "bold innovations of modernism." In terms of technical achievements alone, Todd undoubtedly enjoys a high reputation [85]. Unfortunately, this highly creative artificial nature eventually became a powerful boost to Nazi military action and provided war resources for Hitler's crazy acts. In this sense, Todd's professional activities cast a shadow of "evil" on the project and made himself known as a "devil accomplice" and "war promoter."

(3) Care ethics—"moving with emotion"

"Caring ethics" was first proposed by moral psychologist Carol Gilligan in 1982. In 1984, the feminist educational philosopher Nel Noddings gave the philosophical meaning of the word "care" and further put forward the Feminist Care Ethics and its practical proposition in moral education. On the theoretical construction level, Feminist Care Ethics reshapes a new ethical self by removing the order and hierarchical model established by men and rewrites an ethical discourse that considers women's voices and highlights women's care. In terms of moral value orientation, men prefer fairness and justice, while women prefer care and responsibility. In terms of moral thinking mode, men focus more on individual and logical thinking, while women refer more on relational and situational thinking. As far as the development model of moral judgment is concerned, men follow the development model of individualization, separation, and autonomy, while women follow the development model of self-preservation, self-sacrifice, and nonviolence. Male moral values are related to justice, power, competition, independence, and rule-abiding, while female morality involves generosity, harmony, obedience, and efforts to maintain close relations [86]. The "care" emphasized by Nottinges is not only the life agitation of meeting people in the existing situation but also the acceptance and continuous commitment in attitude, with the belief that only then can caring relationships blossom. The main purpose of Noddings's "care ethics" is to learn how to truly examine yourself and face others, not to suppress internal emotions to endure others due to external constraints, nor to be insensitive and indifferent to the surrounding characters, and to freely use what you have learned to care for others and yourself [87]. Engineering ethics applies it to a broader engineering context in order to realize "engineering for human benefit."

If Cole supports care ethics, he will focus on whether the project will hurt someone and whether it will affect the public welfare. Although Cole and the contractor will obtain the expected economic benefits, the falling glass curtain wall will endanger people's life and safety. As an authority with professional skills, Cole knows the specific situation better than the public and can prevent risks. If Cole fails to report the information, this deception is extremely unfair to the uninformed people. If Cole values the concept of justice and has a feeling of caring for others, it is inevitable to choose to inform the contractor.

A relevant case is the Aswan Dam, completed in the early 1970s. On the surface, the dam brought cheap electricity to the Egyptians, reduced floods and droughts, and irrigated farmland. However, the construction of the Aswan Dam has completely destroyed the ecological balance of the Nile Basin and caused a series of disasters. As the sediment and organic matter of the Nile River are deposited at the bottom of the reservoir, the oases on both sides of the Nile lose hundreds of millions of tons of silt, an important "fertilizer source," and the soil is becoming more and more saline. Due to the lack of sediment supply at the Nile estuary and the inland contraction of the estuarine delta plain, factories, ports, and national defense fortifications are in danger of sinking into the Mediterranean. The lack of salt and organic matter from the land reduced the annual catch of sardine by 18 thousand tons. Due to the barrier of the dam, the living water in the lower reaches of the Nile becomes a relatively static "lake," leading to the spread of schistosomiasis [88]. The lack of "environmental care" and "humanistic care" makes Aswan Dam bring great benefits and disastrous consequences.

(4) Responsibility ethics

As a "forward-looking, self-discipline, caring and holistic" ethics to deal with the negative effects of science and technology, responsibility ethics goes beyond the defect of traditional ethics that takes human beings as the center and ignores the inherent value of nature and life. It provides a value judgment scale for engineers' behavior in the technological world. Traditional ethics is represented by Aristotle's theory of virtue, Christian theory of conscience, Kant theory of obligation, Bentham utilitarianism, and John Rawls's theory of justice. It is a "close ethics" focusing on the moral norms between people. Its moral goal focuses on freedom and virtue, and its moral standard focuses on the internal code of conduct and external norms, which is essentially the care of human goodness and power. In short, the whole traditional ethics is a kind of anthropocentric ethics [89]. However, the unpredictable and destructive consequences caused by technology in the world are becoming increasingly obvious. The principle of moral responsibility in the era of technology urgently needs to break through the limitations of anthropocentrism. Based on the contemporary people's responsibility to the "existing" nature and "future" life, Jonas puts forward the "long-distance ethics," that is, the responsibility ethics, which takes "paying attention to the possibility of future human survival" as the absolute command. It means that we must establish a strong sense of "responsibility." Reasonable attribution of blame is necessary for engineers to bear ethical responsibility. Responsibility ethics expands the ethical relationship from "man and man" to "man and nature and society." It emphasizes the philosophical reflection on the results of scientific and technological progress, the ethical inquiry on the consequences of social development, and the anxiety and quest for the future trend of humankind. Its rise and development are an answer to the new requirements of ethics in today's reality [90].

If Cole agrees with the responsibility ethics, he will examine his own behavior from the perspective of the responsibility principle, putting the practice of professional responsibility in the first place, do his duty to inform the curtain wall contractor about the defect rate of connectors, and explain in detail that the strong hurricane may cause the curtain wall to fall and cause potential casualties.

A relative case is, at 5:52 p.m. on June 29, 1995, more than half of the Sanfeng department store, once a landmark in Seoul, collapsed almost instantaneously. Within 20 s, the 5-story department store collapsed to the 4-story underground, resulting in 502 deaths and 937 injuries. Property losses reached 270 billion Korean Won (about US $216 million). The subsequent cause investigation finally focused on the arbitrary intervention and arbitrariness of the president of the Sanfeng group in the construction project. In order to pursue the increase of structural function, he changed the design drawings without authorization and increased the structural load without permission, which eventually led to the tragedy. The designer, construction manager, and supervision engineer lost their "responsibility" in the accident and did not assume the "bottom line responsibility" that engineers cannot escape. In this case, the design engineers acted non-professionally. The original contractor refused to construct according to the new design. As a result, the president dismissed them and delegated the task to his subordinate construction company. Plus, the dereliction of duty of the design engineer. All of these make the project's hidden danger not be found as soon as possible and removed in time, ultimately bringing shame to the Korean people [91].

(5) Situational ethics—"circumstances determine the truth"

Joseph Fletcher put forward the idea of situational ethics in the 1960s. Situational ethics is also known as environmentalism, contingency theory, contextualism, and even realism. These names point out the core idea of situational ethics: at any time, the actor himself should make behavior choices independently according to the situation faced by the actor—the specific environmental factors and background at that time [92]. The real situation is changing rapidly. If we analyze real-life with rigid ethical rules and dogmas, we will inevitably fall into the dilemma of disconnection between theory and practice. Therefore, Fletcher believes relativism is the most typical cultural feature of the scientific era and contemporary humankind. The ethics under this era's background and cultural heritage also need to adapt to the current situation, get rid of the descriptions of "perfect," "absolute," and "never," and investigate ethics from real and accurate practice. The relativity of situational ethics provides it with a sharp tool to understand difficulties. However, its absoluteness is also controversial. There is only one true morality for situational ethics—"love." Only the commandment of 'love' is absolutely good [92], and "love is the only norm." Situational ethics gives "love" a solid and unshakable position. In view of this, it is considered to have an "absolutist" tendency. L. J. Binkley pointed out: "regardless of Fletcher's initial explanation of his position, there are no principles or rules in his ethics except that we should start from love and try our best to do the greatest amount of good." Alexander Miller simplified it as follows: situational ethics has an absolute component and a computational component. Investigating the absolute "love" and the "calculation" of the real situation can provide effective guidance for practitioners.

If Cole agrees with situational ethics, he will adhere to the highest principle of "love," analyze the ethical problems in different situations and the impact of different ethical choices on various stakeholders and make corresponding choices according to his moral concepts. "Don't do to others what you don't want." Cole will put himself in the position of victims who may encounter the risk of glass falling in the future, use empathy to deal with product quality problems properly, and frankly inform the contractor of potential dangers.

Network morality is typical context ethics. Because it is relatively difficult to restrict and supervise in the network environment, it is unrealistic to require those with low moral standards to "be cautious and independent," reflecting the anomie of network morality. There are angry youths and hackers who break into the restricted area. Online communities' virtual and hidden characteristics make netizens less consider the bad consequences of bad behavior in virtual space in the real world. Human flesh search and irresponsible abuse reflect the essence of some people who lack moral self-discipline and a sense of social responsibility. These behaviors will even become cyber crimes, such as online fraud, spreading computer viruses, etc. From the characteristics of network ethics, on the one hand, as a professional ethics and place situation ethics closely related to information network technology, it reflects the specific requirements of this high-tech on people's moral quality and literacy. On the other hand, as a new type of moral consciousness and code of conduct, it is restricted by a certain economic and political system and cultural tradition and has a certain nationality and particularity. Some real-life moral principles may not apply to the network world. Setting appropriate ethical standards according to the characteristics of the network world is a practice from the perspective of situational ethics [93].

10.4.2 Solutions to Engineering Ethics Problems

The way to solve engineering ethics problems is not unique and established. Engineers should achieve ethical self-discipline, improve the sensitivity of actively discovering ethical problems and insight into in-depth understanding of ethical problems, and make full use of discipline others. It is required to flexibly choose countermeasures according to ethical principles and ethical norms and enhance the judgment of choosing appropriate ways and the execution of solving problems. Self-discipline emphasizes that engineers consciously abide by professional ethics, independently carry out self-discipline and spontaneously take necessary corrective measures. Self-discipline is embodied as follows: engineers should not only be able to internalize ethical norms and form value orientation and moral pursuit. They are also required to have good ethical imagination, broaden ethical vision and surpass the limitations of technical thinking. Disciplining others emphasizes using mandatory formal rules to restrict the arbitrary behavior of engineers. Disciplining others is embodied as follows: engineers should not only reasonably follow and apply ethical norms, that is, pay attention to corporate organizational culture, standardize project management content and discuss ethical norms, but also make good use of consensus and consultation to reach consensus with stakeholders, including goal consensus, procedure consensus, and moral consensus. Consultation is the means to reach consensus, and the consensus is the basis for entering into consultation. The ultimate goal of consensus and consultation is to achieve harmonious coexistence among stakeholders [94]. Based on negotiation ethics, engineering ethics dialogue includes three levels: occupation, public opinion, and system. The professional level dialogue is conducted among engineering and technical personnel and is committed to ensuring the fairness of benefit distribution in specific engineering projects. The dialogue at the level of public opinion is conducted between engineering technicians, the public, and the media. It is committed to real-time supervision of engineering practice at the level of public social opinion. The dialogue at the institutional level is carried out between engineers, technicians, and managers. It is committed to providing institutional guarantees for the effective realization of public interests through institutionalized ways. At present, the research on the measures and strategies to solve engineering ethics problems is relatively scattered, and its effect is not very obvious. Realizing the perfect combination of self-discipline and discipline other is the most effective way for engineers to solve ethical problems.

The earliest method to solve engineering ethics problems originated from the idea of adjusting engineering methods to ethical methods put forward by Mantell, an American Engineering ethicist, in 1960. He believes that ethical problems are almost as complex as engineering problems and are solved by applying principles. The principle of engineering application is the scientific and technological theory, and the principle of the ethical application is ethical theory. The engineering method has achieved great success in solving human material problems. If this method is slightly adjusted to deal with ethical problems, it will also make great progress in solving the problems in human spiritual life. Inspired by Mantel's thought, participants proposed a decision-making model to deal with ethical issues at the 1990 Chicago Conference on Engineering Ethics Education (CEEE). The decision-making process has seven steps: (1) Identify and define ethical issues and be ready to make amendments at any time; (2) Investigate and verify facts; (3) Form alternative solutions and continue to verify facts; (4) Analyze these alternative solutions according to the resources required by alternative solutions and their possible results; (5) Build an ideal choice and persuade or negotiate with others for implementation; (6) Predict the defects or unsatisfactory results of the ideal choice, and take measures to prevent them; (7) Take action (go back to the first and second steps to see if the ethical problem has been solved and check whether any facts have been omitted).

In addition to the seven-step method, foreign engineering ethics has studied and developed a series of other methods for engineering ethics decision-making, such as line drawing and the creative middle way proposed by Harris et al., as well as the priority principle for solving the conflict between various responsibilities proposed by Lunk.

When facing ethical considerations, engineers need to follow three steps. One is to clearly identify the ethical meaning of engineering and analyze the problems of engineering ethics, which is the premise and basis of engineering ethics research. The second is to specifically judge the ethical nature of the project (whether it is good or bad, and how the positive and negative values are) and take corresponding solutions. The third is to test whether the decision is reasonable. Next, we will discuss the three-step solution procedure for ethical problems.

(1) Judging whether a problem is an ethical problem

An important shift in the development of contemporary ethics is the move from meta-ethics to applied ethics. The term "application" in applied ethics refers to the negotiation of universal principles with the particular situation, facts and values, and ends and means, and then reaching a consensus on emerging ethical issues (e.g., cloning technology). Engineering ethics research seeks to improve the ethical environment of the engineering profession with the ultimate goal of helping engineers solve the ethical problems they face. Ethical dilemmas in engineering practice cannot be solved by simply transposing principles; practice ethics begins with the problem, and its primary purpose is to solve it. The focus of practice ethics is on problem situations and the empirical ethics of practitioners and those impacted by stakeholders, emphasizing action-oriented and embedded in "action."

Harold Gortner raised questions to guide decision-makers and provide a reference framework for ethical decision-making [95]. Extending these problems to the engineering field will help engineers identify ethical problems in engineering. The specific contents are as follows. Does this issue produce a significant value conflict? Can the value in the problem be recognized? Or can they be found through other efforts? Analyze and reasonably rank these competing values. The following issues must be examined if we want to achieve perfect ethics. Decrees—what laws can tell us what to do? Philosophical and cultural context—can we find a reasonable response by clarifying philosophical or cultural elements? Profession and professionalism—does professional training help to find solutions? Organizational dynamics—is the relationship with the organization part of the problem? Personal level—what role does "I" play in dealing with ethical dilemmas?

Some scholars also apply the value analysis method to identify engineering ethics problems. First, we should try our best to organize ethical problems and turn them into conflicts of ethical values. Then, the value analysis method is applied to solve the contradiction. In the concentric ring model of value analysis, the core value is arranged in the center, the peripheral value is at the edge, and the authority value is between the two. This value arrangement is the result of consensus. Each culture has consensus core values, and the differences between different cultures are mainly reflected in the core values.

The core issue at this stage is to promote the integration of "professional perspective" and "social perspective." On the one hand, it points out the limitations of a simple "professional perspective" to engineering technicians, finds the problems, and analyzes the specific problems using engineering ethics principles. On the other hand, it provides a "social perspective" to observe problems, so that engineers and technicians can understand the possible consequences and social impact of engineering ethics problems and the methods to solve them. Understanding through interpretation requires the joint efforts of engineers and ethicists.

(2) A comprehensive analysis of the dilemma from the perspective of ethics

Engineers often fall into engineering ethics dilemmas in engineering practice, resulting in many ethical puzzles. How do we balance the cost and benefit of engineering activities? If you choose to carry out the employer's orders, how can the interests of the public be protected? What should engineers do if they expose unethical behavior, face the risk of dismissal and retaliation, and not expose it against professional ethics? According to Lawrence Kohlberg's individual moral development theory, people are at different moral levels, and their moral psychology and behavior are different. People are in different moral situations, and the internal and external conditions they can rely on are also different. To help engineers out of the ethical dilemma, we can consider problems according to the moral level and moral situation of engineers: (1) When engineers face the basic ethical dilemma, they can get rid of the dilemma under the guidance of ethical principles and by using ethical norms. (2) When engineers are faced with areas beyond the consideration of ethical norms, they can adopt the way of ethical negotiation and consensus, which is in line with most people's expectations. (3) When external conditions do not constrain engineers, they need to rely on their own moral self-discipline to regulate personal behavior and play the role of ethical autonomy. These three ways are reflected in combination with the ethical behavior of engineers. An engineer will encounter a variety of ethical puzzles and need to take corresponding ethical treatment methods instead of relying on one model.

Dealing with ethical problems is not rigid but requires creative thinking. Sisera Bock, a philosophy professor at Harvard University, put forward the "Bock Model" for ethical choice, which is divided into three parts. The first step is to ask your conscience, the second is to seek expert advice, and the third is to have an open discussion. The biggest advantage of this model is that when facing ethical difficulties, we should not only examine our hearts but also strive to find flexible methods and make the right moral choice after asking others' opinions. However, because the steps are based on experience, the results often lack a theoretical basis, and the reliability is relatively low. Harvard theologian Ralph Porter proposed the "Porter's Square" for ethical choice. This model proposes to solve the ethical dilemma through four steps to make choices and judgments. The four steps are (1) Understand the facts; (2) Outline the internal values of the decision; (3) Apply relevant philosophical principles; (4) Clearly show a kind of loyalty [96]. Unlike the Bock model, the biggest advantage of this judgment method is not to make choices and decisions blindly but to compare internal values based on understanding the truth, apply appropriate moral principles, and finally judge who they are loyal to from different positions. In Porter's model, actors often cannot choose between values, ethical principles, and loyalty. Therefore, it is often difficult to make moral choices quickly in the face of imminent moral problems.

Mike W. Martin integrated the steps to solve the ethical dilemma. (1) Moral clarity: identify relevant moral values. The most basic step to solving ethical dilemmas is to be aware of them. Only when the moral concept is clear can we better identify the moral problems in engineering, which is also the pertinence of the research on engineering ethical decision-making. (2) Clear concept: clarify key concepts. (3) Know the facts: get relevant information. The main difficulty in solving a moral dilemma is often the uncertainty of facts rather than the value of conflict itself. (4) Understand options: consider all options. At first, the ethical dilemma seems to force engineers into the dilemma of "black or white," but after careful observation, we can often find other options. For example, engineers can resolve conflicts of interest in the following ways. (a) Reject—reject gifts and refuse to disclose confidential information of the enterprise. (b) Give up—give up illegal gains and ethical decision-making power. (c) Resignation—resign as a project manager and leave the enterprise. (d) Do not participate—do not participate in evaluating contractors with a potential relationship with themselves. (e) Disclosure—disclose possible hazards to the public, report corruption and bribery of the project to relevant government departments, and report possible illegal interest exchange in the project to superior leaders. Which method is more appropriate for engineers? It requires engineers to weigh the advantages and disadvantages of the scheme comprehensively. (5) Good reason: make reasonable decisions. If there is no ideal solution, at least one satisfactory solution needs to be sought. Herbert Alexander Simon calls it "satisfying," rather than always looking for the optimal solution.

Arthur Anderson's "situational judgment process" points out seven steps in ethical situational analysis and judgment, as shown below [97]. (1) What are the facts? First collect relevant data, that is, important factors affecting the situation. (2) What are the moral problems? (a) In this case, what problems will involve ethics, not just pure technical problems or other non-ethical problems? (b) Moral problems can also be divided into "personal," "corporate," and "social system." (3) What are the solutions? Using brainstorming method to various problems. (4) Who are the main stakeholders? (5) Evaluate the ethics of each scheme. (a) The moral quality of each programme and the people involved must be assessed. (b) Consider all stakeholders' basic rights and their priorities. (c) consider each scheme's justice side, its fairness and impartiality, and its impact on social justice. In other words, it must be most beneficial to most people. (6) what are the practical limitations? (a) for the most ethical scheme, find out the limitations, difficulties, and risks in practical ability when practicing this scheme. (b) What factors (such as individuals, companies, and society) may restrict the implementation of this best plan? Can these obstacles be overcome? (7) what final decisions should be made? After the decision, specific pre-implementation and post-implementation steps and temporary methods to deal with accidents must be designed. Steps do not mean a one-way, independent movement but an activity carried out jointly and in a repeated pattern. The preliminary discussion of the corresponding moral value and related facts may be followed by the clarity of the concept and the collection of additional information. In turn, they arouse a detailed understanding of the application, value, and related facts and finally make the reasonable answer to the ethical dilemma clear, informed, and fully justified.

In addition, Marcia Hill, Kristin Glaser, and Judy harden [98] put forward an ethical analysis model of feminist views. This model takes into account both rational evaluation and perceptual intuition. It starts by identifying and defining problems, then develops and selects solutions, and provides further reflection through the examination and assessment, which can be used as the starting point for the next problem-solving. Terry L. Cooper thinks about solving ethical dilemmas in seven steps: detection, description, identifying, confirmation, planning, selection, and solution [99]. David B. Resnik explores ethics by structuring problems, collecting information, exploring different opinions, evaluating different opinions, decision-making, and action. The above scholars try to explore the methods to solve the ethical dilemma from the dynamic process perspective, but the significant differences between them are not prominent.

(3) Check whether the decision is reasonable

Carol Lewis considers ethical reasoning and decision as a form of problem-solving. Therefore, a decision checklist was developed (Decision Making Checklist) to promote more responsible and satisfactory solutions, including (1) facts: whether to collect relevant facts and laws? (2) empath: whether to examine your own decisions from the perspective of others? (3) underlying causes: whether to review the impact of decision-makers, others, and external events on ethical issues? (4) stakeholder and responsible: whether to consider the impact of different stakeholders on the problem? (5) Motives and objectives: are the motives and objectives of the decision-makers reviewed? (6) Possible results: what is the possible result of the problem? Who will be affected? (7) Potential harm: who will be hurt by the result of the decision? (8) Participation: consider who will participate in the decision-making and in what form? (9) Long-term frame and anticipated change? (10) Disclosure and publicity: can the decision-making process be disclosed and made public? (11) Appearance and communication: how does the decision's result affect the organization's future? How to face the public? (12) Universality and consistency: decision-makers must measure whether they are consistent with previous policies.

Solving the ethical dilemma is a process similar to engineering creation, which also requires engineers to give full play to their wisdom. Ethics can provide a straightforward answer to a dilemma, but it is not always the case. (1) Norms cannot replace good moral judgment—honest, fair, and responsible moral judgment. (2) Interpreting ethical norms requires good judgment [101]. (3) The formation of good moral judgment is an important part of the formation of engineering experience. (4) Good moral judgment is the main purpose of learning engineering ethics. The rule is like a cookbook. Whether you can make delicious food depends more on the cook's wisdom and understanding.

In addition, we should pay attention to the order of ethical principles to find an appropriate ethical scheme. For example, a common phenomenon in engineering is that "management decision" is more important than "engineering decision"—employers or senior managers suppress and oppose engineers' professional opinions. Joseph Raelin said: "because of the differences in educational background, social environment, values, professional interests, working habits and opinions, there is a natural conflict between management and professional level." This "confrontational relationship" will even lead to employers or managers forcing engineers to do immoral things for the benefit of the company. If engineers dare to report, they will be regarded by enterprises as "traitors" or "heretics" who dig the corner of the organization. In this regard, engineers must stand firm and be loyal to professional judgment and should not make decisions that violate ethical principles at will. Engineers should be aware that they must first be loyal to their profession because the profession is not only a means of livelihood but also bears the special expectations of the public. Since the status of professional is the most important among the multiple identities of engineers, the ethical standard has naturally become the highest standard on which engineers make decisions.

From "beginning" to "end" refers to analyzing the problems to be studied, using Anderson's "dynamic analysis process" or Hill's "feminist analysis perspective" to study how to get rid of ethical dilemmas, and finally using the decision checklist to check whether the selected solution is reasonable. This process can guide in solving ethical problems and help engineering practitioners eliminate ethical dilemmas.

10.5 The Future of Engineering Ethics—General Engineering Ethics in the Perspective of the Engineering Community

10.5.1 The Research of Engineering Ethics Should Adapt to the Trend of the Engineering Community

Ethics research has been moving from narrow to broad in the US and other western countries. Engineers are required to have not only individual responsibility ethics to their employers but also collective ethics among engineers, such as professional justice, information disclosure, intergenerational equity, and social justice [27]. Engineering subject is evolving from a single engineer subject to multiple network subjects. Among multiple subjects, understanding and consensus are achieved through dialogue and negotiation. Altruism and symbiosis are becoming the core values for each member to consider in organizational strategies. The concept of corporate citizenship is gradually becoming popular. These changes all point to one important idea—the engineering community. The engineering community presents more complex organizational characteristics than the traditional single organization. The speed of technological progress and social change puts each member of the engineering community in a more uncertain organizational form and cultural shock, which forces the study of engineering ethics to adapt to it.

First, the complexity and systemic nature of engineering and the comprehensive integration of multiple elements resort to group work, which determines that engineering is not an "individual activity" but a "collective activity." The huge size of the project requires more people to be involved in engineering activities, and the large size of the project organization makes the power controlled by each engineer diluted accordingly; while the power is reduced, the responsibility undertaken by engineers also shrinks sharply, and the phenomenon of engineers using "limited power" as an excuse for not fulfilling their ethical obligations is repeatedly staged. The limits of engineers' ethical responsibilities are divided into direct responsibilities, joint responsibilities, witness responsibilities, and future responsibilities, which are based on causality, resolution, and influence, and are flexibly adjusted according to the role of the three variables [102]. Engineering ethics research can no longer be oriented only to individual engineers but must take the engineering community as a group subject as a new research object.

Secondly, the inherent limitations of engineers themselves determine that engineering ethics research needs to break through the limitations of individual perspectives. Because of their "university education background," engineers may encounter the obstacles to personal development such as "what they learn in university does not match with the current job" and the "inconsistency between knowledge and action." The pressure of survival and adverse social trends easily cause moral collapse. They are receiving blank stares, contempt, and even retaliation from engineering interest groups for adhering to principles and not complicit in the process. These factors shake the foundation of engineers' determination to become "conscientious and diligent gatekeepers." "Knowing," "feeling," "intention," and "action" permeate each other and complement each other, forming an upper spiral system by "knowing, creating a new situation, refining the mind, and practicing." The experience of moral practice is internalized to improve the engineering ethics of "future engineers" [103]. Only by integrating the personal honor and disgrace of engineers into the collaborative development of the engineering community can we find a place for the moral conscience of engineers while ensuring the success of engineering.

Moreover, the complexity of ethical issues in engineering makes it more difficult to solve the problem. Traditionally, the solution to ethical issues in engineering has been attributed to the "dilemma of the individual engineer," but such mechanical thinking does not contribute to the real solution. Complexity, as the understanding of Edmund G. Seebauer and Robert L. Barry, lies in the mixed effects of difficult trade-offs, irreconcilable employment relationships, rising protests from environmental groups, and unpredictable technological influence [104]. Only from the perspective of multiple subjects is it possible to find a solution.

Finally, the territorial character of engineering makes engineering practice socially constructed, as reflected in the social values that permeate engineering design and final products through engineering style, social determination of engineering goals, and optimization of engineering solutions [59]. Examining ethical issues from the perspective of engineering sociology can help promote the development of universal engineering ethics. As Sanhu Li proposed, "For the relationship among science-technology-engineering, technology and science should be viewed from the engineering perspective. At the same time, it is necessary to introduce a social constructivist perspective and start a sociological study on engineering". Engineering should take society as the experimental object, the study of the engineering community has become a trend in the development of engineering ethics [105].

From Joseph R. Herkert's perspective, the debate between the narrow and broad views presents challenges and opportunities for engineering ethics. Herkert tries to focus on the study of the broad perspective based on the narrow perspective and, through the integration of the two perspectives, to obtain a more comprehensive vision to find effective solutions to ethical dilemmas. A deeper investigation into the aspects of engineering ethics from the narrow to the broad perspective in the perspective of the engineering community not only meets the higher demand of "engineering as a social experiment" but also provides engineers with moral sensitivity as an important "software" resource. It can also accelerate the promotion and popularization of engineering ethics awareness in China.

10.5.2 The Formation and Development of the Concept of the Engineering Community

Although engineering activities have been in existence for a long time, engineering communities were formed gradually in modern society. Most of the engineering activities before the first industrial revolution were "temporary" social activities, in which engineering projects were carried out by temporarily conscripting peasants and craftsmen. After the project was completed, the participating farmers and craftsmen had to "return" to their original land or workshops to continue their production activities. At best, there was a "temporary" "engineering community" during this period. Regarding labor division and class stratification, the basic identities of conceivers, commanders, and operators of engineering activities remained peasants, craftsmen, or officials, and no identity differentiation occurred, i.e., only "peasant community" and "official community" existed in a non-strict sense.

With the economic development, production expansion, scientific and technological progress, and clear division of labor in modern society, the artisans in ancient society gradually differentiated and formed different social classes such as workers, engineers, investors, and managers. In the first industrial revolution, as the main production activities of the society, handicraft workshops were replaced by the factory, and machines replaced manual labor. Scientific research results were gradually applied to industrial production, and engineers entered the stage of history as a specialized profession. In the second industrial revolution, advanced industrialized countries built many civil works, and the factory system developed rapidly. Engineers appeared with the continuous development of industrialization and modernization and gradually differentiated from craftsmen or workers.

The engineering community studied in this book focuses on the "engineering activity community," which refers to the virtual organization with various levels, diverse roles, clear division of labor, and multiple interests formed under specific engineering activities to achieve the same engineering goal. It is a "heterogeneous community" composed of investors, engineers, workers, managers, project community residents, and other interest groups. In this regard, Bocong Li made an image analogy: "If engineering activities are compared to a tank or a shovel, the investor can be compared to the fuel tank and fuel, the manager (entrepreneur) to the steering wheel, the engineer to the engine, and the worker to the gun or bucket, each of which is indispensable to the function of the whole machine" [106]. The above analogy is limited to participants who are more relevant to engineering activities, without mentioning other interest groups affected by and/or able to influence engineering, and has not yet fully interpreted the essence of engineering operation from the perspectives of the engineering community. On the one hand, ignoring the interests of any member will definitely lead to unpredictable risks to the project; on the other hand, only by giving full play to the synergy effect of all members can we ensure the project's success. Grasping the increasingly significant giant, social and collective nature of modern engineering can help implement the engineering community's concept.

10.5.3 The Characteristics of Engineering Communities and the New Challenges to the Development of Engineering Ethics

Compared with a traditional single organization, engineering communities present more complex organizational characteristics. The increasing speed of technological progress and social changes put each member of the engineering community in a more uncertain organizational form and cultural impact, all of which force the study of engineering ethics to adapt to them.

(1) In terms of organizational nature, engineering communities belong to social subculture groups, which have the characteristics of "organizational virtualization" such as "temporary project, mobility of members and resource reorganization." The members are wary of each other when sharing core technologies. It is possible to produce a confidence crisis. It is difficult for members to develop a strong organizational identity and a sense of belonging. The engineers intentionally attribute the non-fulfillment of ethical responsibilities to the lack of clear delineation of ethical responsibilities and are willing to position their roles as "instrumentalists."

(2) In terms of motivation mechanism, the motivation of engineering communities to engage in activities originates from the dual needs of survival and social life. On the one hand, the trend of the engineering community as the main body of engineering activities is irreversible, and there is a "fit" between the personal pursuit of engineers (including the needs for achievement, affinity, autonomy, power, etc.) and the vision of engineering community. On the other hand, the engineering community needs to survive better through the project and build a good image of a "conscientious corporate citizen" through the implementation of the project.

(3) In terms of structural stratification, the engineering community conforms to the obvious characteristics of a "flat" organization, such as simplified management level, lean organizational structure, the full delegation of power, unimpeded information sharing, and frequent horizontal contact. However, the change of management mode and organizational function makes members encounter new management obstacles when they strive to become "ethical subjects."

(4) In terms of subject composition, the engineering community is composed of multiple subjects with multiple values, which will lead to inevitable value conflicts if not handled properly: members of the community focus too much on local interests and neglect overall development, and engineering construction result in "tragedy of the commons"; lack of trust in teamwork; inefficient operation; escalation of friction and conflicts intensified. The establishment of mutual trust and mutual benefit mechanism lies in reaching a consensus on the value of the overall situation and the long term; respecting differences, tolerating diversity and treating equally; mutual benefit, reciprocity, and win–win situation; and moderate adjustment of self-worth [107]. In such cases, negotiation and gaming can provide new directions for ethics research.

(5) On the path of recognition, the engineering community seeks internal recognition and external affirmation of the engineering community, but there are inevitably

contradictions in the role-playing and responsibility-taking resort to internal identity and external affirmation. Each member's pursuit of external affirmation is reflected in the "guardian responsibility" to society, which needs to take public welfare and sustainable development of society as the basic guide. It will conflict with the assumption of professional responsibility, causing role ambiguity or even role loss, resulting in the lack of responsibility.

(6) In terms of institutional goals, the engineering community aims to win markets, seek social fulfillment, and transform what exists in the world into something usable for human beings. It consists of three levels of objectives: first, the community must not harm nature in a way that threatens the survival of human beings while achieving its profitability goals; second, the community must make artificial nature meet the spiritual needs of human beings and realize "poetic habitat"; third, the community must embody "people-oriented," care about the "existence" of those who exist, and realize "building for the majority."

10.5.4 The Obstacle to Engineering Community Operation Reveals the Lack of Ethics in China's Engineering Construction

The rapid development of China's economy is closely related to engineering construction. It not only brings new engineering ethical problems to the members of the engineering community but also puts the traditional moral concept to an unprecedented test of how to deal with "people no longer behave traditionally." The established moral system has been losing ground under the impact of the economic wave. The wail of "the times are changing" did not only fail to impact the social morale positively but led to the prevalence of competitive survival philosophy such as "predatory ideology." The following national conditions are all inescapable obstacles to the healthy functioning of the engineering community.

(1) Human relations in Chinese society follow the "ripple model." Xiaotong Fei pointed out that the interpersonal communication pattern in China is a "self-centered" and "differential pattern" from near to far [108], in which relationships, human emotions, and face are the three key factors. A relationship "implies a friendship of continuous interest exchange" [109]; human emotion is the result of an exchange, acquired and strengthened in the process of reciprocating about repaying and owing. The face is the power acquired in the association of relationships and can result from the occurrence of no exchange [110]. Through the operation of relationship, favor, and face, Chinese people obtain invaluable social resources, non-institutional social support and protection, and daily authority that overpower people at the cost of rules, rationality, and system [110]. This tradition permeates every corner of society, and the engineering community cannot be immune from it. In dealing with interest disputes, the role of morality and law is flexible and flexible according to the relationship between intimacy and distance, and its harm is reflected in three aspects: first, the

division of interpersonal relations is not conducive to the integration of resources; secondly, it causes interpersonal conflicts and contradictions, which is not conducive to the internal stability of the community; finally, it is extremely harmful to the survival of vulnerable groups and is not conducive to social harmony. Relationships, favor, and face, as a strategy for people to fight for their own interests, contain the behavior of "treating people as a means rather than an end," which is a great challenge to ethical principles.

(2) Traditional Chinese culture has been quite derogatory to "engineering." Feudal era idea of "political supremacy" caused the deep-rooted thinking of officialdom [111], which formed the social culture of emphasizing officials and farming, less emphasizing industry and commerce, emphasizing Taoism, and less emphasizing technology. The social status of artisans (engineers) was relatively low. In modern society, influenced by the concept that "engineering is only an appendage of science and technology," the status of engineering as an independent discipline is often questioned. The contribution made by engineers is often attributed to scientists, and engineering construction is hardly regarded as a practical activity with independent innovation and unique creativity. These tendencies to belittle engineering activities and engineers reflect that the recognition of engineering activities in society has yet to be improved. It caused low motivation of each member within the community, lack of fulfillment in their work, weak sense of individual responsibility, and insufficient motivation for engineering innovation. All of these have greatly weakened the intrinsic power of engineering creation.

(3) Social transformation affects the transformation of people's beliefs and value systems, increasing the difficulty of benign community development. According to Weber, the rise of rationalism and its comprehensive development in the modern West made people's material life extremely rich, and people pursued wealth growth with more vigor, but their spiritual world became more and more empty [112]. The resulting materialism and hedonism led to the spread of "egoism" in all values, and "ego" became the spiritual fulcrum of human beings. It further leads to value emptiness and lack of beliefs. It causes role disorientation. Members lose the clear positioning of their roles driven by interests, making it difficult to build independent standards for judging right and wrong, good and evil, beauty and ugliness, and to form a true community of interests.

(4) The continued effectiveness of the new economic reform requires improving ethical system construction. Along with the great economic achievements, moral deficiency has also shown a large increase, especially in the engineering field, which is regarded by the general public as a high corruption area. Suppose corruption and non-ethical behaviors are not curtailed as soon as possible, and the social dissatisfaction caused by "sub-moral" or even "immoral" behaviors is stopped. In that case, it will cause the following: Not only the "virtual organization" of the engineering community, which should not be virtual, will exist in name only, but even the success of engineering activities will become "moon in water, flower in the mirror." For example, the local government vigorously promotes "land finance" and forces farmers to sell their land and homes at low prices to become "displaced people." In the process of land expropriation, means such as bullying the weak, illegal forced

demolition, abuse of tyranny, and employing gangs to commit violence is adopted [113], thus triggering a large number of mass incidents against land expropriation and demolition violently. Such unreasonable planning and utilization of land resources not only seriously violate the basic requirements of "environmental ethics" but also ignores the people-oriented "care ethics," not to mention the sustainable development of respecting, following, and complying with "intergenerational ethics"?

For another example, the "second generation" of migrant workers (originally farmers) put forward the appeal of "citizenization of migrant workers". At the same time, the city took the way of "economic acceptance and social exclusion" to respond. In the sociological sense, the "second generation" of migrant workers is a "third pattern" outside the unique urban–rural dualistic pattern of China with no hometown and no future and a real "amphibious man" between urban and rural areas [114]. As an indispensable and important part of the engineering community, migrant workers are crucial to the community's sustainable development. The survival dilemma they encounter questions whether each subject occupying a strong position in the community really follows the three principles of "people-oriented," "respect for life," and "everyone is equal." The concern for the living conditions, the emphasis on the quality of life and the respect for the value of life coincide with the basic value scale contained in engineering ethics—the subjectivity scale.

The inherent characteristics of the engineering community combined with China's national conditions not only help the promotion of engineering ethics in China but also contain some restrictive factors hindering the development of engineering ethics. Engineering ethics rooted in and developed in Chinese ideological and cultural characteristics have become a necessary measure to solve the current development dilemma and open up the prospect of future development. From the perspective of the subject trend of thought, contemporary China is facing the situation of the coexistence of Marxist culture, Chinese traditional culture, and Western culture, so we should absorb the advantages of the three ethical and cultural spirits and complement the shortcomings of the single cultural trend of thought: let the Marxist ethical culture play a leading role in the political and ideological fields; let the spirit of freedom and justice in western culture inspire the economic area; let the humanistic spirit of traditional ethical culture play an active role in personal life, family life, professional life and public social life [115].

In engineering culture, the development of Chinese engineering ethics has the cultural foundation for integrating ancient and modern cultures. It also has the opportunity to integrate sources from China and foreign countries, which has contributed to the perfection of the theoretical engineering management system of "people-oriented, unity of heaven and man, collaborative innovation and harmony." "People-oriented" answers the questions of "for whom" and "relying on whom" of engineering activities; "Unity of heaven and man" elaborates on the harmonious relationship between engineering and nature, engineering and society. "Collaborative innovation" eemphaizes resource sharing, group effort, optimal development, and innovative progress. "Harmony" means the foundation of belief and objectives of building engineering management theoretical system. From the perspective of social and cultural development, "contemporary cultural change in China is not an isolated process, it is an integral part of the overall change of Chinese contemporary society, and a projection and reflection of the overall social change" [117], which will have a great impact on engineering practice. Therefore, at the beginning of its establishment and development, engineering ethics should conform to the direction of social and cultural changes to be more inclusive and adaptable. As William Wulf said, "contemporary engineering practice is undergoing profound changes, bringing macro-ethical issues for the engineering community that has not been considered in the past. These problems arise from the increasing difficulty for humans to foresee all the behaviors of the systems they build, including the catastrophic consequences. As a result, engineering will become a process that requires closer interaction with society [118]. The development of engineering ethics in China needs to fit the trend of coexistence of main directions of thinking, grasp the advantages of integration of engineering culture, and follow the development trend of social culture to develop localized characteristic ethics.

10.5.5 General Engineering Ethics from the Perspective of the Engineering Community

According to Harris, "One of the values of the study of engineering ethics is that it promotes a responsible practice of engineering." It suggests that exploring "how engineers become virtuous" and addressing "how to train engineers to voluntarily choose to be responsible" are the basic tasks of engineering ethics. In the narrow sense, engineering ethics answers these questions, but "engineer ethics" is not equal to engineering ethics, and the development of engineering ethics needs to break through the fence of the inherent limitations of the narrow perspective. Engineering ethics in the US and other western countries are in the process of shifting from the narrow sense to the broad sense. Engineers should not only fulfill the ethics of personal responsibility to employers but also pay attention to the ethics of collective responsibility of engineer groups in terms of professional justice, information disclosure, intergenerational equity, and social justice [73]. Specifically, engineering subjects are evolving from single engineer subjects to multiple network subjects. Understanding and consensus are reached among multiple subjects through dialogue and negotiation. Altruism and symbiosis become the core values for each member to consider in organizational strategies. The concept of corporate citizenship is gradually popularized. In this context, China should also pay more attention to engineering ethics from a macro perspective, such as the engineering community's perspective, manifested in the following seven aspects.

(1) The focus of engineering ethics extends from narrow ethical issues to general ethical issues

The study of engineering ethics began with "professional ethics of engineers" [66] and the exploration of narrow ethical issues such as "individual professional ethics," "relations among engineers," and "loyalty to employers." They have built the bridge from traditional ethics to engineering ethics. However, R. Devon directly and sharply criticizes the traditional individual ethics approach of focusing on the "engineer's dilemma." It is too limited to the perspective of engineers' professional norms. It can only be regarded as the starting point of engineering ethics research and must be supplemented by the broader social ethics approach; as such, a complete engineering ethics system can be built. The purpose of engineering to create welfare for society makes the object of engineering ethics extend to macro ethical issues such as "public welfare" and "sustainable development of society." It was clearly stipulated in the code of ethical conduct of the National Society of Professional Engineers (NSPE): "concern for the safety, health and welfare of the public is the priority of engineering activities." In addition, with the emergence of many non-traditional issues, it is a natural demand for the development of engineering ethics to cross the traditional boundary of "professional ethics" and enter the new frontier of ethics research.

(2) The time-domain concerned by engineering ethics extends from "present" to "future"

It includes two aspects: The first aspect is rational development and utilization of nature. Exploiting resources against the laws of nature is considered against the purpose of engineering for the welfare of humankind. Engineers must follow a rational value system to explore nature. It means that we must gradually move from "anthropocentrism" or "life-centrism" to "ecocentrism", and be able to use natural resources rationally. Secondly, intergenerational equity should be considered based on intra-generational equity, and present and future generations' living conditions and rights should be balanced [119]. In this sense, utilitarianism means to "meet the most people's happiness demands"; emotion means to "consider the interests of future generations"; liberal means having "fair treatment to this generation and future generation based on recognizing the rights of future generation" [120]. The time-domain of engineering ethics must be expanded from contemporary people to the future generation. "Meet the needs of the present generation without harming the needs of the future generation" and extend the object of ethical care to the future generation, which is conducive to the sustainable development of human society.

(3) The subjects involved in engineering ethics evolve from engineers to the core stakeholders in the engineering community

Engineering decision-making is the core of engineering activities. In technical matters requiring engineering expertise, engineers are obliged to make autonomous decisions, i.e., Proper Engineering Decision (PED). Managers are obliged to act as decision-makers in matters involving organizational welfare, i.e., Proper Management Decision (PMD). However, when faced with the ambiguous boundaries required by ethical norms, i.e., "the scope of public health and safety that engineers should protect," the profit-oriented decisions made by managers may conflict with those made by engineers based on scientific laws. The power structure of the organization often makes typical management decisions prevail. The arbitrary act of "leadership decision" is repeatedly performed: managers either take over the decision-making function of engineers or force engineers to take unethical actions. To solve the contradiction between the single engineering decision-making body and the engineering activities that involve multiple elements and interests, it is necessary to enhance the engineering decision-making power of core stakeholders in the engineering community and encourage public participation. Only by doing this can we promote the decision-making work and show higher ethical standards.

(4) The action concept of engineering ethics expands from the concern for the realization of the function of artificial nature to the effect of the implementation of the whole life cycle

Engineering activities are associated with great risks, and the role of engineering ethics is to make ethical considerations permeate the whole process of engineering construction to create more desirable engineering products [121]. It requires a three-fold transformation: first, the design concept of "preventive ethics" proposed by Ryder, i.e., to uphold the principles of foresight, initiative, and care, to take the initiative to do our part to benefit human beings and protect nature with the future behavior in mind, and to be alert to the possible hazards of scientific and technological actions or to reduce the adverse consequences of risks [122], to achieve feed-forward control of engineering. Secondly, engineering personnel needs to change the concept of action, that is, "low energy consumption, low emission, and low pollution" advocated by the low-carbon economy and "reduce waste, reuse, recycle, reorganize and rethink" advocated by the circular economy as a guide to action. Thirdly, the triple performance evaluation model of "economy, environment and society" under the systems thinking is adopted to extend the focus of engineering ethics to the stage after the completion of the project and "make reasonable ethical decisions to avoid more and more serious problems" [123].

(5) The research method of engineering ethics is to introduce a macro-level approach on the basis of a micro-level approach

Since the birth of engineering ethics, the case study method of typical real-life events and the theoretical, analytical approach involving concepts, norms, and principles of engineering practice activities have been dominant [28]. Due to the particularity of the case, the case study method has limited universality and applicability. Due to internal conflict, engineering ethics norms easily lead to ethical dilemmas in the face of the practical situation, leading to constant questioning. In addition, the twentieth century, as "the first era redefined by technology" [124], has triggered the major ethical issue of "human alienation," which also makes engineering ethical issues bi-directional, cross-sectional, pluralistic, and complex. Science, technology, society, and engineering permeate each other. Engineering ethics research must pay attention to them and other generalized problems. Science and Technology Studies (STS) and other generalized methods are imperative. By considering more social and cultural background of ethical issues in engineering practice under the interpretation mode of "social perspective," refining new ethical principles according to new situations and problems [126], can help engineering ethics to open the technical black box and distinguish ethical issues in engineering design.

(6) The academic basis of engineering ethics has evolved through technical ethics—professional ethics—social ethics

The core of technological ethics is whether engineering activities using technology are related to ethical issues. It advocates that engineering is not a value-free problem-solving process but a decision-making process with a value load, in which moral problems permeate [9, 15]. Professional ethics focuses on the ethical responsibility of individual engineers. "Loyalty" has become an important "professional ethics principle" of engineers [126], and engineers bear professional responsibilities by their professional consciousness. Social ethics emphasizes that engineering is a social scale experiment with human beings, which involves the welfare of the public and society. The academic basis has been shifted: from the "technological ethics—the use of technology in engineering activities" to the "professional ethics, with professional knowledge and professional consciousness to fulfill its responsibility" to "social ethics, pay attention to the social benefits of the project." It highlights that the engineering community members should consider society as the focus, consciously assume social responsibility, and protect the common human homeland.

(7) The influence range of engineering ethics spreads from region to global

It is reflected in three trends. First, technological advances triggered by engineering are changing the way people live. In this regard, Robert E. McGinn summarized six points: direct extension of ability, qualitative innovation, reduction or elimination of risk, improvement of function, substitution, and provision of means to express inner life. Technology and engineering have gradually formed a mutual benefitting symbiotic relationship—science and technology are the theoretical support of engineering, and the construction demand of engineering promotes technological innovation. Second, in the international context, when engineering personnel carries out engineering activities in countries with different cultural traditions and levels of economic and technological development, they will encounter the ethical dilemma of cultural conflicts; thus, it is difficult to establish values. Third, the impact of engineering activities is not limited to the region, and problems such as environmental pollution and military technology can transcend national boundaries and spread to the world.

The essence of engineering is a kind of collective material existence. The narrow and broad ethical problems in engineering are often intertwined. The success or failure of engineering practice is no longer tied to the engineering profession itself but is determined by social construction. It presents a new challenge to the narrow sense of engineering ethics, which emphasizes that "engineers are most responsible to employers": how to balance the interests of all parties involved in engineering and follow the "logic of freedom" to achieve sustainable development of human society? The engineering community is a "diverse and heterogeneous structure" that must be socially responsible and environmentally responsible to ensure the project achieves a balance between short-term interests and long-term needs, to promote gradually move to generalized study on engineering ethics, namely, across the border of the traditional of "work ethics." It must pay more attention to (1) the overall relationship between engineering and society; (2) the professional responsibilities of engineers in a broader social context; (3) the influence of engineering decisions at a social and political level; (4) the project's efforts in providing a quality and sustainable environment for future generations. The construction of general engineering ethics from the perspective of the engineering community can not only strengthen the engineers' duty to build an increasingly better life for society but also glue engineering and ethics into a "seamless net" to achieve ethics and excellence in parallel.

References

1. Aristotle. (1994). *Nicoma ethics* (p. 27) (translated by Miao, L.). China Renmin University Press.
2. Li, S. (2003). *Research on engineering ethics and its main problems*. Graduate School of Chinese Academy of Social Sciences.
3. Xu, S. (1963). *Shuo Wen Jie Zi*. Zhonghua Book Company.
4. Anonymous. (1986). *Rites-Le Ji*. People's Literature Publishing House.
5. Li, X. (2003). Ethical thinking of campus environment planning and design. *Southern Architecture, 4*, 68–70.
6. Homer. (1994). *Homer's epic poem* (translated by Luo, N., & Wang, H.). People's Literature Publishing House.
7. Li, S. (2006). Two approaches to engineering ethics research. *Ethics Research, 26*(6), 31–35.
8. Martin, M., & Schinzinger, R. (1996). *Engineering ethics*.
9. Wang, J. (2015). On the integration of engineering and ethics. *Journal of Engineering Management, 1*, 23–27.
10. Wang, J. (2013). Engineering ethics research from the perspective of the engineering community. *China Engineering Science, 11*, 97–102.
11. Li, S. (2008). *Introduction to engineering ethics*. China Social Sciences Press.
12. Xiao, P. (1999). *Engineering ethics*. China Railway Press.
13. Oxford English-Chinese Encyclopedia Dictionary Editorial Department. (1985). Oxford English-Chinese Encyclopedia Dictionary. Encyclopedia Culture Company.
14. Wang, J. (2010). The practical wisdom of engineers from the perspective of ethical behavior constitution. *Dialectics of Nature, 26*(12), 42–47.
15. Martin, M., & Schinzinger, R. (2010). *Ethics in engineering* (translated by Li, S.). Capital Normal University Press.
16. Wang, Z. (2009). The meaning and crisis of Disenchantment-Marx Weber's view of disenchantment and its impact. *Hunan Social Sciences, 4*, 1–8.
17. Wen, X., & Shi, Y. (2005). Between technology and morality: Reflection and breakthrough of the dilemma of cultural disenchantment. *Journal of University of Electronic Science and Technology (Social Science Edition), 1*, 37–41.
18. Pinkus, R. L., Pinkus, R. L. B., Shuman, L. J., Hummon, N. P., & Wolfe, H. (1997). *Engineering ethics: Balancing cost, schedule, and risk* (p. 20). Cambridge University Press.
19. Li, B. (2008). Absolute command ethics and coordination ethics. *Ethics Research, 5*, 42–48.
20. Goulet, D. (1996). *New discipline: Development ethics* (translated by Gao, Y., et al.) (p. 29). Social Science Literature Press.
21. Ning, X. (2006). *An analysis of the engineering talent* (p. 42). Northeastern University.
22. Zhang, H., & Hu, X. (2007). Well-being and responsibility—A review of American engineering ethics. *Philosophical Development, 8*, 58–62.
23. Qi, Y. (2010). *Research on ethical rules of engineering decision-making*. Dalian University of Technology.

24. Harris Jr., C. E., Pritchard, M. S., Rabins, M. J., James, R., & Englehardt, E. (2013). Engineering ethics: Concepts and cases. *Cengage Learning*, 17–18.
25. Tang, L., & Tian, P. (2007). Japanese engineering ethics. *Journal of East Liaoning University, 9*(1), 29–32.
26. Herkert, J. R. (2017). Future directions in engineering ethics research: Microethics, macroethics and the role of professional societies. In *Engineering ethics* (pp. 551–562). Routledge.
27. Guo, F., & Wang, X. (2009). Construction of engineering ethics in China: Background, goals and countermeasures. *Journal of Huazhong University of Science and Technology (Social Science Edition), 23*(4), 116–121.
28. Ladd, J. (1980). The quest for a code of professional ethics: An intellectual and moral confusion. In R. Chalk, M. S. Frankel, S. B. Chafer, et al. (Eds.), *AAAS professional ethics project: Professional ethics* (pp. 154–159). AAAS.
29. Zhang, H., & Qian, W. (2010). The focus of American engineering ethics education and the contemporary trend. *Higher Engineering Education Research, 2*(31–34), 46.
30. Jiang, H. (2009). *Ethical issues in engineering*. Dalian University of Technology.
31. Felix Weber, M. (1998). *Academic and politics* (translated by Feng, K.) (p. 107). Life·Reading·Xinzhi Sanlian Bookstore.
32. Jonas, H. (1985). The imperative of responsibility. In *Search of an ethics for the technological age*. The University of Chicago Press.
33. Wang, G. (2005). The ethical turn of German technological philosophy. *Philosophical Research, 5*, 97–98.
34. Zhu, B. (2006). Ethical responsibility of engineering activities. *Ethics Research, 6*, 36–41.
35. Habermas, J. (1990). *Moral consciousness and communicative action*. MIT Press.
36. German Institute of Engineers. (2002). Basic principles of engineering ethics. In *Engineering Technology Philosophy*. Dalian University of Technology Press.
37. Luegenbiehl, H. C. (2004). Ethical autonomy and engineering in a cross-cultural context. *Techné: Research in Philosophy and Technology, 8*(1), 57–78.
38. Arkhangelski. (1987). Professional ethics and other specialized forms of ethics. *Philosophy Translation Series, 4*, 27–34.
39. Wan, C. (2007). A review of Soviet Russian engineering ethics research. *Journal of Yanshan University (Philosophy and Social Sciences Edition), 8*(1), 133–139.
40. Wan, C. (2004). *Russian technical philosophy research* (pp. 171–175). Northeastern University Press.
41. Chen, W., & Zou, Z. (2008). The regularity of the generation and development of Chinese traditional science and technology ethics. *Ethics Research, 2*, 89–94.
42. Xu, J. (2003). *The evolution of technology from Western ethics of science and technology to explore the value of Dimension*. Wuhan University of Technology.
43. Chen, W. (2008). *Research on Chinese traditional scientific and technological ethics*. Hunan Normal University.
44. Huang, K., & Chen, G. (2004). The ethical function of traditional Chinese architecture. *Architecture, 22*(4), 3–7.
45. Anonymous. (1991). *Huang Di Zhai Jing*. Liaoning Education Press.
46. Chen, W., & Liu, L. (2011). A review of the value of Chinese traditional science and technology ethics. *Ethics Research, 1*, 63–66.
47. Xu, S. (2001). Review and prospect of research on science and technology ethics in China. *Journal of Wuhan University of Science and Technology (Social Science Edition), 3*(4), 1–3.
48. Rao, X. (1999). Discussion on engineering ethics teaching. *General Education Quarterly, 6*(3), 62.
49. Li, X. (2011). *Engineering ethics research from the safety culture perspective* (pp. 9–11). Shanxi University of Finance and Economics.
50. Huang, L., & Chen, A. (1989). *Technology ethics-theory and practice*. Xi'an Jiaotong University Press.

51. Cao, N. (1998). Will the technology finally get out of control? Thoughts generated from the victory of "Dark Blue." *Philosophical Research, 2*, 11–16.
52. Harris Jr., C. E., Pritchard, M. S., Rabins, M. J., James, R., & Englehardt, E. (2013). Engineering ethics: Concepts and cases. Cengage Learning.
53. Cong, H. (2006). The current situation and prospects of engineering ethics. *Journal of Huazhong University of Science and Technology (Social Science Edition), 4*, 76–81.
54. Wang, J. (2010). "Truth", "goodness", and "beauty" of modern engineering from the perspective of ethical thinking. *Morality and Civilization, 2*, 101–105.
55. Li, B. (2006). Several questions about the object and scope of engineering ethics-three talks about some questions about engineering ethics. *Ethics Research, 6*, 24–30.
56. Tang, L. (2007). *American Engineering Ethics Research*. Northeastern University Press.
57. Beder, S. (1998). *The new engineer: Management and professional responsibility in a changing world*. Macmillan Education AU.
58. Gan, S. (2000). Technology ethics: A controversial subject. *Philosophical Development, 10*, 5–8.
59. Durbin, P. T. (Ed.). (1991). *Critical perspectives on nonacademic science and engineering* (Vol. 4). Lehigh University Press.
60. Zou, S. (1987). *Technology and philosophy of technology* (p. 355). Knowledge Press.
61. Mitcham, C. (1989). Ethics and technology. *Research in Philosophy and Technology, 9*, 25.
62. Levitt, N. (2003). *Trapped Prometheus* (translated by Dai, J.). Nanjing University Press.
63. Shulman, E. (1995). *Science and technology civilization and human future* (translated by Li, X.). Oriental Publishing House.
64. Jaspers, K. (1953). *The origin and goal of history*. Routledge and Kegan Paul.
65. Li, B. (2002). *Introduction to engineering philosophy*. Elephant Press.
66. Durkheim, E. (2013). *Professional ethics and civic morals*. Routledge.
67. Martin, M., & Schinzinger, R. (1996). *Ethics in engineering* (3rd ed.). The McGraw-Hill Companies.
68. Baker, U. (2004). *Risk society*.
69. Jiang, Z. (2001). *Research on construction ethics and professional ethics*. Chaoyang University of Science and Technology.
70. Wang, H. (1994). *Ethics teaching design integrated into various engineering courses*. National Science Council Research Project Report.
71. Lin, T. (2001). Construction of engineering ethics environmental Taiwan-21 September Chichi earthquake building earthquake victims of reflection. Yishou University Technical Report, ST903.
72. Li, S. (2006). Analysis of the reasons for the indifference of engineering ethics. *Journal of Beijing Institute of Technology, 6*, 93–97.
73. Li, B. (2010). Micro, meso and macro engineering ethical issues—the fifth talk on engineering ethics. *Ethics Research, 4*, 25–30.
74. Bunge. (1993). Value judgment and moral judgment of science and technology. *Philosophy Translation Series, 3*, 5–10.
75. Wang, X. (2008). *American engineering education certification research from the perspective of Professionalism*. Huazhong University of Science and Technology.
76. Winner, L. (1990). Engineering ethics and political imagination. In *Broad and narrow interpretations of philosophy of technology* (pp. 53–64). Springer.
77. Davis, M. (1998). *Thinking like an engineer: Studies in the ethics of a profession*.
78. Sidgwick, H. (2019). The methods of ethics. Good Press.
79. Ferrer, O., et al. (2005). *Business ethics* (translated by Chen, Y.). Tsinghua University Press.
80. Tilow, J., et al. (2008). *Ethics and life* (translated by Cheng, L., et al.). World Book Publishing Company.
81. Rachels, J., & Rachels, S. (2012). *The elements of moral philosophy 7e*. McGraw Hill.
82. Li, R. (1999). *Confucian bioethics*. Ehu Magazine.
83. Sandel, M. J. (Ed.). (2007). *Justice: A reader*. Oxford University Press on Demand.
84. Aristotle. (2010). *Nicole ethics* (translated by Deng, A.). People's Publishing House.

85. Shirer, W. L. (1991). *Rise and fall of the Third Reich*. Random House.
86. Hu, J. (2011). Moral "gender"-based on the perspective of feminist care ethics. *Yunnan Social Sciences, 6*, 20–25.
87. Li, J. (2010). The power of ethics and aesthetic character education: A preliminary exploration with Nel Noddings. *Aesthetic Education Bimonthly, 177*, 66–75.
88. Shi, S. (2007). *Much of our publicity about the Aswan Dam is misleading*. http://it.sohu.com/20070321/n248858664.shtml.2007-03-21/2014-05-11
89. Wolf, J. C., & Jonas, H. (1996). Eine naturphilosophische Begründung der Ethik. A. Hügli & P. Lübcke. *Philosophie im*, 20.
90. Mao, Y. (2003). Western applied ethics highlighting "responsibility"—A review of western responsibility ethics. *Philosophical Development, 9*(20–24), 42.
91. He, J. (1995). Thoughts after the collapse of the Mitutoyo department store in South Korea. *Construction Technology, 3*, 38–40.
92. Fletcher, J. F. (1997). *Situation ethics: The new morality*. Westminster John Knox Press.
93. Xu, E., Zhang, J., & Li, J. (2012). Research on the correlation of information technology values and the development of network morality. *Educational Technology Research, 12*, 34–40.
94. Wang, J. (2011). Ethical responsibility in engineering activities and its realization mechanism. *Morality and Civilization, 2*, 101–105.
95. Gortner, H. F. (1991). *Ethics for public managers* (Vol. 272). Praeger.
96. Wilkins, L., Painter, C., & Patterson, P. (2021). *Media ethics: Issues and cases*. Rowman & Littlefield.
97. Zhan, D. (1997). *The interactive relationship between ethical decision, emotional influence and virtue cultivation*. Furen University Press.
98. Hill, M., Glaser, K., & Harden, J. (2016). A feminist model for ethical decision making. In *Learning from our mistakes: Difficulties and failures in feminist therapy* (pp. 101–121). Routledge.
99. Cooper, T. L. (2012). *The responsible administrator: An approach to ethics for the administrative role*. Wiley.
100. Resnik, D. B. (2005). *The ethics of science: An introduction*. Routledge.
101. Davis, M. (1999). Professional responsibility: Just following the rules? *Business and Professional Ethics Journal, 18*(1), 65–87.
102. Wang, J. (2013). Analysis of the blame limit of engineers' ethical responsibility under the context of attention. *Dialectics of Nature, 29*(10), 38–43.
103. Wang, J. (2011). The unity of "informed behavior" in the infiltration engineering ethics teaching. *Modern University Education, 4*, 100–105.
104. Seebauer, E. G., & Barry, R. L. (2000). *Fundamentals of ethics for scientists and engineers*.
105. Li, B. (2008). Research on engineering community and development of engineering sociology—The third research of "engineering community." *Dialectics of Nature, 30*(1), 63.
106. Zhang, X. (2008). The nature of the engineering community. *Dialectics of Nature, 12*, 43–47.
107. Hu, M. (2008). On value consensus. *Philosophical Research, 7*, 96–102.
108. Fei, X. (1985). *Rural China*. Life Reading Xinzhi Sanlian Bookstore.
109. Wong, Y. H., & Chan, R. Y. K. (1999). Relationship marketing in China: Guanxi, favouritism and adaptation. *Journal of Business Ethics, 22*(2), 107–118.
110. Zhai, X. (2004). Reproduction of human feelings, face, and power-the way of social exchange in a rational society. *Sociological Research, 5*, 48–57.
111. Zhou, J. (2005). On the negative impact of traditional culture on science and technology. *Journal of Wuhan University of Science and Technology (Social Science Edition), 7*(1), 81–84.
112. Weber, M. (1992). *Protestant ethics and the spirit of capitalism* (translated by Yu, X., & Chen, W.). Life Reading Xinzhi Sanlian Publishing House.
113. Tong, Z. (2004). Land acquisition, house demolition and rights protection in urban construction. *Journal of Wuhan Institute of Economic Management, 18*(3), 17–21.
114. Liu, C., & Cheng, J. (2008). The citizenization of the second generation of migrant workers: Status quo analysis and progress measurement. *Population Research, 32*(5), 48–57.

115. Kong, R. (2009). Conception of the construction of socialist ethical culture with Chinese characteristics//Shaanxi Federation of Social Sciences. In *Proceedings of the thematic forum on the theory and road of socialism with Chinese characteristics in the 60th anniversary of the 3rd academic annual meeting of Shaanxi Social Sciences* (pp. 240–243). Shaanxi Federation of Social Sciences, Xi'an, Shaanxi, China.
116. He, J. (2013). On the core of engineering management theory. *China Engineering Science, 11*, 4–11.
117. Liu, D. (2004). Contemporary Chinese cultural changes and social mentality evolution. *Academic Exploration, 3*, 102–105.
118. Wulf, W. A. (2004). Engineering ethics and society. *Technology in Society, 26*(2–3), 385–390.
119. Mitcham, C., & Mackey, R. (Eds.). (1983). *Philosophy and technology* (Vol. 80). Simon and Schuster.
120. Chen, X. (2009). *Sustainable development and human development*. People's Publishing House.
121. Liao, X. (2004). On intergenerational equity. *Ethics Research, 4*, 25–31.
122. Li, S. (2010). Paradigm analysis of engineering ethics research. *Journal of Beijing Institute of Technology (Social Science Edition), 12*(3), 101–104.
123. Mitcham, C., & Duval, R. (2000). *Engineering ethics*. Prentice Hall.
124. Gadamer, H. G. (1981). *Reason in the age of science* (Vol. 4, p. 82). MIT Press.
125. Zhu, Q. (2011). *Analysis of engineering ethics from the perspective of practical effectiveness*. Dalian University of Technology.
126. Xu, H., & Cheng, X. (2008). On the ethical dilemma and choice of engineers. *Dialectics of Nature, 24*(8), 52–56.

The manufacturer's authorised representative in the EU is Springer Nature Customer Service Centre GmbH, Europaplatz 3, 69115 Heidelberg, Germany. If you have any concerns regarding our products, please contact ProductSafety@springernature.com

Printed and bound by CPI Group (UK) Ltd, Croydon, CR0 4YY
30/09/2026
02225319-0020